ORIGIN OF LIFE

ORIGIN OF LIFE

Proceedings of the Third ISSOL Meeting and the Sixth ICOL Meeting, Jerusalem, June 22-27, 1980

Edited by

YECHESKEL WOLMAN

Department of Organic Chemistry,
The Hebrew University, Jerusalem

D. REIDEL PUBLISHING COMPANY

DORDRECHT : HOLLAND / BOSTON : U.S.A.
LONDON : ENGLAND

Library of Congress Cataloging in Publication Data

Main entry under title:

Origin of life.

Includes index.
1. Life-origin-Congresses. I. Wolman, Yecheskel, 1935-
II. International society for the study of the origin of life (3rd: 1980: Jerusalem). III. International conference on the origin of life (3rd: 1980: Jerusalem)
QH325.06914 577 81-248 AACR2
ISBN 90-277-1229-8

CIP

Published by D. Reidel Publishing Company,
P.O. Box 17, 3300 AA Dordrecht, Holland.

Sold and distributed in the U.S.A. and Canada
by Kluwer Boston Inc.,
190 Old Derby Street, Hingham, MA 02043, U.S.A.

In all other countries, sold and distributed
by Kluwer Academic Publishers Group,
P.O. Box 322, 3300 AH Dordrecht, Holland.

D. Reidel Publishing Company is a member of the Kluwer Group.

Printed in The Netherlands

TABLE OF CONTENTS

PREFACE

This volume is a record of the 6th International Conference on the Origins of Life and the 3rd Meeting of the International Society for the Study of the Origins of Life. The conference was held under the auspices of the Israel Academy of Sciences and Humanities at Jerusalem from June 22nd to June 27th 1980.

A few weeks prior to the conference, Academician Aleksander Ivanovich Oparin passed away. Oparin, the father and founder of the study of the origins of life, proposed over 50 years ago that modern biological molecules had abidogical origins in the past, thus the beginning of life on Earth was preceded by a long period of abiogenic molecular evolution. Oparin was planning to report on his latest work in the opening session of the meeting - "Natural Selection: A Leading Factor in Transition from the Non-Living Matter to Life". This lecture will never be delivered.

In Hebrew we say of those who have died "may their memory be bound with the bonds of eternal life". For Aleksander Ivanovich Oparin those words have particular significance, for surely his pioneering work will endure as long as the spirit of scientific enquiry prevails. This meeting was dedicated to the memory of Aleksander Ivanovich Oparin.

The main topics which were covered in the conference were: organic molecules in the interstellar space and their relevance to the study of the origins of life; meteorite analysis; abiotic synthesis of small molecules; polymerization reactions in the prebiotic era; the origin of optical activity; origin of the genetic code; early biochemical evolution; fossils and microfossils.

One evening was dedicated to the memory of Aharon Katzir-Katchalski, a scholar, a teacher and a friend who lost his life in a terrorist attack seven years ago. In the framework of that evening Ilya Prigonine discussed irreversibility and evolution followed by a panel discussion concerning new avenues in the study of the origins of life.

Another panel discussion was devoted to Origins of Life as a curriculum in education.

One of the main problems that was felt by most of the participants is the barrier that exists between people from different disciplines which are working in the field. There is a lack of communication, astronomers speak differently from geologists and chemists do not always understand the biologist's

Y. Wolman (ed.), Origin of Life, xi–xii.

language. We have to do all that we can in order to break down the barriers and build communication channels among ourselves. The next breakthrough in this area of research will result from a cooperative effort of all of us - astronomers, astrophysicists, physicists, chemists, geologists, biologists. Such a cooperation can take place only when there is a common language between all the people involved.

In order to bring all participants to some common ground no parallel sessions were held. Two poster sessions were held in which a short summary (two to three minutes) of each paper was given before the formal presentation of the posters. The presentation was followed by a discussion between the authors and the participants, moderated by the chairperson. We hope that these arrangements contributed somewhat towards the breaking down of the barriers.

I am indebted to the members of the local organizing committee - M. Paecht-Horowitz, E. Gil-Av and M. Halmann, to the members of the international advisory committee - G. Ponnamperuma, J.W. Schopf and R.S. Young, as well as to all colleagues and friends in Israel and abroad who helped make this conference a success.

Special thanks go to the sponsoring bodies - The Israel Academy of Sciences and Humanities, The Weizmann Institute of Science and the Hebrew University of Jerusalem.

Jerusalem, August 1980 Y. Wolman

LIST OF PARTICIPANTS

ABELL, P. University of Cambridge, Cambridge, England, CB2 3EW
BALTSCHEFFSKY, H. University of Stockholm, 10691 Stockholm, Sweden
BANIN, A. The Hebrew University, Rehovot, Israel
BAR-NUN, A. Tel Aviv University, Tel Aviv, Israel
BLOCH, M.R. Ben Gurion University, Beersheva, Israel
BLOMBERG, C. Royal Inst. of Technology, S-10044, Stockholm, Sweden
BOSSARD, A. Université Paris val de Marne, 94010 Creteil Cedex, France
BOSCHKE, F.L. Springer-Verlag, 6900 Heidelberg, Germany
BRACK, A. Centre de Biophysique Moleculaire, C.N.R.S., 45045 Orleans Cedex, France
BROWN, R.D. Monash University, Clayton, Victoria, Australia 3168
BUVET, R. Université de Paris val de Marne, 94010 Creteil, Paris, France
CHANG, S. Ames Research Center, Moffet Field, California 94035, U.S.A.
CULLMANN, G. 19 Boulevard de la Gardnette, 31000 Toulouse, France
DECKER, P. Tieraertztliche Hochschule, D-3000 Hannover, F.R.G.
DELSEMME, A.H. The University of Toledo, Toledo, Ohio 43606, U.S.A.
DeVINCENZI, D. N.A.S.A., H.Q., Washington, D.C. 20546, U.S.A.
DENES, F. "P. PONI" Institute of Macromolecular Chemistry, 6600 Jassy, Romania
DOWLER, M. San Diego State University, San Diego, California 92182, U.S.A.
EGAMI, F. Mitsubishi-Kasei Inst. of Life Sciences, Tokyo, Japan
EIRICH, F.R. Polytechnic Inst. of New York, Brooklyn, N.Y. 11201, U.S.A.
FRIEDMAN, E. Florida State University, Tallahassee, Florida 32306, U.S.A.
FERRIS, J. Rensselaer Polytechnic Inst., Troy, New York 12181, U.S.A.
FOX, S.W. University of Miami, Florida 33143, U.S.A.
GIL-AV, E.. The Weizmann Institute of Science, Rehovot, Israel
GREEN, B.S. The Weizmann Institute of Science, Rehovot, Israel
GINER-SOROLLA, A. Sloan Kettering Inst., Rye, New York 10580, U.S.A.
GOLDBERG, R. 12 Fifeshire Road, Toronto, Ontario, M2L 265, Canada
HALMANN, M. The Weizmann Institute of Science, Rehovot, Israel
HARADA, K. The University of Tsukuba, Sakura-Mura, Ibaraki, 300-31 Japan
HILL, R. The Open University, Milton Keynes, MK7 6AA, England
IRVINE, W.M. Onsala Space Observatory, S-430 34 Onsala, Sweden

ISHIDA, M. Kyoto University, Sennan-gun, Osaka 590-04, Japan
ISHIGAMI, M. Minamikawachi-Machi, Kawachi-Gun, Japan 329-04
JOHANSSON, K.L.V. Astronomical Observatory, Uppsala University, Uppsala, Sweden
KHARE, B.N. Cornell University, Ithaca, New York 14853, U.S.A.
KINJO, M. Minamikawachi-Machi, Kawachi-Gun, Tochigi-Ken, Japan 329-04
KOKUFUTA, E. University of Tsukuba, Sakuramura, Niiharigun, Ibaraki 305, Japan
LABOUYGUES, J.-M. George Gamow Institute, 63000 Clermont-Ferrand, France
LACEY, J.C, Jr. University of Alabama in Birmingham, Birmingham, Alabama 35204, U.S.A.
LAWLESS, J. Ames Research Center, Moffet Field, California 94035 U.S.A.
LAHAV, M. The Weizmann Institute of Science, Rehovot, Israel
LAHAV, N. The Hebrew University, Rehovot, Israel
LESCHINE, S. University of Massachusetts, Amherst 01003, U.S.A.
LEPONT, P. 96 rue de Val-de-Saire, 50100 Cherbourg, France
LODEN, L.O. Morabergsuaegen 9, F-13300 Saltsjoebaden, Sweden
MATTHEWS, N. University of Illinois, Chicago, Illinois 60680, U.S.A.
MILLER, L. University of California San Diego, La Jolla, California 92093, U.S.A.
MITSUHIKO, A. Kyoto University, Kumatori, Osaka 590, Japan
MORIMOTO, S. Tokushima University, Tokushima 770, Japan
MOUREY, D. Université Paris val de Marne, 94010 Creteil Cedex, France
MULLINS, D.W. University of Alabama in Birmingham, Birmingham, Alabama 35294, U.S.A.
NAGYVARY, J.J. Texas A & M University, College Station, Texas 77843, U.S.A.
NISSENBAUM, A. The Weizmann Institute of Science, Rehovot, Israel
NODA, H. University of Tokyo, Hongo, Tokyo 113, Japan
NOGUCHI, T. Tokyo College of Economics, Tokyo 185, Japan
ORGEL, L.E. The Salk Institute for Biological Studies, San Diego California 92130, U.S.A.
PAECHT-HOROWITZ, M. The Hebrew University, Rehovot, Israel
PAPAGIANNIS, M.D. Boston University, Boston Mass. 02215, U.S.A.
PINCOCK, R.E. University of British Columbia, Vancouver, B.C., V67 1Y6, Canada
PHILLIPS, H. 1820 Avenida del Mundo, Coronado, California 92118, U.S.A.
PONNAMPERUMA, C. University of Maryland, Maryland 20742, U.S.A.
PRIGOGINE, I. The University of Texas at Austin, Austin, Texas 78712, U.S.A.
RAMBLER, M. Boston University, Boston Mass. 02215, U.S.A.
RAULIN, F. Université Paris val de Marne, 94010 Creteil Cedex, France

REIN, R. Roswell Park Memorial Inst., Buffalo, New York 14263, U.S.A.
RICH, A. University of Michigan, Ann Arbor, Michigan, U.S.A.
SAYGIN, O. Celler Str. 12, Hannover, F.R.G.
SCHWARTZ, A. The University of Nijmegen, Nijmegen, The Netherlands
SCHWARTZ, R. Georgetown University Medical Center, Washington, D.C. 20007, U.S.A.
SECKBACH, J. The Hebrew University, Jerusalem, Israel
SHIMIZU, M. University of Tokyo, Meguroku, Tokyo 135, Japan
SINGH, T.D. Bhaktivedanta Institute, Juhu, Bombay 400 049, India
STOKS, P. University of Nijmegen, Nijmegen, The Netherlands
THIEMANN, W. Bravereiweg 18, D2804, Lilienthal, F.R.G.
TOTOLIN, M. "P. PONI" Institute of Macromolecular Chemistry, 6600 Jassy, Romania
VANCE, B.E. D. Reidel Publishing Co. Ltd., Dordrecht, The Netherlands
VISSER, C.M. University of Groningen, Groningen, The Netherlands
VASILESCU, D. Université de Nice, 06034 Nice, Cedex, France
WEINER, S. The Weizmann Institute of Science, Rehovot, Israel
WHITE, D.H. University of Santa Clara, Santa Clara, California 95053, U.S.A.
WOLMAN, Y. The Hebrew University, Jerusalem, Israel
WEINTRAUB, R. 408 Brooks Avenue, Raleigh, NC27607, U.S.A.

INTERSTELLAR MOLECULES AND THE ORIGIN OF LIFE

Professor R.D. Brown

Department of Chemistry,
Monash University,
Wellington Road,
Clayton, Victoria, Australia.

It is perhaps appropriate to remind the meeting of the International Society for the Study of the Origin of Life that to many scientists its activities could be described as replacing the old myths and legends of the origin of the universe and life by twentieth century myths and legends of the origin of the universe and life. We tend to smile rather more about the old legends than the new legends but perhaps at a meeting like this in another century or two they will be smiling about many of the thoughts that we are discussing in the next few days. But let me start by reminding you of just one of the many old legends - a legend of the origin of the universe or specifically the Milky Way. It is portrayed beautifully in a famous painting by Tintoretto. The legend concerned Zeus who was not a very faithful husband and his son Hercules was not that of his wife but of a mortal. He wished to render his son immortal for which he needed to suckle his son at the breast of a goddess, namely his wife. His wife was not too pleased with all of this and in the event milk was spilt! The milk became the Milky Way.

It is a charming but amusing legend to us. It might be called the 'marital infidelity' theory of the origin of the universe. We have passed on from such things to more modern legends whether we talk about the universe or the creation of life.

One can trace the start of the most popular current legends to a letter by Darwin to Hooker in 1871: "...But if (and oh what a big if) we could conceive in some warm little pond with all sorts of ammonia and phosphoric salts, - light, heat, electricity and etc. present, that a protein compound was chemically formed, ready to undergo still more complex changes...."

Y. Wolman (ed.), Origin of Life, 1–9.

The modern work starts from proposals of a chemical, prebiotic evolution on the early Earth by Oparin (1). Experimentally it is typified by the classic experiment of Miller and Urey, in which the simulated atmosphere mixture of methane, water, ammonia and hydrogen was passed through a simulated lightning discharge and a dilute solution ultimately was isolated, which contained a surprisingly small number of compounds in reasonable yield (2). Before the event chemists might have predicted that a vast array of chemical compounds would be formed. In fact not only were very few compounds formed in reasonable quantities but among them was a noticeable proportion of biologically significant molecules. This of course follows the epoch-making concept of Oparin that one had to have the prebiotic molecules formed in the Earth's atmosphere in order to start the whole process of chemical evolution on the surface of the young Earth, leading up to the first primitive living things and biological evolution.

Although that is a very interesting point of view and is the orthodox point of view, I shall discuss the unorthodox point of view which is that life did not originate in the chemical evolutionary sense of the young planet Earth but instead arrived on the planet from elsewhere.

There are at least two questions that have to be asked of the orthodox view for comparison with the alternative point of view:

Firstly what was the nature of the original atmosphere of the primitive Earth after it had cooled down to the biological range of temperatures? From the time of Oparin it has been fasionable to assume that it was reducing, that the carbon was in the form of methane, the nitrogen in the form of ammonia, and so forth, and that molecular hydrogen was present. Geologists and other Earth scientists do not seem to be so convinced of this, many believing that by the time the Earth cooled down the atmosphere was oxidising. It may have been as oxidised as CO_2 and N_2, with a very low partial pressure of molecular hydrogen.

If the atmosphere was not sufficiently reducing this poses problems for chemical evolution in the early atmosphere. The desired chemical precursors such as aminoacids do not seem to be produced by energetic events in an oxidized atmosphere. It may be important to clear up this by further careful studies.

The second point that requires a good deal of examination is: how do you get an adequate concentration of your pre-biotic molecules to go to the next stage of condensation and ultimately bio-polymers? It is again traditional to think in terms of the product of a lightning discharge being washed into a shallow pool by a convenient shower of rain, then relying on the hot sun to concentrate that shallow pool. But there was no ozone layer around the young Earth and so surface compounds received a constant flux of ultraviolet light with sufficient energy rapidly to photolyse organic molecules. For example, an astrophysical calculation some years ago showed that even the most stable molecule, namely carbon monoxide, has a half life of about a hundred years in the normal interstellar flux well away from the solar system (3). The half life close to the Sun when not protected by an ozone layer would be very much shorter than that.

Let us turn now to consider the alternative hypothesis of the Origin of Life. We must start by considering the way in which the solar system was born and scrutinize especially the chemical changes leading up to the formation of planet Earth. It is now widely accepted among astronomers that new stars and planetary systems are a product of the collapse and fragmentation of huge amounts of gas and dust called dark nebulae (4). There are a number of these objects distributed around the Milky Way galaxy, many having other objects closely associated with them, such as bright nebulae. The bright nebulae are regions of gaseous ionised hydrogen, the ionization being produced by embedded very hot stars. They constitute attractive objects when photographed through optical telescopes. Some nebulae have infrared sources within them that represent warmer parts of the cloud and perhaps are incipient stars. Yet other objects such as the Coal Sack nebula in the southern skies are clouds of very cold gas and dust, not associated with hotter objects, the temperature being as low as 10K or less.

It is not very profitable to study dark nebulae with optical telescopes because we can observe little more than the extent to which the dark nebula attenuates starlight from more distant stars beyond it. However in the radio, microwave and millimetre ranges of wavelength the dust does not impede the passage of radiation. By using radio telescopes we have been able to get valuable information about the composition of these nebulae, especially by being able to observe spectral lines in the emitted radiation. These include lines of the rotational spectra of various molecules. By comparing the line frequencies with those of spectra that have been analysed in the laboratory we are able to identify a growing list of molecules as constituents of dark nebulae . The accompanying table lists molecules that have been identified in this way up to the present. Nearly all of these have been detected in a very large object near the centre of our Galaxy referred to by astronomers as Sagittarius B2. This appears to be the most massive dark nebula in our galaxy. Most of the molecules have also been detected in a smaller object much nearer the Sun, part of the great Orion nebula. Smaller lists of molecules have been detected in many other dark nebula within our Galaxy and some have even been detected in other nearby galaxies.

TABLE 1

Interstellar Molecules - June 1980

H_2	HCO^+	HCN	HCO_2H	CH_3-O-CH_3
CH	N_2H^+	HNC	HC≡C-CN	C_2H_5OH
CH^+	HCO	NH_3	HC≡C-C≡C-CN	CH_2=CH-CN
CN	C_3N	H_2O	HC≡C-C≡C-C≡C-CN	HCO_2CH_3
CO	C_4H	H_2CO	H-(C≡C)$_4$-CN	NH_2CN
CS	C_2H	H_2CS	CH_3-C≡CH	CH_3-C≡C-CN
SiO		H_2CNH	CH_3CN	C_2H_5CN
SiS		OCS	$HCONH_2$	CH_2=C=O
SO		H_2S	CH_3OH	HNCO
NS		SO_2	CH_3CHO	HNCS
OH		HNO	CH_3NH_2	CH_3SH
NO	HC≡CH			
	CH_4			

It is probably unwise to assume that the molecules listed in Table 1 are the only chemical constituents of the nebulae. More realistically they represent the ones detectable with telescopes at the presently attainable sensitivity. Similarly those dark nebulae for which the list of molecules is shorter no doubt contain many more molecules that have not yet been detected. Although there have been attempts to see whether the chemical composition of different clouds differ significantly, the results obtained so far are not conclusive. It seems reasonable to assume at present that all dark nebulae contain somewhat similar collections of molecules, the detailed composition no doubt steadily varying throughout the life history of the molecular cloud from its early days as a tenuous whisp, with molecules just starting to form, to its death-throes when it is congealing into a new star cluster.

It will be noticed in the list of detected interstellar molecules that many of these are species which we classify as short-lived in the laboratory. In the dark nebulae their lifetime is considerable because the gas pressures are extraordinary low by terrestrial standards - in the range 10^2 to 10^6 molecules per cc. This has posed a technical problem in identifying some of the observed spectral lines. It is wellnigh impossible to identify an interstellar species until the spectrum has been accurately observed in the laboratory. Thus a substantial effort in the field of investigation of interstellar molecules has been put into the measurement of rotational spectra of transient species. Various spectroscopic groups around the world have contributed. Table 2 summarizes many of the studies of transient species of interest in the investigation of interstellar molecules. All of the work listed in the table was based on the methods of microwave spectroscopy that enable very precise line frequencies to be obtained.

TABLE 2

Short-Lived Species

OH	Sanders, Schawlow, Dousmanis, Townes	1953	free space cell, Zeeman mod., discharge in H_2O vap.
CS	Mockler and Bird	1955	X-band cell, Stark mod., 60Hz discharge in CS_2.
SO	Powell and Lide	1964	split wave guide cell, RF discharge: O + OCS.
SiS	Hoeft	1965	high temperature cell, FeS + Si at $1050^o - 1300^o$.
SiO	" Torring	1968	high temperature cell, Si + SiO_2 at 1800°.
NS	Amano, Saito, Hirota	1969	parallel plate cell, Stark mod., RF discharge $N_2 + SCl_2$
HCO	Saito	1972	parallel plate cell, Stark mod., H_2CO + product of discharge on CF_4.
H_2CNH	Johnson and Lovas	1972	parallel plate, Stark mod., pyrolysis of CH_3NH_2.
HNO	Saito, Takagi	1973	parallel plate, Stark mod., H atom +NO.
HCO^+	Woods, Dixon, Saykally, Szanto	1975	plasma cell, video det., CO + H_2.
HNC	Blackman, Brown, Godfrey, Gunn	1976	parallel plate, Stark mod., pyrolysis of HCN.
	Saykally, Szanto Anderson, Woods		plasma cell, source mod.,
	Creswell, Pearson, Winnewisser, Winnewisser		free space cell, video det., active N + CH_3Br.
N_2H^+	Saykally, Dixon, Anderson, Szanto, Woods	1976	plasma cell, source mod., $N_2 + H_2$.

Some of these studies were made possible by the preceding work of Professor Alan Carrington and his colleagues who pioneered the study of transient species by gas phase electron spin resonance spectroscopy (5) which gave line frequencies not quite of the precision desired by radio astronomers but made the subsequent investigation by laboratory microwave spectroscopy much easier. We must remember however that the dark nebulae no doubt contain a number of other transient species that have not yet been identified because the requisite laboratory spectroscopy has not been successful. This particularly applies to molecular ions because these present a formidable challenge to laboratory microwave spectroscopy. Indeed so far the only successful spectroscopic measurements of this kind have been achieved by R. Claude Woods and colleagues who have observed transitions for CO^+, HCO^+ N_2H^+. One of the important current pursuits in microwave spectroscopy research is to develop methods that will be more generally applicable to the study of molecular ions.

Another category of interstellar molecule, one that would be of interest in connection with the origin of life, is that of molecules such as aminoacids, the DNA and RNA bases and other rather involatile biomolecules. The problem here is to produce a sufficient sample of vapour to enable the laboratory study of the microwave spectrum to be performed. We have put considerable efforts into this kind of laboratory study in my laboratories. By building several spectrometer cells particularly to deal with the problem of compounds of low volatility we succeeded first in observing and analysing the spectrum of urea (6) and subsequently the spectrum of the simplest aminoacid - glycine (7). Neither of these molecules proved to be detectable when we searched for their signals on radiotelescopes but at least we discovered on the one hand that urea in the gas phase is indeed a planar molecule as chemists had suspected and that the form of glycine detected by us in the gas phase is the conformed labelled "4" in the accompanying diagram. It appears to contain a cyclic

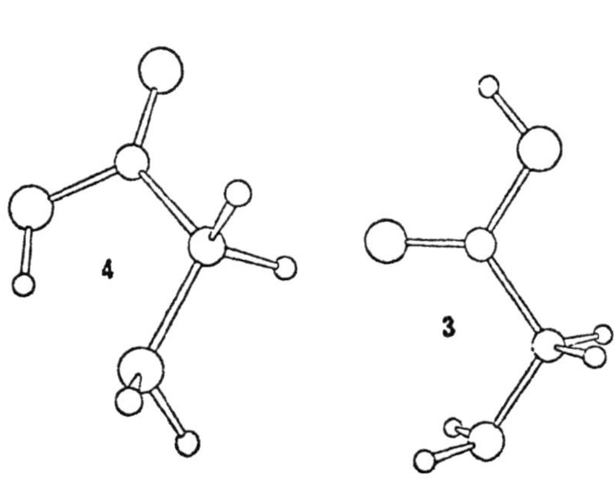

hydrogen bond that confers slightly greater stability on it than other alternative conformers although there is another conformer (labelled "3" in the diagram) that is expected to be of virtually the same stability in the gas phase. Indeed we have seen some additional spectral lines in the laboratory that we believe are to be attributed to that second conformer. The lines are weaker because it appears to have a much smaller dipole moment and line strengths depend on the square of the dipole moment of the molecule concerned. It seems likely that when the sensitivity of radiotelescopes is improved still further that these important biomolecules will in due course be detected in dark nebulae.

Along with the discovery of interstellar molecules came various theories about the way in which the molecules are formed from the original atomic materials of the universe (8). We do not have a complete understanding of the physical conditions that prevail within dark nebulae and particularly know all too little about the nature of the dust. For these reasons and others it is premature to regard any of the current theories as definitive although we appear to be recognising several general types of chemical reaction as playing an important role. But for the purposes of discussion of the origin of life the most significant feature is that many relatively complex molecules have been detected as being already present in the nebula well in advance of its collapse to form new planetary systems. We have little doubt that the chemical complexity of the cloud would steadily increase further as the collapse progresses to form new stars.

During the collapse of the cloud the temperature does not initially greatly increase because the heat of collapse tends to be steadily radiated away. However in the latter stages when the density becomes high enough a steady increase in temperature occurs. Ultimately in the specific spots where new stars are forming the temperature can become high enough to make nuclear processess feasible and thus produce bright stars. But other regions of the cloud, including parts from which some of the planets, meteors and comets are formed, do not experience high enough temperatures to disrupt the complex molecular constituents. We have direct evidence of this from the geochemical examination of some meteorites which contain material that was never heated about 100C during its formation (9). Such meteorites, called carbonaceous chondrites, are considered to be some of the best fossil relics of the original nebula from which our Solar System congealed.

Others of this conference will make reference to the chemical analysis of the contents of certain carbonaceous chondrites, particularly the detection of molecules of biological significance.

Such discoveries have reinforced the concept that much of the required prebiological molecules would have been provided to the surface of the young planet Earth through falling meteorites. Indeed it is known that the infall of meteorites was particularly heavy on all of the inner planets in the earlier history of the Solar System, especially in the first 500,000,000 years of its history.

Thus the rival of the primaeval "dilute soup" is some freshly fallen carbonaceous chondrites that fell on some favourable section of the Earth's surface such as perhaps a swampy or muddy section of land. There is still the problem of the concentration of appropriate molecular species to enable condensation-type reactions to proceed and produce primitive proteinoid or nucleic acid moieties but I personally do not feel that either category of conjecture on the origin of life is particularly persuasive on this aspect of the overall evolutionary story.

The further evolutionary process, which must have been accompanied at some relatively early stage by the emergence of appropriate chirality, also is not readily understood through presently available evidence. Indeed I feel that the chiralty problem is still one of the most challenging in our scientific studies of the origin of life on Earth. But rather than pursuing this line of speculation I would like to turn to another recent speculation that has caused controversy and indeed usually severe criticism from people holding more orthodox views of the subject. I refer to the recent suggestion of Hoyle and Wickramasinghe that life on Earth originated from the showering down on the planet of micrometeorites that are residual cometary dust (10).

They propose that life had in fact evolved in cometary nuclei up to the stage of bacteria and viruses, so much so that they go further and propose that some of the major epidemics in historical times are to be attributed to the arrival of bacteria and viruses in the form of minute dust particles containing such microbiological material and originating in cometary nuclei.

Of the many criticisms that have been levelled at this proposal one of the first that can be raised from the astronomical point of view is to question whether the conditions that prevail in cometary nuclei could be conducive to the progression of chemical evolution from the stage of interstellar molecules to the stage of organised entities like bacteria. Among the various requirements are the presence of liquid water. However recent calculations to be referred to in this conference have demonstrated that it does seem plausible for water to occur in the liquid form at least in larger cometary nuclei because of the effect of heating by radioactive decay of ^{26}Al (11).

It is not my purpose to continue analysing in greater detail the several rival hypothesis that have now been presented for the origin of life on Earth. Perhaps one should think primarily in terms of processes taking place in the atmosphere of the primitive young planet or perhaps one should look towards the original dark nebulae for our antecedents or possibly the crucial action took place in cometary nuclei. In this whole field of discussion of the origin of life it is easy to get to the point of circular arguments. We must exert every endeavour to evaluate available evidence as objectively and dispassionately as possible. It is perhaps appropriate to end with the words of someone of great wisdom whose philosophy is summarised in stanza 27 of his best known work. I refer to the Rubaiyat of Omar Khayyam which runs as follows:

> "Myself when young did eagerly frequent
> Doctor and Saint, and heard great Argument
> About it and about: but evermore
> Came out by the same Door as in I went."

REFERENCES

(1) Oparin, A.I. address to Moscow Botanical Society 1922; 'The Origin of Life' (Moscow, 1924), translation by Ann Synge in: Bernal, J.D., 'The Origin of Life' (Weidenfeld and Nicholson, London, 1967).

(2) Miller, S.L., Science, 117, 528 (1953); J. Amer. Chem. Soc., 77, 2357 (1955).

(3) Stief, L.J., Donn, B., Glocker, S., Gentieu, E.P. and Metall, J.E., Astrophys. J., 171, 21 (1972).

(4) 'Protostars and Planets' edited by T. Gehrels (Univ. of Arizona, Tucson ,1978); 'The Origin of the Solar System' edited by S.F. Dennott (Wiley, Chichester, 1978).

(5) Carrington, A. 'Microwave Spectroscopy of Free Radicals' (Academic, London, 1974).

(6) Brown, R.D., Godfrey, P.D. and Storey, J., J. Molec. Spectr., 58, 445 (1975).

(7) Brown, R.D., Godfrey, P.D., Storey, J.W.V. and Bassez,M.P., J. Chem. Soc., Chem. Comm., 1978, 547 (1978).

(8) Watson, W.D., Rev. Mod. Phys., 48, 573 (1976); Ann.Rev. Astron. and Astrophys., 16, 585 (1978).

(9) Anders, E., Hayatsu, R. and Studier, M.H., Science, 182, 781 (1973).

(10) Hoyle, F., and Wickramasinghe, N.C., 'Diseases from Space' (Dent, London, 1979).

(11) Irvine, W.M., Leschine, S.B., and Schloerb, F.P., Nature, 283, 748 (1980).

MOLECULES IN INTERSTELLAR CLOUDS

W. M. Irvine, Å. Hjalmarson, O.E.H. Rydbeck

Onsala Space Observatory, Chalmers University
of Technology, S-439 00 Onsala, Sweden

Abstract. The physical and chemical state of interstellar clouds is reviewed, including recent investigations at the Onsala Space Observatory. The Orion A region is chosen as an example of a "giant" molecular cloud and hence a formation site for massive stars. The Taurus Molecular Clouds, in contrast, may be the future birthplace of solar-type stars. Molecular species identified in these regions are tabulated.

Introduction. Although the Orion "nebula" was discovered as early as 1610, it took astronomers a long time to accept that such phenomena were clouds of ionized interstellar gas (established spectroscopically in 1864), yet longer before the existence of solid particles in such clouds was shown (first two decades of the twentieth century), and longer still before the ubiquitous nature of the interstellar material was proven (1930´s) (e.g., (1,2)). Today, thanks in large part to radio astronomical molecular spectroscopy, it is recognized that the interstellar medium is clumped into clouds of varying mass, temperature and density, with the relative abundance of gas and particles being generally well correlated. The chemical nature of the particles (usually referred to as "grains" or "dust") remains a matter of considerable dispute, with suggestions for the constituents ranging from diamonds to freeze-dried bacteria (3), but current dogma favors the presence of silicates and possibly of graphite, silicon carbide, "ices" or reduced volatiles such as H_2O and CH_4, and perhaps more complex organics (e.g., (4,5)).

The physical and chemical state of the gas phase is much

Y. Wolman (ed.), Origin of Life, 11–17.

easier to ascertain observationally. The first molecular species (the radicals CH, CN and CH^+) were identified optically about 1940, but the development of radio astronomical techniques has allowed a tremendous expansion in our knowledge of interstellar clouds. Radio methods are valuable, since at the characteristic cloud temperatures and densities almost all molecules are too cold to emit optical or near-infrared radiation, and hence they could previously only be detected at the latter wavelengths by the absorption features produced in the light from background stars. Such stars could not be observed at all through moderately dense clouds because of the obscuring grains. In contrast, the lower energy rotational transitions of molecules are excited at cloud temperatures, and the emitted radio frequency radiation is not absorbed by the grains, thus permitting astronomers to see deep into such dense clouds and for the first time to determine their surprisingly large mass and size.

The interstellar clouds are relevant to the origin of life, since it seems clear that they are the sites of star and hence planet formation. That star formation is an on-going process in our Galaxy is evident from the existence of stars whose luminous lifetime, determined from their mass and rate of energy generation, is not more than a million years. It should be emphasized, however, that the precise sequence of physical and chemical states leading from an interstellar cloud to the actual birth of a star fueled by nuclear reactions is still quite unclear; and that even more uncertainty surrounds the process of planetary formation (6). It is by no means established, for example, that interstellar particles or gas phase molecules survive these formation processes (e.g., (4,7)).

Nonetheless, at the very least the chemistry of the interstellar clouds provides the initial conditions for the cosmogony of planets; and it is certainly conceivable that such material may be transmitted relatively unaltered to the surface of the Earth (from comets, for example (8)). In the following sections we briefly review what is currently known concerning the molecular composition of the gas in two major categories of interstellar clouds.

<u>The Orion molecular cloud</u> is our closest example of a "giant" cloud, and contains such evidence for current star formation as the presence of young stars, luminous sources of infrared radiation (perhaps young stars shrouded in the heated dust and gas from which they formed), glowing gas ionized by the ultraviolet radiation from young massive stars, and radio-emitting molecular masers, which seem to be statistically associated with regions of star formation (9,10). Whether the less massive stars of solar type form predominantly in such clouds, or whether instead their birthplaces are the colder

clouds like TMC-1 discussed below, is not known; but the Orion region does include T Tauri type stars, which may well be the progenitors of solar mass stars (6).

The core of the cloud is a "ridge" of enhanced molecular emission, which (when averaged over dimensions of the order of 10^4 Astronomical Units) has a kinetic temperature of ~75 K and a density of $n = 10^5 - 10^6$ molecules per cc. Note that at these conditions, so far removed from laboratory experience ($n \approx 10^{19}$), the concept of thermal equilibrium has little relevance. Thus, the "temperature" characterizing the distribution of a given molecular species over possible energy levels is not generally the same as the kinetic temperature characterizing the translational velocities of molecules and ions. This must be borne in mind when molecular abundances are deduced or predicted . Within the ridge is a small region with enhanced temperatures, violent gas motions, and perhaps "shock front" chemistry. The gas is primarily molecular hydrogen, but Orion has the greatest variety of molecular species known in an interstellar cloud outside of the Galactic center (Table 1). It appears that gas phase ion-molecule chemistry may be able to account for most of the molecules present. Indeed, such ions as HCO^+, N_2H^+, and their isotopic variants were identified in space before they had been observed in the laboratory. Nonetheless, formation on the interstellar grains, by fragmentation of these grains, in the extended envelopes of stars, or even in "solar nebulae" remain possibilities for H_2 and the larger molecules in Tables 1 and 2 (12,13,14).

We at the Onsala Space Observatory in Sweden have been trying to extend our knowledge of the chemical constitution of interstellar clouds. The observatory now operates the largest radio telescope in the world (20 m diameter) capable of observing at the millimeter wavelengths where most molecular lines are concentrated. Of equal importance in such research is the availability of extremely sensitive receivers, and Chalmers University of Technology (our mother institution) is a world leader in this respect. We have begun a systematic scan across the spectral range accessible with our equipment, a type of program which has not been possible (as a matter of policy) at the facility where most prior radio astronomical studies of molecules have been made, the 11 m telescope of the US National Radio Astronomy Observatory in Arizona. Even though this program is quite new and constitutes only a fraction of our observing program, we have been able to make the first radio astronomical detections of a large number of molecular transitions and have added several new molecules to the known constituents of the Orion Cloud (Table 1) and of the Taurus Clouds (Table 2; see following section). Other species which are believed on chemical grounds to be present in the Orion and Taurus regions, but which do not have accessible millimeter wavelength transitions, include O_2, N_2, CO_2, and HC≡CH.

TABLE 1

Compounds Identified In The Orion Molecular Cloud (cf. 15,11,16).

Simple hydrides, oxides, sulphides, and related molecules:

H_2	CO	H_3CNH_2
OH	CS	$N{\equiv}N^+H$
CH*	OCS*	
H_2O	SiO	
H_2S	SO	
NH_3	SO_2	
CH_4 ?	O_3	

Nitriles, acetylene derivatives, and related molecules:

C≡N	HC≡C-C≡N
HC≡N	C≡CH
H_3C-C≡N	H_3C-C≡C-H
H_3C-CH_2≡CN	HN=C=O*
H_2N-C≡N ?	HN=C

Aldehydes, alcohols, ethers, and related molecules:

H_2C=O	H_3C-O-CH=O*
H-C^+=O	H_3COH
H_3C-CH=O	$(H_3C)_2O$
H_2C=S	

*First reported detection in this source made at Onsala Space Observatory (17). Tentative detections indicated by (?).

For completeness, we should mention that a number of interstellar molecules have been detected to date only in the Galactic center region Sgr B_2 (including NO, NS, HNO, HCO, H_3CSH, HNCS, HCOOH, H_2CNH, H_2CCO, $HCONH_2$, CH_2CHCN, and CH_3CH_2OH). The Orion and Taurus clouds may, however, be more typical of our portion of the Galaxy.

The Taurus Molecular Clouds (TMC-1 and TMC-2) present an interesting contrast to Orion. They are members of the class of cold (kinetic temperature ~10-15 K), dark, relatively dense ($n \cong 10^4 - 10^5$ mols per cc) interstellar clouds which appear to lack internal energy sources such as young stars or protostars. They do, however, contain clumps which appear to be of approximately solar mass, and it has been suggested that such clouds are possibly the birthplaces of stars like the sun (18,19,20). Whether because they are relatively nearby and hence easy to observe, because of their (unknown) age, or possibly for some other reason, the known molecular composition of these clouds is unique. In particular, the heaviest interstellar molecules discovered to date (the cyanopolyynes HC_nN; see Table 2) have been found preferentially here, and, curiously, in the envelope around the envolved star IRC+10216 (21,22). Some of the species listed in Table 1 but not Table 2 are almost certainly also present in the latter case, but not yet detected: H_2 is surely the dominant constituent, but has no transitions excited at the temperature of TMC-1; CH_4 and H_2O are very likely present, but are also difficult to observe by radio techniques in low temperature clouds; other molecules have simply not yet been looked for.

TABLE 2

Compounds Identified In The Taurus Molecular Clouds (e.g., 20, 23,24,25)

Simple hydrides, oxides, sulphides, and related molecules:

OH	SO*	$N{\equiv}N^+H$
CH*	CO	
NH_3	CS	

Nitriles, acetylene derivatives, and related molecules:

C≡N	C≡CH
HC≡N	HN=C
C≡C-C≡N*	
HC≡C-C≡N	
HC≡C-C≡C-C≡N	
HC≡C-C≡C-C≡C-C≡N	
HC≡C-C≡C-C≡C-C≡C-C≡N	

Aldehydes and related molecules:

$H_2C{=}O$

$H{-}C^+{=}O$

*First reported detection in this source made at Onsala Space Observatory.

To conclude: such basic building blocks of biologically important monomers as H_2O, NH_3, H_2CO, CO, H_2S, CH_3CN, and H_2 (the latter detected at ultraviolet and infrared wavelengths) are clearly present in interstellar clouds. Slightly more complex species, such as $HCOOCH_3$ (methyl formate) or CH_3CH_2CN, are also observed. The simplest biological monomers themselves, such as glycine, are intrinsically much more difficult to detect: their larger number of degrees of freedom for internal motion results in the molecular population being spread over many more energy levels, thus dilluting the intensity of any particular radio frequency transition; they typically have a low vapor pressure, making it more likely that they will be frozen out on to the interstellar grains, and also making it more difficult to obtain their spectra in the laboratory; and many of the detections of somewhat simpler molecules were close to the limit of currently available sensitivity. Glycine, imidazole and other cyclic compounds have been sought radio astronomically at Onsala and elsewhere (26,27), thus far unsuccessfully. In view of the discovery of such compounds in carbonaceous chondrites (28,29), however, it seems likely that their detection in interstellar clouds may follow future increases in system sensitivity at radio wavelengths. The ultimate identification of specific polymers, in contrast, will be much more difficult.

ACKNOWLEDGEMENTS We are grateful to H. Olofsson for updating a list of detected interstellar molecules by source. Onsala Space Observatory, Chalmers University of Technology, is operated with financial support from the Swedish Natural Research Council. The low noise receiver development, done by the quantum electronics group at the Research Laboratory of Electronics, Chalmers University of Technology, is supported by the Swedish Board for Technical Development. W.I. was partially supported by NASA grant NGL 22-010-023 to the University of Massachusetts, Amherst.

REFERENCES

1 Mezger, P.G. in Interstellar Matter (ed. Astronomical Inst., Univ. Basel) ch. 1 (Geneva Observatory, Switzerland, 1972)
2 Middlehurst, B.M. and Aller, L.H., eds. Nebulae and Interstellar Matter (Univ. Chicago Press, Chicago, 1968)
3 Hoyle, F. and Wickramasinghe, C. Astrophys. Space Sci. 66, 77-90 (1979)
4 Salpeter, E.E. Ann. Revs. Astron. Astrophys. 15, 267-294 (1977)
5 Proc. Workshop on Thermodynamics and Kinetics of Dust Formation in the Space Medium, Astrophys. Space Sci. 65, Nos. 1-2 (1979)
6 Gehrels, T., ed. Protostars and Planets (Univ. Arizona Press, Tucson, 1978)
7 Clayton, D.D. in Protostars and Planets, op. cit., 13-42
8 Irvine, W.M., Leschine, S.B., and Schloerb, F.P. This volume
9 Goudis, C. Astrophys. Space Sci. 35, 409-458 (1975)
10 Rydbeck, O.E.H., Hjalmarson, Å., Rydbeck, G., Elldér, J., Olofsson, H., and Sume, A. Astrophys. J., submitted
11 Mann, A.P.C. and Williams, D.A. Nature 283, 721-725 (1980)
12 Andrew, B.H., ed. Interstellar Molecules (D. Reidel, Holland, 1980)

13 Sagan, C. and Khare, B.N. Nature 277, 102-107 (1979)
14 Anders, E., Hayatsu, R., and Studier, M.H. Astrophys. J. (Lett.) 192, L101-L105 (1974)
15 Cf. Townes, C.H. Observatory 97, 52-73 (1977)
16 Phillips, T.G. and Knapp, G.R. Bull. Amer. Astron. Soc. 12, 440 (1980)
17 Johansson, L.E.B., in preparation
18 Churchwell, E., Winnewisser, G., and Walmsley, C.M. Astron. Astrophys. 67, 139-147 (1978)
19 Myers, P.C., Ho, P.T.P., and Benson, P.J. Astrophys. J. (Lett.) 233, L141-L145 (1979)
20 Avery, L.W. in Interstellar Molecules, op. cit.
21 Winnewisser, G. and Walmsley, C.M. Astrophys. Space Sci. 65, 83-93 (1979)
22 McCabe, E.M., Smith, R.C., and Clegg, R.E.S. Nature 281, 263-266 (1979)
23 Wootten, A., Bozyan, E.P., Garrett, D.B., Loren, R.B., and Snell, R.L. Astrophys. J., submitted
24 Friberg, P., Hjalmarson, Å., Irvine, W.M., and Guélin, M. Astrophys. J. (Lett.), submitted
25 Rydbeck, O.E.H., Irvine, W.M., Hjalmarson, Å., Rydbeck, G., Elldér, J., and Kollberg, E. Astrophys. J. (Lett.) 235, L171-L176 (1980).
26 Brown, R.D., Godfrey, P.D., Storey, J.W.U., Bassez, M.P., Robinson, B.J., Batchelor, R.A., McCulloch, M.G., Rydbeck, O.E.H., and Hjalmarson, Å.G. Month. Not. Roy. Astron. Soc. 186, 5P-8P (1978)
27 Irvine, W.M., Elldér, J., Hjalmarson, Å., Kollberg, E., Rydbeck, O.E.H., Sørensen, G.O., and Bak, B. Astron. Astrophys., to be submitted
28 Anders, E., Hayatsu, R., and Studier, M.H., Science 182, 781-790 (1973)
29 Stoks, P.G. and Schwartz, A.W. Nature 282, 709-710 (1979)

ON THE ORIGIN OF ORGANIC MOLECULES IN INTERSTELLAR SPACE AND SOME OF ITS CONSEQUENCES

Kjell L. V. Johansson

Astronomical Observatory, Uppsala University
Box 515, S-751 20 Uppsala, Sweden

Abstract. The possible sources of complex organic molecules observed in the interstellar space are discussed. It is found that a leakage from lifebearing planets cannot possibly explain the observed amounts of complex organic molecules. They are probably formed in interstellar clouds, either free in space or on the surface of grains. This opens the possibility that planets (also the earth) moving through such clouds may have collected appreciable amounts of complex organic molecules. It would therefore be feasible to perform experiments like the Urey-Miller's, starting with a considerably more complex mixture than the traditional one.

During the last decades a great number of various organic molecules have been found in the interstellar space through observations at radio-wavelengths. What is the source of this great number? Simple organic compounds, as well as inorganic molecules are commonly believed to be supplied by cold stars, like carbon stars. However, more complex molecules are expected to break down at the temperatures prevailing in the atmospheres of these, or hotter stars. In principle one may then suggest that the organic molecules observed are either formed on lifebearing planets from which they have escaped, or formed in certain regions of interstellar clouds, possibly on the surface of grains.

Y. Wolman (ed.), Origin of Life, 19–25.

If the molecules represent an escape from lifebearing planets, it is of interest to calculate the rate of escape and to compare the amount that could have leaked out from the earth during the last four billion years with the amount that is contained in a typical interstellar cloud according to calculations based on actual observations. Simply by dividing the calculated amount of a certain organic molecule observed in the interstellar space with the amount of this molecule that could have escaped, one obtains a rough estimate of the number of inhabited planets that need to be involved to supply one single cloud.

The number of particles per unit volume (N_e) with speeds equal to or greater than the velocity of escape may be calculated by integrating the Maxwellian distribution of speeds

$$N(v)dv = 4\pi N\left(\frac{M}{2\pi kT}\right)^{3/2} v^2 e^{-Mv^2/(2kT)} dv \qquad \ldots (1)$$

which gives the number of particles N(v) with speeds between v and v+dv per unit volume (N=the total number of particles per unit volume, M=the mass of one particle, k=the Boltzmann constant and T=the absolute temperature). Defining $x=v(M/kT)^{1/2}$ this integral may be written

$$N_e = \int_{v_e}^{\infty} N(v)dv = \int_{x_e}^{\infty} 2N(2\pi)^{-1/2} x^2 e^{-x^2/2} dx =$$

$$= 2N\left((2\pi)^{-1/2} \int_{x_e}^{\infty} e^{-x^2/2} dx + (2\pi)^{-1/2} x_e e^{-x_e^2/2}\right) \qquad \ldots (2)$$

where the index e means escape. Depending on the rate and direction of change, dN/dt, of the population N, the number of outleaking molecules is an increasing, constant, or diminishing function of time, t. If the annual change is constant (=k%) the original population N_o will grow to N_t after t years according to the equation

$$N_t = N_o(1 + 0.01k)^t \qquad \ldots (3)$$

If one assumes that the molecule regarded is a constant fraction of the total mass of life on earth, and that this life has reached an equilibrium, i.e. the earth cannot bear notably more life than the present amount, the case of diminishing population does not exist, and the case of increasing population requires a starting population N_o that is less than the present value. Thus, one arrives to the conclusion that the most favourable case is the one where the number of formed molecules is equal to the number of escaping ones, i.e. N_e=dN/dt. The total number of molecules that could have escaped from the earth, (dN/dt)t, is then the fraction N_e/N times the present population multiplied by the time during which the escape has been going on $((N_e/N)Nt)$.

The reasoning above will be illustrated by the following example. Starting with the equation (2) the fraction N_e/N of a few molecules has been calculated for some different temperatures with the terrestrial velocity of escape v_e=11 200 ms^{-1}. The result is given in Table 1.

Then the total amount of present life on earth may be estimated according to the following. The contribution from woods is the area occupied by the trees and bushes times their mean height times the mean density which is 1000 kgm^{-3}. Assuming 4% of the forestal area to be occupied by trees and bushes, an exaggerated mean height of 20 m and extrapolating the Swedish conditions where 50% of the land area is wood, one obtains a total mass of 6×10^{16} kg for the trees and bushes. Regarding the animals, there are 4 billion humans corresponding to about 2×10^{11} kg, and since we are the most common animal of our size the other animals are not expected to change notably the figure obtained for the forests. Inclusion of the contribution from the seas and oceans will most probably lead to a value that is three times larger. Thus, the total mass of all organic life on earth is roughly about 2×10^{17} kg or less. If distributed homogeneously on the surface of the earth it would hold within a shell having a thickness of 0.5 m.

TABLE 1

Fraction of Molecules with Speeds equal to or greater than the Terrestrial Velocity of Escape for Different Temperatures.

	Molecule	H_2	HCN	HCOOH	CH_3CH_2CN
Temp.	Weight	2	27	46	55
300 K		2×10^{-21}	0	0	0
500 K		4×10^{-13}	0	0	0
1000 K		1×10^{-6}	4×10^{-88}	0	0
1500 K		2×10^{-4}	1×10^{-58}	0	0
2000 K		2×10^{-3}	4×10^{-44}	1×10^{-74}	7×10^{-90}

If the fraction of each organic molecule is 1%, the total amount of terrestrial HCN (for example) is 2×10^{14} kg, since 90%

of the organisms is water. Further, if HCN would survive 2000 K, we find from Table 1 that only $4x10^{-44}$ of the total amount has a speed enough high to allow an escape. During the last 4 billion years thus only $4x10^{-44}x2x10^{14}x4x10^{9}=3x10^{-20}$ kg could have leaked out from the earth. Observations of interstellar organic molecules give a typical column density of 10^{14} cm^{-2}, or 10^{18} m^{-2} (see for example Avery et al. (1)). With a cloud size of 1 light year in diameter the number of molecules in the cloud is about $8x10^{49}$ and thus the total mass of the cloud is $4x10^{24}$ kg which is around half the mass of the earth. The number of planets needed to produce the HCN molecules observed in one single cloud is then $4x10^{24}/3x10^{-20}$ or 10^{44}. If all are earthlike their total mass is $10^{44}x3x10^{-6}=4x10^{38}$ solar masses, which of course is totally impossible, since the mass of our Galaxy is only 10^{11} solar masses. For heavier molecules the chances are considerably smaller. We therefore arrive at the conclusion that a leakage from the planetary atmospheres of lifebearing planets cannot possibly explain the presence of organic molecules observed in the interstellar space.

Thus the remaining possibility is that the organic molecules are formed in the interstellar space. Then one has to explain the chemical reactions and to compare the rate of formation with the rate of dissipation due to collisions with photons and particles, or spontaneous dissociation. A study of some of these problems has recently been published by Hollenbach and McKee (2).

Since the interstellar clouds is the raw material in the stellar formation, and thus also at the formation of the solar system, one would expect that there were originally rather complex molecules available at the formation of our solar system. This seems to be supported by observations of the atmospheres of the outer planets and satellites, as well as the comets, where such compounds as HCN and CH_3CN have been observed. However, one cannot exclude the possibility that these have been formed at later stages. If so, there is still a good chance that the earth has collected a great deal of organic molecules when passing through interstellar clouds during its (and the whole solar system's) motion around the centre of our Galaxy. As the distance to the galactic centre is around 10 kpc, the length of the orbit of the solar system is 63 kpc. A plausible number of interstellar clouds is 5 kpc^{-1} (or more) in most directions of the line of sight in the galactic plane. This means that there are at least 300 clouds distributed along the orbit of the solar system. Assuming that the orbital velocities at 10 kpc from the galactic centre are normally distributed with a mean value of 250 kms^{-1} and a standard deviation of 20 kms^{-1} one obtaines that 4% or 12 of the clouds ought to have velocities deviating less than 1 kms^{-1} from that of the solar system. In the same way one finds that 38% or 115 clouds deviate less than 10 kms^{-1}. Thus around 100 clouds should have a velocity deviating 1-10 kms^{-1} from that of the solar system. During one revolution, corresponding to 250 million years, the earth is expected to pass through at least one cloud, probably more, having a velocity difference in this interval.

With the column density given above, 10^{18} m^{-2}, and the cross section of the earth being 1.3×10^{14} m^2 the number of molecules collected during one single passage is 1.3×10^{32}, which in the case of HCN corresponds to a total mass of 5 700 ton. Otherwise expressed, one finds that each cm^2 has 10^{14} organic molecules to do experiments with, probably even more. As soon as life has formed, it grows rapidly, which is evident from equation (3). As an example, one finds that with the modest increase of 1% per year, it would take less than 5000 years to produce the present total mass of organic life on earth, starting with 1 kg.

To summarize the present paper, we have found that the organic molecules observed in the interstellar space cannot be explained as due to a leakage from lifebearing planets. Thus, the organic molecules observed are most probably formed out there in space. Hence, it is very reasonable to consider the possibility that the surface of the earth, at the time of the start of the development and evolution of organic life, had collected a large amount of more or less complicated organic molecules which may have facilitated or even explicitly made possible the generation of life. It would therefore be feasible to perform experiments of the Urey-Miller kind, starting with a considerably more complex mixture than the traditional one.

ACKNOWLEDGEMENT. I am greatly indepted to Lars Olof Lodén, who suggested the present investigation and has given strong support during the work and many improvements on the final manuscript.

REFERENCES.

(1) Avery, L.W., Broten, N.W., MacLeod, J.M., Oka, T. and Kroto H.W.: Ap. J. (Letters) 205, L173 (1976).

(2) Hollenbach, D. and McKee, C.F.: Ap. J. Suppl. Ser. 41, 555 (1979).

COMETS AND THE ORIGIN OF LIFE

W. M. Irvine

Onsala Space Observatory, Chalmers University of Technology, S-439 00 Onsala, Sweden

S. B. Leschine and F. P. Schloerb

Departments of Microbiology (SBL) and Physics and Astronomy (FPS and WMI), University of Massachusetts, Amherst, MA 01003, USA

Abstract. The possible role of comets in the origin of life is reviewed both con and pro. The former viewpoint stresses our considerable ignorance of the physical and chemical state of comets. The latter position points out the possibility in some comets of an extended thermal history, which might have led to the production of biologically significant macromolecules.

Introduction. Although written records indicate that scholars have studied comets for at least the last three thousand years (1), the nature and origin of these objects remain controversial. This ignorance has not prevented considerable speculation, and comets have been associated with the death of kings, the fall of empires, plagues (2), and more recently with celestial gamma ray sources (3), interstellar masers (4) and the origin of the Earth's volatiles (5). The present general (although not universal (6)) consensus among astronomers is that a comet consists basically of a loose conglomeration of frozen gases with embedded material similar to that found in chondritic meteorites, particularly the carbonaceous chondrites (7,8,9,10). Because of the obvious abundance of volatile material, it is commonly held that comets may be the most pristine samples of the solar nebula out of which the Sun and planets formed (11), although the location where they themselves were formed in this contracting cloud (and even whether some may be interstellar in

Y. Wolman (ed.), Origin of Life, 27–32.

origin (12)) is very uncertain. Because of their elemental composition and possible genetic relationship to carbonaceous meteorites and molecular clouds, both of which are rich in organic molecules (cf. (13)), and in light of recent suggestions concerning a possibly complex thermal history (see below), we may ask whether comets could have played a role in the origin of life. In particular, we shall consider here, from both negative and positive sides, whether comets may have been the sites of chemical and/or biological evolution.

The negative viewpoint is perhaps supported most strongly by the economy of scientific hypotheses (Occam's razor); if it is not necessary to invoke comets to explain the origin and evolution of life, than it is clearly simpler not to do so. Moreover, our lack of knowledge concerning their nature is so great that it is clearly impossible to argue the likelihood that comets played a role in these processes.

Looking more closely at the latter point, we find that direct evidence for both the chemical composition and the size of cometary nuclei (as opposed to the gaseous coma and tail which develop near the Sun) is largely lacking.

TABLE 1

Atomic And Molecular Species Observed In Comets (11)

Organic:	C, C_2, C_3, CH, CN, CO, CS, (HCN),(CH_3CN)
Inorganic:	H, NH, NH_2, O, OH, Si, (H_2O)
Metals:	Na, Ca, Cr, Co, Mn, Fe, Ni, Cu, V
Ions:	CO^+, CO_2^+, CH^+, CN^+, N_2^+, OH^+, H_2O^+
Dust:	Silicates

Of the observed cometary constituents, listed in Table 1, almost all are radicals or ions which are presumably dissociation products of "parent" molecules, which themselves make up the solid nucleus. The exceptions are certainly very interesting to chemical evolutionists (CO, CS, H_2O, HCN, CH_3CN), but in fact the direct observational support for the latter three is rather weak: each was observed only marginally by radio astronomical techniques in only one comet, the apparent velocities differed from those of the optically observed radicals, and subsequent confirmation in the same (for CH_3CN) or later comets has not been achieved (14,15,16). In addition, the presence of (for example) gaseous HCN or H_2O in the coma does not necessarily imply their presence in the nucleus itself; these molecules might well be produced by gas phase chemistry after evaporation of the actual "parent molecules", perhaps by reactions similar to those postulated to explain the existence of these same and other molecules in interstellar space (17).

Of course, these species may have been present in the observed comets, since it is well established that cometary emission is time variable, with "flares" or "outbursts" occurring unpredictably; and it must always be borne in mind that comets may differ considerably from one to another. Some nitrile must be present to explain the production of the CN radical, and circumstantial evidence supports the view that in at least some comets the volatile material is largely H_2O (18,19), which is not surprising simply on the grounds of the cosmic abundance of the elements. Nonetheless, relative abundances in interstellar molecular clouds, which may be the birthplace of comets, are still quite uncertain; and the same holds true for the bulk chemical composition of small bodies at the alternative plausible place of origin, the solar system at approximately the orbits of Uranus and Neptune.

Of probably equal importance with the chemical composition to the possible biological significance of comets is their thermal history. On the traditional view, comets are frozen lumps of ices and "dust", which have been at temperatures $\lesssim$ 100K since their formation (20). Under these conditions it is very difficult to see how any significant chemical evolution could have taken place. According to current theory the solar system is surrounded by a vast cloud of comets at typical distances of 1000 x (radius of Pluto's orbit), and occasionally gravitational perturbations by passing stars will send some of these objects into the inner solar system. Hoyle and Wickramasinghe (21,22) have suggested that solar heating of such comets will lead to melting, the release of chemical energy, and the production of complex molecules; but the lifetime of comets perturbed into orbits where they are periodically heated is exceedingly short ($\lesssim 10^5$ years (23,24)).

On the other hand, Irvine, Leschine and Schloerb (25) have pointed out the possible importance of cometary heating by short-lived radioactive elements such as ^{26}Al, whose presence in the early solar system has been established by analyses of certain meteorites. The time scale for any warm period in the comet's thermal history depends, however, on at least three very uncertain factors: the time it took the comets to form, since if this is longer than the radioactive halflife, there will be little heating of the solid nucleus (26); the thermal parameters (thermal conductivity, heat capacity, mass density) which determine the heat loss and hence rate of cooling; and the physical size of the cometary nucleus, which also profoundly affects the rate of cooling. In consequence of these uncertainties, the time during which liquid water might be available in a comet might be anywhere from essentially zero to greater than 10^9 years.

The positive viewpoint must at our present level of knowledge be an argument of possibility only. Despite our preference for the "simpler" hypothesis of a purely terrestrial origin of life, nature may in fact have been more complex. There may be comets which retained liquid water in their interior (25) for times comparable to the (at most) few hundred million years between the solidification of the Earth's crust and the appearance of life (e.g., (27)). The greater abundance of chemically available volatiles (C,N,O, and particularly H; (11)) in comets relative to the most primitive carbonaceous chondrites, in which amino acids and nucleic acid bases have been found (e.g., (28,29)), then makes it quite plausible that still more complex molecules were produced in comets. And it may be that the cometary environment was more favourable to the polymerization necessary to produce polypeptides, nucleotides or other biologically significant macromolecules than were typical terrestrial environments.

How much further might evolution have gone? And would the result in any case be relevant to terrestrial life? The answer to the first question is simply speculation, although we note that a considerable amount of chemical free energy would be available (30) and that photosynthesis - independent biocommunities apparently exist on Earth (31).

The second question may be reworded into, first, could material be transferred in unaltered form from a comet to the Earth's surface; and, second, how would we ever know? There exists a clear affirmative answer to the first part, since it is well established that meteor showers are the result of the Earth encountering cometary debris and that particles smaller than a few tens of μm diameter are not significantly heated in passing through the atmosphere (32). In fact, the Earth accretes about 10^4 tons of micrometeoroidal material per year; the thermally unaltered fraction of this amount is somewhat uncertain, but is not negligible (8,33). Viewing the same point from a slightly different aspect, if one part in 10^9 (a number, in the spirit of the present paragraph, largely taken out of the air) of this mass were to consist of nucleic acids of molecular weight equal to that of the smallest known self-replicating entities (viroids (34)), then the Earth is subject to a flux of $\sim 10^5$ such molecules per m^2 per year (a unit which, in light of the suggestions by Hoyle and Wickramasinghe (35), might usefully be defined as one sneeze).

But how could an extraterrestrial origin for biomolecules or even life forms be recognized? Unfortunately, our knowledge of chemical and biological evolution is still too limited to answer this question. It is not yet clear whether a comprehensive, self-consistent and purely terrestrial scheme of chemical evolution can be devised. Should it turn out that certain molecules fall outside of plausible models which include the (still quite uncertain) constraints of the primitive environment, it may be reasonable to consider their possible cometary origin. Likewise, the traditionally assumed universality of the genetic code and of a core of common biochemical pathways have been used to argue that all known life is descended from a common ancestral form. Recent evidence that there may be novel metabolic pathways in archaebacteria (36,37) and the discovery of differences in the genetic code employed by mitochondria (38) suggest the kind of evidence which might, at the least, raise the possibility of a more heterogeneous origin for life on Earth.

Hopefully, both future ground-based astronomy and space missions to comets will help to answer the basic questions about the composition, size, and thermal properties of comets; and, ultimately, sample returns will test hypotheses of chemical evolution. It must be borne in mind, however, that although such missions may confirm a link between comets and the origin of life, the heterogeneous nature of comets means that negative findings cannot be easily extrapolated to all comets at all times.

ACKNOWLEDGEMENT. This research was conducted with partial support from NASA grant NGL 22-010-023.

REFERENCES

1 Olivier, C.P. Comets (Williams-Wilkins, Baltimore, 1930)
2. Hellman, C.D. The Comet of 1577: Its Place in the History of Astronomy (Columbia Univ., New York, 1944)
3 Harwit, M. and Salpeter, E.E. Astrophys.J.Lett. 186, L37-L40 (1973)
4 Oppenheimer, M. Nature 254, 677-678 (1975)
5 Chang, S. in Space Missions to Comets (eds. Neugebauer, M., Yeomans, D.K., Brandt, J.C., and Hobbs, R.W.) 59-112 (NASA Conf. Publ. 2089, 1979)
6 Lyttleton, R.A. Quart. J. Roy. Ast. Soc. 18, 213-233 (1977)
7 Whipple, F.L. and Huebner, W.F. Ann. Revs. Astr. Astrophys. 14, 143-172 (1976)
8 Anders, E., Ganapathy, R., Krähenbuhl, U., and Morgan, J.W. Moon 8, 3-24 (1973)
9 Ganapathy, R. and Brownlee, D.E. Science 206, 1075-1077 (1979)
10 Donn, B., Mumma, M., Jackson, W., A'Hearn, M., and Harrington, R. (eds.), The Study of Comets (NASA SP-393, 1976)
11 Delsemme, A.H. in Comets, Asteroids, and Meteorites (ed. Delsemme, A.H.) 3-13 (Univ. Toledo Press, 1977)

12 Witkowski, J.M. in The Motion, Evolution of Orbits, and Origin of Comets (eds. Chebotarev, G.A., Kazimirchak-Polonskaya, E.I., and Marsden, B.G.) 419-425 (D. Reidel, Dordrecht-Holland, 1972)
13 Irvine, W.M., Hjalmarson, Å., and Rydbeck, O.E.H. in This Volume
14 Schloerb, F.P., Irvine, W.M., and Robinson, S.E. Icarus 38, 392-397 (1979)
15 Simon, M.N., Owen, T., and Simon, M. Nature 252, 666 (1974)
16 Ekelund, L., Irvine, W.M., Andersson, Ch., Schloerb, F.P., and Robinson, S.E. To be published
17 Oppenheimer, M. Astrophys. J. 225, 1083-1089 (1978)
18 Delsemme, A.H. in The Study of Comets (eds. Donn, B., Mumma, M., Jackson, W., A'Hearn, M., and Harrington, R.) 711-737 (NASA SP-393, 1976)
19 Delsemme, A.H. Mém. Soc. Roy. Sci. Liège (ser. 6), 9, 135-145 (1976)
20 Whipple, F.L. and Stefanik, R.P. in Nature et Orgine des Comètes 33-52 (Univ. Liège, 1966)
21 Hoyle, F. Mercury 7, 2-7 (1978)
22 Hoyle, F. and Wickramasinghe, C. New Scient. 76, 402-404 (1977)
23 Sekanina, Z. Astron. J. 74, 1223-1234 (1969)
24 Dobrovol'skij, O.V. in The Motion, Evolution of Orbits and Origin of Comets, op.cit., 352-355
25 Irvine, W.M., Leschine, S.B., and Schloerb, F.P. Nature 283, 748-749 (1980)
26 A point emphasized by F. Whipple, private communication (1979)
27 Nisbet, E. Nature 284, 395-396 (1980)
28 Anders, E., Hayatsu, R., and Studier, M-H. Science 182, 781-790 (1973)
29 Stoks, P.G. and Schwartz, A.W. Nature 282, 709-710 (1979)
30 Anders, E. Ann. N.Y. Acad. Sci. 108, 514-533 (1963)
31 Karl, D.M., Wirsen, C.O., and Jannasch, H.W. Science 207, 1345-1347 (1980)
32 Whipple, F.L. Proc. Nat. Acad. Sci. USA, 37, 19-30 (1951)
33 Hughes, D.W. in Cosmic Dust (ed. McDonnell, J.A.M.) Ch. 3 (J. Wiley, New York, 1978)
34 Diener, T.O. Ann. Rev. Microbiol. 28, 23-39 (1974)
35 Hoyle, F. and Wickramasinghe, C. Diseases from Space (J.M. Dent, London, 1979)
36 Woese, C.R., Magrum, L.J., and Fox, G.E. J. Molec. Evol. 11, 245-252 (1978)
37 Bayley, S.T. and Morton, R.A. in Strategies of Microbial Life in Extreme Environments (ed. Shilo, M.) 109-124 (Berlin: Dahlem Konferenzen, 1979)
38 Hall, B.D. Nature 282, 129-130 (1979)

NATURE AND ORIGIN OF ORGANIC MOLECULES IN COMETS

A.H. Delsemme

Jet Propulsion Laboratory, California Institute of Technology (On Sabbatical Leave from the Department of Physics & Astronomy, University of Toledo, Toledo, Ohio)

The organic molecules sublimating from the cometary nucleus either represent the "frost" of interstellar molecules that condensed onto those interstellar grains that have later accreted into comets, or they represent the "snows" that condensed onto silicate grains during the cooling phase of the presolar nebula. Even if one does not accept the speculations of Hoyle and Wickramashinghe (1977), the fact that prebiotic chemistry could have been brought from the interstellar space or from the solar nebula to the Earth by comets is an intriguing possibility that cannot be ruled out at this stage.

The hazy head and the long tail of a comet, come from the permanent loss of gas and dust by a tiny object, the cometary nucleus. The nucleus size is usually from 1 to 10 km, and it is believed to be formed by an icy conglomerate of dust cemented by snows of water and of more volatile gases, that decay in the solar heat (1). Periodic comets decay fast: they can survive 100 to 300 passages that is, typically, a few thousand to a few hundred thousand years. Their decay implies the existence of a steady source of "new" comets to maintain the observed population. Indeed, about 100 different "new" comets have appeared during the last century, at a steady rate of about one per year (2). From the narrow distribution of their binding energies (per unit mass), Oort (3) demonstrated that their previous perihelion was outside of the planetary system; the reason of their sudden appearance was that their orbit had been perturbed at aphelion by the passage of nearby stars. He deduced from their observed flux rate that they come from a large reservoir of about 10^{11} comets (the Oort's cloud), which is gravitationally bound to the sun.

Y. Wolman (ed.), Origin of Life, 33–42.

The periodic comets result from the capture, by planetary perturbations, of some of the "new" comets; whereas the "new" comets seem to have been stored in the deep freeze of the Oort's cloud since the beginning of the solar system. However, since no plausible scenario has ever been proposed to make these comets "in situ" at those large distances, it is widely believed that they were made in the protosolar nebula, then ejected into Oort's cloud by the very processes that have been at work to make the solar system (4). Scenarios on the origin of the solar system suggest that comets may be those planetesimals due to the accretion of particles of ice and dust that had sedimented from the protosolar nebula into thin equatorial rings. Those planetesimals seem to have played a fundamental, although still obscure role, in the formation of the planets; and a small fraction of them seem to have been ejected into and preserved within the Oort's cloud. Their ices may be either the condensates from the solar nebula, or the remnants of frosty interstellar grains that have followed suit in the collapse of the protosolar nebula and finally sedimented into its rings.

Spectra of the cometary head and tail reveal the chemical nature of comets only very incompletely, because the interaction of the cometary nucleus with the solar wind and the solar light is a complex phenomenon. The dust grains present in the tail contain a small fraction of carbon in the form of graphite, carbynes or organic compounds. Solar abundances for metals are suggested by those spectra obtained from meteors coming from cometary orbits (7) as well as by that extraterrestrial dust that has been collected in the upper atmosphere (8). The gas that has sublimated from the snows goes on expanding isotropically at the velocity of sound in a cold spherical "exosphere" centered on the nucleus. The gravity of a comet is so weak that it cannot keep any permanent atmosphere around the nucleus. The solar ultraviolet dissociates and ionizes in different ways the different types of molecules in the expanding gas. Near the nucleus, the gas density is large enough (typically 10^{14} molecules cm^{-3}); therefore, collisions may happen between molecules and their dissociation and ionization by-products, up to distances of the order of ten thousand kilometers from the nucleus. These collisions induce chemical reactions, mainly fast reactions between ions and molecules. Therefore, chemical kinetics prevails (9) and transforms water vapor and the other "parent" molecules stemming from the nucleus (whose nature remains very much uncertain) into the sequence of ions and molecules that are observed in the spectra of cometary heads and tails (Table 1).

TABLE 1

Observed In Cometary Spectra

Organic:	C	C_2	C_3	CH	CN	CO	CS	HCN	CH_3CN
Inorganic:	H	NH	NH_2	O	OH	H_2O	S		
Metals:	Na	K	Ca	V	Mn	Fe	Co	Ni	Cu
Ions:	C^+	CO^+	CO_2^+	CH^+	H_2O^+	OH^+	Ca^+	N_2^+	CN^+
Dust:	Silicates (infrared reflection bands)								

Later, at distances from ten thousand to one million kilometers from the nucleus, there are no collisions anymore because of gas dilution, but radicals are still dissociated by the solar ultraviolet until they become atoms. The atoms become ionized much later. In particular, the resonance lines of carbon, oxygen and hydrogen are observed up to very large distances of the nucleus, in the vacuum ultraviolet, by rockets or space telescopes in orbit around the Earth (10). If the nitrogen line has not been detected so far, it probably is for mere technical reasons. The production rates of C, O, H, completed by the rates of CN and NH that give a lower limit for the rate of N, provide an approximate elementary analysis of the volatile fraction, averaged over four recent comets (11). Combining these data with the dust-over-gas mass ratio, established for two comets only (12), the mean elementary abundance can be reconstructed in % of the silicon abundance (13). It is found that oxygen (present for 2/3 in gases and for 1/3 in the silicate dust) must approximately reach solar abundance; whereas only 1/4 of the solar abundance is reached so far in the observed carbon. However, there is some indication from the infrared properties of dust (5) that it is not a dielectric as pure as silicates are; a small amount of carbon or nonvolatile hydrocarbons mixed up with silicates could easily double the total amount of carbon to 1/2 the solar abundance.

The most spectacular result is, however, the large depletion of hydrogen: its solar abundance is depleted by a factor of 2000 in comets! The balance of the elements in the observed molecules suggest that hydrogen comes exclusively from those molecules like H_2O, HCN or CH_3CN where it is bound to other atoms. There is not enough hydrogen to make room for other undetected molecules containing hydrogen. The redox ratio H/O clearly is in the vicinity of 2, which does not give much leeway for other reduced compounds and suggests,apart from water, a chemistry very much like that of interstellar molecules. When the previous results (uncertain as they are) are compared with elementary abundances in the most primitive meteorites (13), it appears that comets contain from three to ten times more volatile H, C, N, O compounds than the C I carbonaceous chondrites, which contain the largest volatile fraction of all known meteorites (Table 2).

TABLE 2
Elementary Abundances (in atom numbers)

Element	Sun	Comets	C I Chondrites
H	30,000	15	1.5
C	13	3 to 6	0.7
N	3	> 0.1	0.05
O	21	21	7.5
Si	1	1	1
All Metals	~2	~2	~2

Although the chemistry of the cometary nucleus is still very poorly known, present results suggest specific features that were totally unsuspected. For the first time three stable molecules have been observed by radio telescopes in the cometary head, namely H_2O, HCN, and CH_3CN (13). From the production rates of CN deduced from optical spectra in many comets, the cyanides account for about 1% of the total; whereas water is known by numerous circumstantial data to be the primary volatile constituent in many comets. No mechanism has been suggested so far (for instance, a chain of charge-exchange reactions in the coma) (14) that could produce HCN and CH_3CN in the right amounts from other parent molecules. The existence of CO in the recently observed vacuum ultraviolet spectra, as well as of CO^+ and CO_2^+ in the spectra of many comets, suggests that CO and/or CO_2 are two molecules that are likely to be present in the nucleus at a rather large abundance (10% range). The red forbidden line of oxygen that is so strong in many cometary spectra is likely to come at least for 75% from the dissociation of CO_2 into CO+O (11). Combining all these data with the crude elementary analysis mentioned previously (13) obtained from the resonance lines of H, C, and O atoms in the vacuum ultraviolet, it is possible to construct a heuristic model (15) of the molecular compounds present in the cometary nucleus. This model, completed with S and CS data (16) is shown in Table 3.

TABLE 3
Heuristic Models of Cometary Chemistry

Model I (Whipple 1952)		Model II (Delsemme 1980)	
Solar Abundances Thermal Equilibrium Low Temperature (in mass units)		Observational data from four comets (Delsemme 1978) plus S and CS from a fifth one (this paper)	
Silicates	34	Silicates	34
Water	34	Water	30
CH_4	26	CO_2 + CO	6
NH_3	6	S + CS	≤1
		HCN, CH_3CN and parents of C_2, C_3, etc.	1
TOTAL	100	TOTAL OBSERVED	≤72
Dust/Gas ratio	0.5	Dust/Gas ratio (from total gas)	0.9

If the mixture from which comets were made contained all elements in their solar (= cosmic) abundances and were cooled off at thermodynamic equilibrium, it would contain 34% water snow, 26% methane and 6% ammonia (strictly speaking ammonia would also make ammonium hydroxide with some of the water, and some methane would be imprisoned in solid clathrate hydrates). The major difference that appears at first glance is that compounds seem to be less reduced and more oxidized than expected: instead of CH_4 we observe CO and CO_2; instead of NH_3 we observe HCN and CH_3CN. Altogether, it is clear from the elementary abundances that the ratio (H/O ratio) shows an excess of oxygen rather than of hydrogen (after all the possible water has been produced).

Since equilibrium thermodynamic calculations have been used with success for meteorites (at least in a first approximation) to predict high-and-moderate-temperature condensates and the different fractionations observed, Delsemme and Rud (17) have tried to establish whether and under which equilibrium conditions the previous chemical model could be obtained from a solar mixture more or less depleted in hydrogen. It was found first that the solar C/O ratio had to be at least as large as 0.66. This is slightly higher than the nominal recent assessment of the solar

C/O ratio (18) but largely within its error bars (0.6 ± 0.2). It was then possible to duplicate the model, only by assuming that the hydrogen has been drastically depleted by an unknown process in the solar nebula. A thermal equilibrium between 550 and 660 K and at a pressure of 10^{-4} atm (typical of solar nebula models) has to be quenched and the condensates collected between 90 and 100 K. A typical condensate, collected on grains at 96 K was

78% H_2O 21% CO_2 0.8% HCN 0.002% CH_3CN

Graphite condensation constitutes a sink for the carbon present in the gas phase; therefore its appearance at thermal equilibrium suppresses quickly all possibilities of making enough CO_2, HCN and CH_3CN to eventually condense; fortunately, it is known that graphite formation has a very slow kinetics. In this model, the missing carbon remains in uncondensed CO in the primeval nebula, but this discussion suggest that some carbon could be incorporated as graphite in the silicate dust, as also suggested by Ney's infrared reflection spectra of cometary heads.

The major discrepancy of the equilibrium computations is the very small amount of CH_3CN as compared with that of the heuristic model. Larger production rates of CH_3CN could be found only in nonequilibrium situations, similar to the Fisher-Tropsch-type reactions invoked by Anders (19) to explain higher hydrocarbons in carbonaceous chondrites. However, the drastic depletion of hydrogen requested by our analyses and obvious from the crudest data of the Lyman α halos, could be understood in an entirely different context. There is not much doubt that interstellar grains had to follow suit in the contraction of the primitive solar nebula. The inside of the nebula was obviously very much heated by its final contraction, therefore some of the grains were certainly processed by heat or even vaporized. However, those particles that have eventually accreted in the outer rings of this nebula may not have been heated to temperatures much larger than 60 to 80 K, where even their icy mantles could have been preserved.

If comets were formed there from (quasi) unprocessed interstellar grains, the kinetics of the ion-molecular reactions for the formation of their icy mantles in interstellar space may have induced the drastic hydrogen depletion suggested by our results. Let's clarify this matter somewhat further. Even in dense H_2 interstellar clouds, the gas density remains low enough to restrict gas reactions to binary collisions. The low temperature then restricts these binary collisions to the ion-molecule reactions, because they have practically no activation energy to overcome. Since at these low temperatures (typically 10 K) entropy plays no role to speak of, all effective ion-molecule reactions must also be exothermic. The chemistry must presumably be initiated by cosmic-ray ionizations of the most abundant species like H_2 and He; but because of its high ionization potential He^+

provides efficient exothermic reactions for the production of abundant C^+, N^+, O^+, which are the basis for the complex organic chemistry of the dense clouds, whereas H_2^+ quickly disappears into HCO^+. The abundant C^+ does not react exothermically with H_2, which may partially explain the apparent depletion of H in all the subsequent organic molecules. Alternately, surface reactions on cold interstellar grains (although they are less well understood) may also play an important role, and here the fact that hydrogen does not stick easily to grains may also induce a large hydrogen depletion in the final molecular products.

Whatever the origin of this drastic hydrogen depletion, there is no doubt that it has been observed in comets (Table 4). The drastic change in the oxydoreduction ratio H/O of the primeval mixture is much more favorable for a prebiotic chemistry to develop. Of course, stoechiometry is not a sufficient condition, but it makes things easier.

TABLE 4
Elemental Abundances, in Number of Atoms

	H	C	N	O	H/O
Solar nebula (cosmic	30,000	13	3	21	1400
Comets (volatile fraction)	15	3	~0.2	10	~1.5
Life (protoplasm)	27	2.1	0.3	10	2.7

The place where prebiotic chemistry got started is still very controversial, but amino-acids in meteorites demonstrate that it has started much earlier than in the terrestrial oceans. Carbonaceous chondrites, in particular C I chondrites, may be extremely close to outgassed cometary material (20). Anders et al. (19) have repeatedly mentioned that all amino-acids and nucleic bases identified in carbonaceous chondrites can be explained by Fischer-Tropsch-type reactions at 360-400 K and at 10^{-5} atm in the solar nebula. Since most interstellar molecules could also be synthetized by this type of reaction, Herbig's (21) idea is still well alive, namely that prestellar nebulas, scattered away by stellar winds, could be the origin of interstellar molecules. We may be lured by this idea only because we understand classical thermodynamics better than the catalytic chemistry on cold interstellar grains. However, at cold temperatures, the actual temperature is less and less important to change the equilibrium, because the entropy effect narrows, to vanish near the absolute zero; whereas chemical reactivity does not vanish contrarily to the classical picture, because of a molecular tunneling effect (22). Since all equilibria are displaced to the exothermic side, even for very complex molecules, it is even conceivable that all prebiotic molecules could be produced on interstellar grains, although this view has been contested recently (23).

Whether prebiotic chemistry in comets could evolve through and beyond simple amino-acids and nucleic bases seems to mainly depend on the early history of thermal evolution of the cometary nucleus. Severe prerequisites are introduced if liquid water is deemed indispensable. Among the heat sources, we can eliminate gravitational energy because the average gravity field at the nuclear surface is of the order of 10^{-4} g; we can also eliminate the thermal decay in the solar heat: Not only time scales are too short (10^5 years at most) but also most of the vaporizations are sublimations; their steady state temperature is in the vicinity of 200 K and varies with a very small fractional power of heliocentric distance (24).

We are left with extinct or long-lived radioactivities. Whipple and Stefanik (25) have shown that the cosmic abundances of U^{238}, U^{235}, Th^{232} and K^{40} could conceivably raise a typical 5 km radius nucleus to central temperatures of 10 to 20 K. This is clearly not enough, but the situation has changed with the discovery of residues of extinct Al^{26} in carbonaceous chondrites (26). The abundances deduced from the isotopic anomalies of the Allende meteorite would be large enough to much exceed the melting point of water and even that of silicates in cometary cores. The steady state surface temperature needed to radiate the heat away would never be much more than 100 K for a sphere of 5 km radius, losing only the very volatile gases (like CO or CH_4) from the surface layers. The temperature gradient would also produce a migration of the volatiles towards the surface, and therefore a radial gradient in their concentrations. The lifetime of Al^{26} is such that during cooling, a pond of liquid water could percolate in the core for up to a few million years, for the largest cometary sizes. This seems enough for an interesting polymer chemistry to develop, but the size of cometary bodies is decidedly too small to achieve characteristic times of the order of 10^8 years, for the transient period during which a small pond of water could be kept in the liquid state in the core of the nucleus. Of course, this assessment depends on the value of several uncertain parameters. For instance, there were about 10^{12} comets in the early Oort's cloud, 4.5 billion years ago. We have seen 600 of them. If we extrapolate their size distribution to larger sizes, the number of cometary bodies of 100 to 10,000 km could still be several orders of magnitude larger than that of the planets (if there is no cutoff mechanisms to limit their size). Alternately, if the insulation properties of the icy conglomerate are much better than expected, we might have had in our solar system, millions of cometary bodies keeping a central pond of liquid water for the first 10^8 years of their existence, but these vistas are highly speculative.

Even if one does not accept the speculations of Hoyle and Wickramasinghe (1977), the fact that prebiotic chemistry could have been brought from the interstellar space to the Earth by comets still is an intriguing possibility. The terrestrial planets are likely to have been covered by a veneer of H, C, N, O molecules due to an early bombardment of comets. Indeed, Safronov's (27) model explains the origin of Oort's cloud by the ejection of comets from the region of the giant planets as a natural by-product of their accretion; this implies a large fraction of comets ejected on unstable orbits in collision course with the terrestrial planets, and may explain why the Earth and Venus contain two to three times as much carbon as predicted by the classical equilibrium-condensation model of the solar nebula (28) although other explanations are possible (29).

We still do not know how our biosphere was formed or whether the prebiotic chemistry was born in interstellar space. One day, comets may tell us the answer.

ACKNOWLEDGEMENT

NSF Grant AST-79-14789 and NASA grant NSG-7301 from the Planetary Atmosphere Program are gratefully acknowledged.

REFERENCES

1. Whipple, F.L. Astrophys. J. 111, 375 (1950). Delsemme, A.H. A.H., Comets, Asteroids, Meteorites, edit. A.H. Delsemme, publ. Univ. of Toledo Bookstore (1978), p. 8.
2. Marsden, B.G., Sekanina, Z., Everhart, E., Astron. J. 83, 64 (1978).
3. Oort, J., Bull. Astron. Inst. Netherlands, 11, 91 (1950).
4. Cameron, A.G.W., The Origin of the Solar System, edit. S.F. Dermott, publ. Wiley and Sons, New York, (1978), p. 49.
5. Ney, E.P., Astrophys. J. Letters, 189, L 141 (1974). Ney, E.P., Icarus 23, 551 (1974).
6. Slaughter, C.D., Astron. J. 74, 929 (1969).
7. Millman, P.M., Comets, Asteroids, Meteorites, edit. A.H. Delsemme, publ. Univ. of Toledo Bookstore (1977) p. 127.
8. Brownlee, D.E., Rajan, R.S., Tomandl, D.A., Comets, Asteroids, Meteorites, edit, A.H. Delsemme, publ. Univ. of Toledo Bookstore (1977) p. 137.
9. Oppenheimer, M., Astrophys. J., 196, 251 (1975).

10. Opal, C.B., Carruthers, G.R., Astrophys. J., 211, 294 (1977). Feldman, P.D., Tanacs, P.Z., Fastie, W.G., Donn, B., Science 185, 705 (1974).
11. Delsemme, A.H., Comets, Asteroids, Meteorites, edit. A.H. Delsemme, publ. Univ. of Toledo Bookstore (1977) p. 8. Delsemme, A.H., Combi, M.R., Astrophys. J., 228, 300 (1979).
12. Finson, M.L., Probstein, R.G., Astrophys. J., 154, 353 and 327 (1968). Sekanina, Z., Miller, F.D., Science, 179, 565 (1973).
13. Delsemme, A.H., Comets, Asteroids, Meteorites, edit. A.H. Delsemme, publ. Univ. of Toledo Bookstore, (1977) p. 8.
14. Huebner, W.F., Comets, Asteroids, Meteorites, edit. A.H. Delsemme, publ. Univ. of Toledo Bookstore (1977) p. 57.
15. Delsemme, A.H., Proc. Welch Conference, XXI Cosmochemistry, edit. W.O. Milligan, publ. Welch Foundation, Houston (1978) p. 113.
16. Jackson, M.W., Rahe, J., Donn, B., Smith, A.M, Keller, H.U., Benevenuti, P., Delsemme, A.H., Owen, T., Astron. Astrophys. 73, L7 (1979). Smith, A.M., Stecher, T.P., and Casswell, L., Astrophys. J. (1980) in press.
17. Delsemme, A.H., Rud, D.A., Comets, Asteroids, Meteorites, edit. A.H. Delsemme, publ. Univ. of Toledo Bookstore (1977) p. 529.
18. Ross, J.E., Aller, L.H., Science, 191, 1223 (1976).
19. Anders, E., Hayatsu, R., Studier, M.H., Origins of Life, 5, 57 (1974). Anders, E., Proc. Welch Conf. XXI Cosmochemistry, edit. W.O. Milligan, publ. Welch Foundation, Houston (1978) p. 183.
20. Delsemme, A.H., Icarus, 24, 95 (1975).
21. Herbig, G.H., Mem. Soc. Roy. Sci. Liege 5th ser., 19, 13 (1970).
22. Goldanskii, V., Ann. Rev. Phys. Chem., 27, 85 (1976). Goldanskii, V., Proc. Welch Conf. on Cosmochemistry, edit. W.O. Milligan, publ. Welch Foundation, Houston (1978) p. 170.
23. Mukhin, L.M. and Gerasimov, M.V. Origins of Life 10, 61 (1980).
24. Delsemme, A.H., Nature & Origin of Comets, Coll. Internat. Astrophys., Liége, U. Liége (1966).
25. Whipple, F.L., Stefanik, R.P., Nature et Origine des Comètes, Univ. Liége (1966).
26. Lee, T., Papanastassiou, D.A., and Wasserburg, G.J., Astrophys. J., 211, L 107 (1977).
27. Safronov, V.S., Comets, Asteroids, Meteorites, edit. A.H. Delsemme, publ. Univ. of Toledo Bookstore (1977) p. 483.
28. Lewis, J.S., Earth Planet Sci. Lett., 15, 286 (1972).
29. Lewis, J.S., Barshay, S.S., Noyes, B., Icarus 37, 190 (1979).

LIQUID WATER ON A PLANET OVER COSMIC PERIODS

Michael D. Papagiannis

Department of Astronomy, Boston University
Boston, Massachusetts 02215, USA

ABSTRACT From the equation that determines the average temperature of a planet, it is readily seen that even small changes in the distance, albedo and greenhouse effect of the planet, if not properly compensated, can elliminate liquid water from the surface of the planet either through runaway glaciation or through a runaway greenhouse effect. Since active life, at least as we know it, requires the presence of liquid water, and since its evolution to higher intelligence seems to require billions of years, it follows that the ability of planets to maintain liquid water over cosmic periods is probably the critical factor that determines the abundance of advanced civilizations in the Cosmos.

The average temperature $T_e \simeq 290\ ^oK$ of the earth is given by the expression (1,2)

$$T_e = T_s \left(\frac{R_s}{D} \right)^{\frac{1}{2}} \left[\frac{(1-A) \quad (1+G)}{4} \right]^{\frac{1}{4}} \qquad (1)$$

where $T_s \simeq 5{,}750\ ^oK$ is the effective temperature of the sun, $R_s = 7\times10^{10}$ cm the radius of the sun, $D = 1.5\times10^{13}$ cm the distance of the earth from the sun, $A \simeq 0.4$ the albedo of the earth, i.e., the fraction of solar radiation reflected back into space, and $G \simeq 1.4$ a factor that describes the greenhouse effect of the earth's atmosphere produced primarily by its content in CO_2.

It is estimated (3) that during an ice age the average temperature of the earth decreases by only about 2-3 ^oK, i.e., by about 1%, which corresponds to roughly a 5% change in the albedo from A = 0.40 to A = 0.42. It is obvious, therefore, that a significant but still small change in any one of the parameters of Eq.1 can produce a change in T_e that might lead

Y. Wolman (ed.), Origin of Life, 43–50.

either to runaway glaciation or to a runaway greenhouse effect. In either case the result would be the elimination of active life, which, at least as we know it, cannot exist in the absence of liquid water. All of the parameters of Eq.1, however, have been changing throughout the past history of the earth. These changes, though, have occurred in such a beautifully orchestrated manner that made it possible for the earth to maintain liquid water on its surface for the past 4.0-4.5 billion years.

The luminosity of the sun, e.g., has increased by about 30% since the formation of the solar system, as the sun ages on the main sequence (4,5). If this change was to suddenly occur today and remained uncompensated, it would raise the temperature of the earth by about 20 ^{o}K which would melt all the ice, would add substantial amounts of CO_2 into the atmosphere from the oceans and would most likely lead to a runaway greenhouse effect.

The greenhouse effect is determined primarily by the CO_2 content of the atmosphere, which presently stands at about 330 parts per million, with water vapor also making a small contribution. Carbon dioxide is easily dissolved in water and as a result the oceans of the earth contain nearly 60 times more CO_2 than the atmosphere. Present estimates suggest that the amount of CO_2 that has been released into the atmosphere over geological times is about 250,000 times its present content in CO_2. Through the weathering of the rocks, however, and the burying of organic compounds produced by living organisms, water and life have managed to take practically all of the CO_2 out, leaving only a tiny amount in the atmosphere. This amount, however, is most critical, because through the greenhouse effect it produces it manages to raise the temperature of the earth by about 50 ^{o}K thus making our planet habitable. Without any CO_2 in the atmosphere, the earth would have been a desolate, totally frozen planet.

The fact that the CO_2 content of the earth's atmosphere was stabilized to about 300 parts/million, i.e., to about 0.0004% of what was released into the atmosphere over geological times, seems like a miracle, especially since any significant increase or decrease of this tiny amount of CO_2 in the atmosphere would have turned the earth either into a frozen desert like Mars or into a scorching inferno like Venus. It should be noted here that life made an important contribution toward this goal by using the process of photosynthesis as a feedback mechanism to stabilize the CO_2 content of the atmosphere at the desirable level. This is philosophically very interesting, but it also implies that the earth might have not been able to avoid the accumulation of CO_2 in the atmosphere and therefore a runaway greenhouse effect, without the early appearance of life and photosynthesis.

The amount of water incorporated in a planet depends critically on the temperature of the region where the planet was formed and hence at the distance from the central star. The fact that water appears to be so abundant on the surface of our planet, should not make us forget that it is actually a rather rare commodity representing considerably less than 0.1% of the total mass of the earth. The preservation of this valuable substance in liquid form over geological times represents a very difficult task, because the presence of liquid water implies also the presence of a substantial amount of water vapor in the atmosphere where it is easily decomposed into hydrogen and oxygen by the ultraviolet radiation of the sun. Hydrogen, being the lightest of the elements, escapes easily from the atmosphere and thus the water present on a planet is slowly diminished and ultimately is completely lost.

The earth managed to preserve its water supply thanks to an effective Cold Trap higher up in the troposphere which condenses the water vapor back into water droplets. These droplets, because of their weight, begin to descend, form clouds and ultimately return to the surface of the earth in the form of rain or snow. In this manner water vapor is not allowed to reach heights above 10 km where the temperature of the atmosphere has dropped down to about 220 ^{o}K. By staying way inside the troposphere, the water molecules are protected from the U.V. rays of the sun by the overlying layers of the atmosphere. This process has become even more effective thanks to the ozone layer, which is formed at heights between 20 and 45 km in the stratosphere and screens the soft end of the U.V. spectrum of the sun. Thus the Cold Trap and the ozone layer keep the water vapor and the solar U.V. rays apart thus preventing the dissociation of water molecules. A strong ozone layer has become possible only thanks to photosynthesis, which produced a free oxygen atmosphere. Here life seems to have made again a significant contribution toward its own survival by helping preserve the water supply of the earth.

The mass of a planet is another very important factor because it determines the future evolution of the planet and its atmosphere. The gravitational forces on a more massive planet are stronger and therefore the internal heating and hence the outgasing of CO_2 and H_2O are more intense. If the earth had been more massive, there would have been more water on its surface but also more CO_2 in its atmosphere, leading probably to an unavoidable runaway greenhouse effect. If, on the other hand, the earth had been considerably smaller, there would have been much less H_2O and CO_2. Also the escape velocity would have been considerably lower, allowing most of the atmospheric gases to escape. Without a substantial atmosphere, however, it is not possible to maintain liquid water on a planet. It appears, that the mass range for a planet capable of maintaining liquid water on its surface over geological periods is quite restricted.

Clouds, with albedos in the 0.25-0.65 range, are responsible for close to one-half of the earth's total albedo because they usually cover nearly 50% of the earth's surface. Areas covered with ice and snow are excellent reflectors with albedos close to 0.8, but they represent less than 2% of the projected area of the earth as seen from the sun. The oceans, on the other hand, that cover nearly 70% of the earth's surface, are very good absorbers with albedos of only about 0.06 (6). Dark soils and vegetation are also good absorbers with albedos as low as 0.1-0.2.

As seen from Eq.1, an increase of the earth's albedo, produced, e.g., by the expansions of glaciation during an ice age, results in a decrease of the average temperature of the earth. This ought to produce a further expansion of the ice coverage, additional lowering of the temperature and thus ultimately lead to runaway glaciation. An increase of aerosols (small particles) in the atmosphere would also increase the albedo with similar results. The earth, however, has always managed to reverse such trends probably by decreasing the cloud coverage (7) when the temperature drops, or possibly through a small change in the color of the oceans (8) conceivably due to different forms of phytoplancton resulting from changes in the temperature and the salinity of the oceans during an ice age.

The albedo of the earth has undergone many long term fluctuations due to many different changes of the earth. The albedo, e.g., is affected by the location of the land masses relative to the equator and the poles, a situation which is continuously changing due to the continental drift. It is also affected by the level of the oceans, which can vary by about 200 m, because when the level decreases during an ice age the land area increases as a substantial part of the continental shelf becomes exposed. It should be mentioned here that the position of the continents on the globe, as well as the rotation of the earth, also play an important role in the circulation of the atmosphere and the oceans which try to equalize the temperatures all around the earth. Convection and circulation effects, day-night temperature differences due to the rotation of the earth and seasonal differences due to the inclination of the earth's axis, and to a lesser degree due to the eccentricity of the earth's orbit, are not explicitly included in Eq.1. They are accounted for, however, indirectly because they affect the ice and cloud coverage and hence the albedo of the earth. They also affect the distribution of temperatures around the globe, which can make a difference in the energy balance because the energy radiated away per unit area is proportional to T^4.

All these factors have been changing continuously throughout the past history of the earth with significant effects on the climate, but fortunately so far without having precipitated a permanent catastrophe. The rotation of the earth, e.g., has been slowing down, from an original period of about 6 hours to the present 24 hours due to the strong tidal forces of the moon. The inclination of the earth's axis has also been changing due to precession and nutation. Precession makes the earth's axis follow the outside surface of a cone with a period of 26,000 years, while nutation continuously changes the inclination of the earth's axis, i.e., the half angle of the cone, between a minimum of 21.8^{o} and a maximum of 24.4^{o} with a period of 42,000 years. At present the inclination of the earth's axis is 23.5^{o}. At maximum inclination the summers become hotter and the winters colder, while at minimum inclination they both become more mild.

Finally, the eccentricity, e, of the earth's orbit varies between a minimum of e = 0 (a perfect circle) and a maximum of e = 0.067 (a more elongated ellipse) with a period of about 102,000 years. The eccentricity of an ellipse is given by the expression

$$e = \frac{A-P}{A+P} \qquad (2)$$

where P is the distance to the sun at perihelion, i.e., at the closest approach to the sun, while A is the distance to the sun at aphelion, i.e., at the farthest point of the orbit from the sun. At present the perihelion occurs in January, but the eccentricity now is so small (e = 0.017) that the summers in the southern hemisphere are not much warmer than the summers of the northern hemisphere which occur at aphelion. When the eccentricity, however, reaches its maximum, the solar constant at perihelion becomes 30% higher than at aphelion with a significant effect on the overall climate of the earth.

$$\frac{S_{per.}}{S_{ap.}} = \frac{S_o/(1-0.067)^2}{S_o/(1+0.067)^2} = \frac{(1.067)^2}{(0.933)^2} = 1.3 \qquad (3)$$

It is believed (3) that the major ice ages, which last 10-100 million years, are due to the continental drift which rearranges the land masses on the surface of the earth, while the smaller advances and retreats of glaciation, which occur with periods of the order of 100,000 years, are probably due to the above mentioned changes in the eccentricity of the earth's orbit and in the inclination of the earth's axis. The last such retreat was completed only about 10,000 years ago.

It is obvious from the above that over geological times a planet like the earth undergoes many changes (rotation period, inclination of axis, eccentricity of orbit, level of oceans, continental drift, chemical and physical evolution of the atmosphere, etc.) which affect the albedo A and the greenhouse effect G and hence the temperature and climate of the planet. It is also clear that many of these changes, if not properly compensated, can become catastrophic for life by eliminating liquid water either through runaway glaciation or through a runaway greenhouse effect. Though the balance between these two extremes seems to be a very delicate one, the earth has miraculously managed so far to survive all these small changes through different feedback mechanisms including those produced by life.

It is quite possible, however, that through mere chance the earth was endowed with the most favorable conditions which have assured the long presence of liquid water and hence the continuation of life on this planet. The earth, e.g., was formed at the right distance from the sun, far enough to have sufficient water but close enough to maintain it in liquid form. It also has the right mass, big enough to maintain an oxygen atmosphere but not too big to have problems with excessive outgasing of CO_2. Finally it has a fast spinning rate to avoid overheating on the sunlit side and an orbit with small eccentricity to avoid temperature extremes at perihelion and aphelion.

In comparison the other members of our solar system satisfy some, but not all of these requirements and as a result have failed to maintain liquid water on a continuous basis on their surfaces. The moon, e.g., is obviously at the right distance but has too small a mass to sustain an atmosphere. Venus, on the other hand, which has a mass very similar (0.82) to the mass of the earth, is now a bone-dry inferno. The reason is that it either started with much less water having formed 28% closer to the sun, or that it lost its water through a leaky Cold Trap, possibly because of its extremely slow spinning rate (243 days) which allowed the overheating of the sunlit hemisphere. Mars, finally, which has a spinning rate (24.6 hours) similar to the earth's, was formed too far from the sun (52% farther than the earth) to have moderate temperatures without a strong greenhouse effect. Its mass, however, is too small (0.11) compared to the mass of the earth to outgas and sustain a sizable atmosphere, without which liquid water would very rapidly evaporate from the surface of a planet. As a result, Mars is now a frozen planet, though Viking photographs indicate that it might have had brief episodes with running water on its surface, possibly after some major volcanic eruptions which produced a temporary CO_2 atmosphere with a strong greenhouse effect (9).

The earth was also fortunate to have a low eccentricity orbit and a small variation in the inclination of its axis, which prevent the development of temperature extremes. Mercury ($e = 0.206$) and Mars ($e = 0.094$) have orbits with much larger eccentricities, while the inclination of the axis of Mars, which currently (24°) is almost identical to that of the earth, varies over a much wider (15°-35°) range. Finally, with the possible exception of tiny, faraway Pluto, the earth was endowed with the most massive, relative to the mass of the planet (0.012), satellite in the solar system. The moon is probably responsible for the stability of some of these parameters. It might also have played an important role in the origin of life (10) through the tides that helped in the formation of prebiotic compounds through dehydration/condensation in small water pools along the shores.

Finally, it is amazing that as the earth's atmosphere continued to evolve, its greenhouse effect kept decreasing at such a rate as to keep pace exactly with the continuously increasing intensity of the solar radiation, which over geological periods has increased by about 30% (4,5). It is also startling that life came along at the right time to change, through photosynthesis, the atmosphere to an oxidizing one, to keep monitoring the CO_2 content of the atmosphere so as to avoid a runaway greenhouse effect even though the sun was getting hotter, and to produce an ozone layer that helped preserve the water supply of our planet.

In summary, it seems amazing that the earth has managed to maintain liquid water on its surface for over 4 billion years inspite of the different changes and the many factors that could have gone wrong. It is quite difficult to compute the actual tolerance of all these parameters (mass, distance, eccentricity, spinning rate, inclination of axis, a large moon, early appearance of life and photosynthesis, etc.) and their interdependence which would assure the continuous presence of liquid water on a planet. It is even more difficult to compute the combined tolerance of all these parameters for different types of stars, though it appears that only stars similar to our sun, i.e., a fraction of probably less than 1% of all the stars in the galaxy (5,11) is capable of providing the conditions needed by a planet (long term stability and sufficient energy) to maintain liquid water on its surface over cosmic periods. It is clear, however, that no other planet or moon of our solar system has managed to maintain liquid water on its surface over its entire history, while preliminary studies of this complicated problem indicate that the tolerance of some of these planetary parameters, such as the distance from the sun (12), is only in the few percent range.

It is possible, therefore, that of the stars similar to our sun, only a very small fraction, possibly as small as one per million (13), might have a planet capable of sustaining liquid water on its surface over billions of years. This would make a big difference in our chances of finding other technological civilizations in our galaxy, because the critical factor in their emergence seems to be not the origin of life, which in the case of the earth occurred in only 0.1-0.7 billion years after the formation of the oceans, but rather the extremely slow evolution of life toward high intelligence, which in our case took 3.5-4.0 billion years, i.e., 5-40 times longer than the appearance of life. This indicates that given favorable conditions, the most important of which is probably liquid water, life as we know it can originate with relative ease. It is the painfully slow evolution of life into high intelligence, however, and the difficulty of most planets to maintain these favorable conditions for billions of years, that probably limit the number of planets in our galaxy where an advanced technological civilization can appear.

In conclusion, if some day we might conclude that we are one of the very few if not the only advanced civilization in our galaxy, we should not be terribly surprised. The concurrence of favorable circumstances (including the earth's mass, distance, spinning rate, eccentricity of orbit, inclination of axis, as well as the early appearance of life and photosynthesis and the type of star our sun is) that allowed our planet to provide a hospitable environment to life for billions of years, so that life could evolve into a technolocical society, might indeed represent an extremely rare phenomenon in the galaxy.

REFERENCES

1. Papagiannis M.D., "Space Physics and Space Astronomy", Gordon and Breach, New York, 1972.
2. Papagiannis M.D., "Search for Life in the Universe", Astr. Contr. Boston Univ. Ser.II, No.70, 1979.
3. Press F. and R. Siever, "Earth", W.H. Freeman & Co., San Francisco, 1974.
4. Ulrich R.K., Science, 190, 619, 1975.
5. Hart M.H., Icarus, 37, 1979.
6. Payne R.E., J. Atm. Sci., 29, 959, 1972.
7. Weare B.C. and F.M. Snell, J. Atm. Sci., 31, 1725, 1974.
8. Papagiannis M.D., EOS, Trans. A.G.U., 61, 255, 1980.
9. Cess R.D., V. Ramanathan and T. Owen, Icarus, 41, 159, 1980.
10. Ponnamperuma C., "The Origin of Life", Thames and Hudson, London, 1972.
11. Pollard E.G., Am. Sci., 67, 653, 1979.
12. Hart M.H., Icarus, 33, 1978.
13. Papagiannis M.D., in "Origin of Life" ed. by H. Noda, Center Acad. Publ. Japan, Tokyo 1978.

ORGANIC ANALYSIS OF THE ANTARCTIC CARBONACEOUS CHONDRITES

R.K. Kotra, Akira Shimoyama,[1] Cyril Ponnamperuma

Laboratory of Chemical Evolution, Dept. of Chemistry
University of Maryland, College Park, MD. 20742.

P.E. Hare

Carnegie Institution of Washington, Wash., D.C. 20008

K. Yanai

National Inst. of Polar Research, Tokyo, Japan

In the search for extraterrestrial evidence of chemical evolution, organic analysis of carbonaceous chondrites has proven to be very fruitful. Recently, the discovery of a large number of meteorites in Antarctica has provided new opportunities to further our knowledge of meteorite composition and, hence, the conditions during the formative stages of the solar system. We have acquired two carbonaceous chondrites for analysis and have identified amino acids of extraterrestrial origin in the Allan Hills (77306) and the Yamato (74662). Chromatographic analysis combined with mass spectrometry indicates the presence of more than 25 different amino acids. More importantly, these Antarctic meteorites appear to be uncontaminated by terrestrial organic matter.

In the search for extraterrestrial evidence for chemical evolution carbonaceous chondrites have proved to be the only source of complex organic matter. The lunar samples contained

---

[1]present address: Mining College, Akita University, Akita 010, Japan

Y. Wolman (ed.), Origin of Life, 51–57.

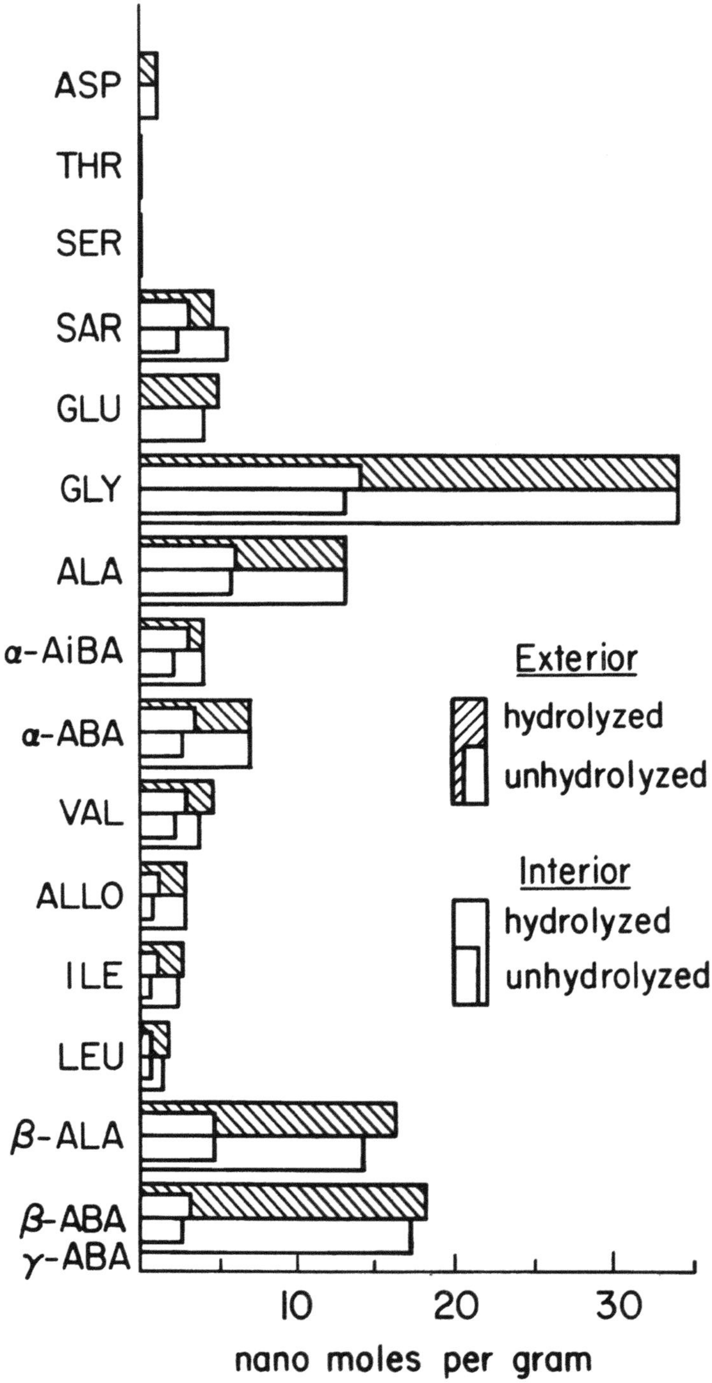

Figure 1. Amino Acids in the Yamato (74662) carbonaceous chondrite.

only a trace of organic matter (1), and Martian soil revealed almost none (2). Comets and asteroids have yet to be studied in detail. The detection of both protein and non-protein amino acids and the equal abundance of the D and L enantiomers of the amino acids in the Murchison meteorite confirmed the extra-terrestrial abiotic origin of organic compounds in carbonaceous chondrites (3). The Murray, the Orgueil and the Mighei also contained indigenous organic compounds (4,5,6,7,8,9).

Our information comes from a handful of meteorites, and we are also confronted by sample heterogenerity and terrestrial contamination in some samples of the above meteorites. Thus, for a better understanding of the organic synthesis in the early solar system we need additional information from other meteorites. Recent expeditions to Antarctica have returned with a large number of meteorites (10,11,12). Some of these are carbonaceous chondrites which may have been protected from terrestrial contamination by the environment. We acquired two of these, the Allan Hills (77306) and the Yamato (74662), for organic analysis. These C2 meteorites have been analyzed for amino acids and hydrocarbons.

The meteorite samples were processed in a Class 100 clean room prior to chemical analysis. The analytical procedures are given in detail elsewhere (13,14). The hydrolyzed Allan Hills extracts were analyzed by ion-exchange chromatography (IEC) for quantitation, and by gas chromatography (GC) for enantiomer ratios. The unhydrolyzed and hydrolyzed Yamato extracts were analyzed by ICE, GC and gas chromatography combined with mass spectrometry (GCMS) for confirmation of identifications. For comparative purposes a sample of the Murchison was also analyzed.

Table 1 and Figure 1 show the quantities of amino acids found in these two Antarctic carbonaceous chondrites. Fifteen were detected in the Yamato and twenty were detected in the Allan Hills based on retention times. The most abundant are glycine and alanine. Both protein and non-protein amino acids are present. Upon hydrolysis there is an increase in the quantities of amino acids. The most important observation here is that the quantities in the exterior and interior portions of these meteorites are almost identical.

Gas chromatography of the derivatized samples showed that the abundance of the two enantiomers of several amino acids is nearly equal in both the meteorites for the exterior and the interior. Figure 2 shows selected portions of the chromatograms of the Yamato interior. Alanine, glutamic and aspartic acids show near equal abundance of their enantiomers.

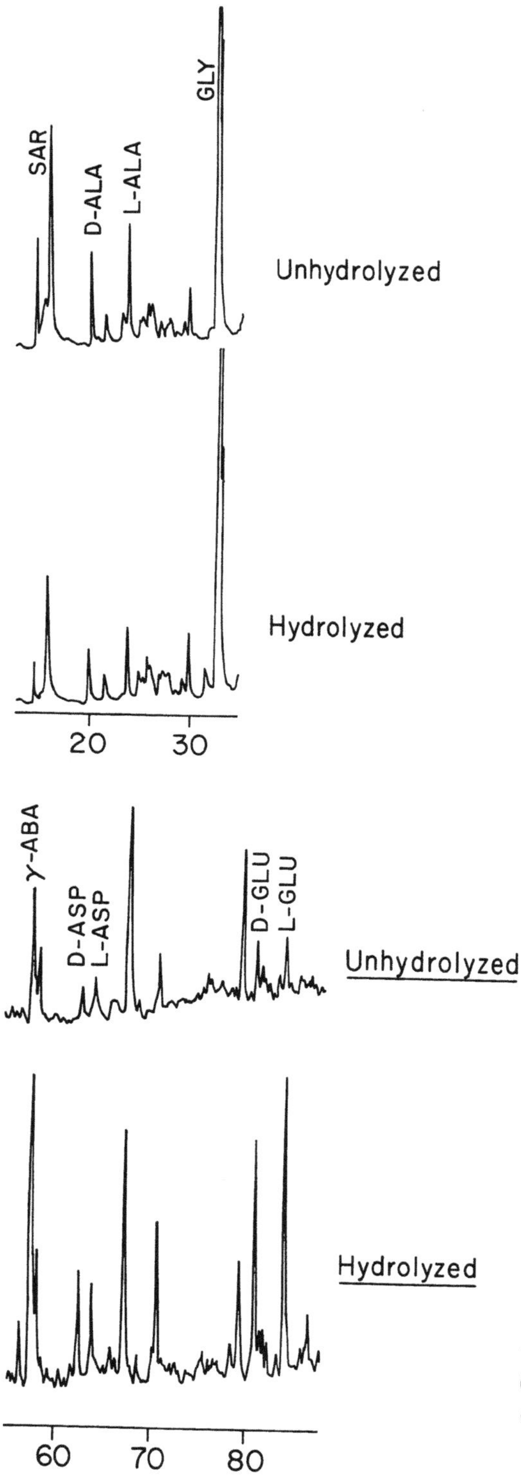

Figure 2.
Gas Chromatogram of Amino acids of the interior portion of the Yamato (74662) carbonaceous chondrite.

TABLE 1

units: nanomoles/gm

Amino Acid Abundances in the Allan Hills and the Murchison Meteorites Based on Ion-Exchange Chromatography

Amino Acid	Allan Hills Exterior	Allan Hills Interior	Murchison Exterior	Murchison Interior
Aspartic Acid	2.1	1.4	33.3	3.6
Threonine	0.3	0.3	13.3	1.6
Serine	0.5	0.4	14.4	1.9
Glutamic Acid	1.1	1.2	48.7	13.0
Glycine	14.1	7.3	101.1	37.0
Alanine	2.8	1.8	54.1	15.2
α-Aminoisobutyric Acid	1.3	1.4	29.0	43.1
α-Aminobutyric Acid	5.5	0.9	16.5	7.4
Valine	0.3	0.4	28.1	10.7
Methionine	0.3	0.2	3.7	1.9
Alloisoleucine	0.6	0.5	10.6	3.3
Isoleucine	0.1	0.2	8.1	0.6
Leucine	0.2	0.2	12.0	2.0
Norleucine	0.3	0.3	1.2	0.7
Tyrosine	0.2	0.3	1.3	0.7
Phenylalanine	0.1	0.1	4.2	0.8
β-Aminobutyric Acid	1.6	1.3	5.0	3.2
β-Alanine	3.4	1.8	14.1	7.8
β-Aminoisobutyric Acid	1.3	1.2	9.0	3.4
γ-Aminobutyric Acid	1.6	3.0	16.2	23.3

The same derivatives were also analyzed by a Hewlett-Packard 5992 GC-MS and the retention time-based identifications were confirmed by utilizing selected ion monitoring. Standard amino acids were derivatized and their fragmentation patterns were obtained prior to meteorite analysis. Table 2 shows the amino acids confirmed in the Yamato by retention times and selected ion monitoring. The Allan Hills could not be analyzed by GC-MS due to small sample size. However, since the techniques of analysis used were identical the mass spectral confirmations of the Yamato may be extended to the Allan Hills.

TABLE 2

Amino Acids from the Yamato (74662) Carbonaceous Chondrite as Confirmed by Selected Ion Monitoring and Retention Times

DEFINITE	TENTATIVE
D-ALANINE	D-VALINE
L-ALANINE	L-VALINE
D-α-AMINO-n-BUTYRIC ACID	D-ALLOISOLEUCINE
L-α-AMINO-n-BUTYRIC ACID	L-ISOLEULINE
GLYCINE	D-β-AMINOISOBUTYRIC ACID
D-NORVALINE	L-β-AMINOISOBUTYRIC ACID
L-NORVALINE	D-NORLEUCINE
β-ALANINE	L-NORLEUCINE
γ-AMINO-n-BUTYRIC ACID	D-LEUCINE
D-ASPARTIC ACID	L-LEUCINE
L-ASPARTIC ACID	D-or L-LYSINE
D-GLUTAMIC ACID	
L-GLUTAMIC ACID	

The amino acid assemblage, abundances, and the enantiomer ratios in these chrondrites indicate an abiotic extraterrestrial nature. These samples may have been in the Antarctic ice for a long time as the age of the blue ice around the Yamato mountains is approximately 250,000 years (11). Yet the distribution patterns do not indicate any terrestrial contamination. The quantities found in the Yamato are comparable to those found earlier in the Murchison and the Murray, all of which are C2 meteorites. The Allan Hills, also a C2, contains about one fifth to one tenth the quantity. Weathering may not have altered these significantly.

Thus, the two Antarctic meteorites we have examined seem to be the least contaminated and possibly the least altered meteorites examined to date. Hence, the organic content in these may represent an assemblage closest to that of carbonaceous chondrites in space. Experiments are now in progress to study the hydrocarbon content of these samples.

ACKNOWLEDGEMENTS

This work was supported by NSF Grants DPP 7706993 and DPP 79009-91 and NASA Grant NGR 21-00-317. The authors thank J.J. Kemper, University of Maryland, for editorial assistance with this manuscript.

REFERENCES

1. Gehrke, C.W., Zumwalt, R.W., Kuo, K., Ponnamperuma, C., Shimoyama, A., Origins of Life, 6:540 (1975)
2. Biemann, K., Oró, J., Toulmin, P., Orgel, L.E., Nier, A.O., Simmonds, D.G., Flory, D., Diaz, A.V., Rushneck, D.R., Biller, J.A. Science, 194:72 (1976)
3. Ponnamperuma, C. Ann. N.Y. Acad. Sci. 194:56 (1972)
4. Lawless, J.G., Kvenvolden, K.A., Peterson, E., Ponnamperuma, C., Moore, C., Science, 173:626 (1971)
5. Oró, J., Gibert, J., Lichtenstein, H., Wikstron, S., Flory, D.A. Nature, 230:105 (1971)
6. Oró, J., Nakaparksin, S., Lichtenstein, H., Gil-Av, E. Nature, 230:107 (1971)
7. Cronin, J.R., Moore, C.B. Science, 172:1327 (1971)
8. Lawless, J.G., Kvenvolden, K.A., Peterson, E., Ponnamperuma, C., Jarosewich, E. Nature, 236:66 (1972)
9. Buhl, P.H., Ph.D. Thesis, University of Maryland (1972)
10. Yoshida, M., Ando, H., Omoto, K., Naruse, R., Ageta, Y. Nankyoku Shiryo (Antarctic Rec.), 39:62 (1971)
11. Cassidy, W.A., Olsen, E., Yani, K. Science, 198:727 (1977)
12. Yanai, K., Mem. Natl. Inst. Polar Res., Spec. Issue 8:1 (1978)
13. Kotra, R.K., Shimoyama, A., Ponnamperuma, C., Hare, P.E. J. Mol. Evol., 13:179 (1979)
14. Shimoyama, A., Ponnamperuma, C., Yanai, K. Nature, 282:394 (1979)

NITROGEN COMPOUNDS IN CARBONACEOUS METEORITES: A REASSESSMENT

Peter G Stoks and Alan W Schwartz

Exobiology Department, Faculty of Science, The University, Nijmegen, The Netherlands.

A reinvestigation of the presence of *N*-heterocyclic compounds in the Murray, Murchison and Orgueil carbonaceous meteorites has been undertaken in our laboratory. The positive identification of the pyrimidine uracil and the purines adenine, guanine, xanthine and hypoxanthine is reported. The *s*-triazines melamine, ammeline, ammelide and cyanuric acid could not be detected in any meteorite, above a detection limit of 50 ppb.
These results, together with laboratory data on the formation of *s*-triazines, are discussed in relation to proposed mechanisms of formation of organic compounds in carbonaceous meteorites.

Meteorites are generally believed to be remnants of the early history of the Solar System (1,2,3). The carbonaceous, or carbon containing meteorites as a subclass are of great value to the study of the origin of Life, since the analysis of their indigenous organic content may provide valuable information about processes that have occurred in the Solar Nebula, as well as about chemical evolution on the primitive Earth.
Detailed analyses of the extractable organic material present in carbonaceous meteorites have revealed, (to list but a few examples) the presence of hydrocarbons (4,5), amino acids (6,7), carboxylic acids (8,9) and *N*-heterocyclics (10,11,12,13,14,15). Although different analyses may show variations, certain quantitative and qualitative features within several compound classes appear to be characteristic for a specific meteorite, or even for different meteorites belonging to the same type

Y. Wolman (ed.), Origin of Life, 59–64.

(2,7,8,16). However, a marked discrepancy exists regarding the distribution of *N*-heterocyclics, even within fragments of the same meteorite. Hayatsu (12) and Hayatsu *et al.* (13,14) identified several purines and *s*-triazines in extracts from the Orgueil and Murchison meteorites. In contrast, Folsome *et al.* (10,11) identified only 4-hydroxypyrimidine derivatives in extracts from these two meteorites, as well as in extracts from the Murray meteorite. Van der Velden and Schwartz (15) identified purines only, in extracts from the Murchison meteorite. In addition, they obtained strong evidence that the hydroxypyrimidines reported by Folsome *et al.* were due to artefacts.

Fisher-Tropsch type reactions, (catalytic reactions of H_2, CO and NH_3), have been proposed as the source of organic material in the Solar Nebula and in meteorites (1,5,17). Although distribution patterns of hydrocarbons in Fisher-Tropsch type products and in carbonaceous meteorites show a marked correlation, which apparently cannot be explained by Miller-Urey type reactions (1,5,17,18), the observed discrepancies in *N*-heterocyclic content of carbonaceous meteorites (10,11,12,13, 14,15,19) have made a correlation with Fisher-Tropsch type products rather difficult. In view of the importance of purines and pyrimidines in theories of the origin of Life, as well of *N*-heterocyclics in general for theories on the origin of organic material in meteorites, we have reanalyzed samples of the Murchison meteorite, as well as new samples of the Orgueil and Murray meteorites, using a combination of analytical techniques. We have attempted to explain the discrepancies observed in the literature, thus enabling a more reliable correlation with proposed mechanisms of formation, such as Fisher-Tropsch type reactions (20).

In brief, extraction procedures involved sonication at $60^{o}C$ in benzene, water and formic acid successively, and hydrolysis of the water and formic acid extracts, as well as the residues after formic acid extraction, in 3*M* HCl ($110^{o}C$, 18hr). Extracts were purified on charcoal, AG50WX8(H^+) and C18 reversed phase columns and were analyzed by cation and anion exclusion HPLC. Samples of the Allende carbonaceous meteorite served as analytical blanks for all stages of the work. The presence in extracts of the Murchison, as well as the Murray and Orgueil meteorites, of the biologically important purines adenine, guanine, xanthine and hypoxanthine, as well as the pyrimidine uracil was indicated from liquid chromatographic analyses. Subsequent mass spectrometric analysis of isolated peaks from the LC columns confirmed these identifications in all instances (19,20). Table 1 summarizes, qualitatively, the analytical results for these meteorites obtained by different research groups. The meteorite extracts were analyzed for *s*-triazines by gas chromatography, using a technique described elsewhere (21). No *s*-triazines could be detected in any sample, above a detection limit of 50 ppb. In agreement with Van der Velden and Schwartz (15), we were unable to identify the 4-hydroxy-pyrimidines previously reported (10,11) in samples from these same meteorites, above background levels of about 10 ppb.

TABLE 1

N-heterocyclic compounds identified in carbonaceous meteorites

references	1C,11	14	15,19,20
analytical techniques	GCMS	MS	HPLC, MS, GC
adenine	–	+	+
guanine	–	+	+
xanthine	–	–	+
hypoxanthine	–	–	+
uracil	–	–	+
thymine	–	–	–
cytosine	–	–	–
4-OH-pyrimidines	+	–	–
s-triazines	–	+	–

As suggested by Van der Velden and Schwartz (15) these pyrimidines seem to be artefacts, synthesized from impurities present in the silylation mixture (*N*,*O*-bis(trimethylsilyl)-acetamide/pyridine), used for gas chromatographic analyses. De Vries and Stoks (unpublished work) have shown that even with a carefully purified derivatization agent these same compounds are produced under a variety of conditions, albeit in lesser abundance. A possible mechanism to account for the formation of such compounds is shown below.

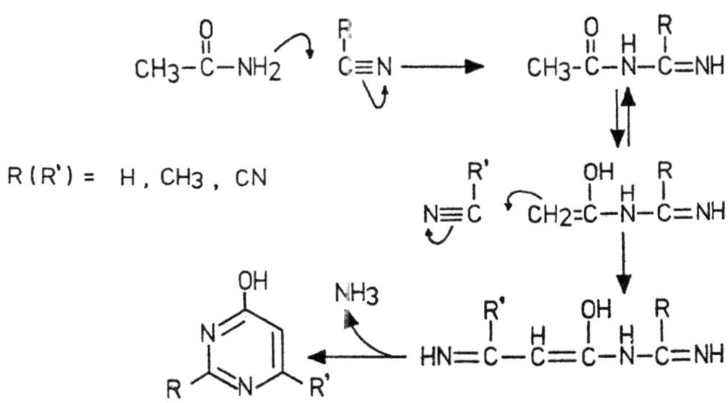

A similar mechanism has been suggested for the formation of 4-substituted 2,6-dimethylpyrimidines (22).

Discrepancies regarding the *s*-triazines may also be due to artefacts of the experimental conditions applied. *s*-Triazines may easily be synthesized from simple starting materials under a variety of conditions (23). We have found that guanylurea produced *s*-triazines when refluxed in acetic anhydride, or when heated in a mass spectrometer. Guanylurea has been identified in meteorite extracts in concentrations up to 270 ppm (13,14). In addition, *s*-triazines have only been identified in these same meteorite extracts following such manipulations (12,13,14). The formation of *s*-triazines from guanylurea is shown below.

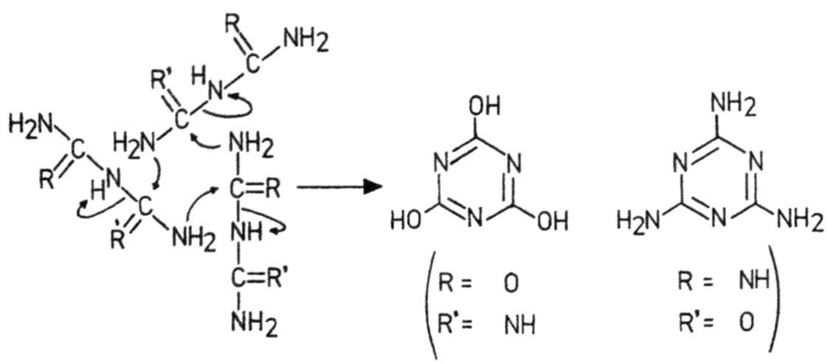

By-products in this triazine synthesis (urea, guanidine) may participate in the reaction as well, giving rise to additional *s*-triazines upon trimerization and cyclization. It can be seen from the above scheme that the two possible amino,hydroxy-*s*-triazines are produced from different combinations of R and R'.

The observed discrepancies in the distribution of *N*-heterocyclics now seem to be largely explainable on the basis of artefacts of the analytical procedures, although insufficient sensitivity of analysis, as well as compositional heterogeneity of compound distribution have probably also been important. It should be noted in this regard, that the HPLC analysis we have employed for purines and pyrimidines is one to two orders of magnitude more sensitive than previously applied techniques.

Both purines and pyrimidines have been identified in products from Fisher-Tropsch type reactions, under conditions suggested to have occurred in the Solar Nebula (1,17,24,25). However, compound distributions appear to be difficult to reproduce and, in addition, slightly different reaction conditions seem to give completely different results. We have recently suggested (19), that the occurrence of uracil in carbonaceous meteorites tends to support the possible relevance of Fisher-Tropsch type reactions for the Solar Nebula. However, this must now be balanced against the absence of detectable amounts of *s*-triazines in carbonaceous meteorites, which are reportedly synthesized in Fisher-Tropsch type reactions in large quantities under several sets of conditions (13,24,25). It would, however, be desirable to replicate these syntheses under controlled conditions and with the analytical precautions which have been applied to the analysis of meteorites.

Although many arguments favour the Fisher-Tropsch hypothesis as a source of hydrocarbons (1,5,17,18), we think that the evidence regarding *N*-heterocyclic compounds is less conclusive. Other mechanisms are also clearly of interest. It has been noted (26), that there is a resemblance between the distribution patterns of purines found in carbonaceous meteorites and those formed by hydrolysis of the products of HCN oligomerization. We have made a preliminary attempt to investigate this possible relationship further by screening our meteorite extracts for 4,5-dihydroxypyrimidine, which is one of the most abundant *N*-heterocyclic products from HCN thus far identified, as well as the minor product 5-hydroxyuracil (27,28). We have been unable to detect either of these compounds in concentrations higher than the background level of 25ppb. However, recoveries of these compounds have not yet been determined and further work will be necessary to clarify this point.

REFERENCES

1 Anders E, Hayatsu R. and Studier M.H. *Science* 182: 781 (1973)
2 Hayes J.M. *Geochim. Cosmochim. Acta* 31: 1395 (1967)
3 Nagy B. *Carbonaceous Meteorites*. Elsevier, Amsterdam (1975)
4 Orő J, Gibert J, Lichtenstein H, Wilkstrom S. and Flory D.A. *Nature* 230: 105 (1971)
5 Studier M.H, Hayatsu R. and Anders E. *Geochim. Cosmochim. Acta* 36: 189 (1972)
6 Kvenvolden K.A, Lawless J, Pering K, Peterson E, Flores J, Ponnamperuma C, Kaplan I.R. and Moore C. *Nature* 228: 923 (1970)
7 Cronin J.R, Pizzarello S. and Moore C.B. *Science* 206: 335 (1979)
8 Yuen G.U. and Kvenvolden K.A. *Nature* 246: 301 (1973)
9 Lawless J.G, Zeitman B, Pereira W.E, Summons R.E. and Duffield A.M. *Nature* 251: 40 (1974)
10 Folsome C.E, Lawless J.G, Romiez M. and Ponnamperuma C. *Nature* 232: 108 (1971)
11 Folsome C.E, Lawless J.G, Romiez M. and Ponnamperuma C. *Geochim. Cosmochim. Acta* 37: 455 (1973)
12 Hayatsu R. *Science* 146: 1291 (1964)

13 Hayatsu R, Studier M.H, Oda A, Fuse K. and Anders E. *Geochim. Cosmochim. Acta* 32: 175 (1968)
14 Hayatsu R, Studier M.H, Moore L.P. and Anders E. *Geochim. Cosmochim. Acta* 39: 471 (1975)
15 Van der Velden W. and Schwartz A.W. *Geochim. Cosmochim. Acta* 41: 961 (1977)
16 Cronin J.R. and Moore C.B. *Science* 172: 1327 (1971)
17 Hayatsu R, Matsuoka S, Scott R.G, Studier M.H. and Anders E. *Geochim. Cosmochim. Acta* 41: 1325 (1977)
18 Studier M.H, Hayatsu R. and Anders E. *Geochim. Cosmochim. Acta* 32: 151 (1968)
19 Stoks P.G. and Schwartz A.W. *Nature* 282: 709 (1979)
20 Stoks P.G. and Schwartz A.W. *Submitted for publication*
21 Stoks P.G. and Schwartz A.W. *J. Chromatogr.* 168: 455 (1979)
22 Forsberg J.H, Balasubramanian T.M. and Spaziano V.T. *J. Chem. Soc. Chem. Comm.* 1060 (1976)
23 Smolin E.M. and Rapoport L. *s-Triazines and Derivatives* Weisberger A. Ed. Interscience Publishers, Inc. New York (1959)
24 Hayatsu R, Studier M.H, Matsuoka S. and Anders E. *Geochim. Cosmochim. Acta* 36: 555 (1972)
25 Yang C.C. and Orő J. In: *Chemical Evolution and the Origin of Life* Buvet R. and Ponnamperuma C. Eds. North-Holland Publishing Cy. Amsterdam (1971)
26 Schwartz A.W. In: *Marine Organic Chemistry* Duursma E.K. and Dawson R. Eds. Elsevier *In Press*
27 Ferris J.P, Joshi P.C. and Lawless J.G. *BioSystems* 9: 81 (1977)
28 Ferris J.P, Joshi P.C, Edelson E.H. and Lawless J.G. *J. Mol. Evol.* 11: 293 (1978)

ABIOTIC ORGANIC SYNTHESIS IN SPACE

M.R. BLOCH and H.L. WIRTH

Max-Planck-Institut für Kernphysik, Heidelberg and Research and Development Authority, Ben-Gurion University, Beer Sheba

A model of iron meteorite formation at low temperature (250°C) in space (e.g. comets) has been experimentally investigated. A $Fe(CO)_5$ + $Ni(CO)_4$ + H_2S gas stream was heated (180°C) between the poles of a permanent magnet. Iron-nickel alloys similar to natural meteorites were formed (Kamazit,Taenit enveloping Troilite and Pentlandit). In addition graphite and hydrocarbons were tentatively recognized. This finding is supported by technical literature on the Fischer-Tropsch synthesis where the decomposition of metalcarbonyls in presence of H_2 and NH_3 at 240°C leads to hydrocarbons and primary amines. It is assumed that in a magnetic field the streaming gas-mixture might react to form some assymetric, not completely racemic compounds. Such compounds could have reached earth with impacting comets and meteorites.

The chemists at the Max-Planck-Institute for Nuclear Physics in Heidelberg were challenged by W. Gentner in 1969 to propose a new working hypothesis for meteorite formation in nature. The problem was to allow high melting iron and nickel alloys to solidify together with low-melting and temperature sensitive compounds either simultaneously or afterwards. Carbonates, silicatehydrates, sulfides and organic compounds with dissiminated alloys of iron and nickel are frequently found in meteorites.

The challenge was taken up (Bloch, Wirth 1971, Bloch, Müller 1971,1973) by proposing iron-carbonyl and nickel-carbonyl type of compounds as the source

Y. Wolman (ed.), Origin of Life, 65–71.

of the nickel and iron-alloys karmacite and taenite. They are at 20°C clear, colorless liquids with a vapor pressure similar to water. It can be shown that by decomposing these carbonyls at 150-200°C metal bodies are made which have several characteristics also found in iron meteorites. They show shrink crevices similar to those in meteorites which up to now were interpreted as shock phenomena. They can easily be polished to give a metallic reflection and they also show locally changing composition in iron and nickel-rich zones giving "M" like scan diagrams for nickel and iron content similar to those found with meteorites. At present these are supposed to be caused by diffusional demixing homogeneous alloys in the course of millions of years. In both meteorites and carbonyl derived iron-nickel alloys free carbon can be found as graphite.

Hübner (1970) suggested that iron carbonyls are present in comets. According to Whipple (1968) comets have a nucleus consisting of a "dirty snowball" some kilometers in diameter. He also suspects that these snowballs have been agglomerated from cosmic ice dust clouds before the creation of the sun (Voelk 1979).

It has been known for a long time (Mond-Langer 1890) that metal carbonyl vapors decompose to metal globules and carbonmonoxide at 200 °C. At the suggestion of the late O. Müller a stream of iron carbonyl $Fe(CO)_5$ and nickel carbonyl $Ni(CO)_4$ gas mixtures were passed through the space between the poles of a permanent magnet at 180 °C, Figure 1.
It was found that the growing metal globules associate to strings which have great similarity in form and chemical constitution to "swaddling kamazit. Debye-Scherrer diagrams show them to be almost identical with the metal phase of certain meteorites (Hoba, Sikote) Figure 2.

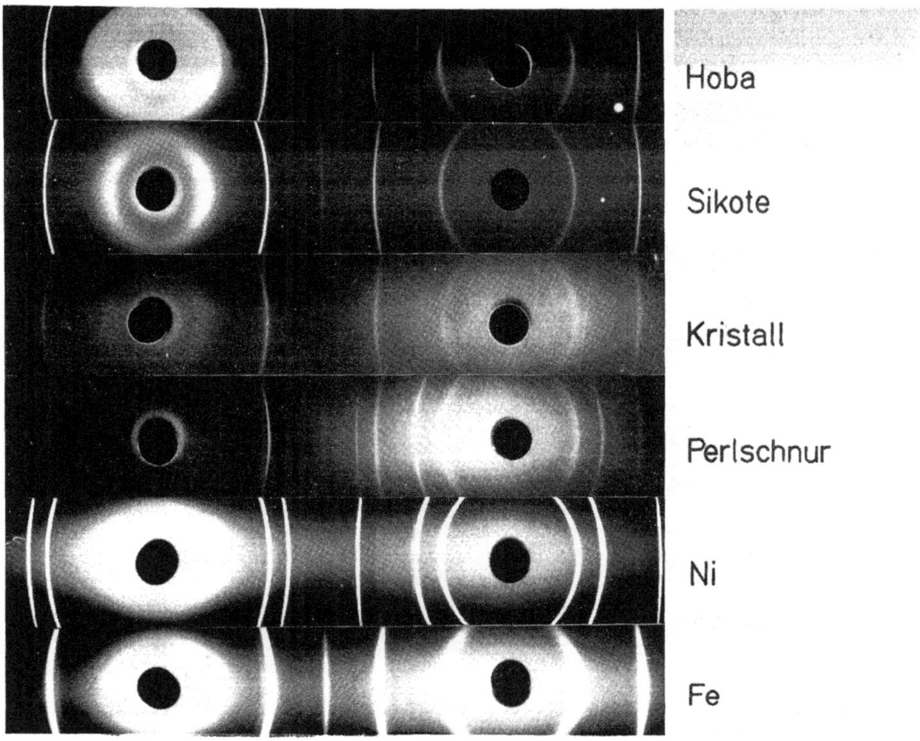

The globular strings show when scanned by microsonde the typical "M" form of the nickel and iron content of kamazit and taenit sequences as in Widmannstaett configuration (Bloch,Müller,Nagel 1979) Figure 3.

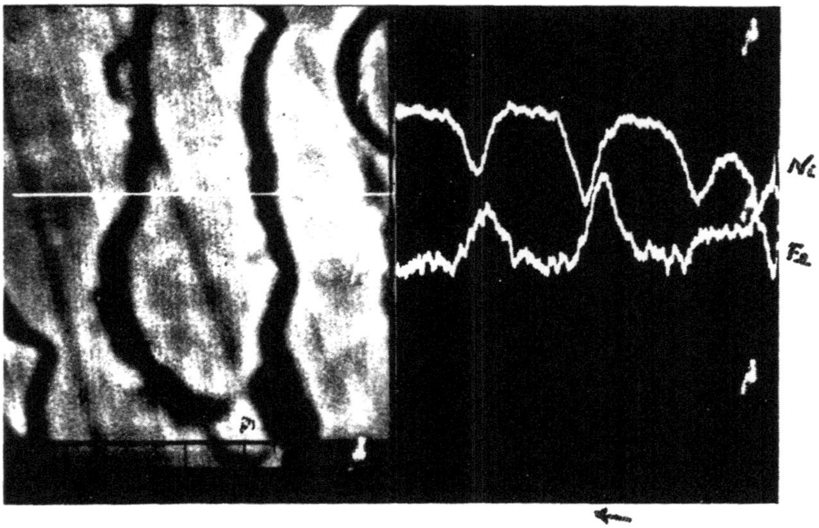

It was further established that when the carbonyl gases contain added H_2S and CS_2 gas, iron and nickel sulfides globules (troilite and pentlandite) form which envelope the kamazite taenite phases or are themselves enveloped by them much as seen in meteorites. Figure 4.

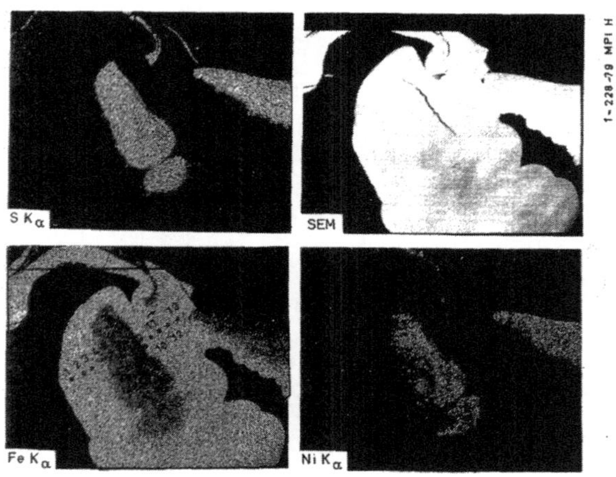

Nos.	Ni	Fe	S	P
2	.8	97	.02	.05
4	40	53	.7	.04
6	35	40	20	.02
8	38	31	20	.02
10	5	95	.03	.02
12	.40	97	.03	.03

Table in %

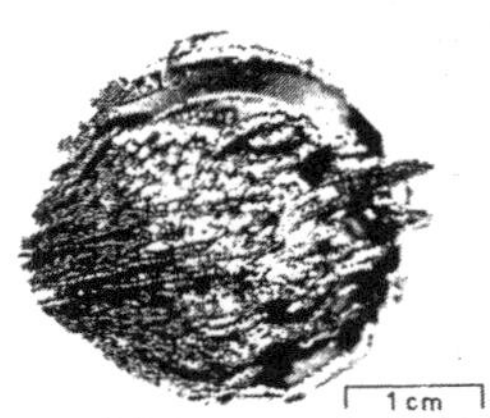

globular strings

Considering that H_2S is probably present in the"dirty ice"of comets together with many other hydrides, the CO of metal carbonyls can react in a Fischer-Tropsch type reaction to form hydrocarbons and their oxigen compounds. In fact this is claimed to be a commercially viable process working at 250°C for producing fuel(IG-Farben 1928). A later patent (Ruhr-Chemie 1949) claims the production of primary amines by adding small quantities of NH_3 to a Katalyst-hydrogen-carbonmonoxide system. This makes it probable that the low temperature decomposition of iron and nickel-carbonyls in the presence of amo-ia will lead to amino acids. We have started experi-ments to verify the claims of the patents mentioned and we have already got indications of their accuracy.

It seems very probable that organic substances and graphite can not only be made in space according to the two mechanisms proposed by Studier and Miller-Urey. They proposed high energy reactions (Urey-Miller 1952) and quasi equilibrium reactions (Studier, Hayatsu and Anders 1965);however, also the Fischer-Tropsch reaction is a possibility involving nickel and iron carbonyls at low temperature.

Many hydrocarbons have been found in carbonaceous chondrites containing disseminated metal and metal sulphide phases in silicate matrix (Nagy 1968).

We have further put forward the hypothesis that not only the hydrides of the type of H_2O, NH_3, CH_4, HCl and H_2S and metal carbonyls are present as ice in comets and cosmic ice dust, but that SiH_4, alkali and earth alkali hydrides can also coexist with H_2O ice at very low temperature, $\sim 10^{\circ}$K,for 10^9 years. However, as soon as a comet comes near the sun the hydride carbonyl mixture heats up and starts to lose H_2 (Biermann 1978) and volatiles which move partly into space and partly to the still cooler rather porous mass deep in the comet body where they will condense in the cavities of the "dirty snow". This fractional destillation into the interior will deposit different carbonyls at different points, mostly near the centre of the rotating comet body (Whipple 1979). So it can be suggested that under the influence of temperature gradients iron, nickel and other carbonyl forming metals can collect more or less separately in the inner parts of bodies which are much too small for seperation effects due to gravity.

When the separately collected metal carbonyls finally come to be heated up by radiation, impact or gravity compression and by exothermic reaction between H_2O and other hydrides metals, inorganic and organic compounds are formed similar to what is actually found in meteorites. For instance:Figure 5.

$$\underbrace{H_2O + SiH_4 + MgH_2 + H_2S + Fe(CO)_5 + Ni(CO)_4}$$

$$\downarrow$$

$$10\,^{\circ}K \rightarrow 100\,^{\circ}C \rightarrow 1500\,^{\circ}C \text{ (exothermic reaction)}$$

$$\downarrow$$

$$\overbrace{\text{Fe,Mg-Silicate; Fe,Ni-Metal; Fe,Ni-sulfide; } C_xH_y\text{; } H_2.}$$

Such reactions are exotherme and the cold unequilibrated mixture is a store of considerable energy which leads to a silicate hydrate matrix. If this happens in a gas stream when crossing a magnetic field (Scott 1967) the synthesis of some assymmetric molecules might be biased in favor of one or the other isomere and so leading to optically active organic substances. We are planning experiments to test this hypothesis.

ACKNOWLEDGEMENT

We are grateful to Professor H. Gil-Av of the Weizmann Institute for careful discussion and important literature references

REFERENCES

(1) Bierman,L. and Michael, K.W.: The Moon and the Planets, 18, 477, (1978)

(2) Bloch, M.R. and Wirth, H.: The Meteoritical Society 34th Meeting 1971, Experiments for iron-meteorites Simulation

(3) Bloch, M.R. and Müller, O.: An Alternative Mode for the Formation of Iron-Meteorites, Earth Planet Sci.Lett. 12, 134 (1971)

(4) Bloch, M.R. and Müller, O.: Low -temperature Meteorite Formation: Some Simulation Experiments, Meteoritics 8, 327, (1973)
(5) Bloch, M.R., Müller, O., Nagel, K.: A Low-temperature Formation for Nickel-Suphide, Meteoritics 14, 353 (1979)
(6) Huebner, W.F.: Dust from Cometary Nuclei, Astron. and Astrophys. 5, 286 (1970)
(7) Mond, Langer(1890),see:Remy,H.:Lehrbuch d. Anorgan. Chemie II, Leipzig 1961
(8) Nagy,B.: Carbonaceous Meteorites , Endevaour XXVII, 101 (1968)
(9) Scott, G.G., Sturner, H.W., Williamson, R.M.: Gas Torque Anomaly in Weak Magnetic Fields, Phys. Rev.158, No.1, 118, (1967)
(10) Studier, M.H., Hayatsu, Anders,E.: Organic Compounds in Carbonaceous Chondrites, Science 140, 1455, (1965)
(11) Ruhr-Chemie DBP 904891(1949) und IG-Farben EP 270705 (1928)
(12) Urey, H.C., The Planets (Yale Univ. Press, 1952)
(13) Völk, H.J.: Sizes of Protostellar Grains Embedded into Comets, Proc. of a Workshop on Cometary Missions, Bamberg 111-119, 1979, Astronomisches Institut der Universität Erlangen.
(14) Whipple, F.L.: Earth,Moon and Planets. Harvard Press 1968 and Whipple, F.L.: The Spin of Comets, Scientific American, March 1978, pp. 88-96

FORMATION OF PREBIOTIC PRECURSORS FROM MODEL REDUCING ATMOSPHERES: ROLE OF HYDROGEN ESCAPE

D. MOUREY, F. RAULIN and G. TOUPANCE

Laboratoire de Physicochimie de l'Environnement
Université Paris Val de Marne
94010 CRETEIL CEDEX-FRANCE

It is generally accepted that the amount of molecular hydrogen in the atmosphere of the Earth, must have always been small, on account of its escape. Now, in classical experiments simulating the chemical evolution of model reducing atmospheres, large quantities of hydrogen are generally formed and accumulate. The results of these experiments must be therefore reassessed if we want to apply them to the modelisation of the chemical evolution of terrestrial planets. For this reason, we have carried out sytematic experiments with and without hydrogen escape simulation. The influence of this parameter :
-On the evolution of the atmosphere from a CH_4-H_2O containing atmosphere to a CO_2 - H_2O containing atmosphere.
-On the formation of atmospheric organic precursors such as aldehydes, is reported and discussed.

In classical MILLER'S type of experiments on the chemical evolution of model reducing atmosphere, molecular hydrogen is the most abundant product found in the remaining gases(1). It is obtained by craking of starting molecules containing H atoms,mainly : CH_4, NH_3 and H_2O. Such H_2 accumulation does not fit with plausible models of the primitive atmosphere of the Earth (2 and 3). In fact, the partial pressure of hydrogen in the atmosphere of the Earth must have been always low in account of its escape (4). The influence of excess of H_2 on the synthesis of organic compounds from CH_4 - NH_3 - H_2 and CO_2-N_2-H_2 mixtures submitted to a silent electric discharge or to U.V. irradiation has been previously theoretically discussed (5). In order to estimate more precisely the influence of hydrogen escape on the chemical evolution of a model atmosphere, we have

Y. Wolman (ed.), Origin of Life, 73–82.

set up an experimental device simulating the escape of hydrogen derivated from the device used by BUVET et al(6) in CERES program. In CERES experiments, aqueous solutions were always present ; in addition, these studies were mainly concerned with the aqueous phase. The results of the evolution of the gas phase were only qualitatively reported. The authors showed that CO_2 appears in the gas phase jointly with a decrease of CO. In this paper we report the results of a set of experiments on the chemical evolution of CH_4(50torr) - H_2O (100 torr) gaseous mixtures with and without H_2 escape.
The main objective of this work is to estimate qualitatively and quantitatively the influence of H_2 escape :
- on the evolution of the atmosphere from a CH_4 - H_2O rich medium to a CO_2 - CO - H_2O rich medium,
- on the synthesis of organic compounds.

As a first step of our program, the experiments have been performed without liquid water.

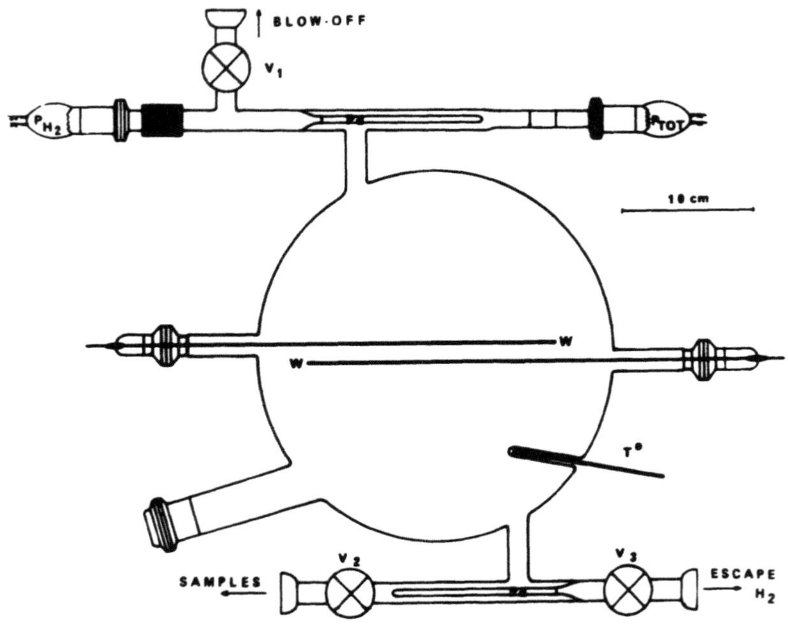

Figure 1 : Experimental Device

Figure 1 Shows the apparatus used for these studies. It consists in a 10 liters spherical pyrex flask with two tungsten electrodes. The spark discharges are produced by a Tesla coil with a high frequency voltage (22 000 volts). In this apparatus two palladium membranes, heated up to 245°C and connected to a vacuum line, simulate the hydrogen escape. Two membranes transducers indicate continuously the total pressure and the H_2 partial pressure. The whole device is heated to 70°C to prevent any water condensation. On account of the slow rate H_2 diffusion through the Pd membranes, the sparking is operated only 1 minute each 10 minutes in order to limit the hydrogen partial pressure to a sufficiently low value. To estimate the possible catalytic effect of palladium, a similar apparatus has been designed without palladium membranes. For systematic studies of model atmosphere, three distinct experiments are carried out. Two experiments are performed in the reactor with palladium membranes : one with hydrogen escape (V_1 and V_3 open). and the other without hydrogen escape (V_1 and V_3 turned of). In these two separate runs, the temperature ot Pd fingers is the same. The third run is performed in the reactor without Pd. The temperature of the flask and the sparking program are identical in these 3 runs. Periodically samples are taken, from the reactor and analyzed :

- hydrocarbons (C_2 to C_4) are analyzed by gas chromatography on a 3m x 2mm i.d. stainless steel column packed with PORAPAK Q, 80-100 mesh, using a flame ionization detector.
- CH_4, CO, H_2O and CO_2 are also analyzed by gas chromatography on a 2m x 2mm i.d. stainless steel column packed with CARBOSIEVE S, 100 - 120 mesh, using a T.C. detector.

The O - containing organics are analyzed after collecting the gaseous sample in a glass flask and dissolving in water :

- formaldehyde is analyzed by colorimetric titration using chromotropic acid method (7).
- ketones, aldehydes, alcohols are analyzed by gas chromatography on a 4m x 2mm i.d. stainless steel column packed with PORAPAK Q, 100-120 mesh, using a flame ionization detector.

Figure 2 (a - b - c) presents the evolution of CH_4, H_2, CO, CO_2, and H_2O. The partial pressure of these compounds is plotted as a function of effective sparking time. The results of the experiments carried out without hydrogen escape are plotted on the figure 2a and 2b. No significant differences are observed between the results of the experiment carried out in the reactor with palladium membranes(fig. 2a) and the results of the experiment carried out in the reactor without palladium membranes (fig. 2b.). On account of this comparison, we can conclude that the catalytic effect of Pd on the evolution of the major compounds such as CH_4, CO, CO_2 and H_2O is negligible. Thus, the differences observed between the behaviour ot these coumpounds during the experiment with hydrogen escape simulation (fig. 2c.) and the experiment without hydrogen escape simulation (fig. 2a) can be mainly attibuted to hydrogen escape.
In the experiments perfomed without hydrogen escape, H_2 becomes rapidly the more abundant product. About 60% of initial amount of hydrogen atoms is converted into molecular hydrogen

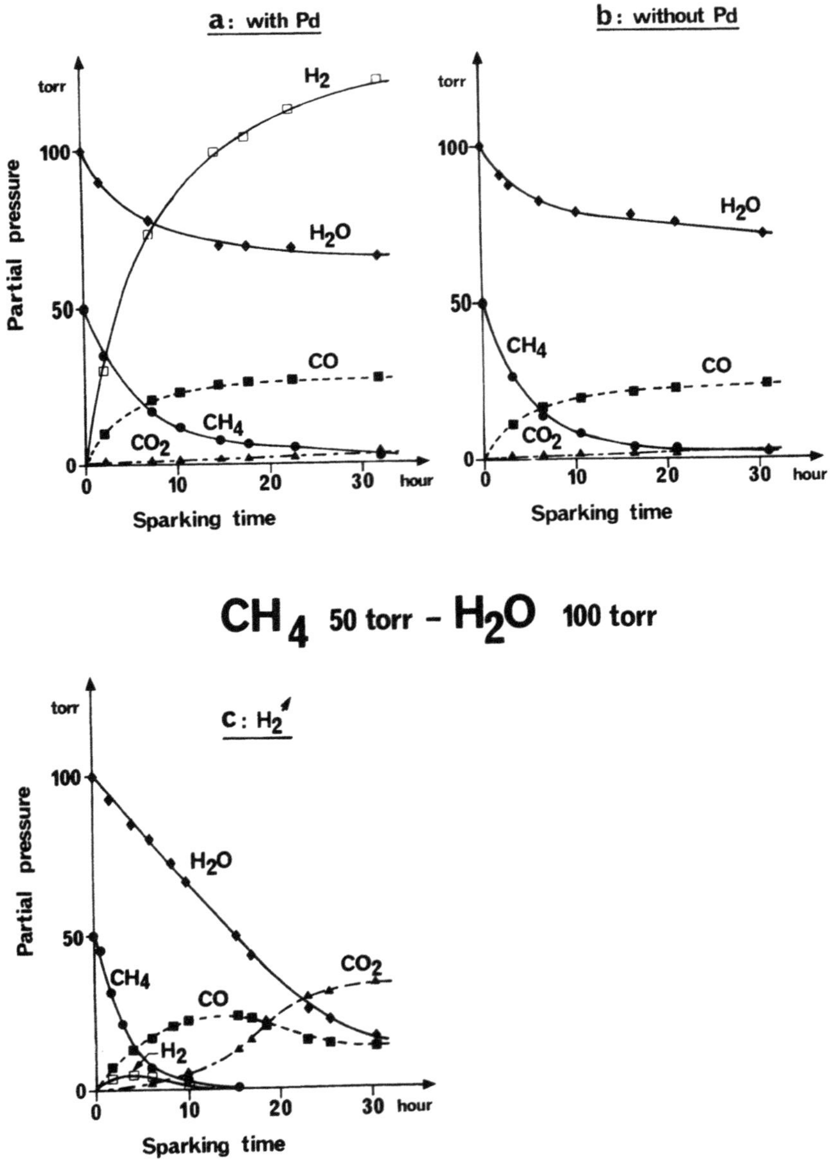

Figure 2 : Partial pressure of major gases as a function of sparking time. a : without H_2 escape with Pd; b : id. without Pd; c : with H_2 escape.

The major product of the oxidation of methane is carbon monoxide. This is consistent with MILLER's results (1) on the evolution of CH_4 - NH_3 - H_2O mixture. CO_2 is present in lower quantities than CO and the ratio $\frac{CO}{CO_2}$ is about 7 at the end of our experiment. This value is comparable with the value of about 9.5 obtained by TOUPANCE (8) with a CH_4 - H_2O (1 :2) gaseous mixture (total pressure 20 torr) submitted to a silent electrical discharge. An important result is the observation that a steady concentration is reached by CO, CO_2, H_2,H_2O and CH_4. The presence of such a steady state could be probably explained if we assume that an equilibrium between these coumpounds exists at the high temperature where the reactions occur.

Figure 2c shows that the steady state reached in the experiment with hydrogen escape is very different. In this run, the partial pressure of residual hydrogen is lower that 5 torr. The curve corresponding to the partial pressure of CH_4 shows that CH_4 is strongly destroyed by the discharge. After 15 hours of sparking,the CH_4 pressure is down to about 0.5 torr and it continues to decrease with time. In the experiment performed without H_2 escape, the decrease of methane is less important since the partial pressure of CH_4 is about 3 torr after 32hours of sparking. This points out the influence of H_2 partial pressure on the rate of disappearance of CH_4. These differences suggest that the accumulation of hydrogen in run (a) increases the role of mechanisms giving back CH_4. During the first 15 hours sparking the evolution of CO is about the same in both runs. But after about 15 hours sparking, the amount of CO in the experiment with H_2 escape decreases jointly with an increase of CO_2. Then carbon dioxide becomes more abundant than CO. In addition, we can observe that H_2O decreases rapidly. Results are quite different in the experiments without hydrogen escape since pressure of CO and H_2O does not decrease with time. These differences can be explained by the following elementary model:

$$H_2O \xrightarrow[\text{elec. disch.}]{k_1} OH + H \qquad [1]$$

$$OH + H_2 \xrightarrow{k_2} H_2O + H \qquad [2]$$

$$OH + CO \xrightarrow{k_3} CO_2 + H \qquad [3]$$

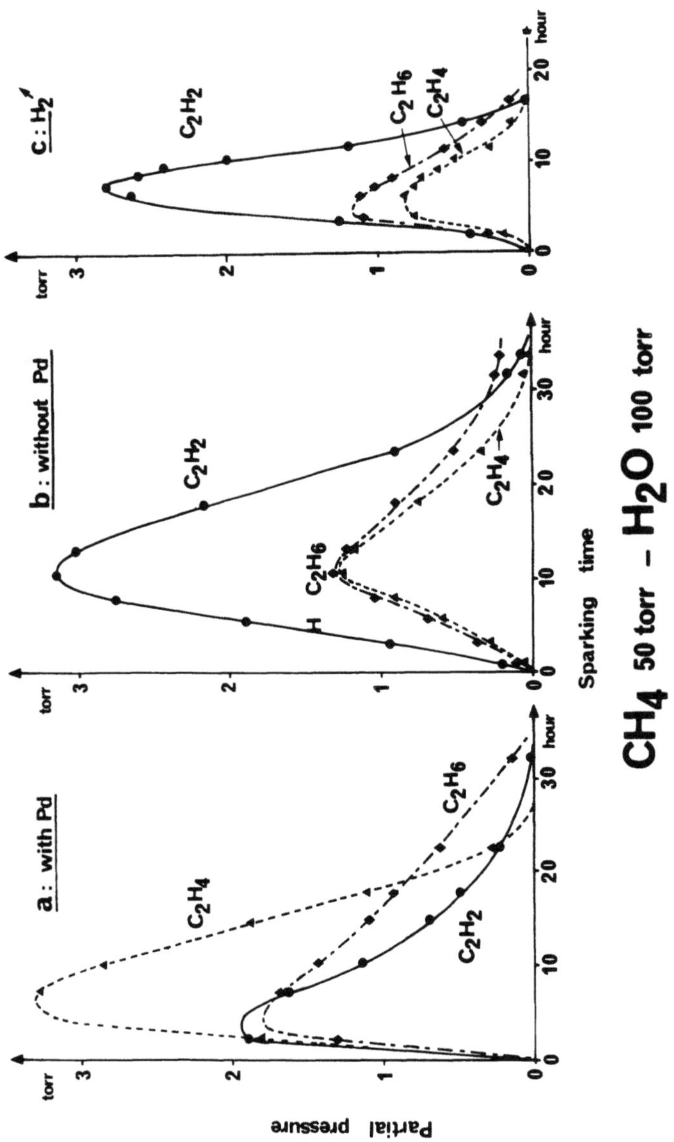

Figure 3 : Partial pressure of C_2-hydrocarbons as a function of sparking time. a : without H_2 escape with Pd; b: id° without Pd; c: with H_2 escape.

The reactions occuring in electrical discharges are often described by high temperature processes. According to MILLER's results obtained with a similar source of energy (1), we can assume that the temperature of the discharge is about 1000K. At 1000K, the values of k_2 and k_3 are such that $k_2 \sim 7\ k_3$ (9). At the end of runs carried out without H_2 escape, the molar fraction of H_2 relative to CO is about 6. So it comes $V_2 \sim 40\ V_3$. Thus the production of CO_2 by [3] is mostly inhibited by the back synthesis of H_2O by
In experiment carried out with H_2 escape, the molar fraction of H_2 relative to CO reaches a value of about 0.02 after 15 hours of sparking. From this value, we can calculate that reaction [3] in ten times faster than reaction [2]. Thus, the synthesis of CO_2 predominate over the back synthesis of H_2O and consequently the amount of CO decreases. If we assume that the rate of craking of H_2O by [1] is almost the same in all the experiments, the amount of remaining H_2O must also decrease.
This model is in a good agreement with the experimental results.

The results of analysis of C_2 - hydrocarbons are reported in the figure 3a-b-c. In the three cases, unsatured hydrocarbons are more abundant than satured hydrocarbons. An identical result is observed for the C_3 and C_4 - hydrocarbons. Acetylene is the major hydrocarbon produced in the experiments achieved with hydrogen escape (fig.3c) and also in the experiment achieved without palladium membranes (fig. 3b). In both cases, it reaches a maximum partial pressure of about 3 torr after about 10 hours of discharges. On the contrary, in the experiment carried out without hydrogen escape in the reactor with Pd membrane (fig. 3a), the major synthesized hydrocarbon is C_2H_4 whose maximum partial pressure is about 3 torr. In addition, the amount of C_2H_2 (maximum partial pressure : about 2 torr) is quite smaller here than in the two others experiments. This is probably related to the hydrogenizing reaction of acetylene on palladium, giving rise to ethylene formation.
The influence of palladium on the behaviour of C_2 hydrocarbons point out the limitation of our simulation of hydrogen escape. Nevertheless, the results related to the C_2 hydrocarbons show that the rate of disappearance of these compounds is strongly increased by the escape of hydrogen. The same resulte is found for the evolution of C_3 and C_4 hydrocarbons.

Sparking of mixtures CH_4 - H_2O gives rises also to several oxygen containing compounds (10). Systematic analysis of the samples shows the presence in the gaseous mixtures of alcohols : mainly methanol, ethanol, and very small amounts of 1 and 2 propanol. Aldehydes and ketones are also present : methanal, ethanal and butanone have been identified on the obtained gas chromatograms. Propanal and propanone are co-eluted in our chromatographic conditions, so that it is not pos-

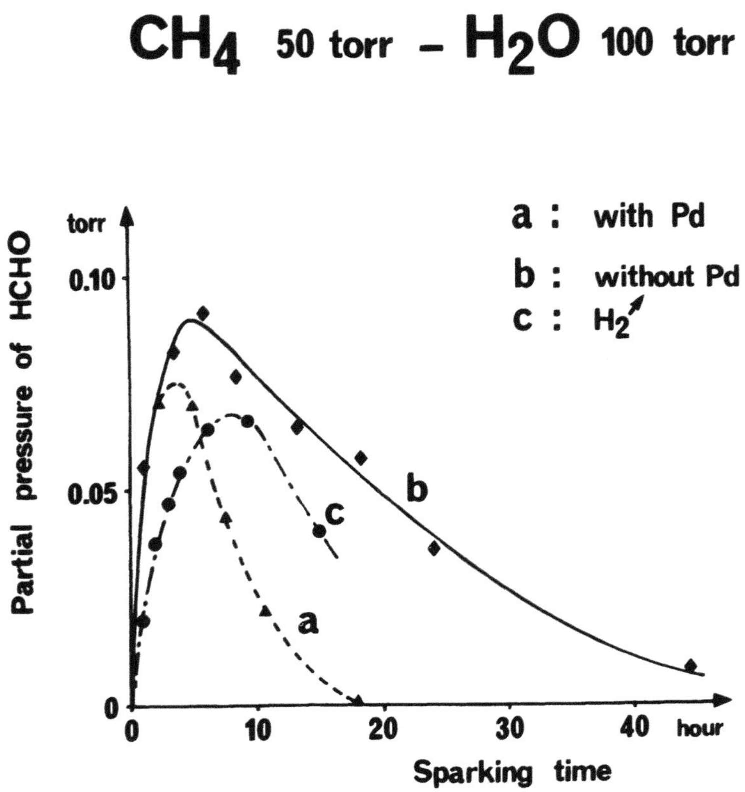

Figure 4: Partial pressure of HCHO as a function of sparking time. a: without H_2 escape, with Pd; b: id° without Pd; c: with H_2 escape.

sible to know the contribution of each in the corresponding peak, always present on the chromatograms. Quantitative analysis has been performed for most of these compounds. Fig.4 shows the variation of partial pressure of synthesized HCHO as a function of sparking time, for the 3 different runs.

The quantity of formaldehyde reaches a maximum value around 7 to 9.10^{-2} torr, after 5 to 7 hours of sparking. Then the partial pressure of HCHO decreases with the sparking time. In the run (b), performed without Pd membranes, the amount of HCHO decreases slower than in runs a and c. This relative stability of HCHO in run b may be explained by catalytic hydrogenizing of HCHO on palladium. The catalytic effect is not observed in run c, despite the presence of Pd because in the experiment carried out with the H_2 escape, the amount of hydrogen is always very small. A similar catalytic effect is found in the case of the other aldehydes but none in the case of alcohols. Despite the palladium effect, comparison between the run carried out without Pd and the run carried out with hydrogen escape shows that the rates of disappearance of the O - containing compounds are lightly increased by the hydrogen escape. The same effect has been found for hydrocarbons. This suggests that O-containing compounds are mainly synthesized from hydrocarbons.

In conclusion, several results can be pointed out from the present study.
1. The stability of a CH_4 rich atmosphere is strongly decreased by hydrogen escape.
The synthesized hydrocarbons are destroyed much more quickely when hydrogen escape is simulated. In this case, the initial CH_4 rich atmosphere evolves toward an oxidized atmosphere containing mainly CO_2 with small quantities of CO. Without H_2 escape the initial atmosphere evolves toward a CO rich atmosphere containing small quantities of CO_2).
2. In all the experiments, the quantities of synthesized atmospheric precursors decrease strongly during the oxidation of the medium.
3. No important effect of hydrogen escape on the synthesis of O-containing organic compounds is observed. O-containing organics are synthesized in quite low quantity. Howewer their production should be influenced mainly by the presence of a liquid phase(11). The high solubility of these compounds in liquid water should allow their accumulation in the so-called primitive soup.
Another serie of experiments including liquid water, is now in progess in the laboratory, in order to study the influence of the presence of a liquid phase on the evolution of the volatile compounds formed in the gaseous phase.
Another important precursor plays also a crutial role in prebio-

tic chemistry : HCN (14). Studies are now in progress, in the laboratory on the possible influence of H_2 escape on the evolution of N - containing atmospheres.

ACKNOWLEDGEMENT

We thank A. Bossard for his comments. This work has been supported partly by CNRS grant INAG 3721 and CNES.

REFERENCES

1. Miller S.L.,J.Amer. Chem. Soc.,77,2351,(1955).

2. Holland H.D., "Petrologic studies : A volume in honor of A.F. Buddington", A.E. Engel, H. James and B.F. Leonard Edit., Geological Society of America, New-York, (1962), p.447.

3. Hart M.H., Icarus,33, 23,(1978).

4. Mc Govern W. J. Atmos. Sci., 26, 623, (1969).

5. Toupance G.,Mourey D.,Raulin F., Proceedings of the international conference "Evolution of planetary atmospheres and and climatology of the Earth", CNES édit. Toulouse, (1978), p. 31

6. Buvet R., Stoetzel F, Origins of life, 7, 400, (1975).

7. Bricker C.E., Johnson H.R., Ind. Eng. Chem. 17, 400, (1945).

8. Toupance G, Thèse de doctorat d'Etat, Université Paris 6, (1973).

9. Hampson Jr. R.F., Garvin D., N.B.S. SP-513,(1977).

10. The synthesis of O-containing compounds from sparking of CH_4 - H_2O mixtures has previously been reported by : Allen W.V.,Ponnamperuma C., Curr.Modern Biol.,1, 24 (1967). However the authors analyzed only monocarboxylic acids.

11. Sergio R., Thorton J.D., J. Appl. Chem., 17, 325, (1967)

12. For review see : Miller.S.L., Urey H.C., Oro J., Mol. Evol. 9,59, (1976).

ORGANIC CHEMICAL EVOLUTION OF REDUCING MODEL OF THE ATMOSPHERE OF THE PRIMITIVE EARTH. ROLE OF UV LIGHT AND ELECTRIC DISCHARGES

A. BOSSARD, F. RAULIN, D. MOUREY and
G. TOUPANCE

Laboratoire Physicochimie de l'Environnement
Univ. Paris Val de Marne, 94000 Creteil. France

The purpose of this paper is to compare the role of UV light and electric discharges, the two most important sources of energy on the primitive earth, in the synthesis of organic compounds out of a reducing model of that atmosphere. Since Miller's experiments in 1953, most of the experimental simulations have been performed with electric discharges, and it is frequently assumed that UV radiation would give similar results.
In order to check this assumption we have performed both experimental simultations in our laboratory. Experimental results indicate that this assumption is wrong in a large extent. Our four main conclusions are :

1 - Unlike electric discharges, UV light is not an efficient source for producing unsaturated carbon chains.

2 - UV light is efficient for producing nitriles in CH_4 - NH_3 mixtures when the molar ratio of NH_3 is very low while electric discharges need a higher molar mixing ratio of NH_3.

3 - UV light is not able to produce nitriles from CH_4 - N_2 mixtures while electric discharges produce diversified nitriles from these mixtures.

4 - UV light is not efficient for producing aldehydes from CH_4 - H_2O mixtures, while electric discharges seem to be able to produce them more efficiently.

The two most important sources of energy on the primitive earth have been the ultraviolet light from the sun (41 cal. cm^{-2} yr^{-1} for wavelengths lower than 200 nm (1)) and the electric discharges from the thunderstorms (4 cal. cm^{-2}. yr^{-1} (1)).

Y. Wolman (ed.), Origin of Life, 83–92.

TABLE 1

Partial Pressures (Torr) of the Different Gases Employed in the Reviewed Experiments[a].

Source of Energy		CH_4	CH_4-N_2	CH_4-NH_3	CH_4-H_2O
UV Light: λ (nm)	123.6	10 (12)	no detail (13)	20-20 (14)	
	125-170		4 - 4 (15)		
	147	9 (9)	5 - 5 (16)	9 - 1 (9)	9-1 (9,10)
	184.9		230-200[b] (17) T = 345 K	95-30[c] (18)	430-240 (17) T = 345 K
Electric Discharges	spark (static system)	300 (19)	50-100 (19) T = 325 K	190-190 (21)	50-100 (11,19) T = 345 K
			140-620 (20)		
	corona (flow system)	20 (7)	10-10 (7,8)	18-2 (7,8)	18 - 2 (7)

a: Temperature is given, when different from ambient.

b: Composition of this mixture: CH_4(230)-N_2(200)-H_2O(240).

c: Composition of this mixture: CH_4(95)-NH_3(30)-H_2(474)-He(60).

Since Miller's first experiments in 1953 (2) simulating the transformation under electric discharges of reducing gaseous mixtures equilibrated with an aqueous phase, an important number of experiments have been performed with reducing models of the atmosphere of the primitive earth. Among them, only four (3-4-5-6) have been performed with UV light. Because UV light and electric discharges have produced similar results, principally with respect to aminoacids analysis, it is frequently assumed that the two sources of energy have played a similar role in the primitive earth. However it is easy to calculate that, for most of the plausible reducing models of the atmosphere, the absorbtion of far UV radiations from the sun (i.e. 190 nm and less) occurs in the middle and upper atmosphere (10 km and more) - i.e. at low pressure and far from any liquid water phase.

Consequently the results of Miller's type of experiments using far UV radiation are not directly significant for prebiotic chemistry. The respective role of the electric discharges and of UV light must be discussed again on the basis of the results of the experiments using only a gaseous phase. The objective of this paper is to present a first comparison of the relative importance of these two sources of energy in the atmospheric production of organic precursors of biomonomers.

Table 1 summarizes the principal experimental conditions which have been used in our experiments and in virtually all previous experiments of this type performed by others. Our experiments with UV light and corona discharges were carried out with diversified compositions of different types of mixtures. Table 1 includes only these which best simulate the conditions of the primitive earth. In particular in the case of the experiments with CH_4 - NH_3 mixtures we have chosen those which contain low NH_3 molar ratio because it is generally assumed that, if NH_3 were to have been present in the primitive atmosphere of the earth, it would have always been a minor component.

The results of all these experiments indicate the following four main differences between the actions of UV light and electric discharges on reducing models of the atmosphere of the primitive earth.

The first important difference concerns the possibility of synthesizing saturated or unsaturated carbon chains. We have summarized in table 2 the values of the ratios of unsaturated C_2 or C_3 hydrocarbons to saturated C_2 or C_3 hydrocarbons, which can be calculated from the available data. This table clearly shows that, when the UV light is the source of energy, the production of ethane and propane is more important than the production of their unsaturated homologues. This effect is far more

TABLE 2

Ratios of Synthesized Unsaturated C_2 or C_3 Hydrocarbons to Synthesised Saturated C_2 or C_3 Hydrocarbons.

Source of Energy \ Gaseous Mixture		CH_4		CH_4-N_2		CH_4-NH_3		CH_4-H_2O	
		C_2	C_3	C_2	C_3	C_2	C_3	C_2	C_3
UV Light; λ nm	123.6	0.3 (12)	0.01			< 0,05 (14)			
UV Light; λ nm	147	0,6 (9)	0,1	0,5 (16)	0.2	0,05 (9)	< 0,01	0,03 (9,10)	< 0,01
Electric Discharges	Spark (static system)	10 (19)	25	10 (19)	10			3 (11,19)	3
Electric Discharges	Corona (flow system)	1,3 (7)	0,7	1,3 (7,8)		1.3 (7,8)		2 (7)	

These ratios have been calculated, for the photochemical experiment, from the value of the initial quantum yield, and, for the electric discharges experiments, from the value of the initial rate of production.

important when the mixture contains water vapor or ammonia, or when the number of carbon of the chain increases. On the other hand electric discharges are very efficient producers of unsaturated carbon chains from all these reducing mixtures. In the case of sparking experiments, among the major hydrocarbons produced, are acetylene, ethylene, propylene but also very unsaturated species like allene or biacetylene. In addition the synthesis of benzene has been established in sparking experiments while it has never yet been detected in experiments using UV light. The possibility of synthesis of those unsaturated carbon chains must be important for further chemical reactions of prebiotic interest. For instance,the synthesis of **acrylonitrile or** cyanoacetylene from CH_4 - NH_3 or CH_4 - N_2 mixtures submitted to electric discharges (cf table 3 which will be discussed below) must be reasonably attributed to the presence of the large quantities of ethylene and acetylene which are produced in the gaseous medium.

The second important difference which may be observed concerns the production of N-containing organics from C-N-H systems. Table 3 summarizes the main qualitative results from experiments with CH_4 - NH_3 and CH_4 - N_2 mixtures. We have established that UV irradiation (147 nm) of CH_4 - NH_3 mixtures leads to the formation of HCN, CH_3CN and C_2H_5CN (8,22). At a longer wavelength (184.9 nm), Ferris and Chen (18) have also shown that the production of HCN occurs from such a mixture. As mentionned above it is generally admitted that ammonia would always have been a minor component of the primitive atmosphere of the earth. For this reason, we have studied the possibility of synthesis of nitriles in low NH_3 molar ratio mixtures (9) and we have shown that the quantum yields of formation of those compounds increase when NH_3 molar ratio decreases. We have also established that the synthesis of amines is not observed in those mixtures poor in NH_3, probably because of their destruction by the incident UV light.

The efficient production of HCN from UV irradiation of model atmospheres containing low partial pressures of NH_3 is significant for prebiotic chemistry because this compound is not easily destroyed by the UV radiations reaching its zone of synthesis. Consequently there is a high probability that HCN would have reached the surface of the ocean and dissolved. In addition, Mizutani et al (23) have shown that the main product resulting from the UV irradiation of gaseous HCN is cyanogen, which should also be a possible precursor of biomonomers.
The production of nitriles from CH_4 - NH_3 mixtures using electric discharges has been established by Ponnamperuma et al (21) and by Toupance et al (7,8). The latter have also shown (7,8) that synthesis of nitrile may occur for low NH_3 molar ratio.

TABLE 3

N-Containing Organics Synthesized from gaseous mixtures of CH_4 - NH_3 and CH_4 - N_2.

Source of Energy \ Gaseous Mixtures		CH_4 - NH_3	CH_4 - N_2	
UV Light; λ nm	123.6		no nitrile (13)	
	125-170		HCN (15)	
	147	HCN CH_3CN (9) C_2H_5CN	no nitrile (16)	
	184.9	HCN (18)	no N-containing organics (17)	
Electric Discharges	Spark (static system)	HCN CH_3CN (21) C_2H_5CN aminonitriles	HCN (19) CH_3CN C_2H_5CN $CH_2{=}CHCN$ $CH{\equiv}CCN$ $(CN)_2$	HCN (20) $CH{\equiv}CCN$
	Corona (flow system)	HCN (7,8) CH_3CN C_2H_5CN $CH_2{=}CHCN$	HCN (7,8) CH_3CN C_2H_5CN $CH_2{=}CHCN$ $CH{\equiv}CCN$ $(CN)_2$	

The previous results are relevant to the geological period when NH_3 may have been present as a minor component. The following are relevant to the period when NH_3 was absent. Table 3 shows clearly the important differences obtained from photochemical and electric discharge experiments performed with CH_4 - N_2 mixtures. The production of a nitrile (HCN) has been reported only once in photochemical experiments (15) and with a low quantum yield. Several attempts to replicate this result have failed (13). Unlike UV light, electric discharges (spark or corona) have been found to be very efficient in producing varied nitriles, HCN, CH_3CN, C_2H_5CN, C_2H_3CN, C_2HCN, $(CN)_2$. In particular, it is the only way that has been found to produce cyanoacetylene (8,19,20). The experimental data from CH_4-N_2 mixtures demonstrate that UV light would have never been an efficient source of energy for producing nitriles, while electric discharges seem to have played a very important role in the production of triple bonds between C and N atoms.

The final difference between UV light and electric discharges with respect to chemical evolution concerns the production of O-containing organics. Table 4 summarizes the results obtained from experiments on CH_4 - H_2O mixtures. Mourey et al (11) have shown that the sparking of CH_4 - H_2O mixtures produces C_1 to C_3 alcohols, C_1 to C_3 aldehydes and C_3 to C_4 ketones. Because thunderstorms take place in the lowest layers of the atmosphere near the surface of the oceans, most of these compounds may have dissolved in the "primitive soup" and thus may have taken part in the synthesis of biomonomers. As in the case of sparking experiments, the UV irradiation of CH_4 - H_2O mixtures leads to the formation of aldehydes, ketones and alcohols at 184.9 nm (17) or to the formation of ketones and HCHO at 147 nm (9,10). Nevertheless, as mentionned above and discussed in detail elsewhere (10), the synthesis of these compounds takes place in layers of the atmosphere above 10 km where pressures are low. These compounds absorb intensively near UV light from the sun (200 nm $< \lambda <$ 350 nm) and hence would tend to be destroyed and converted into CO in the atmosphere by these radiations. The final result of the action of the UV light from the sun, upon CH_4 - H_2O atmosphere, is mainly the conversion of CH_4 to carbon oxides. It is very unlikely that O-containing organics would be produced in this way.

To conclude this review, the present experimental data lead us to believe that, despite the fact that electric discharges was a less abundant source of energy than UV light in the primitive earth, it would have played an important role in the synthesis of atmospheric organic precursors of biomonomers. Its role was probably more important than that of UV light from the sun. The most important contribution of UV light to the evolution of the primitive earth has been surely the more or less rapid

TABLE 4

O-Containing Organics Synthesized from gaseous mixtures of CH_4 and H_2O.

UV Light	147 nm	Ketones HCHO	(9-10)
	184,9 nm	Aldehydes Ketones Alcohols	(17)
Electric Discharges	Spark Discharges — Static System	Aldehydes Ketones Alcohols	(11,19)
	Corona Discharges — Flow System	HCHO CH_3CHO CH_3OH	(7)

transformation of a CH_4 - N_2 - H_2O atmosphere to a CO_2-CO-N_2-H_2O atmosphere.

ACKNOWLEDGEMENT. The authors wish to thank R. Swensson for his help during the preparation of this manuscript. This work has been supported by the French Space Agency (CNES).

REFERENCES

(1) MILLER S.L. and ORGEL L.E., "The Origins of Life on the Earth", Prentice-Hall Inc., New Jersey, (1974), 55.

(2) MILLER S.L., Science, 117, 528, (1953)

(3) GROTH W.E. and Von WEYSSENHOFF H., Planet. Space Sci., 2 79, (1960).

(4) DODONOVA N.Y. and SIDOROVA A.I., Biofizika, 6, 149, (1961).

(5) SAGAN C. and KHARE B.N., Science 173, 417, (1971).

(6) JOSHI P.C. and PATHAK H.D., J. Brit. Interplanet Soc., 28 90, (1975).

(7) TOUPANCE G., These de Doctorat d'Etat ès Sciences Physiques, Université Paris 6, (1973).

(8) TOUPANCE G., RAULIN F. and BUVET R., Origins of Life, 6, 83, (1975).

(9) BOSSARD A., These 3ème cycle Chimie-Physique, Université Paris 6, (1979).

(10) BOSSARD A. and TOUPANCE G., Communication at the 6th ICOL Conference, Jerusalem (Israel), (1980).

(11) MOUREY D., RAULIN F. and TOUPANCE G., Communication at the 6th ICOL Conference, Jerusalem (Israel), (1980).

(12) HELLNER L. and VERMEIL C., J. Chim. Phys., 67, 221, (1970). and HELLNER L., Thése de 3ème cycle Chimie Physique, Faculté des Sciences de Paris, (1969).

(13) CHANG S., SCATTERGOOD T., ARONOWITZ S. and FLORES J., Communication at the NASA workshop on "The Saturn System", NASA Conference Publication n°2068, RESTON (Virginia), 161, (1978).

(14) GARDNER E.P. and Mac NESBY J.R., J. Photochem. accepted for publication, (1980).

(15) DODONOVA N.Ya., Russ. J. Phys. Chem., 40, 523, (1966).

(16) BOSSARD A., unpublished results.

(17) FERRIS J.P. and CHEN C.T., J.Amer. Chem. Soc., 97, 2962, (1975).

(18) FERRIS J.P. and CHEN C.T., Nature, 258, 587, (1975).

(19) MOUREY D., to be published.

(20) SANCHEZ R.A., FERRIS J.P. and ORGEL L.E., Science, 154, 784, (1966).

(21) PONNAMPERUMA C. and WOELLER F., Curr. Modern Biol., 1, 156, (1967).

(22) TOUPANCE G., BOSSARD A. and RAULIN F., Origins of Life, 8, 259, (1977).

(23) MIZUTANI H., MIKUNI H., TAKAHASI M. and NODA H., Origins of Life, 6, 513, (1975).

(24) FERRIS J.P., SANCHEZ R.A. and ORGEL L.E., J. Mol. Biol., 33, 693, (1968).

FAR UV PHOTOLYSIS OF METHANE-WATER GASEOUS MIXTURES AND THE PREBIOTIC SYNTHESIS OF ALDEHYDES.

Alain BOSSARD and Gérard TOUPANCE

Laboratoire de Physicochimie de l'Environnement
Université Paris Val de Marne - Avenue du Général
de Gaulle - 94010 Créteil Cédex - France.

Far UV photolysis of model reducing atmosphere (CH_4-H_2O) has been performed in order to test the possibilities of organic synthesis on the primitive earth. The only detected organic products are hydrocarbons, ketones and formaldehyde. The results are compared with those of a previous study which evidenced the synthesis of aldehydes, alcohols and ketones. The prebiotic significance of these results are discussed.

The nature of the primitive atmosphere of the earth has been the subject of a lot of discussions since the MILLER-UREY hypothesis (1). However no final answer has been established yet and various types of atmosphere, ranging from CH_4-N_2-NH_3-H_2O to CO-CO_2-N_2-H_2O must be studied with regard to the possibilities of prebiotic synthesis of organic compounds. The results we report here concern the evolution of methane-water gaseous mixtures under UV light and principally the conditions of synthesis of O-containing organics.
UV light from the sun has been the most important energy source on the primitive earth but it has been rarely employed (2-3-4-5) to test the possibilities of organic synthesis from reducing model of the primitive atmosphere,,mainly owing to experimental difficulties. We would like to point out that for three of these experiments performed with far UV light (2-3-4) the gaseous mixture was equilibrated with a liquid water solution and for all these experiments the pressure of the irradiated gas phase was high, in the order of magnitude of one atmosphere.
As a matter of fact, it is easy to calculate that for most of the possible model atmospheres, the absorption of far UV radiations of the sun, let us say 190 nm and less, occurs in the middle and upper layers of the atmosphere, at pressures ranging from 1 torr

Y. Wolman (ed.), Origin of Life, 93–100.

SCHEME 1

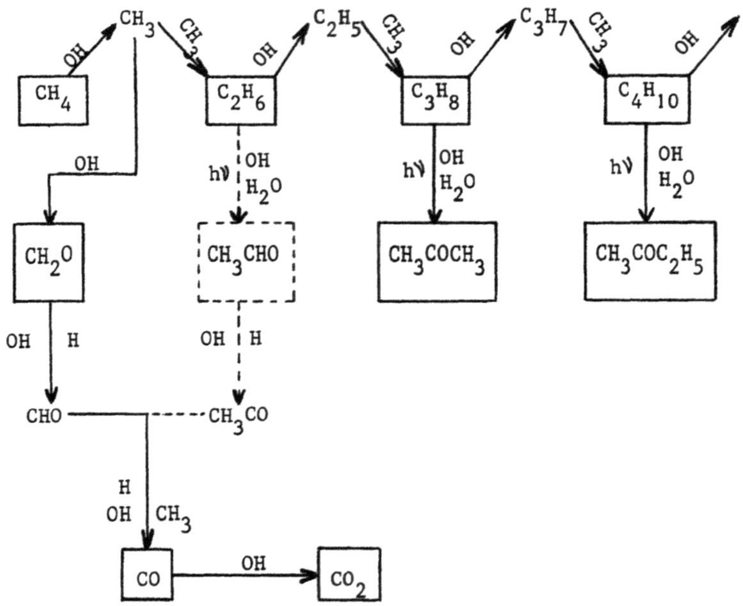

TABLE 1

Experimental Studies of the Photochemistry of Methane-Water Gaseous Mixtures Performed before 1979.

	STIEF et al (7)	FERRIS et al (5)
Total Pressure (torr)	21	660
Molar ratio of water vapor	0.86	0.36
Wavelength used (nm)	123.6 147 145-185	184.9
Products Detected	CO C_2H_6 (8) CO_2	Aldehydes Ketones (9) Alcohols

to hundred of torr. Consequently far UV photochemical evolution of the primitive earth is mostly relevant to atmospheric chemistry so that experiments using a liquid phase previously reported (2-3-4) must be considered as very bad simulations of prebiotic photochemistry.
In order to test directly in more plausible conditions the possibilities of organic synthesis in these atmospheric layers of the primitive earth, we have studied the far UV photolysis of CH_4-H_2O gaseous mixtures at low pressure. In addition, the experimental knowledge of this photochemistry may be very useful for problems relevant to primitive Mars or to Titan, as it has been recently claimed (6).

The study of the photochemistry of that type of mixture has been reported only twice until today (5,7) and only once (5) with the objective of modeling the chemical evolution of the atmosphere of the primitive earth. We summarize in table 1 the experimental conditions used in these experiments and also the obtained results.

In our study, CH_4-H_2O gaseous mixtures of different compositions under a total pressure of 10 torr have been irradiated at 147 nm (low pressure xenon lamp). The apparatus and experimental procedure used for this study have been previously described elsewhere (10). The experiments have been carried out at ambient temperature with a lamp emitting 8.10^{15} photons.s^{-1}.
Table 2 summarizes the results of the analyses after irradiation of 3 types of mixture (water molar ratio : 0.1, 0.5 and 0.8). The major synthesized products are always CO, CO_2 and C_2H_6. This fits with the results previously reported by STIEF et al (7) at the same wavelength and at similar pressure (table 1). Search for H_2 has not been performed. The formation of molecular oxygen has been searched for by gas chromatography (G.C.) on molecular sieve support ; the results were always negative.
The synthesized hydrocarbons consist mainly of saturated hydrocarbons. Ethane represents more than 95 % of the C_2 hydrocarbons, propane and butanes represent respectively more than 99 % of the C_3 and of the C_4 hydrocarbons. Butanes consist of equal amounts of iso and normal. Trace amounts of C_5, C_6 and C_7 are also synthesized. The increase of the molar ratio of water vapor in the mixture decreases the synthesis of hydrocarbons. This inhibitory effect is much more important for the unsaturated hydrocarbons than for the saturated ones.
The only O-containing organics that we have detected are ketones and formaldehyde. The absence of acetaldehyde, methanol, ethanol and dimethyl-ether, has been established using two GC supports (Porapak Q and chromosorb 103) in the limits of detectability of the GC analysis (see table 2).Search for carboxylic acids hasn't been performed.
Ketones have been clearly identified by GC analysis on three

TABLE 2

Quantities of Carbon Oxides and Organic Products (nanomole) Synthesized after 30mn Irradiation at 147nm of Methane-Water Gaseous Mixtures.

Molar Ratio of H_2O	0.1	0.5	0.8
CO	600	950	400
CO_2	250	900	1500
C_2	2500	1640	460
C_3	260	190	30
C_4	80	49	8
Acetone	55	105	50
Butanone	45	49	16
Formaldehyde	20	12	<5
Acetaldehyde	<5	<5	<5
Methanol	<5	<5	<5
Ethanol	<5	<5	<5
Di-methyl-ether	<1	<1	<1

types of support (Porapak Q, chromosorb 103, Durapak OPN/Porasil C). They consist mainly of acetone and butanone, which are synthesized respectively with quantum yields of the same order of magnitude than those of C_3 or C_4 hydrocarbons.
2-Pentanone, 3-Pentanone and 3-methyl-2-butanone have also been detected in the range of few nanomoles.
Formaldehyde has been detected by colorimetric method using 2-Hydrazinobenzothiazole reagent (11-12). The limit of detectability for this compound is 5 nanomoles, with our experimental procedure.

In order to explain our results, we have carefully discussed the occurence of many elementary reactions in our conditions. In fact, photochemistry, in the system we use, becomes rapidly very complicated because of the absorption of a significant part of the incident light by the products, principally by the hydrocarbons. The initial photochemical step is the dissociation of water into OH et H radicals. The dissociation into O and H_2 is a very minor process at 147 nm (13) and may be neglected. The absorption of CH_4 is weak in comparison with the absorption of H_2O and is probably not significant to explain the observed results.
The main chemical step which governs the syntheses is the reaction between hydroxyl radical and methane to produce methyl radicals. From these radicals (CH_3 and OH) ethane and formaldehyde are expected to be the primary photoproducts from the reactions:

$$CH_3 + CH_3 \longrightarrow C_2H_6 \qquad (a)$$
$$CH_3 + OH \longrightarrow CH_2O + H_2 \qquad (b)$$

These two primary photoproducts are also easily attacked by the hydroxyl radical. Ethane produces ethyl radicals and leads to the formation of higher hydrocarbons by reactions similar to (a). Formaldehyde produces formyl radicals which, by further attack of H, OH or CH_3, leads to the formation of carbon monoxide. Formation of carbon dioxide must be attributed to the reaction between CO and OH. The synthesis of relatively important quantities of acetone and butamone may be interpreted by the photolysis of propane and butane which occurs very rapidly.Photolysis of alkanes at 147 nm leads principally to the formation of alkylidenes radicals (14 a). For example, in propane and n-butane photolysis the main primary processes are :

$$CH_3CH_2CH_3 \xrightarrow{147\ nm} CH_3CCH_3 + H_2$$
$$CH_3CH_2C_2H_5 \longrightarrow CH_3CC_2H_5 + H_2$$

These radicals may induce directly the formation of ketones through reactions with H_2O or OH such as :

$$CH_3CCH_3 \ + \ OH\ (or\ H_2O) \qquad CH_3COCH_3 + H\ (or\ H_2)$$

In fact, through this reaction pathway, acetaldehyde would also be produced from ethane. The absence of acetaldehyde in our experiments is probably due to the rapid decomposition of that product into acetyl radical by the attack of H, OH or CH_3. The acetyl radicals,as the formyl radicals, may lead themselves to the formation of CO. The scheme 1 appears to us to be representative of the photochemistry of CH_4-H_2O mixtures at 147 nm.

The results we report here, about the synthesis O-containing organics, differ obviously from those reported by Ferris et al (5). The major product, in their experiments,is formaldehyde, while it is acetone and butanone in our experiments. They also reported the synthesis of alcohols while we do not have detected any. These differences may be explained principally by our use of a different wavelength and different pressure. In the experiments by Ferris et al the synthesis of saturated hydrocarbons occurs certainly by the same way than in ours. Nevertheless, these compounds can't be photolyzed by the 184.9 nm radiations because they don't absorb this wavelength (14b). Thus the formation of ketones through this way is impossible. This fact may surely explain that Ferris et al find a yield of synthesis of ketones lower than what we find. The differences concerning the synthesis of alcohols may be explained through the differences of pressure used in the Ferris et al experiments and in ours.Practically, the only way for producing methanol is the reaction $CH_3 + OH + M \rightarrow CH_3OH + M$. This reaction competes with that of formation of CH_2O ($CH_3 + OH \rightarrow CH_2O + H_2$). These two reactions have quite the same rate constant at atmospheric pressure (15-16) and consequently quite the same probability; thus the production of CH_3OH and CH_2O may occur simultaneously.
At low pressure, the reaction of formation of methanol is surely unfavoured because of its important exothermicity and consequently the reaction of formation of CH_2O is favoured. The absence of CH_3OH in our mixtures after irradiation may be explained by this consideration. The most important product for prebiotic chemistry is undoubtedly formaldehyde. Ferris et al (5) have calculated from their results a value of 2.10^{-4} as initial quantum yield of production of CH_2O. From our results we can calculate respectively values of 8.10^{-4} and 5.10^{-4} for mixtures of 0.1 and 0.5 water molar ratios . These three values are in the same order of magnitude. In fact Ferris et al experiments have shown that the concentration of detected HCHO corresponds to a steady state concentration (the molar fraction is about 8.10^{-7}). We have also shown by irradiating mixtures containing 10 % of water vapor during 1/2 h and 1 h that HCHO reaches a steady state concentration (molar fraction about $7\ 10^{-5}$) at least during those times of irradiation. Differences between these two steady state concentrations may be partially explained by the occurrence of a longer wavelength (254 nm) emitted intensively by the UV lamp used by Ferris et al. This radiation can effectively destroy the organic photoproducts because of their important absorptions at this wavelength. We have calculated that in our experiments, photo-destruction of HCHO by incident UV light is negligeable.

This last remark is very important because of its prebiotic significance. Effectively the atmospheric synthesis of organic products from CH_4-H_2O atmosphere would have occured on the primitive earth under the intensive irradiation of near UV light from the sun. For instance formaldehyde absorbs intensively UV light in the range 200-360 nm. (14c). This absorption leads almost always to the production of CO with a quantum yield higher than 0.5 (14c). In the primitive atmosphere the UV energy available for the synthesis of formaldehyde from methane and water would have been, at the best, the hundredth of that which could destroy it. So, the main results of the action of UV light is the oxidation of CH_4 to carbon oxides as we have observed. Ferris et al have calculated from the value of their initial quantum yield that a synthesis of about 10^{11} moles of HCHO per year would have been assured by UV light. The prebiotic importance of that synthesis depends on the conservation of that product by its dissolution into the primitive ocean. It is very difficult to appreciate this rate of dissolution, nevertheless, it seems unlikely that a major part of the photoproduced HCHO would have been preserved from photodestruction. Consequently the maximum value of photosynthesized HCHO on the primitive earth must be lower than 10^{10} moles per year.
From the results obtained by Mourey et-al (17), on sparking CH_4-H_2O gaseous mixtures, we have calculated that the production of one mole of HCHO needs an energy of about 8 10^8 calories. Since the energy available on the primitive earth from electric discharges was 4 cal. cm^{-2} . yr^{-1} (18a) it can be calculated that electric discharges would have produced about 2.5 10^{10} moles of HCHO per year. The synthesis of HCHO under electric discharges would have occured mostly near the surface of the ocean and consequently the probability that HCHO dissolves would have been important. Finally electric discharges would have been probably more efficient for producing formaldehyde on the primitive earth than UV light.

The synthesis of formaldehyde or the primitive earth is thought to be essential for chemical evolution : it can participate in numerous prebiotic reactions and its self-condensation is the only way known to produce sugars. Nevertheless this production of sugars needs surely a concentration of formaldehyde in the primitive ocean higher than 0.01 M (18 b). It is pratically impossible that this concentration would have been reached up with the quantities synthesized from CH_4-H_2O atmospheres as well with electric discharges as with UV light. Synthesis of the important quantities of HCHO required for producing sugars must be searched by other ways which have not yet been reported.

ACKNOWLEDGEMENT - This study has been supported by the french space agency CNES.

REFERENCES

1. Miller, S.L., J. Amer. Chem. Soc., 77, 2351 (1955)
2. Groth, W.E. and Von Weyssenhoff, H., Planet. Space Sci., 2 , 79 (1960).
3. Dodonova, M. Ya. and Sidorova, A.I., Biofizika, 6 (2), 149, (1961)
4. Joshi, P.C. and Pathak, H.D., J. Brit. Interplanet. Soc., 28 (2), 90 (1975)
5. Ferris, J.P. and Chen, C.T., J. Amer. Chem. Soc., 97 (11), 2962 (1975)
6. Yung, Y.L. and Pinto, J.L., Nature, 273, 730 (1978)
7. Stief, L.J., De Carlo, V.J. and Hillman, J.J., J. Chem. Phys., 43 (7), 2490 (1965)
8. Search for other products has not been undertaken.
9. The analytical procedure doesn't permit the detection of volatile gases CO, CO_2, light hydrocarbons. The major products are formaldehyde and acetone.
10. Raulin F., Bossard A., Toupance G. and Ponnamperuma C., Icarus, 38, 358 (1979)
11. Sawicki E. and Hauser T.R., Anal. Chem., 32, 1434 (1960)
12. Josimovic L., Anal. Chim. Acta, 62, 210, (1972)
13. Okabe H., "Photochemistry of small molecules", John Wiley and sons, New York, (1978), 203.
14. CALVERT J.G. and PITTS Jr J.N., "Photochemistry", John Wiley and sons, New York (1966)
a - p 497, b - p 493, c - p 368 - 371
15. Greenberg R.I. and Heincklen J., Int. J. Chem. Kinet., IV, 417, (1972)
16. Fenimore C.P., "Twelfth Symposium on Combustion", (1969) from Westbrook C.K., Combust. Sci. Technol., 20, 5 (1979)
17. Mourey D., Raulin F. and Toupance G., Communication at the 6th ICOL Conference, Jerusalem (Israel) (1980)
18. Miller S.L. and Orgel L.F. , "The Origins of Life on the Earth", Prentice Hall Inc., Englewood Cliffs (New Jersey), (1974), a - p 55, b - p 109.

PHOTOLYSIS OF CH_4-NH_3 MIXTURES AND PH_3 AS MODELS FOR THE PHOTOCHEMICAL TRANSFORMATIONS ON THE PRIMITIVE EARTH AND JUPITER

James P. Ferris, John Y. Morimoto and Robert Benson

Department of Chemistry
Rensselaer Polytechnic Institute
Troy, New York 12181

Methane, ammonia and phosphine are some of the possible constituents of the atmospheres of the Jovian planets and their satellites. Photolysis of NH_3 in the presence of CH_4 at 185 nm in the temperature range of 25°C to -100°C results in the decomposition of CH_4. The reaction is inhibited by added H_2 or SF_6. These findings are consistent with the reaction of hot hydrogen atoms with CH_4 to give $CH_3\cdot$. P_2H_4 is the initial product formed by the photolysis of PH_3 at 206 nm. Kinetic studies established that it is the intermediate in the formation of P_4 from PH_3. The potential significance of these reactions to the atmospheric photochemistry of the Jovian planets and moons will be discussed.

Methane, ammonia, phosphine and other hydrides are some of the constituents of the atmospheres of the planets Jupiter, Saturn and Neptune and the moons Ganymede, Callisto, Titan and Triton. These gases are subject to photolysis by the short wavelength ultraviolet light from the sun. For example, the ammonia present in the atmospheres of these planets is photolyzed to nitrogen and hydrogen by light at wavelengths <210 nm. We calculated that the ammonia on Jupiter would have a lifetime of only 6×10^7 years if it were not regenerated by thermal processes in the lower depths of the atmosphere (1). Ammonia regeneration is not possible in the cold atmosphere of Titan and it has been postulated that any ammonia that was present in its primordial atmosphere has been photochemically converted to nitrogen (2,3,4).

Photolysis of phosphine (PH_3) to red phosphorus (P_4) is one of the explanations of the red color of the Great Red Spot and the red coloration in other parts of Jupiter's atmosphere (5).

Y. Wolman (ed.), Origin of Life, 101–105.

The red phosphorus formed is reconverted to phosphine by reaction with hydrogen atoms in the upper atmosphere (6) and by thermal processes in the lower hotter regions of the Jovian atmosphere (5).

In the present study we investigated the photolysis of phosphine and ammonia, the latter in mixtures with methane and hydrogen, to determine the possible roles of these photochemical reactions in the atmospheric chemistry of the Jovian planets and their moons.

In previous studies of the irradiation of mixtures of methane and ammonia it was observed that the photolysis of ammonia resulted in the loss of methane (7,8). Methane loss was also observed when the photolysis was performed in the presence of excess hydrogen, a result which suggested that hot hydrogen atoms did not initiate the reaction of methane. In these previous studies the photolysis was performed with the quartz spiral of the low pressure mercury lamp immersed in the mixture of gases to be irradiated. Even though the temperature of photolysis vessel was controlled by a cooling bath much higher temperatures (150°-200°C) were measured in the photolysis zone next to the quartz coil. Consequently, these experiments were repeated with the gas mixtures in a quartz cell 1 cm away from the low pressure lamp. In this way the thermal effects of the low pressure lamp are minimized and the cell containing the gas mixtures can be conveniently thermostated to a variety of temperatures.

Photolysis of a mixture of about 30 Torr of ammonia and about 90 Torr of methane at room temperature resulted in the loss of about 0.1 mole of methane per mole of ammonia photolyzed. The $\Delta CH_4/\Delta NH_3$ ratio increased to about 0.2 when a mixture of 10 Torr of ammonia and 20 Torr of methane was photolyzed. Further studies in which an initial 1:3 ratio of $NH_3:CH_4$ was maintained but the total pressure of the system was varied indicated that the $\Delta CH_4/\Delta NH_3$ ratio increased to 0.20-0.25 as the total pressure of the gas mixture decreased to zero.

The dependence of the $\Delta CH_4/\Delta NH_3$ ratio on methane pressure was investigated next with the ammonia pressure held constant at about 6 Torr. A constant value of 0.25 was observed at methane pressures greater than or equal to 16 Torr. Subsequent studies were performed using an initial ammonia and methane pressures of 6 Torr and 16 Torr respectively because there was no change in the amount of methane reacting when higher partial pressures of methane were used. The loss of methane is undoubtedly due to the reaction of the photochemically formed hydrogen atoms with methane (equation 1,2).

$$NH_3 \xrightarrow{h\nu} NH_2\cdot + H\cdot \qquad (1)$$

$$CH_4 + H\cdot \longrightarrow CH_3\cdot + H_2 \qquad (2)$$

These studies confirmed that methane was decomposed when ammonia was photolyzed at 20°C but they did not establish if this loss occurred at the much lower temperatures present in the atmospheres of the Jovian planets and moons. The $\Delta CH_4/\Delta NH_3$ ratio remained constant at about 0.2 when mixtures of methane and ammonia (6 Torr and 16 Torr respectively) were photolyzed at -84°C and -107°C. The absence of a temperature effect established that methane loss is not due to the attack of thermal hydrogen atoms on methane. The rate constant of this reaction varies significantly with temperature. It is 1.4×10^5 at 27°C and 0.13 at -98°C ($k = 4.0 \times 10^{13} \exp(-11600/RT)$ cm^3/mol-sec)(9). If thermal hydrogen atoms were responsible for the loss of methane one would expect a significant decrease in the $\Delta CH_4/\Delta NH_3$ ratio as the temperature is decreased to -100°C.

Nonthermal (hot) hydrogen atoms would be expected to be equally as effective at 20°C and -100°C in abstracting hydrogen from methane (equation 2). Photolysis of ammonia with light of wavelength 185 nm results in the formation of hot hydrogen atoms with a kinetic energy of about 48 Kcal/mol. This kinetic energy does not decrease significantly as the temperature is decreased so these hot hydrogen atoms would be expected to be equally as efficient abstracting hydrogen from methane at 20°C or -100°C. A test of a mechanism involving hot hydrogen atoms is the introduction of an inert gas which can thermalize the atoms. Addition of 30 Torr of SF_6 to the methane-ammonia mixture (6 Torr and 16 Torr respectively) resulted in a 50% decrease in the $\Delta CH_4/\Delta NH_3$ ratio. Methane loss is completely supressed when 210 Torr of SF_6 is used. The SF_6 only affected the CH_4 loss and did not change the rate of NH_3 decomposition.

A similar decrease in the $\Delta CH_4/\Delta NH_3$ ratio was observed when hydrogen was added to the photolysis mixture. Hydrogen does not behave as an inert gas in that the rate of ammonia photolysis decreases due to the reaction of $NH_2\cdot$ back to NH_3 (10,11), but the hydrogen does effectively thermalize the hot hydrogen atoms formed by ammonia photolysis.

These findings demonstrate that photolysis of ammonia would result in the reaction of methane on those Jovian planets and moons where the atmospheric pressure is low. Titan and possibly Pluto, Ganymede and Callisto are possible sites for the loss of methane by this mechanism. This photochemical process would be expected to yield higher hydrocarbons, amines, nitriles and HCN (7,8). At present we have the most information about the atmosphere of Titan which contains methane and significant amounts of inert gases (N_2, Ar?) as well as other unknown species which may include higher hydrocarbons, methylamine and HCN (2,3,4). One possible scenario for the evolution of Titan's present atmosphere is the photolysis of its primordial methane-ammonia atmosphere with the resultant formation of more complex hydrocarbons, organic nitrogen compounds, nitrogen and hydrogen. Ammonia photolysis probably did not lead to an appreciable formation of complex organic compounds on the primitive earth and it is probably not a source of these substances on Jupiter and Saturn because the hot hydrogen atoms involved are thermalized by the higher pressures present in that atmospheres of these planets.

We have also investigated the photolysis of phosphine because of its potential significance for the formation of red phosphorus in the atmosphere of Jupiter (5). Diphosphine (P_2H_4) has been identified as the initial photoproduct resulting from the photolysis of phosphine at 206 nm. This is the first time that diphosphine has been shown to be a product so the kinetic schemes proposed previously to explain phosphine photolysis will have to be revised significantly to include the formation of this derivative. Termolecular processes and nonthermal (hot) atom reactions were shown to be unimportant by the absence of an effect of added N_2 or SF_6 on the formation of P_2H_4. That P_2H_4 is formed in the gas phase, and not in a wall reaction, was established by the observation of its formation 300 μs after the flash in flash photolysis studies. Wall reactions were eliminated because this time interval is too short for PH_2 to diffuse to the cell walls and combine to form P_2H_4 and for the P_2H_4 to then diffuse to the path of the analyzing beam passing through the center of the photolysis cell.

Kinetic studies provide insight into the mechanism of phosphine photolysis. The quantum yield (extrapolated to zero reaction time) for phosphine loss is 1.78, for diphosphine formation, 0.80 and for red phosphorus formation is 0.04. The quantum yield for phosphine loss close to 2 suggests the reaction of hydrogen atoms with phosphine is an important process along with phosphine photolysis in diphosphine formation. Similarly the quantum yield for diphosphine formation, which is close to 1, also requires an efficient pathway for formation of PH_2 other than the photolytic expulsion of a hydrogen atom from PH_3.

The efficient formation of diphosphine from phosphine (quantum yield 0.8) and the low quantum yield for red phosphorus formation (0.04 at zero reaction time) indicate the following reaction sequence: $PH_3 \rightarrow P_2H_4 \rightarrow P_4$. The previous reaction schemes which postulated that the disproportionation of PH_2 is an important step in P_4 formation cannot be correct because they do not include the formation of diphosphine. Further studies which further delineate the mechanism of phosphine photolysis will be published elsewhere.

These studies suggest that diphosphine would be expected to be a constituent of the atmospheres of some of the Jovian planets since it is formed efficiently by phosphine photolysis. Most of it will be converted to P_4; however there may be sufficient diphosphine to be detected either by ground based IR measurements or by the Galileo orbiter and probe of the Jovian atmosphere. The formation of diphosphine by the combination of 2 PH_2 radicals suggests the possible formation of PH_2NH_2 by the combination of PH_2 and NH_2. PH_2NH_2 would be expected to decompose to phosphorus paranitride $(PN)_x$, a reddish-brown polymer (13) which may also be responsible for some of the coloration in the atmosphere of Jupiter.

ACKNOWLEDGEMENT. This research was supported by NASA Grant NGR 33-018-148. The purchase of the Cary 219 UV spectrophotometer used in these studies was supported in part by an NIH Biomedical Science Research Support Grant. We thank Professor Robert Strong for the flash photolysis studies of phosphine.

REFERENCES

1. Nicodem, D. E.; Ferris, J. P. Icarus 19, 495 (1973).
2. Atreya, S. K.; Donahue, T. M.; Kuhn, W. R. Science, 201, 611 (1978).
3. Lutz, B. L.; Owen, T.; Cess, R. D. Astrophys. J., 203, 541 (1976).
4. Hunten, D. M., "Chemical Evolution of the Giant Planets," Ponnamperuma, C., ed., Academic Press, N.Y., 1976, 27.
5. Prinn, R. G.; Lewis, J. S. Science, 190, 274 (1975).
6. Howland, G. R.; Harteck, P.; Reeves, R. R. Jr. Icarus, 37, 201 (1979).
7. Ferris, J. P.; Chen, C. Nature, 258, 587 (1975).
8. Ferris, J. P.; Nakagawa, C.; Chen, C. COSPAR Life Sciences and Space Research, XV, 95 (1977).
9. Schofield, K., Planetary Space Sci., 15, 643 (1967).
10. Ferris, J. P.; Nicodem, D. E. Nature, 238, 268 (1972).
11. Ferris, J. P.; Nicodem, D. E. "The Origin of Life and Evolutionary Biochemistry", Dose, K., Fox, S. W., Deborin, G. A., Pavlovskaya, T. E., editors, Plenum, N.Y., 1974, 107.
12. Norrish, R. G.; Oldershaw, G. A. Proc. Roy. Soc. London, Ser. A, 262, 1 (1961).
13. Wiles, D. M.; Winkler, C. A. J. Phys. Chem., 61, 902 (1957).

POSSIBLE ROLE OF PHOSPHINE IN CHEMICAL EVOLUTION

F. Raulin[1] and C. Ponnamperuma[2]

1 : L.P.C.E., Université Paris Val de Marne
Avenue du Général de Gaulle, 94010 Créteil - France
2 : L.C.E., University of Maryland,
College Park, Md. 20742 - USA

Experiments simulating the formation of organic molecules in model reducing atmospheres including phosphine show that a variety of organic molecules can be synthesized. From the experiments using UV light as an energy source, it appears that the presence of phosphine drastically increases the yield of photolysis of methane and, thus, the yield of the formation of organic matter. The presence of phosphine enhances the formation of organophosphorous compounds, such as methyl and dimethylphosphine.

Phosphine has been detected in the jovian atmosphere in 1974 by Ridgway (1). This discovery has been confirmed by Larson et al (2) and more recently by Tokunaga et al (3) and by the IRIS experiment during the Voyager mission to the outer solar system (4). Thus, although there are still some controversies about PH_3 abundance (5), it can be assumed that there are regions in the jovian atmosphere where PH_3 and CH_4 are simultaneously present. This observation shows that PH_3 is a possible source of phosphorus in a reducing planetary atmosphere and it justifies the development of experiments simulating the organic syntheses in model atmospheres including PH_3.

The only simulation experiments of this kind have been carried out in the late sixties by Rabinowitz, who studied electric discharge reactions in gaseous mixtures of phosphine, methane, ammonia and water (6). He showed that the presence of PH_3 gives rise mainly to inorganic phosphorus compounds containing phosphates, pyrophosphates and hypophosphates, and also, in some cases, to organophosphorous compounds with C-O-P bounds.

Y. Wolman (ed.), Origin of Life, 107–114.

Among the energy sources which were available for primordial organic syntheses, U.V. light was probably the most abundant. Several experimental studies of the photochemistry of PH_3 have already been reported, particularly after the discovery of PH_3 in the jovian atmosphere. For example Vera-RUIZ and Rowland (7) have shown the scavenger effect of C_2H_2 and C_2H_4 on the UV photolysis of PH_3 into red phosphorus and the mechanism itself of this photolysis has been studied (8-10). However, so far, no experiment has been performed on the synthesis of organic compounds with UV light in a model reducing atmosphere including PH_3. The availability at the Laboratory of Chemical Evolution (L.C.E.) of a high flux solar light simulator has allowed us to developp such simulation experiments by irradiating gaseous mixtures containing CH_4 and different molar ratios of other gases including PH_3.

The experimental device is described elsewhere in detail (II). The light source, a Giannini high pressure Argon lamp, emitts a continuum covering the range of wavelengths between 100 and 700 nm. In the range 100-200 nm, which is the most favorable for organic syntheses, the flux is about 10^{17} photons/cm^2/sec. The samples are irradiated in 500 ml flasks, including CO_2 samples irradiated foractinometric measurements. Qualitative and quantitativeanalysis are performed by IR spectrophotometry (for CH_4,NH_3,PH_3 and HCN), by Gas chromatography and by Gas chromatography - mass spectrometry.

Preliminary experiments carried out at LCE (12) on a gas mixture of CH_4-NH_3-H_2O-H_2S-PH_3 showed the presence of hydrocarbons (mainly C_2H_6 and C_3H_8) and alcohols (CH_3OH, C_2H_5OH and 2-C_3H_7OH) in the volatile fraction obtained after irradiation. The only detected P-containg compound was metaphosphate, analyzed by paper chromatography in the water extract of the non volatile fraction. Since, in the presence of water, the P-containing products obtained from PH_3 are mainly inorganics, the further systematic studies were carried out on water-free gaseous mixtures, by irradiating different mixtures containing CH_4 (5torr) NH_3,H_2S and PH_3. As CH_4 is the only source of carbon in these mixtures, the quantity of photolyzed methane must give an estimate of the quantity of synthetized organic molecules. For each mixture, the quantity of starting material remaining after irradiation was determined and the value of the relative amount of photolyzed methane was calculated. The lamp emitts a continuum, so that it is not theoretically possible to determine a quantum yield of reaction at a given wavelength. Nevertheless, taking into account the known value of the flux of the lamp in the range 125-200 nm (about 3.10^{17} photons/sec) and the duration of the irradiation, we can estimate an average yield of photolysis of methane.

As shown on figure 1, the yield of photolysis of CH_4 from CH_4-PH_3 mixtures increases with the initial pressure of PH_3. From pure methane (5torr) it is less than a few 10^{-4} molecule/photon. The yield reaches around 10^{-2} molecule/photon from mixtures containing initially 5 torr of CH_4 and 5torr of PH_3. In fact, kinetic studies of the partial pressure of remaining CH_4 and PH_3 as a function of time irradiation show that, for each studied mixture, CH_4 pressure decreases linearly with time. This observation shows that the rate of photolysis of CH_4 is constant with the irradiation time, for a given initial PH_3 pressure. It justifies our use of an average yield of photolysis of CH_4. This behaviour is also consistent with the fact that an important fraction of the initial PH_3 is present in the sample after irradiation. The strong favorable effect of PH_3 on the yield of photolysis of CH_4 may be explained through the process of formation of hot hydrogen atoms H^* from the PH_3 photolysis :

$$PH_3 + h\nu \longrightarrow PH_2 + H^* \qquad [I]$$

CH_4 does not absorb photons of wavelength higher than 150 nm. On the other hand, PH_3 absorbs strongly in the region 150-230 nm, and process [I] is energetically possible below 360 nm, since the bond dissociation energy of $H-PH_2$ is 3.4 ev (13). Thus, in the range 150-230 nm, the incident photon in reaction [I] has an excess energy of between 3.5 and about 5 ev. Because of the mass difference between PH_2 and H, this excess energy is carried almost entirely by the H^* atoms generated by [I]. Thus, the hotest H^* atoms have a sufficient energy to break the $H-CH_3$ bond (the dissociation energy of which is about 4.5 ev) :

$$H^* + CH_4 \longrightarrow CH_3 + H_2 \qquad [2]$$

In order to compare the strong effect of PH_3 on the CH_4 photolysis with the possible effect of H_2S and NH_3, able also to give hot H^* atoms (14) similar studies have been carried out with these two compounds. Figure 2 indicates the lowest and the highest values of the yield of CH_4 photolysis obtained from the several experiments which have been performed on pure CH_4 (5torr) and on mixtures of CH_4 (5torr) - other gas (I torr). From pure methane, only a small fraction of CH_4 is photolyzed giving, as previously found, an average yield smaller than a few 10^{-4} molecule/photon. When I torr of NH_3 is added, the average yield of photolysis of CH_4 increases slightly, up to about 8.10^{-4} molecule/photon. With I torr of H_2S, it increases up to about 2.10^{-3} molecule/photon. Finally, with PH_3, this effect is much greater since the presence of I torr of PH_3 in 5 torr of CH_4 increases the yield up to about $7\ 10^{-3}$ molecule/photon (around 40 % of the initial CH_4 is photolyzed in this last case). The yields of photolysis of methane from mixtures containing CH_4 (5torr) - NH_3 (5 torr) - PH_3 (I torr) ; CH_4 (5torr) - H_2S (I torr) - PH_3 (I torr) and CH_4 (5 torr) - NH_3 (5 torr) - H_2S (I torr) - PH_3 (I torr) reaches also values up to about 10^{-2} molecule/photon.

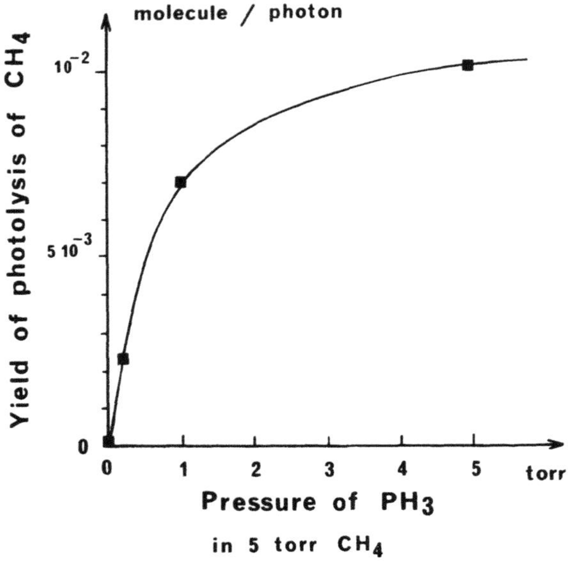

Figure 1 : Average yield of photolysis of CH_4 from mixtures containing initially CH_4 (5torr)-PH_3, as a function of the initial PH_3 pressure.

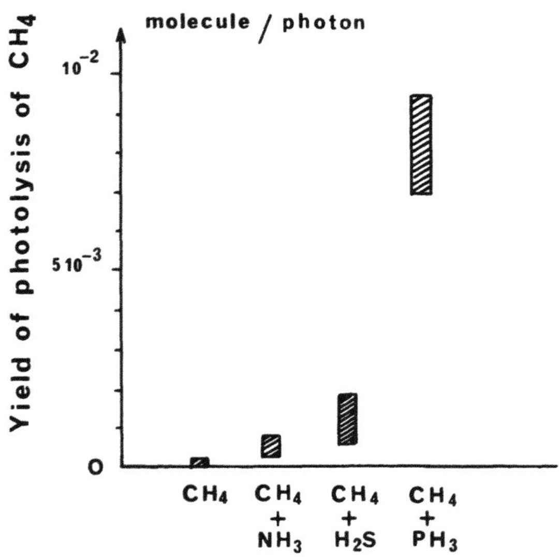

Figure 2 : Average yield of photolysis of CH_4 after 4 hours irradiation of pure methane (5torr) and of mixtures containing 5 torr CH_4 and 1 torr another gas.

The different behaviours of PH_3, NH_3 and H_2S relative to CH_4 photolysis can also be explained by means of hot hydrogen atom reactions. The photolysis of H_2S and NH_3 can produce H^* atoms :

$$H_2S + h\nu \longrightarrow HS + H^* \quad [3]$$
$$NH_3 + h\nu \longrightarrow NH_2 + H^* \quad [4]$$

The bond dissociation energies (13) of H-NH_2 (4.4 ev) and H-SH (3.9 ev) are notably higher than the bond energy of H-PH_2 (3.4 ev), so that the excess energy contained in the incident photon of a given wave length is higher for PH_3 than for H_2S and higher for H_2S than for NH_3. For H_2S, as for PH_3, the excess energy is carried almost entirely by the H^* atom. For NH_3, the mass ratio between NH_2 and H is smaller and the H^* atoms carry a fraction of the excess energy, slightly smaller than in the case of PH_3 or H_2S. Thus, the level of energy of the H^* atoms produced from these three compounds at a given wavelength increases in the following order : NH_3-H_2S-PH_3. With a continuum UV light, PH_3 photolysis can produce more energetic H^* atoms able to break the H-CH_3 bonds than H_2S, and H_2S more than NH_3. This is consistent with the experimental results of figure 2.

Volatile organic compounds produced during the experiments have been systematically sought by analyzing the gaseous mixtures obtained after irradiation. Figure 3 shows gas chromatograms of samples obtained from irradiation of pure CH_4 (5torr) and of a mixture of CH_4 (5torr) - PH_3 (I torr). From pure CH_4, the products are mainly saturated hydrocarbons = ethane, propane, n and iso-butane. The only unsaturated hydrocarbons detected were acetylene and isobutene. Qualitatively, these results are similar to the results obtained with a monochromatic far UV light (15). When I torr of PH_3 is added to 5 torr of CH_4, the same hydrocarbons are formed. In addition, a detectable amount of propene is present and several P-containing compounds are synthetized : mainly methylphosphine and dimethylphosphine. Another P-containing compound (P ?) has been detected but it has not been identified. The alkylphosphines may be formed by a recombination reaction such as :

$$R + PH_2 \longrightarrow RPH_2 \quad [5]$$

or an H abstraction reaction such as :

$$R\text{-}CH + PH_3 \longrightarrow RCH_2PH_2 + H \quad [6]$$

The products obtained from irradiation of CH_4-NH_3-PH_3 mixtures are mainly those formed from the binary mixtures CH_4-NH_3 and CH_4-PH_3, with the addition of an unidentified volatile product which could be a phosphorus and nitrogen containing compound. The results are slightly different in the case of CH_4-PH_3-H_2S mixtures, since the alkylphosphines are no longer detected in the corresponding samples analyzed after irradiation. This is probably due to the acid-base reaction of the alkylphosphines with H_2S. However, the irradiation of the mixture CH_4-PH_3-H_2S-NH_3

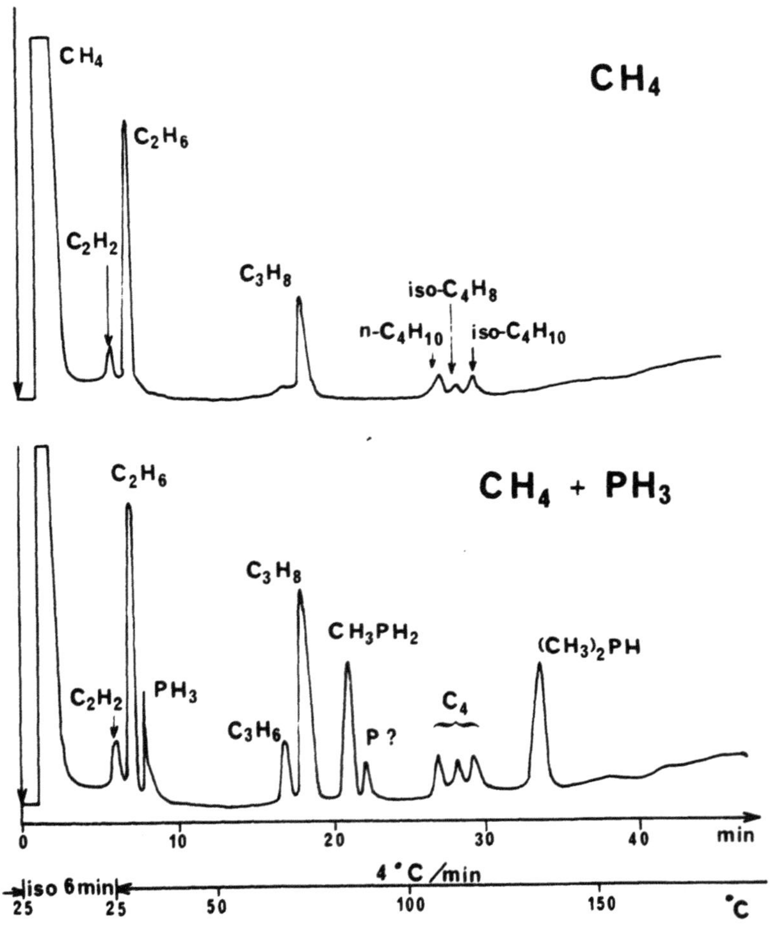

Figure 3 : Gas chromatograms of samples obtained after 4 hours irradiation of CH_4 (5torr) and CH_4 (5torr)-PH_3 (I torr). Column : Porapak Q 1/8" i.d. x 6' long - isothermal 6 min at 25°C; then programmed temperature at 4°C/min up to 200°C. Helium : 30 ml/min. Detector : F.I.D., 10^{-11} a.f.s.

gives rise to most of the compounds formed from the binary mixtures, including methylphosphine and dimethylphosphine. In this last case, the photosynthetized HCN is probably in the form of NH_4CN and H_2S in the form of NH_4SH, so that the phosphines stay in their free form as detected by the gas chromatograph. In addition, an unidentified chromatographic peak is observed, which may correspond to a volatile compound containing S and P atoms.

Some studies have also recently been carried out at LCE on CH_4-PH_3 mixtures submitted to an electrical are discharge (16). The preliminary results seem to show that the presence of PH_3 increases the rate of decomposition of CH_4. This observation could also be explained by means of hot H atoms. In addition, alkylphosphines were also detected in the obtained products.

In conclusion, PH_3 appears to be a good photon acceptor, able to increase drastically the yield of photolysis of methane, probably through reactions involving hot hydrogen atoms. Thus, the presence of PH_3 in a reducing model atmosphere increases notably the yield of synthesis of organic compounds. In addition, PH_3, without changing too much the distribution of the P-free organic products, allows the synthesis of organophosphorus compounds such as alkylphosphines. PH_3 could play such a role in the jovian atmosphere and it could have played such a role in the primordial atmosphere of the earth, assuming that this atmosphere was reducing. From biochemical data, it seems clear, despite the low cosmic abundance of phosphorus and the absence of P-containing compounds in the list of the detected interstellar molecules, that this element has played an important role during the processes of chemical evolution (17,18). PH_3 and the organic phosphines may be some of the P-containing products which have allowed the incorporation of P in chemical evolution.

ACKNOWLEDGEMENT - This report presents the results of one phase of research carried out at the Laboratory of Chemical Evolution. The research has been supported by a NASA contract NGR 21-002-317 and by the NSF and CNRS under the US-French scientific exchange program for F. Raulin. The authors whish to thank A. Bossard and D. Mourey for their comments during the preparation of this manuscript.

REFERENCES

(1) Ridgway S.T., Bull. Amer. Astron. Soc. 6 376 (1974); Ridgway S.T., Wallace L. and Smith G.R., Astrophys. J., 207 1002 (1976).
(2) Larson H.P., Treffers R.R. and Fink A., Astrophys. J., 211 972 (1977).
(3) Tokunaga A.T., Knacke R.F., Ridgway S.T. and Wallace L., Astrophys. J. 232 603 (1979).

(4) Hanel R., Conrath B., Flasar M., Kunde V., Lowman P., Maguire W., Pearl J., Pirraglia J., Samuelson R., Gautier D., Gierasch P., Kumar S., and Ponnamperuma C., Science 204 974 (1979)
(5) Beer R. and Taylor F.W., Icarus 40 189 (1979)
(6) Rabinowitz J., Woeller F., Flores J. and Krebsbach R., Nature 224 796 (1969), Rabinowitz J., Helv. Chem. Acta 53 53 (1970)
(7) Vera Ruiz H.G. and Rowland F.S., Geophys. Res. Lett. 5 407 (1978)
(8) Lee J.H., Michael J.V., Payne W.A., Whytock D.A. and Stief L.J., J. Chem. Phys. 65 3280 (1976).
(9) Howland G.R., Harteck P. and Reeves Jr R.R., Icarus 37 301 (1979).
(10) Ferris J.P. and Benson R., Nature, in press (1980).
(11) Raulin F., Abeyagunawardene S.I., Le Mercier I. and Ponnamperuma C., American Chemical Society Meeting (Section : Organic Chemistry - Identification : 258), Washington D.C. (1979); Raulin F., Abeyagunawardene S.I., Lee A., Le Mercier I., Ponnamperuma C. and Hanel R., to be published.
(12) Yamada M. and Ponnamperuma C., L.C.E. report, University of Maryland (1978).
(13) Okabe M., "Photochemistry of Small Molecules" John Wiley, N.Y. (1978)
(14) See for instance. Hong K.Y., Hong J.H. and Becker R.S., Science, 184 984 (1974).
(15) Toupance G., Bossard A. and Raulin F., origins of Life 8 259 (1977); Bossard A., Thèse de 3ème cycle, spécialité chimie physique, Université Paris VI (1979).
(16) Rabenovets M., Raulin F. and Ponnamperuma C., L.C.E. Report, University of Maryland (1979).
(17) Halmann M., in "The Origin of Life and Evolutionary Biochemistry", Dose K., Fox S.W., Pavlovskaya T.E. and Deborin G.A. Eds, Plenum Press, N.Y., 169 (1974).
(18) Griffith E., Ponnamperuma C. and Gabel N., Origins of Life, 8 71 (1977).

CLAYS AS PREBIOTIC PHOTOCATALYSTS

Lelia M. Coyne,[1] James Lawless,[1] Noam Lahav,[2] Stephen Sutton,[3] and Michael Sweeney[4].

[1]Ames Research Center, NASA, Moffett Field, California, U.S.A. [2]Hebrew University of Jerusalem, Rehovot, Israel, [3]Washington University, Box 1105, St. Louis, Missouri, U.S.A., and [4]University of Santa Clara, Santa Clara, California, U.S.A.

ABSTRACT
Clay minerals catalyze peptide bond formation in fluctuating environments. A number of plausible mechanisms have been proposed and tested. We are exploring the possibility that clays may actually be energizing the reaction by means of electronic excitation, creating mobile or trapped holes and electrons in the lattice. We have discovered that clays emit light upon dehydration. We are testing the correlation between dehydration-induced, or thermoluminescent, processes and the yield of glycine oligomers after treatments known to affect the luminescent yields in an effort to understand the catalytic mechanism.

We are exploring the mechanism by which clays catalyze prebiotic dehydration reactions, in particular, the formation of the peptide bond. Figures 1a and 1b (ref. 1) illustrate our experimental system and protocol. That clays should be singled out for study from among scores of potential prebiotic mineral catalysts has been proposed (2, 3) and rationalized in a number of past publications (4-6). Investigations are revealing that a high degree of selective adsorption of molecular species of prebiotic significance can be provided by clays, consequent to their association with transition metal ions (7, 8). They also have structural features that, in principle, nicely adapt them for absorbing and storing electronic energy. An energized solid state, whether manifested by formation of chemically altered active sites or simply by separated circulating or trapped charges, might be expected

Y. Wolman (ed.), Origin of Life, 115–124.

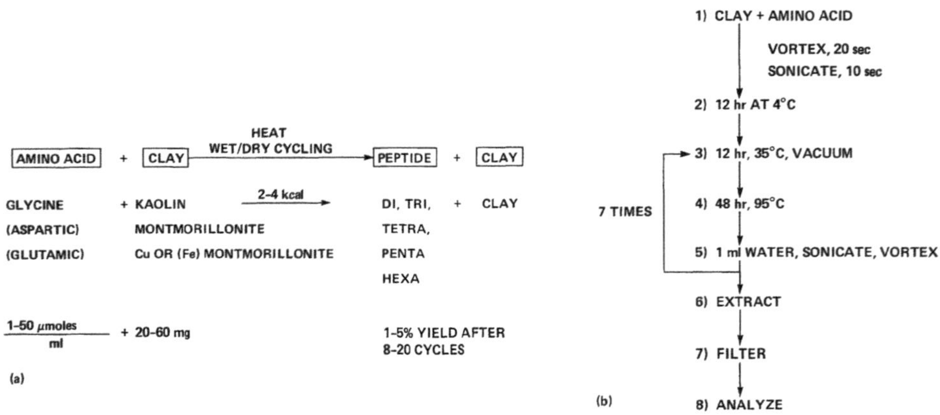

FIGURE 1. Experimental protocol: (a) System; (b) Oligomerization procedure.

to be more catalytically active than one not exposed to a radiation field.

The questions posed in the present study are threefold.

1. Is the catalytic activity of clays and certain other minerals associated with electronic transitions?

2. Are there accompanying luminescent phenomena in the mineral which can be used to explore:
   a. The reactive potential of the mineral?
   b. Monitor the course of the reaction?
   c. Aid in establishing the mechanism?

3. If items (1) and (2) are so, must one see them as contradictions of or as additions to the presently accepted lore of catalysis on these surfaces, or is it possible that the energy can be simply coupled to the classical chemical reaction mechanisms to give them enhanced activity in the presence of an energy source?

The chemical and crystallographic composition of clays has been well characterized, for example, (9, 10); however, actual clay crystals deviate from ideality in many ways. Crystal defects derive from radiation damage (either cosmic or natural decay), crystallization imperfections, mechanical stress, interstitial ions, and adsorbed gases (11). The most distinctive form of lattice defect for clay minerals, however, is that produced by isomorphous substitution. This substitution — which for montmorillonite and many other clays is stoichiometrically equivalent to the exchangeable cations (12), but is more debatably so for kaolin (13, 14) — has been shown to produce interspersed and interposed electronic energy levels in the band structure of the octahedrally substituted material (15). These levels are attributable to the local effect of excess electron density in the environment of the isomorphously substituted cation. A similar effect is predicted for minerals with predominant substitution of Al for Si in the tetrahedral layer.

Defects, generally speaking, can also serve as traps for migrating electrons or holes produced by photo-, radiation-, piezo-, or tribo-excitation mechanisms, as well as from chemical reactions or energy transfer from adsorbed species. Some common electronic phenomena in photoconducting solids are considered in Figure 2 (ref. 16). The charge transfer absorption from isomorphously substituted ions corresponds to transition 2a shown in Figure 2. Any of the common modes of recombination, free electron/free hole, trapped electron/free hole, free electron/trapped hole, may be luminescent, though they are by no means constrained to be. Conducting or luminescent relaxation can be produced by such means as pressure, grinding, friction, or irradiation. The most common is heating, in which case any emitted light is called thermoluminescence (TL).

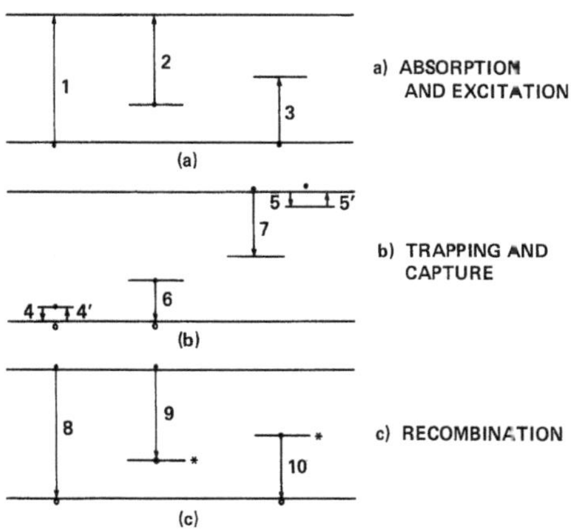

FIGURE 2. Common electronic transitions in photoconductors. (From Bube, Richard, Photochemistry of Solids, John Wiley & Sons, 1960.)

We have made an effort to correlate trends influencing the TL of clay minerals as published in the literature (17) with reaction yields in our cycling system. The TL samples of Nishita (11) were drained of trapped charges by preannealing, then irradiated with gamma rays and the TL measured as a function of temperature. Trends 1-3 noted by Nishita were used as guides in our initial experimental design.

Trend 1: Upon preannealing, the TL tends to shift to lower temperatures, indicating a collapse of the trap depth.

Trend 2: Upon preannealing, the TL capacity of the material is diminished, as evidenced by an overall dimution in the integrated luminescence intensity.

Trend 3: Clay soils bearing calcium oxide or carbonite impurities emit their TL at lower temperatures than do the more pure clay minerals. This could be taken to indicate either a more shallow bulk trap structure or a set of luminescent centers derived more from reactive surface sites than from crystal defects. Moreover, the TL yield for these materials is increased by preannealing rather than diminished. This perhaps implies a dominance of the trap-populating versus trap-draining influence of heat on the charge separation in the material.

Inference: Trends 1 and 3 might imply activation of the material by both preannealing and addition of $CaCO_3$. Trend 2 could push the activity either way depending on the meaning of the decrease in luminescence capacity. Transiently stored energy is not measured by TL, for a delay is deliberately imposed between γ-irradiation and TL determination. This delay avoids detection of energy stored in shallow traps, which to the customary dating application of TL measurements is spurious. However, the importance of the short-term emission may be equal to or greater than that of the high-temperature component of the TL in assessing catalytic activity. Whether it is of equal or greater importance depends on whether the TL is shown to indicate (1) the capacity of the material to separate charge, (2) present content of stored energy in the material, (3) an available source of reaction energy, or (4) the condition of catalytically active sites. Strict correlation should not be expected, however, because individual variation in samples, even of the same mineral type, dominate luminescence behavior, for luminescent effects are more often related to impurities than to the bulk properties of the pure material. Our preliminary cycling results are shown in Figure 3a.

Nishita et al. (18) have also performed measurements of the time dependence of decay of thermoluminescence in a number of γ-irradiated clays. Their results prompted us to preirradiate our clays with γ's in order to check the effect on the glycine oligomerization yield of post-irradiation release of stored energy at ambient temperatures. We irradiated our clays with Co-60 γ-rays to a dose of ~1 megarad. This dose was chosen because it has been demonstrated (19) that it is adequate to saturate preexisting luminescent centers with trapped charges without creating new ones. Thus, enhancement of the yield might more likely be attributable to stored energy than to creation of new catalytically active sites. At present, however, we cannot convincingly differentiate between these alternatives, because the TL sites may not be the catalytic sites. Although gamma-irradiated, unannealed Cu-bentonite showed a significantly enhanced yield, as shown in Figure 3b, kaolinite showed no such enhancement unless it was preannealed before the preirradiation. Because the surface density of the glycine applied to the clay was not adequate to maximize yield/glycine, this lack of enhancement in the unannealed kaolin is not yet considered to be definitive. Furthermore, there appear

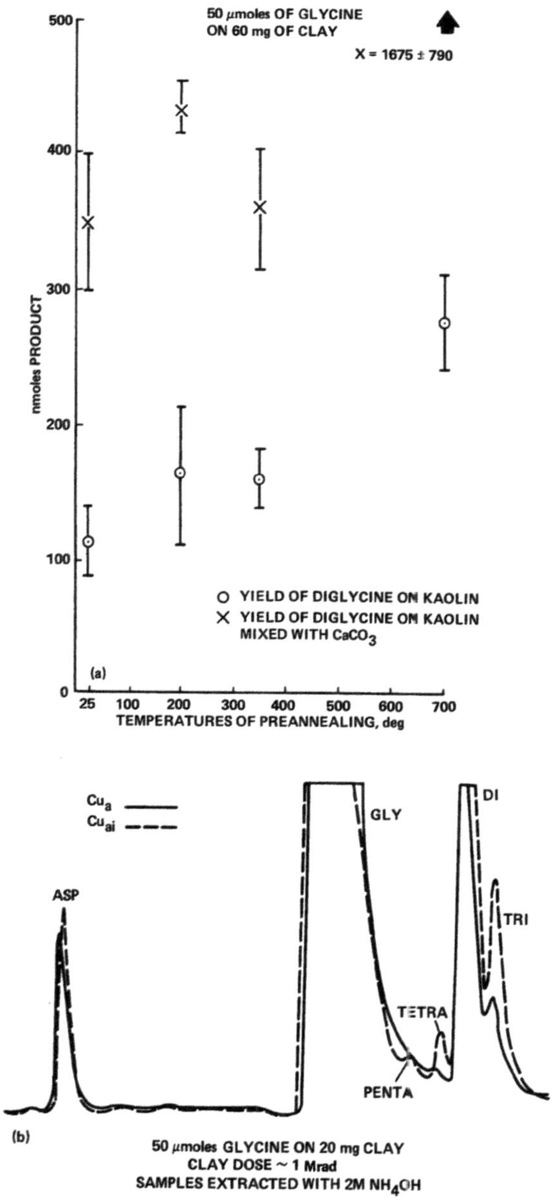

FIGURE 3. Effect of preannealing and preirradiation on oligomer yield: (a) Preannealed clays — yield of diglycine on kaolin and kaolin mixed with $CaCO_3$; (b) Effect of γ preirradiation of Cu-bentonite on oligoglycine yield.

to be multiple degrees of binding strengths or multiple reaction mechanisms. Evidence for heterogeneity of binding sites stems from our observation that extraction by NH_4OH removes a preannealing, preirradiation sensitive component of oligopeptide yield, but extraction by tetramethylammonium chloride yields a component that is independent of preannealing and preirradiation. This feature of fractional extraction may prove to be useful in structural determinations of the interaction between binding site and reactant.

Figure 3 as well as additional incompletely analyzed data gives experimental corroboration of our predictions that glycine oligomerization should be enhanced on preannealed, preirradiated, and $CaCO_3$ treated clays.

Perhaps the most interesting aspect of our efforts to relate catalytic activity and stored energy is a phenomenon discovered by Lahav and Coyne (20) and verified by Sutton (21). Given that freeze-drying procedures produce considerable distortion of the clay platelets, we were led to investigate the possibility that light is released when kaolin and other clay minerals are dried; that is, to look for triboluminescent phenomena. We found that kaolin-water paste will give a delayed burst of photons at the point where the sample appears superficially dry (Figs. 4a, 4b, 4c). Although this light is emitted, with considerably diminished intensities, from samples heated to 300°C, it is barely detectable from samples preannealed to 700°C. Upon standing for 6 months there is some recovery of the luminescent effect from the 300°C sample, but no recovery in the 700°C sample. The spectral distribution of the emitted light is unknown, but is presumed to peak in the blue, by analogy with the TL of this sample and the frequency response of the phototubes used to detect it.

These preannealing treatments, as has been previously seen, are accompanied by enhanced catalytic activity (Fig. 3). The preannealing temperatures used and the partial recovery of the 300°C sample are tentatively being interpreted as a measure of the correlation between the dehydration luminescence and changes in the condition of the structural hydroxyls of the clay on annealing. The reason for this interpretation is that dehydroxylation — partially reversible at 300°C, irreversible at 550°C — is known to accompany annealing (22).

A number of other clay minerals were tested for the dehydration-induced luminescence. A number of them do emit light, but not in a delayed pulse, as for kaolin. A representative sample of this more exponential-like decay is shown in Figure 5. Since kaolin also exhibits a small component of emission from t = 0, the simplest model rationalizing the two types of decay would attribute the delayed peak to a bulk mineral phenomenon and the exponential-like component to emission from sites associated with broken bonds and impurities. This latter component is very sensitive to grinding or sonication or both; it is observed, to a varying extent, in all clay minerals investigated, including pyrophyllite and talc, neither of which has either isomorphous substitution or exchangeable cations. Lack of observation of the delayed component in clays of higher isomorphous substitution or cation exchange capacity (C.E.C.) than kaolin — such as illites, attapulgites, vermiculites, and montmorillonites — can be attributed to the difficulty of obtaining appreciable quantities of dry, somewhat optically transparent films under the conditions of our experiments.

No delayed component was observed from halloysites either, but in this case the material was prepared by grinding a rock form of the mineral. A later experiment with a rock kaolinite indicates that the delayed component is partially interconvertible with the exponential-like decay. The exponential-like component is increased by grinding. Efforts are being made to eliminate an

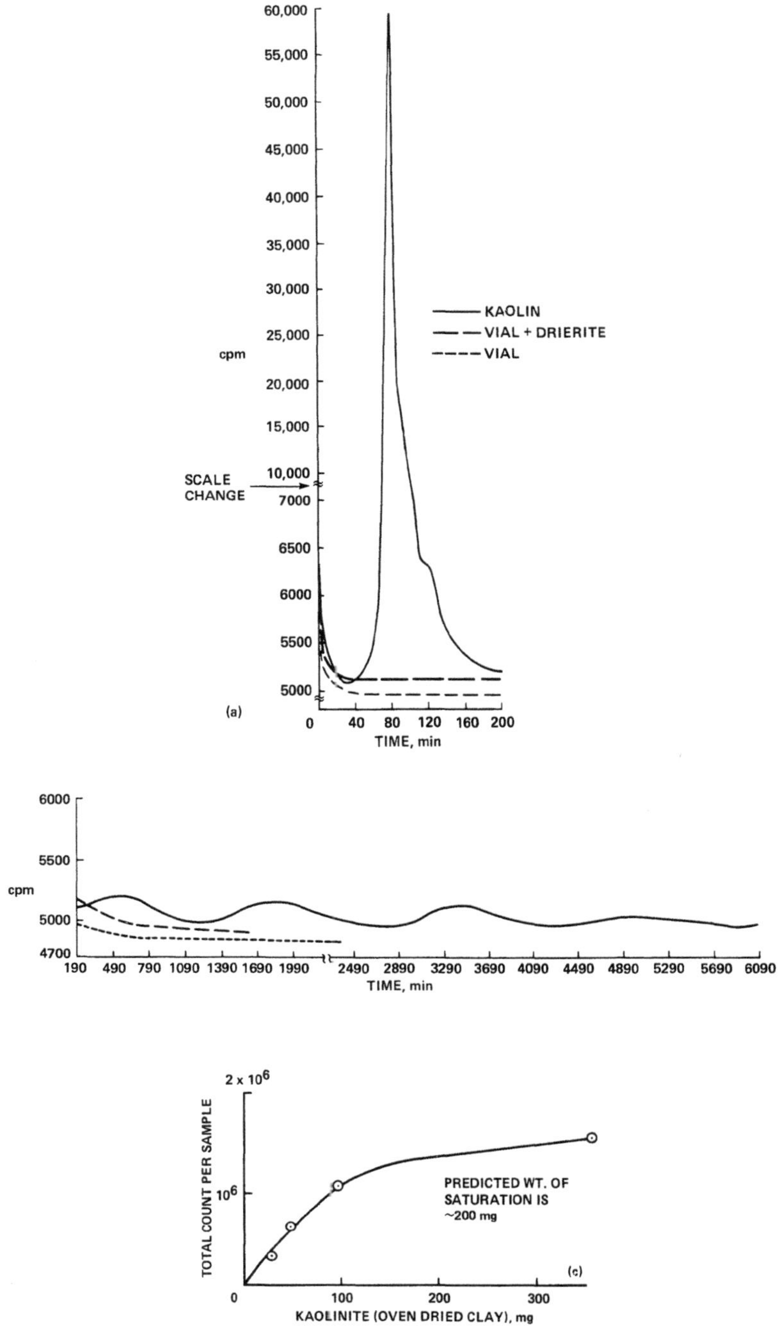

FIGURE 4. Kaolin—dehydration induced luminescence: (a) Kaolin—dehydration-induced luminescence (~100 mg of clay); (b) Kaolin continued; (c) Kaolinite plus silica gel dose response.

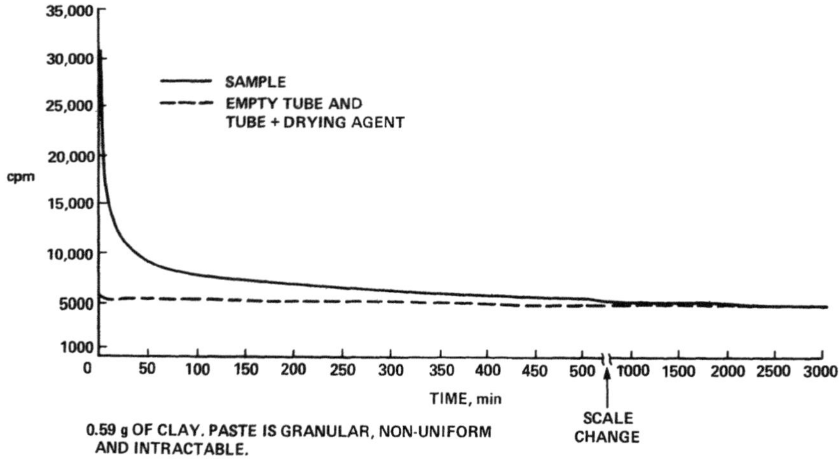

FIGURE 5. Attapulgite — dehydration-induced luminescence.

alternative hypothesis that the kaolin effect simply is unique to this mineral.

The effect is under continued investigation, both with respect to its intrinsic nature and with respect to its degree of correlation with clay reactivity. A number of possible photon-counting artifacts have been eliminated by control experiments and by confirmation by another experimental protocol. Possible artifacts relating to low-level extraneous mineral impurities are under investigation. The effect is not, however, caused by TiO_2, the most likely contaminant for kaolin.

Estimates have been made of the quantitative relationship between isomorphous substitution, observed reaction yield, and the photon yield of TL and dehydration-induced luminescence. The required number of electronic events necessary to produce our observed conversion is shown in Figure 6. From a highly sensitive soil sample, saturated with natural radioactive decay, one can anticipate (10^8 photons/sec mg) $\times$ 10 sec $\times$ 30 mg (1.0/0.01) (collection efficiency) $\simeq 3 \times 10^{13}$ photons. This number of photons is barely adequate to provide events for 1 cycle; it certainly is insufficient for 10 cycles. The clays that we use have been measured by Sutton (21) to produce yields that are less than optimal by a factor of 10^3 to 10^4. As a result of (1) these less than optimal yields, (2) the fact that the spectral output is in the blue and is thus insufficiently energetic to excite glycine, and (3) the observation that the TL photons are available only at temperatures of 300°C and higher, we conclude that the process is not emission activated, either by emission/reabsorption or energy transfer. This conclusion is corroborated by the fact that the

$$\frac{\text{1-50 } \mu\text{moles GLYCINE}}{\text{20-60 mg CLAY}} \times .1 - .5\% \frac{\text{CONVERSION}}{\text{CYCLE}} =$$

$$\boxed{\frac{\text{"EVENTS"}}{\text{CYCLE}} = 6 \times 10^{14} - 6 \times 10^{15}}$$

WHERE EVENTS ARE DEFINED TO BE:
PHOTONS
ELECTRONS
HOLES

ASSUMPTION: ONE CAN EXPECT ONE EVENT FOR EACH EXCHANGEABLE CATION

FOR KAOLIN, AS AN EXAMPLE:

$$\frac{\text{60 mg}}{\text{300 g/mole}} = 2 \times 10^{-4} \text{ moles} = 10^{20} \text{ "MOLECULES" OF CLAY}$$

CATION-EXCHANGE CAPACITIES RANGE FROM $5 \times 10^{-3} - 5 \times 10^{-2}$ equiv./mole

$7 \times 10^{17} - 7 \times 10^{18}$ EVENTS/SAMPLE CYCLE

THUS WE HAVE AN EXCESS OF AVAILABILITY OVER REQUIREMENTS BY $10^2 - 10^4$
(MONTMORILLONITE WILL HAVE AN ADDITIONAL 100-1000 FOLD PRODUCTION)

SPECULATION: THE EVENTS MUST BE REGENERATED EACH CYCLE RATHER THAN BEING RELEASED AS PREVIOUSLY STORED ENERGY, FOR THIS MATERIAL.

FIGURE 6. Number of electronic events necessary to produce observed conversion.

dehydration-induced luminescence does not appear to be significantly quenched by glycine. However, systems such as this one are not constrained by the requirements of classical photochemistry. One photon represents about 75 kcal of energy, whereas we need only 2-4 kcal to overcome the thermodynamic barrier (23) and 20 kcal to overcome the activation barrier (24). Not only is there a plethora of lower depth traps — which, if populated with releasable energy in a regenerable manner, provide the potential for a higher yield at lower temperature — but there exist reaction mechanisms involving excited states that do not depend on quantization of the energy in the form of photons. These mechanisms can be associated with changes in the acidity or redox potential in either the bulk material or in specific sites within it; for example, repulsion between trapped holes and protons can produce super-acidic surface sites. We postulate that photons represent energy wastage, not available energy. Hence, the photons we see are probably not indicative of the tip of an energy iceberg. Instead, they give us two pieces of valuable information — a lower limit on the reactive potential and a potential monitor of reactive potential, reaction progress, or reaction mechanism.

We conclude that clays have luminescent properties that can be related both to structural condition and to catalytic activity. It appears that a more thorough investigation of the solid-state spectroscopic properties of clay minerals will prove invaluable in (1) extending our understanding of the process by which prebiotically important molecules are selected for chemical reaction and (2) in elucidating the mechanisms by which these reactions occur under the prevailing surface and atmospheric conditions which are likely representations of those that prevailed on interstellar grains or the crust of the primitive earth.

REFERENCES

(1) Lahav, N., White, D., and Chang, S. Science 201, 67 (1978).
(2) Bernal, J. D. The Physical Basis of Life, Routledge and Kegan, Paul, Ltd., London, 1951.
(3) Paecht-Horowitz, M. Origins of Life 5, 173 (1974).
(4) Lahav, N. and Chang, S. J. Mol. Evol. 8, 357 (1974).
(5) Chang, S. and Bunch, T. Geochimica, in press.
(6) Banin, A. and Rishpon, J. J. Mol. Evol. 14, 133 (1979).
(7) Lawless, J. and Edelson, E. Proc. of the Comm. on Space Research, in press.
(8) Tunzi, M., Church, F., Mazzurco, J., O'Brien, K. and Lawless, J. Abstract No. 40, 179th ACS Natl. Mtg., Houston, Tex., Mar. 24-27, 1980.
(9) Grim, R. Clay Minerology, McGraw-Hill, 1978.
(10) Brown, G. The X-ray Identification and Crystal Structures of Clay Minerals. Minerological Soc. Clay Minerals Group, London, 1961.
(11) Nishita, H. and Hamilton, M. Soil Science 110, 371 (1970).
(12) Grim, R. op. cit. p. 193-195.
(13) Little, L. IR Spectra of Adsorbed Species, Academic Press, London, 1966, p. 366.
(14) Van Olphen, H. An Introduction to Clay Colloid Chemistry, 2nd Ed., Wiley Interscience, New York, 1963, p. 70.
(15) Aronowitz, S., Coyne, L., Lawless, J. and Rishpon, J. To be presented at ACS Natl. Mtg., San Francisco, Calif., Aug. 1980.
(16) Bube, R. Photochemistry of Solids, John Wiley & Sons, 1960.
(17) Nishita, H. and Hamilton, M. Soil Science 108, 1 (1969).
(18) Nishita, H., Moore, R., Beckman, R. and Hamilton, M. Soil Science 119, 259 (1975).
(19) Sutton, S., Washington University, St. Louis, Mo., personal communication.
(20) Lahav, N., Hebrew University of Jerusalem, and Coyne, L., Ames Research Center, in preparation.
(21) Sutton, S., Washington University, personal communication.
(22) Little, op. cit., chap. 13.
(23) Fox, S. and Dose, K. Molecular Evolution and the Origin of Life, Rev. Ed., Marcel Dekker, N.Y. and Basel, 1977, p. 142.
(24) White, D., Dept. of Chem., University of Santa Clara, Santa Clara, Calif., unpublished data.

CLAY-MEDIATED REACTIONS OF HCN OLIGOMERS. THE EFFECT OF THE OXIDATION STATE OF THE CLAY.

J. P. Ferris, K. W. Alwis, E. H. Edelson, N. Mount and W. J. Hagan, Jr.
Department of Chemistry
Rensselaer Polytechnic Institute
Troy, New York 12181

Montmorillonite clays which contain Fe(III) inhibit the oligomerization of aqueous solutions of HCN. The inhibitory effect is due to the rapid oxidation of DAMN, a key intermediate in HCN oligomerization, by the Fe(III) incorporated into the aluminosilicate lattice of the clay. The Fe(III) oxidizes DAMN to diiminosuccinonitrile (DISN), a compound which is rapidly hydrolyzed to HCN and oxalic acid derivatives. DAMN is not oxidized when Fe(III) in the montmorillonite is reduced with hydrazine. The oxidation state of the clay is an important variable in experiments designed to simulate clay catalysis on the primitive earth.

Hydrogen cyanide condenses in aqueous solution to oligomers which yield purines, pyrimidines and amino acids when hydrolyzed (1). Since clays are postulated to have had an important role in chemical evolution (2), their effect on the formation of HCN oligomers was investigated (3). Montmorillonite clays were found to dramatically inhibit the oligomerization of aqueous solutions of HCN. Yields of colored oligomers, urea and diaminomaleonitrile (DAMN) were all diminished although the rate of cyanide reaction remained essentially unchanged. When ^{14}C-DAMN was reacted with the montmorillonite all of the ^{14}C was accounted for in the supernatant after centrifugation of the mixture, yet no DAMN was present in the supernatant. This finding established that the DAMN is transformed to other products and is not bound to the clay. HCN is a reaction product and two moles are formed for every mole of DAMN which reacted.

Y. Wolman (ed.), Origin of Life, 125–128.

The clay-mediated decomposition of DAMN is similar to the Ni^{+2}-catalyzed reaction of DAMN in which two moles of HCN are formed per mole of DAMN (4). The reaction with montmorillonite is not due to the associated exchangeable cations since the same reaction was observed when a series of homoionic clays (NH_4^+, Na^+, Cu^{2+}, Zn^{2+}, and Ni^{2+}) were used in place of the naturally-occuring montmorillonite.

The role of non-exchangeable cations, present in the aluminosilicate lattice of the montmorillonite, in the reaction with DAMN was investigated. The effect of Fe(III) on the decomposition of DAMN was studied first because it is one of the principal ions replacing aluminum in the clay lattice (5). A 10^{-4} M solution of DAMN is completely decomposed when percolated through a column of Dowex 50 to which Fe(III) is bound. Two moles of HCN were detected in the effluent per mole of DAMN reacting. When the column was washed with excess acid to elute the bound iron salts two equivalents of Fe(II) were detected for every mole of DAMN decomposed. The reduction of Fe(III) to Fe(II) indicated that an oxidation reaction was responsible for the destruction of DAMN. This was confirmed by the observation that a Dowex column to which Fe(II) is bound does not affect the decomposition of a 10^{-4} M solution of DAMN.

The formation of two equivalents of both Fe(II) and HCN is consistent with the oxidation of DAMN to diiminosuccinonitrile (DISN) (6) and the subsequent hydrolysis of the DISN to HCN, NH_3 and oxalic acids (equations 1 and 2) (7,8). This stoichiometry was confirmed by the detection of oxalic acid by its reduction to glycolic acid with magnesium. The formaldehyde released on heating the glycolic acid with sulfuric acid was determined from the intensity of the absorption maximum of its 2,7-dihydroxynaphthalene adduct at 530 nm (9). No formaldehyde was detected when the reduction step was omitted.

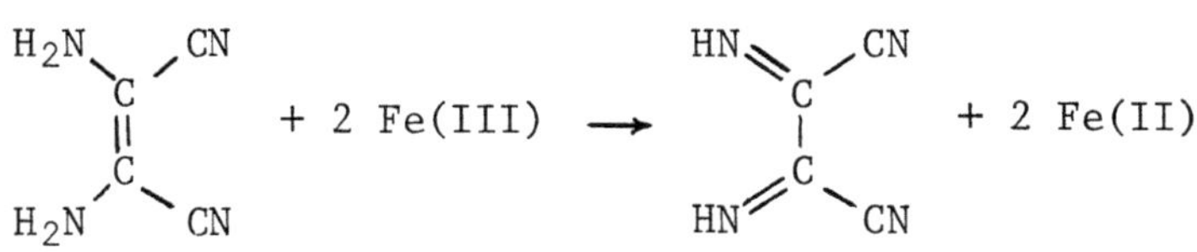

(1)

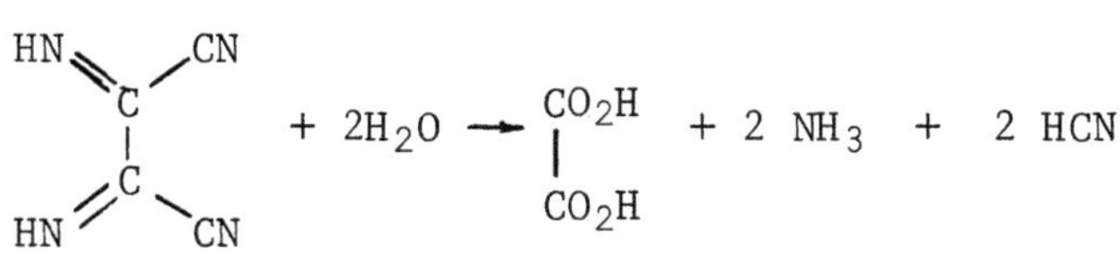

(2)

A similar oxidation of DAMN was observed when aqueous solutions of $Fe(NO_3)_3$ were reacted with DAMN. The course of the reaction differed from that using Fe(III) bound to Dowex in that less than 2 moles of Fe(III) were reduced for every mole of DAMN that reacted. In addition, the oxidation of DAMN was never complete even though excess Fe(III) was present.

It was established that the Fe(III) present in the clay lattice is responsible for the oxidation of DAMN. Reduction of montmorillonite with hydrazine (9) results in the conversion of about 75% of the iron present to Fe(II) (10, 11). Presumably the Fe(III) which is not reduced is in an environment where it is not accessible for reaction with chemical reducing agents. Aqueous solutions of DAMN are not decomposed when mixed with this reduced montmorillonite, a finding consistent with Fe(III) being responsible for the oxidation of DAMN. Possible chemical reactions between the reduced montmorillonite and DAMN are currently being investigated.

The difference in the course of the reaction of oxidized and reduced montmorillonite with DAMN demonstrates that the catalytic effects of clays may be markedly different for different oxidation states. Thus the oxidation level of the clay must be considered as an additional variable in investigations of clays as potential catalysts in the simulations of chemical events on the primitive earth.

It is likely that both oxidized and reduced clays were present on the primitive earth. A preponderance of reduced clays were probably present shortly after the earth formed and weathering first produced clays. The oxidation state increased as O_2 was generated by the photolysis of water and later when photosynthesis became widespread. Consequently the principal chemical processes catalyzed by clay minerals on the primitive earth probably changed as chemical and biological evolution progressed. It has been postulated that minerals were an integral part of the earliest forms of life (12). This study shows that the iron-substituted montmorillonites have the capacity to serve as electron transfer systems. It is possible that they had this function in primitive forms of life.

ACKNOWLEDGEMENTS. This research was supported by NASA Grant NGR 33-018-148 and NSF Grant CHE-76-11000. The purchase of the Cary 219 UV spectrophotometer used in these studies was supported in part by an NIH a Biomedical Sciences Research Support Grant.

REFERENCES

1. Ferris, J. P.; Joshi, P. C.; Edelson, E. H.; Lawless, J. G. J. Molec. Evol., 11, 293 (1978).
2. Bernal, J. P. "The Physical Basis of Life", Routledge and Kegen Paul, London, 1951.
3. Ferris, J. P.; Edelson, E. H.; Mount, N. M.; Sullivan, A. E. J. Mol. Evol., 13, 317 (1979).
4. Ferris, J. P.; Edelson, E. H. J. Org. Chem., 43, 3989 (1978).
5. Grim, R. E.; Güven, N. "Bentonites: Geology, Minerology, Properties and Uses", Elsevier, Amsterdam, 1978, 18.
6. Webster, O. W.; Hartter, D. R.; Begland, R. W.; Sheppard, W. A.; Cairncross, A. J. Org. Chem., 37, 4133 (1972).
7. Begland, R. W.; Hartter, D. R. J. Org. Chem., 37, 4136, (1972).
8. Ferris, J. P.; Ryan, T. J. J. Org. Chem., 38, 3302 (1973).
9. Feigl, F. "Spot Tests in Organic Analysis", Elsevier, Amsterdam, 1956, 353.
10. Solomon, D. H.; Loft, B. C.; Swift, J. D. Clay Minerals, 7, 389 (1968).
11. Tennakoon, D. T. B.; Thomas, J. M.; Tricker, M. J. J. Chem. Soc. Dalton, 2211 (1974).
12. Cairns-Smith, A. G. J. Theoret. Biol., 10, 53 (1966).

CHEMICAL EVOLUTION AND PLATE TECTONICS

Dale E. Ingmanson
Sea Grant
University of California
La Jolla, California

Michael J. Dowler
Dept. of Natural Science
San Diego State University
San Diego, California

Laboratory analysis of Red Sea Atlantis II Deep brine solutions revealed a significant thiocyanate concentration (2.4 x 10^{-5}M). Amino acid analysis revealed large quantities of glycine (about 1 micromole per liter) and no other dissolved free amino acids. This brine and the underlying sediment constitute a reducing, acidic, clay rich, anhydrous to hydrous gradient system that can vary in temperature from 1100°C underlying magma to 63°C water. Since these conditions occur along a sea floor spreading zone there may be an important link between the origin and evolution of the Earth's crust and chemical evolution.

One of the most elusive aspects to the elucidation of the concept of the chemical evolution of life on earth has been verification of the abiotic synthesis of organic molecules in the natural environment. It has long been held that natural environments on earth would not be fruitful ones to investigate due to the ubiquity of life. As a result, the search has centered upon extra-terrestrial environments. This search has led to some successes in finding organic molecules in meteorites(1) and on the moon(2). Still, one feels a certain disquiet due to the lack of evidence in the only environment known to contain life. In this paper, we propose a natural terrestrial environment worthy of analysis for the abiotic synthesis of organic molecules, cite evidence for such synthesis and suggest a new environmental system in which chemical evolution could have occurred.

The Atlantis II Deep brine and underlying sediments should be explored further for evidence of abiotic synthesis of organic

Y. Wolman (ed.), Origin of Life, 129–134.

molecules. The Atlantis II Deep is located in the Red Sea (21° 25′N and 38° 03′E) off Jidda, Saudi Arabia and lies in the rift valley at the axis of a global plate spreading zone. Several characteristics make the Atlantis II Deep interesting from the point of view of those involved in research in chemical evolution: (a) it has been reported sterile(3); (b) the environment is reducing(4); (c) there is no free oxygen(5); (d) the methane and ethane content is 10^3 times that of normal sea water(6); (e) clays such as montmorillonite are present in the sediments underlying the brine(7); (f) the temperature of the brine is 63°C(8); (g) heavy metals important in chemical evolution are abundant(9); and (h) thiocyanate has been reported in the brine (10).

Recently we have completed a study of the amino acid composition of the Atlantis II Deep brine(11). The composition is unique among natural environments on earth. Glycine is present at a concentration of about 1 micromole per liter. Virtually no other amino acids are present; certainly none is present above a concentration of 10 nanomoles per liter. Previous reports of total dissolved amino acids in the ocean suggest that deep water concentrations are 100 nanomoles per liter or less(12) and that surface ocean water contains about 1 micromole per liter with glycine comprising about 0.2 micromoles per liter(13).

We have suggested alternative explanations for the unique amino acid composition of the Atlantis II brine(11). Basically, there are four plausible alternatives.

The most obvious explanation is that there has been biological contamination of the brine and/or our procedures. The brine has been repeatedly found sterile(3). Since the amino acid distribution of our control brine indicates no contamination and bacterial assays of the same Atlantis II Deep brine samples were negative, this explanation seems unlikely. In addition, contamination due to handling shows a characteristic distribution of amino acids that we do not observe(14).

A second possibility is that the glycine bleeds off the Chelex 100 resin which was used in the analysis, a previously reported procedural problem(12). If this were the explanation, comparable glycine concentrations should have been found in the synthetic control brine. This was not found.

A third alternative is that abundant bacteria above the Atlantis II brine decompose organisms and the amino acids leach into the brine where they accumulate. Perhaps there is some chemical compound present in the brine that selectively removes all dissolved amino acids except glycine from the brine or alters other amino acids to form glycine. There has been no such chemical compound reported.

A fourth explanation is that the glycine has been formed abiotically involving a Strecker synthesis from formaldehyde and cyanide. Our previous report of thiocyanate at a concentration of about 2.4×10^{-5}M, a methane concentration of about 6×10^{-2} ml/liter(6) and the present NH_3 in the area of sea floor vents lend support to this explanation. Unfortunately, the volume of our samples is insufficient to collect enough glycine for a C^{12}/C^{13} ratio determination which would shed light on the validity of this explanation.

The evidence for supporting the need for more extensive exploration of the Atlantis II Deep from the point of view of prebiotic chemistry is clear. Since proposals call for commercial mining of the rich heavy metal deposits beneath the brine as early as 1982, we propose that an expedition be planned for 1981, perhaps under the auspices of the ISSOL.

Our research and research pertaining to other hydrogeothermal systems associated with plate tectonic spreading zones, such as the Galapagos area, have led us to propose a relationship between chemical evolution and the evolution of the earth's crust(15).

Many general articles have been published enunciating the global plate tectonic model for crustal evolution(16, 17). These articles describe the characteristics of plate spreading zones. The zones stretch in relatively narrow bands (2 to 20km wide) across 85,000km of the earth's surface. They typically are submarine, volcanically active and extensively fractured. Lava is extruded at relatively frequent irregular intervals along the fractures. In recent years geothermal springs have been reported associated with these regions.

In cross-section from the sea floor to a depth of 15 km one finds a layer of ocean bottom sediments including clays, extrusive volcanic rocks (submarine lava flows), intrusive volcanic rocks and crystalline basement rocks which are mainly iron/magnesium silicates. Molten material and gases come up from the earth's mantle in these regions. Water circulates through rock fractures, gets heated in proximity to molten material and rises along fractures to the sea floor.

The solutions circulating through this geothermal system and the adjacent deposits are of special interest to prebiotic chemistry. The role of concentrating economically important ores by these solutions has been well-documented(18). High concentrations of iron, lead, zinc and magnesium in these solutions have been documented. The role of metal ions, especially lead, in abiotic synthesis has been reported by Orgel and co-workers(19). Finding significant quantities of water containing substantial concentrations of these substances in tide pools, for instance, has been an enigma for researchers in chemical evolution.

These geothermal systems have a thermal gradient from the sea floor to the underlying magma at a temperature of, perhaps 1100°C. Since the geothermal solutions apparently flow intermittently through the fractures in the system, periods of dehydration would be expected. How to account for the range of temperatures and areas for anhydrous and hydrous reactions required for various steps in chemical evolution has also been a problem for those envisioning tide pools or the ocean thermocline as an environment conducive to chemical evolution. However, this environment would seem to be ideal for the occurrence of such processes as those envisioned by Fox and co-workers(20).

The evolutionary importance of iron/sulphur enzyme systems has been noted(21). This environment would be ideal for the evolution of such systems since it is rich in both reduced iron and sulfides(22). In addition, the abundance of clay and related metamorphosed rocks in these regions is consistent with the concept that clays may have played an important role in amino acid synthesis as suggested by Paecht-Horowitz and Katchalsky (23).

One characteristic of the spreading zone geothermal system is high pressure. The role of pressure in chemical evolution experiments is virtually unexplored. There has, however, been extensive experience in industrial production of organic compounds to indicate that this aspect warrants study.

Hydrogeothermal systems located at global plate spreading zones appear to be viable environments for chemical evolution. These systems are large, have an array of conditions conducive to chemical evolution and were apparently present in early Pre-Cambrian time. We envision the possibility of a complete series of chemical processes associated with chemical evolution occurring in this hydrogeothermal system. These processes would include: the conversion of reduced gases such as methane and ammonia to reactive species like formaldehyde and cyanide near the source of magma; their further conversion to important biomonomers like amino acids, sugars, purine and pyrimidine bases and subsequent polymerization and selection involving metallic ion and clay catalysis higher in the system where rock fractures could allow water intrusion. Within the system, temperature could vary from as low as 4°C near the sea floor to 1100°C near molten material. The solutions would be enriched with metallic ions such as lead and zinc. There would be anhydrous conditions created when water is evaporated by high temperatures or squeezed out of clays by compression during earthquakes. Abundant clays are present to supply suitable templates for polymerization. Indeed, these systems appear to embody more of the known conditions required for the extensive variety of chemical reactions presently thought to be needed to account for chemical evolution than other terrestrial environments previously postulated.

We are not the first, by any means, to suggest that the oceans or sea floor may have played an important role in chemical evolution. Weyl(24) and Degens and Matheja(25) in particular have hypothesized on this role. We are suggesting that the validity of their conjectures is enhanced when one envisions a key source of organic molecules in the hydrogeothermal systems of plate spreading zones. We also believe that there may be an important link between the process of the origin and evolution of the Earth's crust and the process of the chemical evolution of life. It is possible to test this hypothesis by more extensive exploration of the geothermal systems at global plate spreading zones.

REFERENCES

1. Kvenholden, K. A., Lawless, J., Pering, K.,Peterson, E., Flores, J. and Ponnamperauma, C.: Nature, 228, p. 923 (1970).
2. Harada, K., Hare, P. E., Windsor, C. R., and Fox, S. W.: Science, 173, p. 433 (1971).
3. Watson, S. W. and Waterbury, J. B.: in "Hot Brines and Recent Heavy Metal Deposits in the Red Sea" (E. T. Degens and D. A. Ross, Eds.), Springer-Verlag, New York (1969), p. 272.
4. Ryan, W. B. F., Thorndike, E. M., Ewing, M. and Ross, D. A.: Ibid., p. 153.
5. Brewer, P. G., Densmore, C. D., Munns, R., and Stanley, R. J.: Ibid., p. 138.
6. Swinnerton, J. W. and Linnenbom, V. J.: Ibid., p. 251.
7. Bischoff, J. L.: Ibid., p. 368.
8. Ross, D., personal communication.
9. Brooks, R. R., Kaplan, I. R., and Peterson, M. N. A.: in "Hot Brines and Recent Heavy Metal Deposits in the Red Sea" (E. T. Degens and D. A. Ross, Eds.), Springer-Verlag, New York (1969), p. 180.
10. Dowler, M. J. and Ingmanson, D. E.: Nature, 279, p. 51 (1979).
11. Ingmanson, D. E. and Dowler, M. J.: manuscript submitted for publication.
12. Lee, C. and Bada, J. L.: Earth Planet. Letters. 26, p. 61 (1975).
13. Clark, M. E., Jackson, G. A., and North, W. J.: Limn. Ocean. 17, p. 749 (1972)
14. Hamilton, P. B.: Nature, 205, p. 284 (1965).

15. Ingmanson, D. E. and Dowler, M. J.: Origins of Life, 8, p. 221 (1977).
16. Dewey, J. F.: Scientific American, 226, May, p. 56 (1972).
17. Alexander, T.: Smithsonian, 5, Jan., p. 30 and Feb., p. 38 (1975).
18. Rona, P. A.: Scientific American, 229, July, p. 86 (1973).
19. Sleeper, H. L., Lohrmann, R. and Orgel, L. E.: J. Mol. Evol. 13, p. 203 (1979).
20. Fox, S. W., Harada, K., Kendrick, J.: Science, 129, p. 1221 (1959).
21. Dickerson, R. D.: Scientific American, 224, Mar., p. 136 (1980).
22. Stephens, J. D. and Wittkop, R. W.: in "Hot Brines and Recent Heavy Metal Deposits in the Red Sea" (E. T. Degens and D. A. Ross, Eds.), Springer-Verlag, New York (1969), p. 441.
23. Paecht-Horowitz, M., Berger, J., and Katchalsky, A.: Nature, 228, p. 636 (1970).
24. Weyl, P. K.: Science, 161, p. 158 (1968).
25. Degens, E. T. and Matheja, J.: J. Brit. Interplanetary Soc., 21, p. 52 (1968).

THE STRECKER SYNTHESIS IN THE PRIMITIVE OCEAN

Stanley L. Miller and James E. Van Trump

Department of Chemistry, University of California, San Diego, La Jolla, California 92093

Abstract

Most prebiotic amino acid syntheses proceed through the amino nitrile (Strecker synthesis). The equilibrium constants for the formation of glycine nitrile and alanine nitrile have measured as well as the kinetics of their hydrolysis to the amides and amino acids. From the data the yields of amino and hydroxy acids can be calculated as a function of pH, temperature and ammonia concentration. It can also be shown that amino acids can be synthesized at high dilutions in the primitive ocean.

There is an extensive literature on the prebiotic synthesis of amino acids, and the precursors to the amino acids are the corresponding amino nitriles in most of these syntheses. This includes prebiotic syntheses which start with aldehydes, ammonia and hydrogen cyanide (Strecker synthesis) produced by a spark or other source of energy, amino nitriles produced from cyanide polymerization, and amino nitriles that may be produced directly in a spark discharge.

Most of these syntheses are done under forced conditions--that is, the use of high concentrations of ammonia, high concentrations of intermediates, hydrolysis conditions that involve high temperatures and/or acid catalysis, and high inputs of energy per liter of simulated primitive earth atmosphere. Although it is possible that some of these forced conditions, such as high concentrations of intermediates, were present at the site of amino acid synthesis on the primitive earth, it is more

Y. Wolman (ed.), Origin of Life, 135–141.

likely that most of the amino acids were synthesized under dilute conditions in the primitive ocean.

In order to understand the time course of prebiotic amino acid synthesis from the precursors, we have determined the equilibrium constants for the formation of glycine and alanine nitriles as well as the kinetics of hydrolysis of these nitriles to the amide and the amide to the acid (1,2). The details of the Strecker synthesis have also been investigated by Commeyras and coworkers (3) who obtained data for acetaldehyde in agreement with those presented here. These data allow us to calculate the ratio of hydroxy acid to amino acid produced from a precursor aldehyde as a function of temperature, pH and ammonia concentration.

The Hydroxy Acid/Amino Acid Ratio

The kinetics scheme is as follows:

$$\mathrm{RCHO + NH_3 + HCN \underset{}{\overset{K_{AN}}{\rightleftharpoons}} RCH(NH_2)CN + H_2O} \qquad (1)$$

$$\mathrm{RCHO + HCN \overset{K_{HN}}{\rightleftharpoons} RCH(OH)CN} \qquad (2)$$

Eqs. 1 and 2 combine to give

$$\mathrm{RCH(OH)CN + NH_3 \overset{K'_{AN}}{\rightleftharpoons} RCH(NH_2)CN + H_2O} \qquad K'_{AN} = \frac{K_{AN}}{K_{HN}}$$

The hydrolysis steps are

$$\mathrm{RCH(NH_2)CN \xrightarrow{k_{1AN}} RCH(NH_2)CONH_2 \xrightarrow{k_{2AN}} RCH(NH_2)COOH}$$

$$\mathrm{RCH(OH)CN \xrightarrow{k_{1HN}} RCH(OH)COHN_2 \xrightarrow{k_{2HN}} RCH(OH)COOH}$$

In discussing the product ratio, we only need to consider the kinetics of the nitrile hydrolyses, since amino nitriles and hydroxy nitriles can interconvert but the amides or the acids cannot. The hydrolysis steps are irreversible.

The rate of formation of the amino amide, which will ultimately be hydrolyzed to the amino acid, is given by

$$+\frac{d(\text{amino amide})}{dt} = k_{1AN}[RCH(NH_2)CN] = k_{1AN}K_{AN}[RCHO][HCN][NH_3]$$

The rate of formation of the hydroxy amide is given by

$$+\frac{d(\text{hydroxy amide})}{dt} = k_{1HN}[RCH(OH)CN] = k_{1HN}K_{HN}[RCHO][HCN]$$

Integrating and taking the product ratio after complete hydrolysis gives

$$\frac{(\text{hydroxy acid})}{(\text{amino acid})} = \frac{k_{1HN}K_{HN}}{k_{1AN}K_{AN}[NH_3]} \tag{3}$$

The values of k_{1AN} and k_{1HN} are the pseudo first order rate constants for the hydrolysis of corresponding nitrile. They are obtained from pH rate profiles which have been determined at a number of temperatures (1).

The values of the rate constants at 0° and 25° and pH 8 are shown in Figs. 1 and 2. The ratio of hydroxy acid/amino acid is shown in Fig. 3 as a function of pH for formaldehyde and for acetaldehyde at 25° and 0° when the total dissolved ammonia $[NH_4^+]+[NH_3]$ is 0.01 M. At higher ammonia concentrations the curves would be shifted down proportionally to the concentration, and low ammonia concentrations would shift the curves up correspondingly. It is apparent that this ratio is not much dependent on temperature or on the pH in the region of interest (pH 6 to 9). Glycine is favored over glycolic acid relative to alanine/lactic acid by a factor of about 4 at a given ammonia concentration. Other amino acids would be expected to have comparably small differences.

If the ammonia concentration in the primitive ocean were too low no amino acids would have been produced; hydroxy acids would have been the sole product. If we make the assumption that it was necessary for the hydroxy acid/amino acid ratio to be less than 1, we can calculate the minimum ammonia concentration in the primitive ocean. These are shown in Table 1. Also shown in Table I are the minimum NH_4^+ calculated from the aspartic acid-fumaric acid equilibrium as well as the upper limit on the NH_4^+ concentration based on the clay mineral ion exchange equilibria (4). These concentrations are close to that proposed to give a greenhouse effect necessary to prevent the earth from freezing (5).

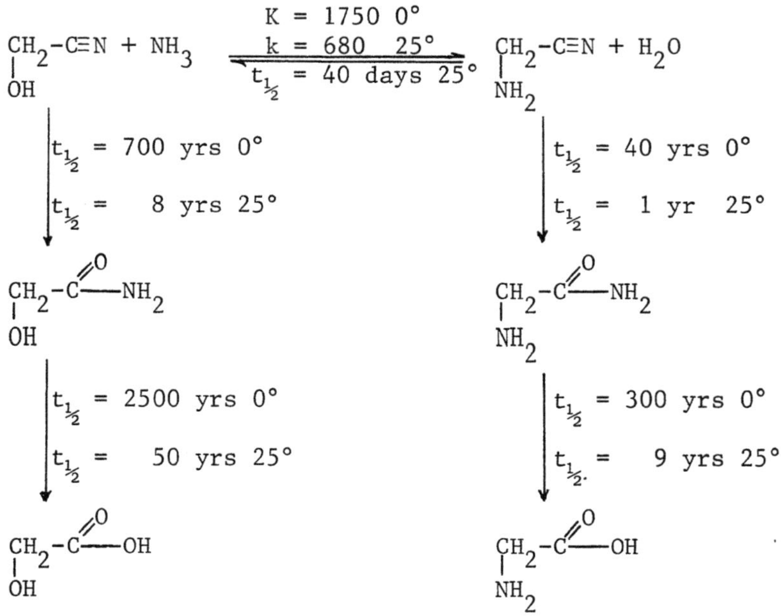

Figure 1. Equalibrium and kinetic data for the formation of glycine and glycolic acid.

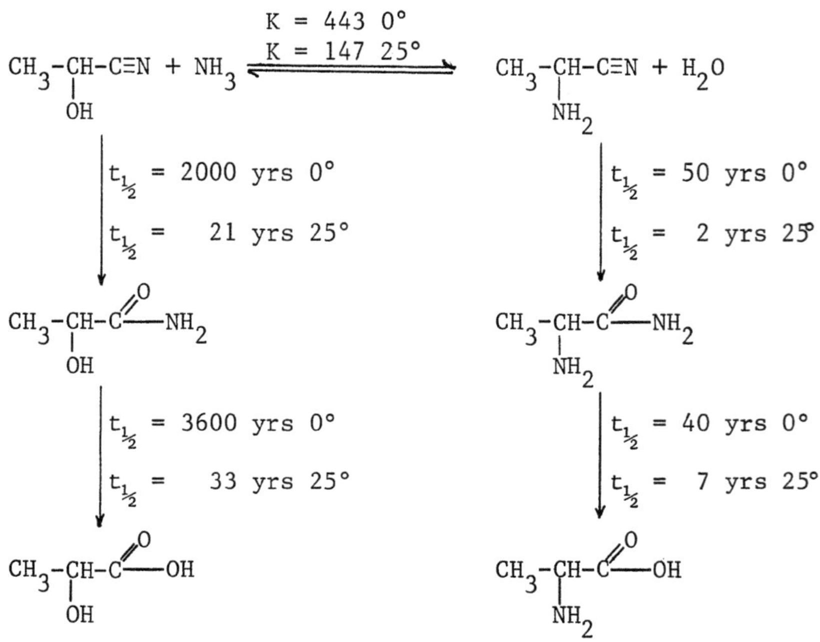

Figure 2. Equilibrium and kinetic data for the formation of alanine and lactic acid.

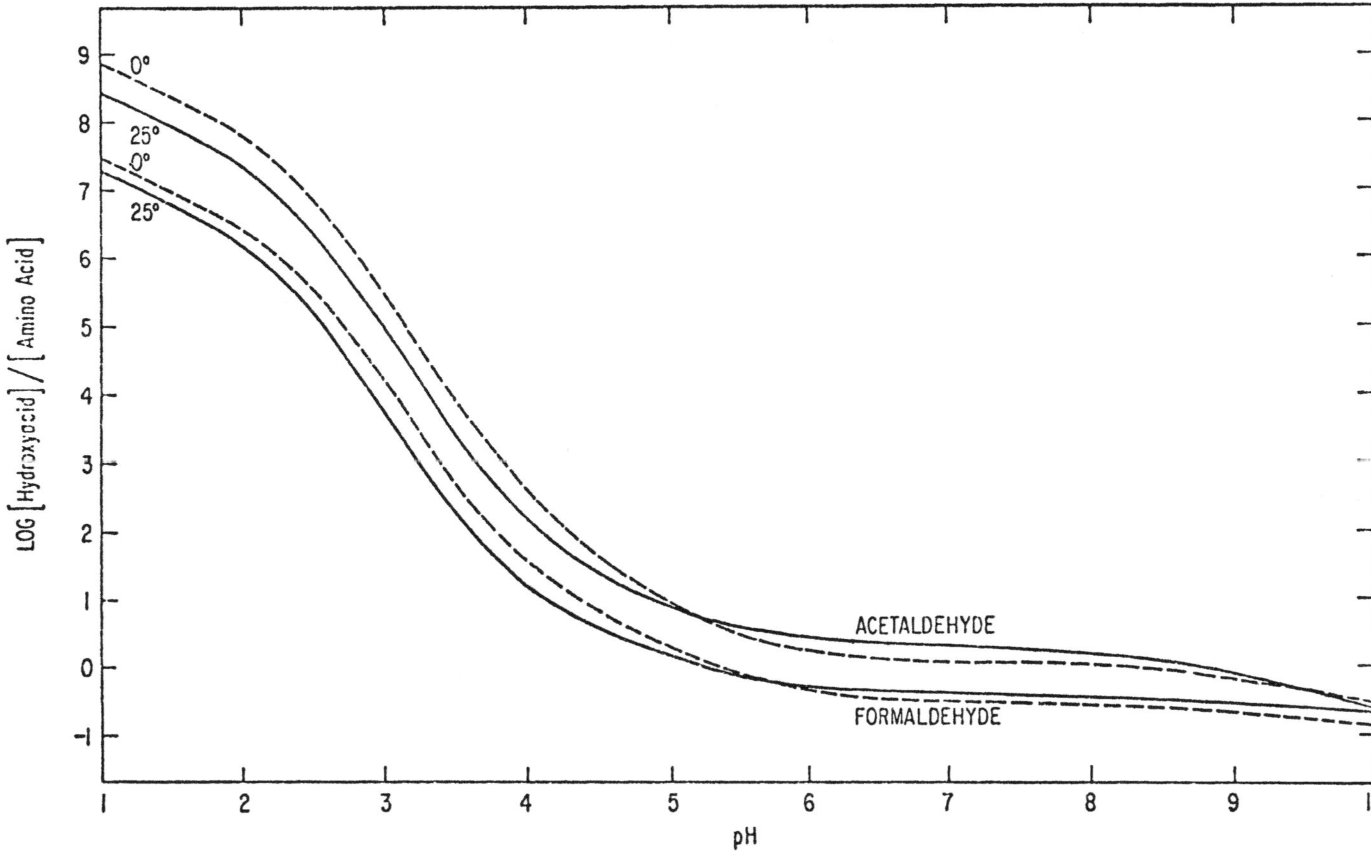

Figure 3. The ratio hydroxy acid/amino acid as a function of pH when $NH_4^+ + NH_3 = 0.01$ M. The dotted line is for 0°C and the solid line is for 25°C.

TABLE 1
Ammonium Ion Concentrations Giving the Indicated Ratio of Products at pH 8 and the Indicated Temperature

	0°	25°	50°
Glycolic acid/glycine = 1	3.1×10^{-3}	4.3×10^{-3}	2.2×10^{-3}
Lactic acid/alanine = 1	1.2×10^{-2}	1.4×10^{-2}	0.46×10^{-2}
Fumaric acid/aspartic acid = 1	1.0×10^{-3}	1.0×10^{-3}	4.8×10^{-3}
Maximum NH_4^+ controlled by clay mineral equilibria	1.0×10^{-2}	1.0×10^{-2}	1.0×10^{-2}

Strecker Synthesis at High Dilutions

Under very dilute conditions, most of the amino and hydroxy nitriles will be dissociated to the aldehyde, HCN and NH_3. Although there will be some amino nitrile that could be hydrolyzed to the amino acid, the hydrolysis of the HCN to formamide and formic acid will be the major process. The kinetics are give by

$$-\frac{d[HCN]}{dt} = k'_{1HCN}[HCN][OH^-] = k_{1HCN}[HCN]$$

This is the loss of HCN to formate. The loss of HCN to amino acids is

$$-\frac{d[HCN]}{dt} = k_{1AN}[RCH(NH_2)CN]$$

When the loss of the HCN to formate equals the loss of HCN to amino acids, we have

$$k_{1HCN}[HCN] = k_{1AN}[RCH(NH_2)CN] = k_{1AN}K_{AN}[RCHO][NH_3][HCN]$$

and

$$[RCHO] = \frac{k_{1HCN}}{k_{1AN}K_{AN}[NH_3]} \qquad (4)$$

At lower aldehyde concentrations than this, HCN goes to formate; at higher RCHO the HCN goes to amino acids. The value of k_{1HCN} can be calculated from the second order rate constants (6).

$$\log k'_{1HCN}(hr^{-1}M^{-1}) = 23.981 - 7022.2/T$$

Taking the conditions as 0° pH = 8, NH_4^+ = 0.01, we have from Eq. 4, $CH_3CHO = 3.0 \times 10^{-7}$ and $H_2CO = 9.1 \times 10^{-10}$. At higher temperatures the aldehyde concentrations are higher.

These considerations show that the Strecker synthesis will synthesize amino and hydroxy acids under quite dilute conditions such as the primitive ocean. Therefore, special concentration mechanisms are not needed for amino acid synthesis.

REFERENCES

(1) Van Trump, J. E., Thesis, University of California, San Diego.

(2) Van Trump, J. E. and Miller, S. L., in preparation.

(3) Bejaud, M., Mion, L. and Commeyras, A., Bull. Soc. Chim. France 1976, 1425 and preceeding papers.

(4) Bada, J. L. and Miller, S. L., Science 159, 423 (1968).

(5) Sagan, C. and Mullen, G., Science 177, 52 (1972).

(6) Sanchez, R. A., Ferris, J. P. and Orgel, L. E., J. Mol. Biol. 30, 223 (1967).

PHOTOASSISTED CARBON DIOXIDE REDUCTION AND FORMATION OF TWO- AND THREE-CARBON COMPOUNDS

M. Halmann, B. Aurian-Blajeni, Isotope Department,
Weizmann Institute of Science, Rehovot, Israel and
S. Bloch, Laboratoire Energétique Biochimique,
Université Paris Val-de-Marne, Créteil, France.

We found that aqueous carbon dioxide, in the presence of inorganic minerals with semiconducting properties, underwent photosensitized reduction by ultraviolet and visible light - the main products being formaldehyde and methanol. Effective photoactive materials included naturally occurring minerals such as nontronite, anatase, wolframite, molybdenite, minium, cinnabar and hematite. No appreciable carbon dioxide reduction was observed in the presence of bentonite. The heterogenous photoreduction of carbon dioxide by natural semiconducting minerals could be a precursor of plant photosynthesis. Photosynthetic condensation of dilute aqueous formaldehyde solutions to glyoxal and malonaldehyde was obtained by UV-irradiation in the absence of oxygen. The malonaldehyde concentration reached its maximum after several hours and then declined. The known condensation reactions of malonaldehyde with urea or guanidines to form hydroxy- or amino-pyrimidines may be a potential prebiotic route to pyrimidines.

The present study is intended to indicate pathways for nonbiological photosynthesis, using visible and near ultraviolet light, in the presence of naturally occurring minerals as photoactive agents. In recent work, we showed that semiconducting solids, by irradiation with light more energetic than their bandgap, are able to effect the reduction of aqueous carbon dioxide to organic products: formic acid, formaldehyde, methanol and methane (1,2). The heterogenous photoassisted reaction thus provides a new model of prebiological photosynthesis, which may be considered a potential precursor to the photosynthetic fixation of carbon dioxide by plants. The photosynthetic reduction of carbon dioxide in either heterogenous or homogeneous systems

Y. Wolman (ed.), Origin of Life, 143–150.

was also reported recently by other investigators (3-6). The reductive fixation of molecular nitrogen in a similar reaction, by UV-irradiation of nitrogen and water in the presence of titanium dioxide was shown to yield ammonia and hydrazine (7). Molecular nitrogen was also fixed by an oxidative photocatalytic reaction, by near-UV irradiation in the presence of titanium dioxide-producing nitric oxide (8). Irradiation of aqueous solutions of formaldehyde containing ammonium nitrate was found to yield glyoxal, presumably by dimerization of two formyl radicals CHO formed in the initial photochemical step (9). Also, UV-irradiation of dilute formaldehyde produces traces of ribose and deoxyribose (10). In the absence of oxygen, irradiation of dilute aqueous formaldehyde resulted in the formation of glyoxal and malonaldehyde (11).

Measurements of the solar radiation incident on Mars and on the other Inner Planets (12) indicate favorable conditions for photosynthetic reactions. Iron-rich clays similar to montmorillonite and nontronite, which belong to the smectite group, have been identified as the most abundant component of the soil of Mars (13). In the fine material on the surface of Mars, the most abundant elements are silicon and iron, with concentrations of 45±5 wt % SiO_2 and 19±3 wt % Fe_2O_3 (14). The "Labeled Release" experiments carried out during the Viking Lander tests indicated some carbon assimilation of ^{14}C-CO_2 (15). Laboratory simulation of these experiments were achieved using ferric iron saturated clays (16).

Nontronite (originating from Manito, Washington, U.S.A.) was crushed in a mortar and activated by heating in an evacuated quartz tube at 0.001 torr at 600°C for 6 hours.

Optical Properties of nontronite and of other minerals were determined by diffuse reflectance spectroscopy. Results of such spectra for several materials are shown in Figure 1. For nontronite, a comparison of the reflectance spectrum with the photoacoustic spectrum (Fig. 1b) indicates that nontronite starts to absorb visible light below 700 nm, and absorbs very intensely below 500 nm.

Irradiations. High-pressure mercury lamps of the immersion type (Hanau Q 81 or Q 718) were used. Irradiations were carried out in double-walled reactors (17), in which the lamp was immersed in the center and surrounded by circulating thermostated cooling water. The outer jacket contained the reaction mixture, and was flushed with either CO_2 or argon gas from a sintered glass inlet at the bottom. For the irradiation of aqueous carbon dioxide, the photochemical reactor was constructed of borosilicate glass (to filter out the UV-light at wavelengths below 300 nm), while for irradiation of aqueous formaldehyde, the

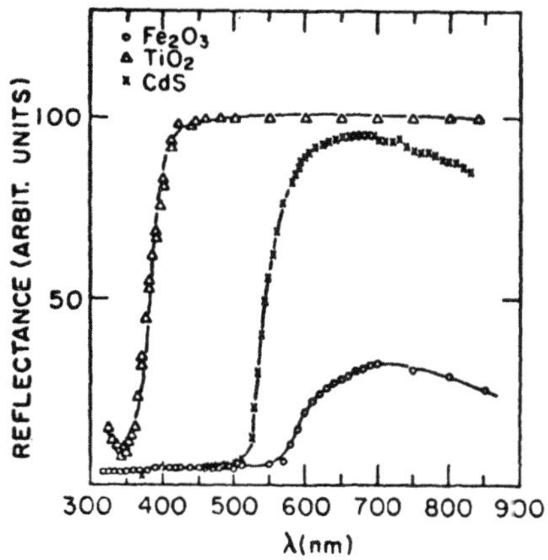

Fig. 1a Diffuse reflectance spectra of ferric oxide, titanium oxide and cadmium sulfide.

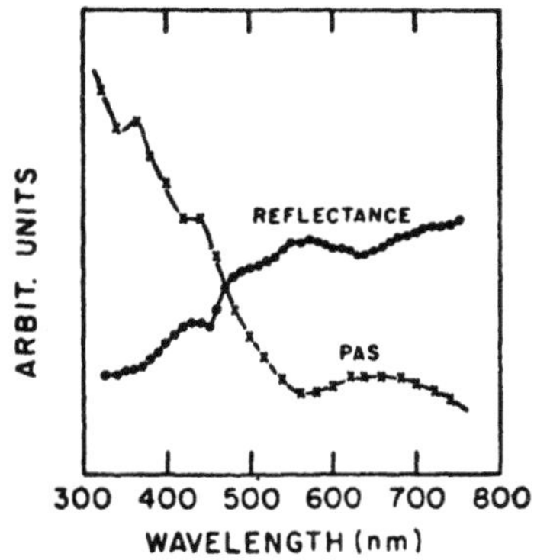

Fig. 1b Comparison between the diffuse reflectance spectrum and the photoacoustic spectrum (PAS) of nontronite (the PAS normalized against carbon black).

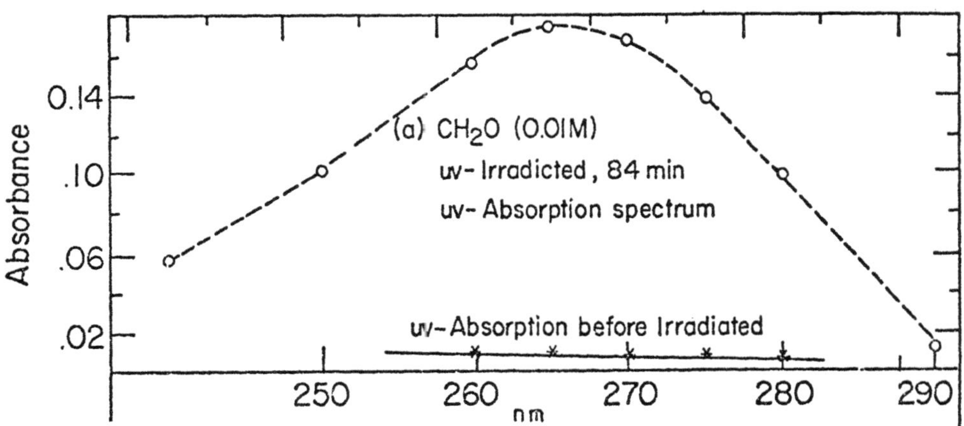

Fig. 2. Absorption spectrum of UV-Irradiated aqueous formaldehyde (0.01M) in 0.01M $NaHCO_3$, pH 8.3.

TABLE 1. Formaldehyde and Methanol Production by the Photoassisted Reduction of Aqueous Carbon Dioxide.

Mineral	Bandgap eV	Irradiation Conditions	Light Source	Irradiated area cm^2	Production Rate µmoles h^{-1}	
					CH_2O	Ch_3OH
Nontronite		Suspension	a	25	0.06	1.2
Bentonite		Suspension	a	25	0	0
TiO_2 (anatase)	3.3	Suspension	a	25	0.32	2.0
WO_3 (wolframite)	2.7	Coated Surface	c	3,300	43.3	2.7
		Suspension	a	25	0.2	5.1
MoS_2 (molybdenite)	1.8	Coated Surface	c	2,400	5	19
Pb_3O_4 (minimum)	2.1	Suspension	a	25	<0.001	0.55
		Coated Surface	b	368	<0.001	1.8
HgS (cinnabar) coated on Ba TiO_3	2.1	Suspension	a	25	0.3	0.29
Fe_2O_3 (hematite)	2.2	Suspension	a	25	<0.001	3.0
		Coated Surface	a	25	<0.001	0.03

a = 70 W Hg-Lamp ; b = 500 W Hg-Lamp ; c = Sunlight.

inner tubes were built of quartz-tubing (to provide exposure also to the short UV below 300 nm).

Analysis of formaldehyde was carried out by the colorimetric reaction with chromotropic acid (18). Methanol was determined by gas chromatography using a column of Porapak Q (length 4 m, I.D. 6 mm). Glyoxal and malonaldehyde were determined by the colorimetric reactions with thiobarbituric acid, with absorption maxima at 450 and 530 nm, respectively (19). Malonaldehyde was also assayed by its intense UV-absorption maximum at 266 nm in alkaline media.

Aqueous carbon dioxide saturated suspensions or surfaces of the minerals were irradiated either with high-pressure mercury lamps or with sunlight. The results of some typical experiments are shown in Table 1. The results with nontronite were poorly reproducible, and varied with different samples of the mineral. More consistent data were obtained with the synthetic minerals tested. The semiconductor bandgaps were taken from the literature (20), and were also determined by diffuse reflectance spectroscopy.

The photosynthetic condensation of formaldehyde to glyoxal, CHO-CHO, and to malonaldehyde, $CH_2(CHO)_2$, proceeds in a stepwise reaction, with initial formation of glyoxal. The production of malonaldehyde is delayed; its concentration reaches a maximum only after several hours and then declines. The results of one experiment are demonstrated in Figure 2, showing the UV-absorption of aqueous formaldehyde (0.01M, in 0.01M $NaHCO_3$), before and after 84 minutes of irradiation with a 70W high-pressure Hg-lamp. The 265 nm absorption maximum is due to malonaldehyde

The photosynthetic reduction of aqueous carbon dioxide to formic acid, formaldehyde, methanol and methane involves the conversion of light energy into chemically stored energy, with the standard enthalpies $\Delta H°$ and Gibbs free energies $\Delta G°$ given below (21,22):

	$\Delta H°$ kJ/mole	$\Delta G°$ kJ/mole
$CO_2(g) + H_2O(\ell) \rightarrow HCOOH(\ell) + 1/2\ O_2(g)$	270	286
$CO_2(g) + H_2O(\ell) \rightarrow H_2CO(g) + O_2(g)$	563	522
$CO_2(g) + 2H_2O(\ell) \rightarrow CH_3OH(\ell) + 3/2\ O_2(g)$	727	703
$CO_2(g) + 2H_2O(\ell) \rightarrow CH_4(g) + 2O_2(g)$	890	818

As we have shown above, this reduction occurs with the visible light of the solar spectrum and uses naturally occuring minerals. It is obvious to consider that such reactions may have proceeded also on the Primitive Earth, on which then (as now) visible light was the most abundant energy source. Such photosynthetic oxidation-reduction reactions may be the initial

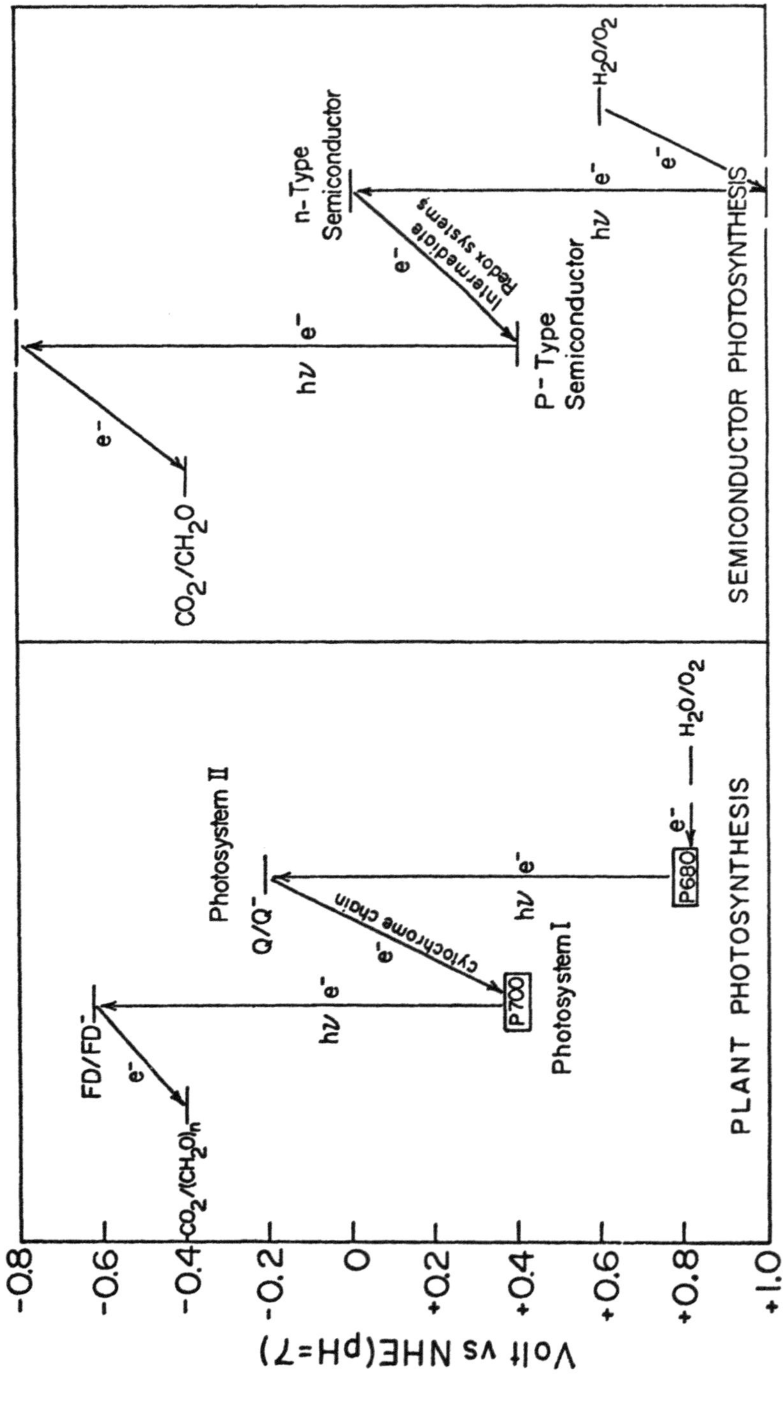

Fig. 3. Comparison of photosynthesis in plants and with semiconductors.

process of producing simultaneously energy-rich reduced carbon compounds and molecular oxygen.

One hypothesis accepted by many geologists is that the Primitive Earth atmosphere was partially oxidized, due to outgassing of rocks, consisting mainly of CO_2, H_2O, N_2 - as on Mars or Venus now. Photosynthetic surface reactions on semiconducting materials may lead to formaldehyde, which may undergo further photosynthetic condensation to glyoxal and malonaldehyde. Malonaldehyde readily condenses even in dilute aqueous solutions with urea or guanidines, producing hydroxypyrimidines or aminopyrimidines (23). The known mutagenicity of malonaldehyde may be due to such condensation reactions (24).

An alternative scenario is a completely reduced atmosphere of methane, ammonia and water. The photosynthetic formation of aminoacids was achieved by irradiation of methane, ammonia (or ammonium chloride) and water in the presence of platinized titanium oxide with near UV-light. The product mixture contained glycine, alanine, serine, aspartic acid and glutamic acid (25). The overall reaction was

$$2CH_4 + NH_3 + 2H_2O \rightarrow H_2NCH_2CO_2H + 5H_2 \qquad \Delta G^\circ = 231.8 \text{ kJ/mole}$$

This reaction too is photosynthetic, involving a net gain of energy.

A formal analogy of semiconductor-assisted photosynthesis with that in living plant can be represented as in Figure 3.

Thus, photosynthetic reactions with visible light, in the presence of photoactive minerals of semiconducting properties, may provide models for prebiological carbon and nitrogen fixation in both oxidized and reduced atmospheres.

ACKNOWLEDGEMENT: This study was supported in part by a grant from the United States-Israel Binational Science Foundation (BSF), Jerusalem, Israel, and from NASA.

REFERENCES

1. Halmann, M., Nature, 275, 115 (1978).
2. Halmann, M. and Aurian-Blajeni, B., Proceed. 2nd E.C. Photovoltaic Solar Energy Conf., Editors, Van Overstraeten, R. and Palz, W., Berlin (West), Reidel Publ., Dordrecht, 1979, p. 682.
3. Hemminger, J.C., Carr, R. and Somorjai, G.A., Chem. Phys. Lett., 57, 100 (1978).
4. Inoue, T., Fujishima, A., Konishi, S. and Honda, K., Nature, 277, 637 (1979).

5. Fruge, D.R., Fong, G.D. and Fong, F.K., J. Amer. Chem. Soc., 101, 3694 (1979).
6. Åkermark, B., Eklund-Westlin, U., Baekström, P. and Löf, R., Acta Chem. Scand., B 34, 27 (1980).
7. Schrauzer, G.N. and Guth, T.D., J. Amer. Chem. Soc., 99, 7189 (1977).
8. Bickley, R.I. and Vishwanathan, V., Nature, 280, 306 (1979).
9. Pavlovskaga, T.E. and Telegina, T.A., Origins of Life, 5, 303 (1974).
10. Ponnamperuma, C. and Mariner, R., Radiat. Res. 19, 183 (1963).
11. Halmann, M. and Bloch, S., BioSyst., 11, 227 (1979).
12. Levine, J.S., Kraemer, D.R. and Kuhn, W.R., Icarus, 31, 136 (1977).
13. Klein, H.P., J. Geophys. Res., 82, 4677 (1977); Icarus, 34, 666 (1978).
14. Snyder, C.W., J. Geophys. Res., 84, 8487 (1979).
15. Horowitz, N.M., Hobby, G.L. and Hubbard, J.S., Science, 194, 1321 (1976).
16. Banin, A. and Rishpon, J., Life Sciences and Space Research, Proceed. Conf. Innsbruck, Austria, 1978.
17. Trachtman, M. and Halmann, M., J. Chem. Soc., Perkin Trans. II, 132 (1977).
18. Burton, R.M., in: Methods in Enzymology, Vol. III, Editors, Colowick, S.P. and Kaplan, N.O., Academic Press, New York, 1957, p. 247.
19. Morre, J., Ann. Chim. 4, 227 (1969).
20. Strehlow, W.H. and Cook, E.L., J. Phys. Chem. Ref. Data, 2, 163 (1973).
21. Bolton, J.R., Science 202, 705 (1978).
22. Aurian-Blajeni, B., Halmann, M. and Manassen, J., Solar Energy, in press (1980).
23. King, T.P., Biochemistry, 5, 3454 (1966).
24. Mukai, F.H. and Goldstein, B.D., Science, 191, 868 (1976).
25. Reiche, H. and Bard, A.J., J. Amer. Chem. Soc., 101, 3127 (1979).

MELANOIDIN POLYMERS AS POSSIBLE OXYGEN SINKS IN THE PRE-BIOTIC OCEANS

Andrei Serban and Arie Nissenbaum

Geoscience Group, Isotope Department, Weizmann Institute of Science, Rehovot, Israel

A prerequisite for abiotic organic synthesis is an anoxic environment. Several mechanisms for oxygen removal were proposed among which reactions with ferrous iron are considered to be the most important. We are proposing here a supporting system for an "in situ" sink for oxygen. Carbohydrates and amino acids, presumably present in the "prebiotic soup" might interact through a Maillard reaction in order to form humic acid like compounds - melanoidins. Laboratory experiments show that the melanoidin synthesis is characterized by a rapid incorporation of the dissolved molecular oxygen into the final condensation product. This reaction may have been important during the main stage of abiotic synthesis of organic matter, but its importance probably declined when the non-biological production of organic matter diminished.

It is generally accepted that the earth's early atmosphere was a reducing one (1). According to most authors a lack, or at least, a depletion of oxygen in the pre-biotic atmosphere is a requirement both for the production and for the preservation of considerable amounts of low molecular weight organic compounds from which, according to the Oparin-Haldane model, living systems will eventually emerge.

The source of oxygen in the atmosphere has been a subject of debate between supporters of origin by photosynthesis (2) and proponents of origin by photodissociation of H_2O (3). It is quite possible, however, that even if plant systems are the major source of atmospheric oxygen, small amounts of oxygen were always produced by photodissociation. Several mechanisms for the removal of this oxygen can be postulated. The most widely accepted

Y. Wolman (ed.), Origin of Life, 151–156.

hypothesis is that ferrous iron may have been responsible for regulating the concentration of atmospheric oxygen (4).

In the present report we propose an additional sink for oxygen which may have been operative during the stage of abiotic formation of organic matter.

This hypothesis is based on the observation that in recent aquatic environments one of the major pathways of removal of dissolved organic carbon from the oceans or freshwater bodies, is through the non-biological condensation of amino acids and carbohydrates to form humic substances. It was proposed (5) that similar reactions may have been operative in pre-biotic oceans and were responsible for the scavenging of soluble organics from the water column into the sediments. If so, then any interaction of this diagenetic pathway with the oxygen cycle may have a bearing on oxygen regulation in the early atmosphere.

In order to test this hypothesis we have investigated the reaction of melanoidins (brown polymers formed by the Maillard reaction between carbohydrates and amino acids) with molecular oxygen.

Melanoidins were synthesized in aqueous solutions of D-glucose (1 or 2M) and an amino acid (L-Glycine or L-Lysine, 1 or 2M). The solutions were buffered at a slightly alkaline pH using either solid calcium carbonate or 0.1M borate buffer, pH 8.6. The reaction was conducted in a reaction vessel fitted with a screw cap with a silicon rubber septum.

At various time intervals, samples of the gas phase from the reaction vessels were taken with a gas-tight syringe and injected into a Perkin Elmer Sigma 2B gas chromatograph equipped with a Molecular Sieve 5A, 60/80 mesh, 6 ft x 1/8" SS column. The chromatographic conditions were: column temperature - 50°C, carrier gas-helium at 30 ml/min, detector-thermal conductivity, sample size - 0.1 cc. The oxygen concentration was obtained from a Sigma 10B Chromatography Data Station connected on-line with the gas chromatograph. The analytical results of oxygen concentration were plotted against reaction time by a best fit curve using computer smoothing subroutine which computes the cubic splines. The visible-ultraviolet spectra were recorded using a Cary Model 118 double beam spectrophotometer.

In all the reaction mixtures investigated, the oxygen concentration in the gas phase above the solution decreases as the reaction proceeds toward completion. The oxygen consumption rate is strongly dependent on temperature as well as upon the nature of the amino acid (Fig. 1). This suggests a possible direct incorporation of the oxygen into the reaction product, since the

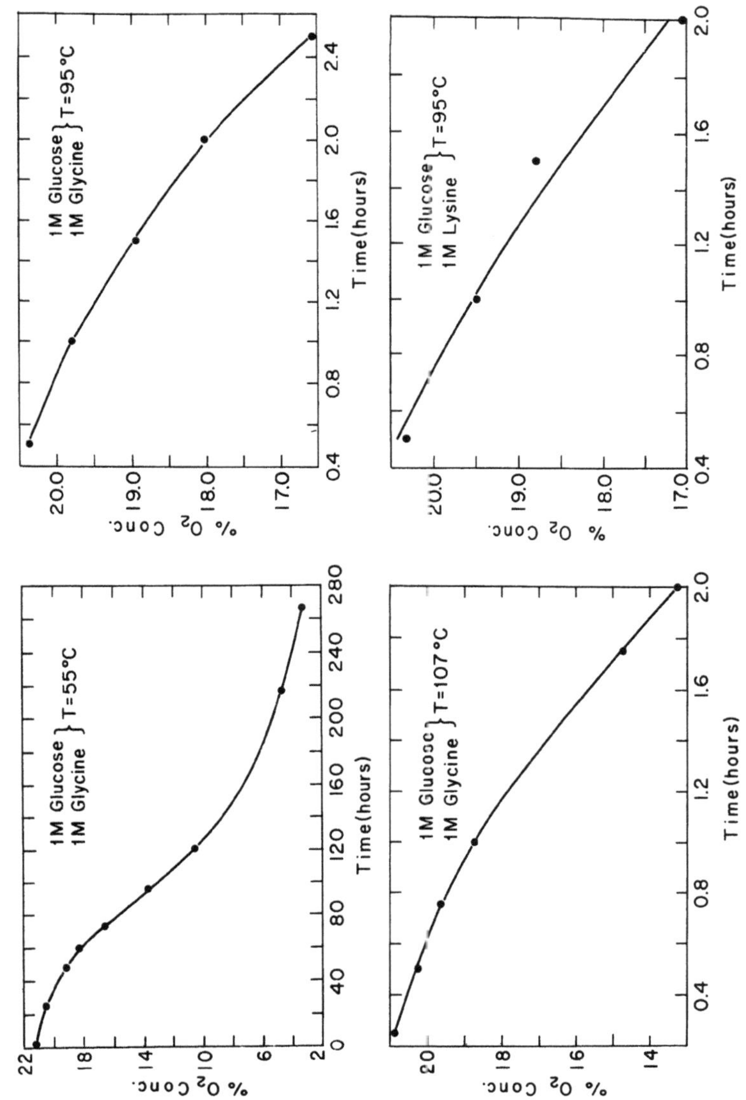

Figure 1 - Oxygen consumption during melanoidin formation

rate of melanoidin formation, has been shown to be strongly dependent upon both temperature and nature of the amino acid (6).

As can be seen from Fig. 1, the oxygen consumption rate does not follow a simple reaction pathway but rather one involving several intermediates, resulting in a S-shaped curve. Several such intermediates have been either isolated or postulated by many other studies (6). It is possible to assume that oxygen may react with one or more intermediates formed during the reaction, each one with a different kinetics. This explains the increase in the oxygen disappearance rate after a certain period of time (which may coincide with formation of intermediate (s)) and for the return to the initial rate toward the completion of the reaction. This assumption is also based on the appearance of an intermediate which absorbs at 295 nm (Fig. 2) and which disappears slowly toward the end of the reaction. The appearance of a UV absorbing intermediate coincides with the increase in the oxygen consumption rate.

Melanoidins are known to form over a wide range of temperatures, with an increased rate at higher temperatures. For oxygen removal during melanoidin synthesis we have observed a change in the reaction pathway which is temperature dependent. When the temperature is raised from 55°C to 85°C the initial gently-sloping part of the curve is reduced while at higher temperatures the curve began to approximate straight line (Fig. 1).

The amounts of oxygen consumed during melanoidin synthesis is given in Table 1.

TABLE 1
Consumption of Oxygen During Synthesis of Melanoidins

Reactants* volume (ml)	(conc.) (molar)	Reaction vessel vol. (ml)	Temp. (°C)	Reaction time (hours)	O_2 consumed**
Glu:Gly	10 (1)	63	55	265	0.06
Glu:Gly	10 (1)	63	85	76	0.22
Glu:Gly	5 (1)	34	95	2	1.85
Glu:Gly	6 (1)	34	107	1.75	3.41
Glu:Lys	5 (1)	34	95	1.50	2.11
Glu:Lys	1 (2)	3.5	72	1.83	1.02

* The reactants are in equimolar amounts.

** μg/mg C in reaction mixture x hr^{-1}.

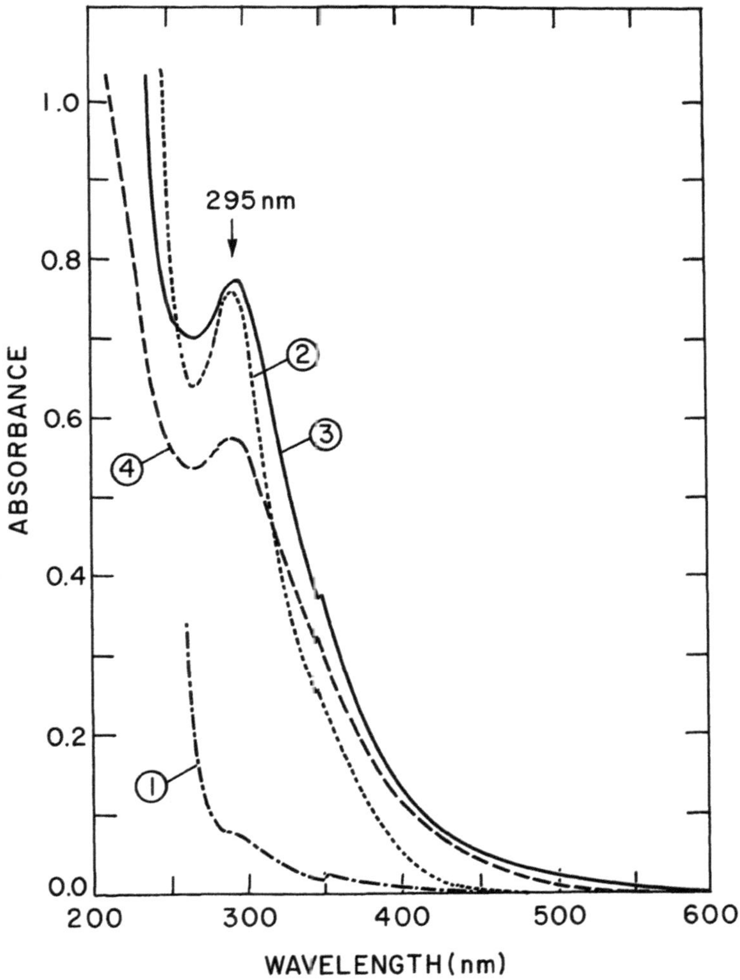

Figure 2 - UV - VIS absorption spectra of Glycine-Glucose mixtures at 55°c at various reaction times. (1) 2 hours. (2) 25 hours. (3) 121 hours. (4) 267 hours.

We have shown that condensation of amino acids with carbohydrates to form melanoidin polymers is accompanied by removal of free oxygen from the reaction mixture. It is probable that humic substances, which are a major component of organic matter in recent aquatic environment, are formed from similar reactions where the starting materials are derived from the decomposition of plankton (7). If so, then the browning reaction which occurs in modern oceans may be responsible for removing dissolved oxygen from waters which, due to stagnant conditions, may not replenish their oxygen from the atmospheric reservoir (8).

However, similar scavenging of oxygen could occur if amino acids (or any other amines) and carbohydrates (or other aldehydes) were supplied to the aqueous millieu by an abiotic source. Thus, prebiotic formation of such monomers should inevitably lead to the formation of melanoidin-like polymers. These would then be removed into the sediment leaving behind an ocean with a relatively low concentration of dissolved organics (5). During condensation, molecular oxygen, which may have been formed by photocatalytic decomposition of water, will be fairly efficiently scavenged.

However, once the main stage of abiotic formation of organic matter is over, the role of this particular scavenging mechanism diminishes considerably, since the residence time of the organic polymers in the water column is geologically short. Subsequently, other mechanisms, such as the oxidation of ferrous iron, may serve as regulators of molecular oxygen concentration. It is also probable, after the development of extensive marine biota which will supply by its decomposition the necessary reactants, that this reaction may again be instrumental under local and specialized conditions in the formation of anoxic enviornments.

REFERENCES

(1) Berkner, L.V. and Marshall, L.C.: Discussions Faraday Soc., 37, 122 (1964).
(2) Cloud, P.E.: Paleobiology 2, 351 (1976).
(3) Towe, K.M.: Nature 274, 657 (1978).
(4) Cloud, P.E.: Econ. Geol. 68, 1135 (1973).
(5) Nissenbaum, A.: Origin of Life 7, 413 (1976).
(6) Hodge, J.E.: Agr. Food Chem. 1, 928 (1963).
(7) Nissenbaum, A. and Kaplan, I.R.: Limn. Ocean. 17, 570 (1972).
(8) Nissenbaum, A.: Advances in Organic Geochemistry, ed. B. Tissot and F. Bienner, Edition Technip, Paris, 1974, 37.

FORMATION OF ENERGY RICH PHOSPHATE IN FENTON'S REACTION

Oemer Saygin and Peter Decker

Chemical Institute, Veterinary School, Hannover, FRG.

Mechanism and origin of the ubiquitous biological phosphorylation stays obscure since the discovery of ATP by Lohmann in 1929. Asking whether the extreme reactivity of radical intermediates could not explain such reactions, we found that up to 30% of aqueous inorganic phosphate is bound into an energy-rich oxidation product of 2-methylimidazole by oxidation with H_2O_2 or O_2 using Fenton's reagent. These results emphasize, that coupled energy and group transfer and self-organization requires open systems ("bioids") driven by an adequate energy source like redox or photo-reactions, and that transfer of all, molecules (in chemical reactions), energy and information are complementary and inseparable aspects of a generalized Darwinian evolution i.e. of accumulation of information over time in such systems.

Three aspects of evolution.

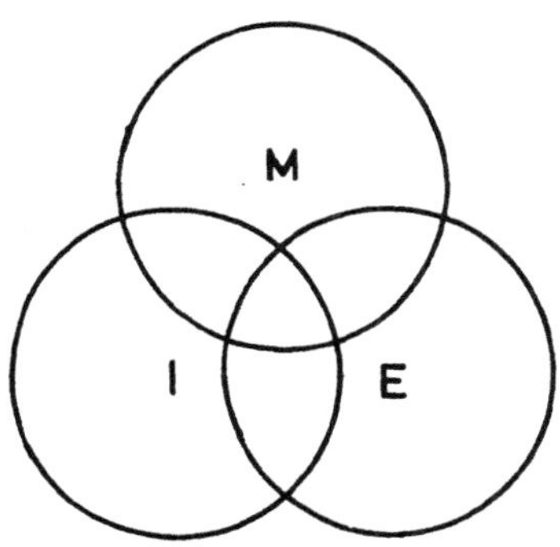

Matter, information and energy are three complementary aspects involved in all "bioids" (1-5), chemical systems capable of a generalized Darwinian evolution i.e. of acquisition, conservation and accumulation of information. In more complex systems molecules and energy may be replaced by more general entities as "elements" and "driving force" whithout changing the formal structure and

Y. Wolman (ed.), Origin of Life, 157–164.

function, resulting in fields as cybernetics or synergetics (6), which can describe biological interactions between cells and/or organs and ecological,sociological and economical interactions between individuals.

In chemical systems catalysis is the most prominent feature involving information: catalysis may be considered as information transfer, as a "translation" of the catalysts concentration into a reaction rate, linking different points of a material network. Conspicuously, standardized catalysts, protein enzymes produced by the nucleoprotic mechanism, represent the most prominent feature of terrestrial life.

Feedback loops in such relations may produce autocatalysis and more complex behaviour like bistability, excitability (7) oscillations and spatial dissipative structures as studied by Prigogine in the "Brusselator" (8), in the Zhabotinsky (9) reaction or by Seelig (10) and Decker (11,12) in bioids which may produce optical activity by kinetic bistability.

Self-organization and energy transfer.

However, all such systems, including the many abiogenic syntheses considered as examples of chemical "evolution" require an energy source in order to become eligible as elements of models where evolution means more than in "evolution of hydrochloric acid", namely a process of acquisition and conservation over time of "information" in an open system. This requires at least two steady states ("species") and one transition ("mutation" by selection of the more stable one) representing the acquisition of one bit, the minimum amount of information. Decker proposed the term "Bioid" (1) for an open system fulfilling those minimum requirements.

As the sole known chemical system which could acquire at least one bit of information by switching into a selfreproducing autocatalytic reacting state we studied in 1969 the autocatalytic formation of sugars from formaldehyde. Formaldehyde could be formed steadily from a methane atmosphere, and methane could become recycled by fermentation of biomass in a reducing atmosphere. Moreover we considered as a possible additional feedback loop, a second bit of information, the self-enhancement of formaldehyde formation through photooxidation of methane by water,evolving H_2 (an "inverse assimilation" (4,13,14)), sensitized by some products of a nonenzymatic browning (Maillard-) reaction of the sugars with ammonia and amino acids, which could couple the system to solar energy. Among the products of such reactions are "melanoidins" which may play a role as sensitizers or precursors of such, as ligands modifying and enhancing the catalytic actions of metal ions, or as precursors of kerogen-like polymers similar to those found in sediments.

Binding of phosphate.

Therefore we became interested in possible interactions of such systems with energy sources, and, considering the ubiquity of chemical energy transfer and reactions in biochemistry, in possible spontaneous reactions of this kind.

Although ATP is formed at almost every energetically suitable metabolic step, and the problem is open since the days of Lohmann's discovery of ATP (15), the mechanism of such phosphate binding reactions remains rather obscure. Vectorial processes (16) presently may play a role in many such processes. However, since cells, membranes and enzymes hardly were the primary features of self-organization, phosphorylation should have originated in homogenous systems.

Hypotheses in origins of life research often lack the bioid requirement of a steady energy source and considered formations involving dehydration orthophosphates by heath or by the use of already energy rich specific reagents (17-18). More interesting in our context are reactions involving coupling of phosphate binding to hydrolytic resp. phosphorolytic cleavage of cyanogen compounds which may be claimed to have occured in a prebiotic scenario (19). An alternative, suggested by bioreactions is a coupling of phosphate binding to redox reactions. The idea of trivalent phosphorous has been rejected because of the instability of its compounds in the prebiotic environment (20). The other alternative, binding of phosphate during oxidation of a substrate, has been explored in several examples by Th. Wieland and Baeuerlein (21). Generally Lewis acid-base-reactions are considered in such mechanisms (22,18), whereas Wieland's systems also may involve radical intermediates.

Since photo-reactions (23) also involve radical-like excited states, as a simplified approach we screened the oxidation of different substrates by a modified Fenton's reagent as a source of radical intermediates which, possibly, could scavenge phosphate or acylate ions into energy rich compounds. Instead of the reducing formol condensation mixture we used ascorbic and EDTA as a defined co-reductant and compexing agent respectively.

In a typical experiment the complete system contained the following components (μmoles in 1 ml of total volume) at pH 4-6 and $0^{o}C$: $Na_2HPO_4 \cdot 2H_2O$ 100; $FeSO_4 \cdot 7H_2O$ 4; ascorbic acid 40; EDTA 20 and substrate 40. The solution was cooled with ice and the reaction was initiated by the addition of 50 μl of 30% H_2O_2. After two minutes the phosphate determination was started.

Inorganic phosphate was removed by extraction as the molybdate complex with isopropyl acetate and bound phosphate was then determined colorimetrically using an additional extraction after hydrolysis of the aqueous layer (24).

With imidazole as substrate in dilute solutions 0.9% of inorganic phosphate was transformed into an organic form. Omission of each, phosphate, imidazole, Fe/EDTA or H_2O_2 suppressed its formation. In contrast, omission of ascorbic acid only reduced the yield to 0.2% and its role as reductant could by partly replaced by an increase of Fe(II). Similarly, EDTA is not essential and could be replaced by other complexing agents. Co(II), Mn(II) or V(IV) could not replace Fe(II); however, Cu(II) was as active as Fe(II) (in the absence of EDTA, since the Cu complex is not reduced by ascorbate.

Among alternative substrates, 2-methylimidazole gave by far the best yields - up to 32% with a more concentrated reaction mixture. 1-methylimidazole, 4-hydroxymethylimidazole, histamine and pyrimidine also gave positive results (Table 1). No measurable effect was found with imidazole-4-acrylic acid, histidine, adenine, xanthine, guanine, uracil, pyrazole, pyrazine, 1,2,4 triazole, acetonitril, imidazolinthion,acetamid, cyanamid, dicyanamid, N,N dimethylformamid, oxamid, ethyleneurea.

Maximum yield is obtained after about 2 minutes, followed by a slow hydrolytic decay: Half-life at 25°C (35°C) was ca 1 h (20 min) in 1 N acid and ca. 5 min at 25°C and pH 4.5. Maximal yields were obtained between pH 5 and 7 depending on the respective substrate.

Using 2-methylimidazole at pH 6 and 0°C we studied the effect of different variables on yields; with the exception of the respective component shown as variable in Fig. 1, the concentrations were (μmol/ml): inorganic phosphate 300; 2-methylimidazole 475; ascorbic acid 250; Fe(II) 25; EDTA 50; hydrogen peroxyd 2000. Only in (C) Fe(II) and EDTA are varied simultaneously at a constant ratio of 1:2. Inorganic phosphate and 2-methylimidazole as variables show linear dependence in small concentrations and saturation at higher concentrations. With Fe(II)-EDTA as variable, saturation was reached at very low concentrations, indicating the catalytic role of iron. Only in the case of ascorbic acid a maximum was observed. To understand the decrease of the yield at higher concentrations, in additional experiments we added ethanol or tert.-butanol (250 μmoles per ml) and observed also a decrease of the yield (50%). Since alkohols are known radical trapping agents, the observed decrease of the yield at higher concentrations of ascorbic acid can be explained by trapping the OH radicals.

In other experiments we tried to replace hydrogen peroxide

Table 1

Bound phosphate formation from imidazole

System	Yield of bound phosphate A (µmol/ml)	A (%)	B (µmol/ml)	B (%)
complete system	0.9	0.9	23.2	8.0
imidazole omitted	0.0			
phosphate omitted	0.0			
Fe(II)-EDTA omitted	0.0			
H_2O_2 omitted	0.0			
ascorbic acid omitted	0.2	0.2		
ascorbic acid omitted and initial Fe(II)-EDTA concentration doubled	0.5	0.5		
EDTA replaced by equimolar:				
oxalic acid	0.8	0.8		
ethylene-triamine 1,1,4,7,7 penta acetic acid	0.9	0.9		
Fe(II) replaced by equimolar:				
Mn(II)	0.0			
Co(II)	0.0			
V(IV)	0.0			
Cu(II) and EDTA omitted	0.9	0.9		
imidazole replaced by equimolar:				
1-methylimidazole	0.4	0.4	3.2	1.1
2-methylimidazole	1.4	1.4	95.0	32.7
4-hydroxymethylimidazole	0.3	0.3		
histamine	0.2	0.2		
pyrimidine	0.2	0.2		
imidazole replaced by 2-methylimidazole and 1 h bubbeling with air instead of H_2O_2	0.7	0.7		

System A contained (µmol/ml): inorganic phosphate 100, imidazole 40, Fe(II) 4, EDTA 20, ascorbic acid 40, H_2O_2 1700; System B: inorganic phosphate 300, imidazole 800, Fe(II) 25, EDTA 50, ascorbic acid 250, H_2O_2 2000. For reaction conditions see text. Yield (%) based on inorganic phosphate.

by O_2. Using 2-methylimidazole as substrate, after one hour of bubbling air through the solution (whithout H_2O_2) easily detectable amounts of bound phosphate were obtained (Table 1).

By elution of the reaction mixture on ion exchange columns (Dowex 1x8 in acetate form) with destilled water two fractions of bound phosphate were obtained between the elution volumes of

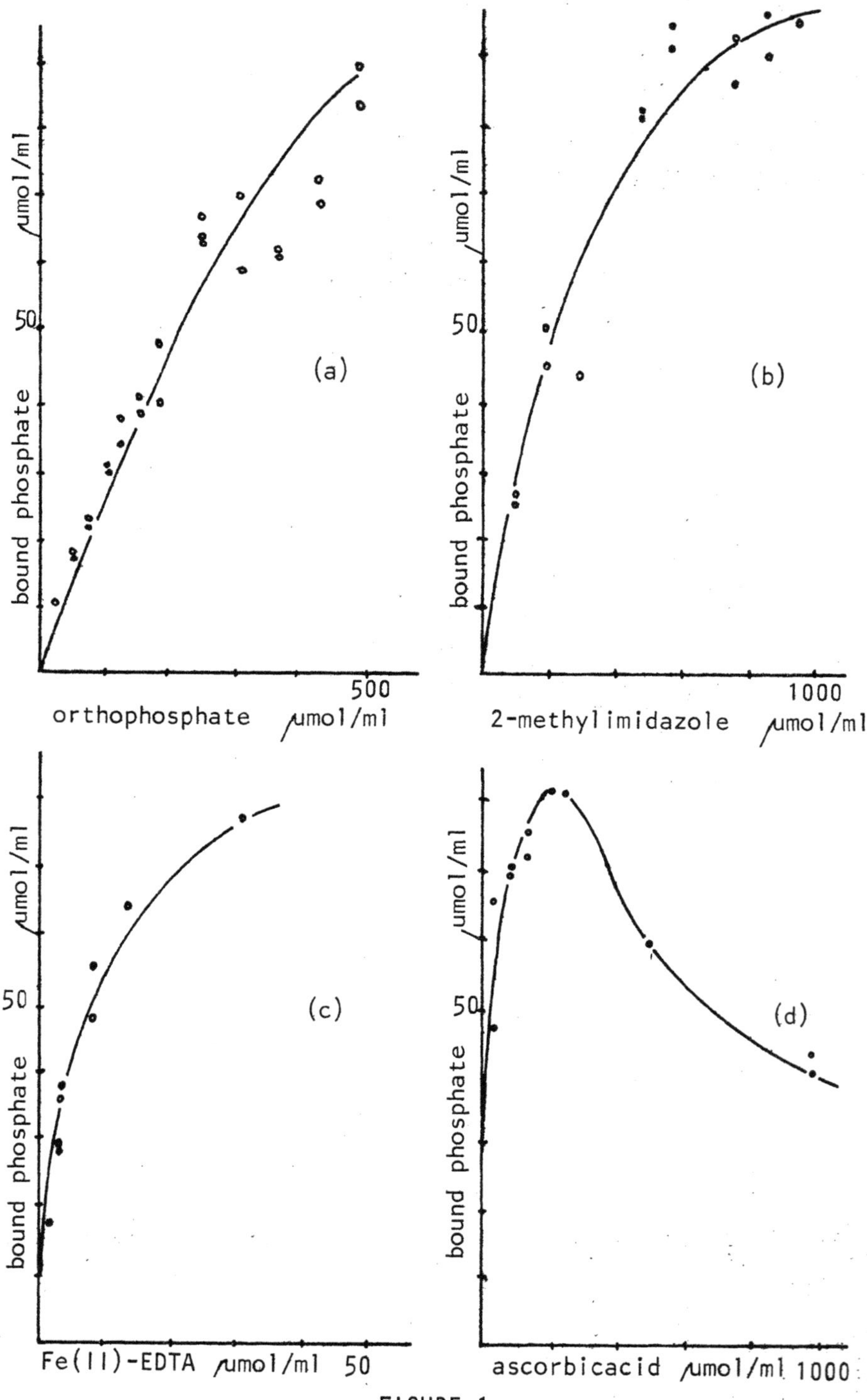

FIGURE 1
Bound Phosphate Formation Using 2-Methylimidazole

40-80 and 125-200 ml; inorganic phosphate remains absorbed on the column under these conditions. After freeze-drying the product was obtained by precipitation with ethanol or acetone from water giving white crystals, m.p. 130°C (decomposition).

Conclusions.

The investigation of the crystallized product so far has shown that it is an alcali labile phosphoric acid ester of a 4,5-dihydroxyimidazoline (25) in contrast to the acid-labile N-phospho-imidazole derivatives described as intermediates in ATP-formation in polar organic solvents (26). Since we obtained the same crystalline product using UV and sensitizers known to induce the formation of singlet oxygen (23), it appears that this novel kind of oxidative phosphorylation also involves $^1O_2$ (which may be formed from the reduction of metal complexes by the $O_2^{-\cdot}$-radical (27)) and not, as we expected, through radical intermediates like those which occur e.g. in anaerobic photosensitized oxidations of EDTA coupled with H_2-evolution (28,14). However, the reaction which we found with our screening test for bound phosphate anyway appears remarkable because of its ability to scavenge so much inorganic phosphate from aqueous solution.

If we consider self-organization in chemically reacting open systems as relevant to origins of life, reactions as potential elements of such systems appear more relevant than all "biochemical" compounds arising through "abiogenic synthesis". Among reactions, those involving coupling between energy and group transfer deserve special interest. The principle of group activation through redox reactions involving radical intermediates might occur frequently enough to justify further search using appropriate screening tests as analytic tools in what we proposed as "Hannover program" (which hopefully will spread to other places and which of course already is commonplace to many researchers) (2,3): the search and experimental reproduction of the first steps of chemical self-organization in open systems ("bioids") which produced the steady supply of defined and energy rich reactants necessary for the formation of ordered polymers allowing individuation and the deciding step in the evolution of terrestrial life - the standardization of catalyst production through the nucleic acid system.

REFERENCES

(1) Decker, P. and Speidel, A., Z.Naturforsch. 27b,257(1972). (Bioids 1).

(2) Decker, P. and Heidmann, W. in: H.Noda, Ed., "Origin of Life", Proc. 2. ISSOL Mtg., Jap.Sci.Soc.Press, Tokyo, pp. 617-624, 1978.

(3) Decker, P., Ann. New York Acad.Sci. 316, pp. 236-250, (1979) (Bioids VII).
(4) Decker, P., in: DFG-Kolloquium "Evolution in Planetenatmosphären", Schliersee, FRG, 2.-4.Okt. 1979, in press.
(5) Decker, P., Evolution in offenen Systemen: prize essay submitted to the Bavarian Academy of Sciences; reprints(170pp) available from Documentation Center on the Origin of Life, Universitätsbibliothek, Ulm, FRG.; cf. Decker, P., Nachr. Chem. Tech. 23, 167 (1975).
(6) Haken, H., Naturwissensch. 67, 121 (1980).
(7) Decker, P. and Saygin, Ö., Z.Naturforsch. 34c, 649-51(1979).
(8) Glansdorff, P. and Prigogine, I., Thermodynamics of Structure, Stability and Fluctuations, Wiley-Interscience, New York, 1971.
(9) Zaikin, A.N. and Zhabotinsky, A.M., Nature 225, 535 (1970).
(10) Seelig, F.F., J.Theor.Biol. 31, 355 (1970).
(11) Decker, P., Nature New Biol. 241, 72 (1973); J.Mol.Evol. 2, 137 (1973), 4, 49 (1975); Origins of Life 6, 211 (1975) (Bioids III-VI).
(12) Decker, P., in: D.C. Walker, Ed., Origins of Optical Activity in Nature, Elsevier, Amsterdam, pp. 109-124, 1979.
(13) Decker, P., Pure and Applied Geophysics 112, 865 (1974).
(14) Decker, P., in: H.Noda, Ed., "Origin of Life", Proc. 2. ISSOL Mtg., Jap.Sci.Soc.Press, Tokyo, pp.631-637, 1978.
(15) Lohmann, K., Naturwissensch. 17, 624 (1929).
(16) Mitchell, P., Eur. J. Biochem. 95, 1 (1979).
(17) Lohrmann, R. and Orgel, L.E., Science 171, 490 (1971); Ponnamperuma, C. and Chang, S., in "Molecular Evolution I", Buvet, R. and Ponnamperuma, C. eds., North Holland p. 216, 1971.
(18) Schramm, G., Groetsch, H. and Pollman, W., Angew.Chem.Int. Ed. 1, 1 (1962);Bruice, Th. and Bencovic, S.J., Bioorganic Mechanisms, Vol. II, p. 88-98,Benjamin, New York, 1966.
(19) Degani, Ch. and Halmann, M., in: "Molecular EvolutionI", R.Buvet and C.Ponnamperuma eds.,North Holland, p.224, 1971.
(20) Schwartz, A.W., in: "Molecular Evolution", Rohlfing, D.and Oparin, A.I. eds., Plenum Press, New York, p.129, 1972.
(21) Wieland, Th. and Bäuerlein, E., Naturwissensch. 54, 80 (1967).
(22) Clark, V.M., Hutchinson, D.W., Kirby, A.J. and Warren, S.G., Angew. Chem. 76, 704 (1964).
(23) Decker, P. and Saygin, Ö., this Congress
(24) Saygin, Ö., Z.Naturforsch., in Press
(25) Decker, P. and al., in Preparation
(26) Brinigar, W.S., Knaff, D.B. and Wang, I.H., Biochemistry 6, 36 (1967).
(27) Czypski, G. and Ilan, Y.A., Photochem. Photobiol. 28, 652 (1978).
(28) Krasna, A.I., Photochem. Photobiol. 31, 75 (1980).

ABIOTIC SYNTHESIS OF PHOSPHORIC ESTERS OF MONOSACCHARIDES ACCORDING TO THE „COLD MODEL"

C.I.Simionescu, S.Dumitriu and Al.Constantinescu

Polytechnic Institute, 6600 Jassy, Romania

It is investigated, according to the „cold model", the abiotic synthesis of some phosphoric esters of monosaccharides by cold plasma decomposition of methane-water mixtures in the presence of apatite.This high frequence electric discharge induced the formation of a complex reaction product,containing alcohols,aldehydes,ketones and acids.

The results suggested that aldehydes (especially formaldehyde) take part in the condensation reaction (at alkaline pH values) yielding sugars,which in the presence of the apatite participate in the phosphorylation process yielding phosphoric esters of monosaccharides.

The simulation of the abiotic syntheses supposed to have taken place on the earth from a small number of simple compounds (methane,ammonia,water) under the action of different forms of energy represents an important investigation field as part of the studies devoted to the origin of life.Precursors of living matter were thus obtained,such as aminoacids,sugars, purine and pyrimidine bases,porphyrins etc.

In the present paper the abiotic synthesis of some phosphoric esters of monosaccharides which played an important role in the process of vegetal kingdom occurence is reported.These compounds participate to the photosynthesis (glycerolaldehyde-3-phosphate,ri-

Y. Wolman (ed.), Origin of Life, 165–172.

bose-5-phosphate),metabolism of starch and of other polysaccharides (glucose-1-phosphate,glucose-6-phosfate,fructose-6-phosphate) as well as to the respiration (Krebs cycle) and fermentation processes.

Some phosphoric esters obtained by abiotic phosphorylation reactions are listed in Table 1.

Table 1 Phosphoric esters obtained by phosphorylation abiotic syntheses

Phosphoric ester	Obtaining conditions	Literature
D-glucose-1-phosphate	D-glucose + orthophosphate + cyanogen	/1/
Saccharose-6'-phosphate	Saccharose-2,3,4,6,1',3',4'-heptaacetate-cyanoethylphosphate + DCC in pyridine	/2/
Phosphorilated derivatives of D-fruc-	D-fructose + orthophosphate + cyanogen	/3/
Phosphorilated uridine	Uridine + ammonium oxalate + hydroxyapatite	/4/

The experimental model of our studies,reffered to as „the cold model" employs the primary atmosphere composed of methane,ammonia and water and the high-frequency electric discharge as an energy source.The reaction products (simple molecules and activated species) are cooled and retained on the ice surface, where subsequent transformations take place.The ice is the first matrix which contributed to the accumulation and evolution of the initial compounds protecting them against destructive actions /5/.

The role played by some rocks and clays,when came into contact with the primary ocean,must also be taken into account.Due to their adsorptive and catalytical properties they contributed to the accumulation of some compounds,initiation of some reactions and introduction of new chemical elements necessary to life.At the same time they constituted the structural matrix for perfecting some macromolecular formations /6/.In case of the „cold model" the apatite was also used as a phosphorus source /7/.

The experimental equipment for the decomposition

of the methane-water mixtures in cold plasma is illustrated in Fig.1.

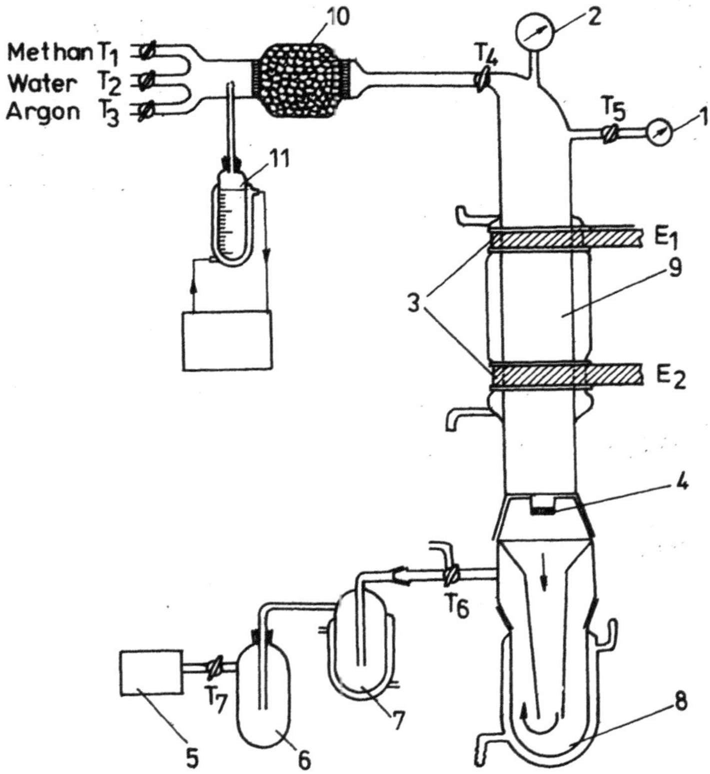

Figure 1. Experimental equipment.
1-Mano-vacuummeter; 2-Vacuummeter; 3-Isolating teflon rings; 4-Frite for catalyst; 5-Vacuum pumpe; 6,7,8-Collecting vessel for reaction products; 9-Reaction space; 10-Room for gases mixture; 11-Vials with water; T_1-T_7-Stopcocks; E_1,E_2-High frequency electrodes.

After leaving the discharge zone the reaction products are collected on either apatite or other phosphorus compounds (apatite carbonate,sodium pirophosphate) in a glass vial cooled at -170°C.The apatite was subjected previously to the extraction in water,in a Soxlet apparatus for 48 hours,followed by extraction in carbon tetrachloride and finally dried and irradiated with UV light for 24 hours.The third principal component,ammonia,was introduced into the collecting vessel as solutions of different concentrations.After synthesis,the collecting vessel was brought to the room temperature and then allowed to stay exposed to the solar radiations for 10-12 days. At the same time,imediatlly after thawing,some samples were irradiated with UV light from a bacteriostatic lamp of 500 W,with a wide spectrum.The reaction products were identified by means of thin-layer chromatography as well as of gas-chromatography.By thin-layer chromatography,the detection of monosaccharides and disaccharides as intermediate products in obtaining the phosphoric esters was followed. Polygram Sil G (Machery-Negel Co.) plates covered with silica gel and the n-butanol - methylacetate - isopropanol - acetic acid - water (35:100:60) mixture as eluent were used.The plates were developed with aniline - diphenylamine - phosphoric acid (5:5:1) mixture /8/.The same compounds were gas-chromatographed and previously sililated according to the literature method /9/.The monosaccharides and disaccharides were separated by means of a Carlo-Erba GV chromatograph with a column of SE-30 (5%) deposited on silanized Chromosorb W under the following conditions: $t_{col.}$ = 210°C (for monosacharides) and 250°C (for disaccharides).The phosphoric esters of the monosaccharides were separated by thin-layer chromatography with plates of the same type using n-butanol - acetic acid - water (74:19:50) mixture as eluent and the ammonium molibdate solution - conc. hydrogen chloride - perchloric acid - water (25:2:5:68) mixture as developer.After drying at 110°C the plates were introduced into a H_2S stream freshly generated /8/.The phosphoric esters were sililated by the method described by W.W.Wells and cowork./10/.The column of SE 30 (5%) deposited on Gas-Chrom P were used for gas-chromatography.

The decomposition of the methane-water mixtures in cold plasma leads to a great number of compounds resulted by the recombination reactions of the radicals formed.By the gas-chromatographic analysis of the liquid phase resulted after thawing aldehydes, ketones,alcohols and acids were detected (Fig.2)/11-12/.Among these compounds the formic aldehyde plays an essential role in the abiotic synthese of sugars. The formic aldehyde is concentrated by adsorption on the apatite surface participating then to condensation reactions with formation of monosaccharides.The synthesis conditions for monosaccharides were reported in another paper /13/.The sugars identified by gas-chromatography are listed in Table 2.

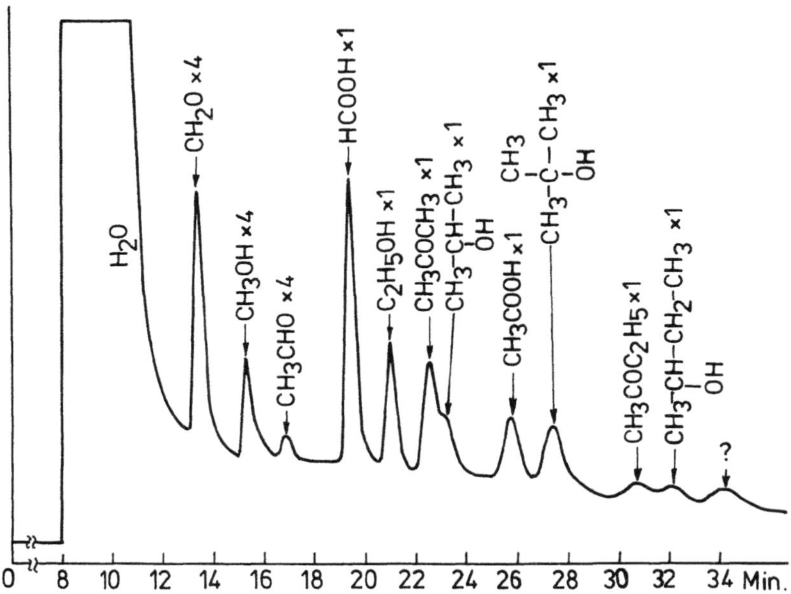

Figure 2. Gas-chromatogram of liquid phase from the cold plasma decomposition of methan-water mixtures.

Table 2. Sugars synthesid by decomposition of the methane-water mixture in cold plasma.

pH	D-Erythrose	D-Xylose	D-Ribose	D-Glucose	D-Manose	D-Galactose
-	-	+	-	-	β+	++
2	-	++	-	α++	β+	-
8	-	+	-	α+	α+, β+	-
10	+	-	+	α+	α+, β+	-

+ present in small quantities; ++ present in big quantities; α and β refer to the two anomere forms.

The sugar formation is noticed to be promoted by a strongly alkaline medium.By subsequent experiments the formation of disaccharides was made evident,the maltose and cellobiose being identified by thin-layer chromatography (Fig.3).The A_1,A_2 and A_3 syntheses were carried only in the presence of apatite,apatite and a 5% ammonia solution and apatite and a 10% ammonia solution, respectively.

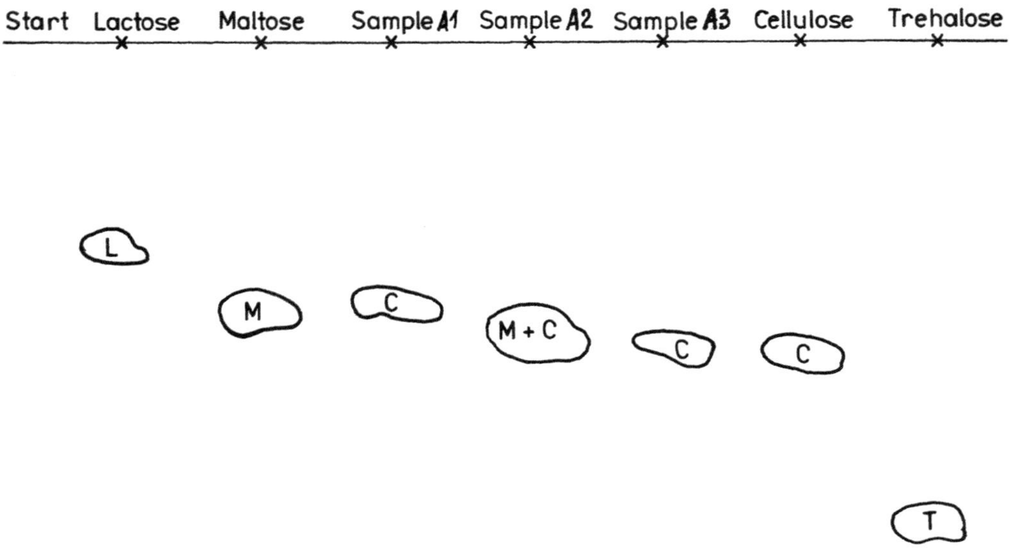

Figure 3. Thin-layer chromatogram of disaccharides synthesized by cold plasma decomposition in methan-water-apatite system at pH = 8, realized with ammonia solution.

It may be noticed that in the A_2 synthesis a greater amount of disaccharides (maltose and cellobiose) formed. By increasing the methane/water mole ratio and under the influence of ammonia which enhances the catalytical activity of apatite, the amount of formic aldehyde increases, resulting in increasing amounts of sugars and an increased number of the reactions between the components within the system.

By using the same reaction system, namely methane -water - apatite some phosphoric esters of monosaccharides were synthesized and detected by gas-chromatography (Table 3). The date in Table 3 show the pentoses to form in a greater amount than hexoses. The esterification takes place preferentially at the primary hydroxy group which is more reactive (2.24 μg/ml glucoso-6-phosphate and 0.70 μg/ml glucoso-1-phosphate). As regards the behaviour toward the UV irradiation, the riboso-5-phosphate and manoso-1-phosphate are noticed to show a rather reduced stability while the formation reaction of hexose-6- and -1-phosphate are accelerated by the UV radiation.

Table 3. Concentration of phosphoric esters of monosaccharides in the reaction mixtures synthesized with the methane - water - apatite system.

Phosphoric ester	Unirradiated UV sample, (μg/ml)	Irradiated UV sample, (μg/ml)	Sample obtained with a modified high frequency generator, (μg/ml)
Riboso-5-phosphate	4.39	1.83	3.09
Manoso-1-phosphate	2.52	traces	traces
Glucoso-6-phosphate	2.24	5.00	2.83
Glucoso-1-phosphate	0.70	1.65	1.97

The frequency and the power of the high-frequency generator influence directly upon the concentration of phosphoric esters in the reaction mixture.An increase in these two characteristics of the generator results in a decrease in the pentose concentration with an increase in the hexose concentration.

Witth the „cold model" proposed by us we obtained from the components of primary atmosphere,in the presence of the above mentioned energy sources, compounds whose structural complexity increases continously from precursors to biomonomers.

REFERENCES

1.Degani Ch., Halman M., J.Chem.Soc., C, 1971, 1459, (1971).
2.Buchanan J.E., Cummerson D.A.,Turner D.M., Carbohyd. Res., 21, 283 (1972).
3.Degani Ch., Kewatsuji M., Halman M., J.Mol.Evol., 6,51 (1975).
4.Schwartz A.W., Biochim.Biophys.Acta, 281, 477 (1972).
5.Simionescu C.I., Mora R., Simionescu B.C., Bioelectrochem.Bioenerg., 5, 1 (1978).
6.Burton F.G., Neuman M.W., Neuman W.F., Curr.Mod. Biol., 3, 20 (1969).

7.Schwartz A.W., Molecular Evolution: Prebiological and Biological, Rohlfing D.L. and Oparin A.I. eds., Plenum, New York 1971, p.129.
8.Tamaş V.P., Iohan F., Cromatografie in strat subţire, Ed.Tehnică, Bucureşti, 1971, p.240.
9.Sweeley C.C., Bentley R., Makita M., Wells W.W., J.Am.Chem.Soc., 85, 2497 (1963).
10.Wells W.W., Katagi T., Bentley R., Sweeley C.C., Biochim.Biophys.Acta, 82, 408 (1964).
11.Simionescu C.I., Dumitriu S., Popa V.I., Bulacovschi V., Simionescu B.C., Rev.Roum.Chim., 23, 89 (1976).
12.Simionescu C.I., Dumitriu S., Popa V.I., Z.Naturforsch., 31b, 466 (1976).
13.Simionescu C.I., Dumitriu S., Bulacovschi V., Popa V.I., Cellulose Chem.Technol., 12, 143 (1978).

SYNTHESIS AND DEGRADATION OF AMINO ACIDS BY CONTACT GLOW DISCHARGE ELECTROLYSIS, A POSSIBLE ROUTE FOR PREBIOTIC FORMATION OF BIO-ORGANIC COMPOUNDS

Kaoru Harada, Jun-ichi Terasawa and Hiromi Gunji

Department of Chemistry, The University of Tsukuba,
Sakura-mura, Niihari-gun, Ibaraki, 305, Japan

Abstract: Oxidative degradations of β- and γ-amino acids by contact glow discharge electrolysis (CGDE) were carried out. It was found that stepwise oxidative reactions proceeded during the reactions. Applying the unusual and powerful oxidation method, several hydroxyamino acids and amino alcohols were oxidized and the time courses of the amino acid formation reactions were studied. Aliphatic amines were also oxidized to the corresponding amino acids.

1. INTRODUCTION

Contact glow discharge electrolysis (CGDE) is a type of chemical reaction carried out by means of an electric discharge between an aqueous solution containing substrates and an electrode in contact with the solution [1,2]. In this reaction, the solution which contain chemical substances is regarded as an electrode. The ions produced by the electric discharge are accelerated by the electric field, and the ions generate radicals in the solution by hitting water molecules. The primarily formed radicals react with the substrates to form reactive species and the activated molecules react with each other to form bio-organic compounds. Therefore, the reaction is quite different from that employing electric discharge between metal electrodes in gases under relatively reduced pressure.

Few CGDE studies on organic compounds have been reported. The CGDE is a new type of chemical reaction in organic compounds and the CGDE could also be considered as a simulation of a lightning striking on the primitive hydrosphere which contains a various organic and inorganic compounds.

Y. Wolman (ed.), Origin of Life, 173–180.

In the previous investigations on the CGDE in this laboratory, several studies on the amino acid formations were carried out. These are: ammination of carboxylic acids [3], carboxylation of aliphatic amines [4], formation of amino acids from ammonium bicarbonate or ammonium formate [5], formation of amino acids from elemental carbon using aqueous ammonia [6], formation of amino acids from unsaturated carboxylic acids [7], carboxylation of 2-pyrrolidone [8], amination of aliphatic nitriles and cyanization of aliphatic amines [9]. The CGDE could also be used for dehydrogenation reaction. Dihydropyrimidines (dihydrouracil, dihydrothymine and dihydroorotic acid) were converted to its corresponding dehydrogenated pyrimidines [10].

2. RESULTS AND DISCUSSION

In this investigation, oxidative degradation of 1) β- and γ-amino acids, 2) hydroxy amino acids, 3) amino alcohols and 4) amines were carried out.

2.1 Oxidation of β- and γ-amino acids

The degradation of β-alanine (β-Ala), γ-amino butyric acid (γ-NH_2-BA) and other β- and γ-amino acids were carried out [11]. From the results obtained, it was found that the α-methylene bound to the carboxyl group was apparently eliminated stepwise by CGDE. In order to study the reaction pathway of the unusual degradation reaction by CGDE, the following hypothesis has been made to explain the degradation. In the aquoeus solution, the primary product of CGDE would be the hydroxyl radical, as in the irradiation of electrone to the aqueous solution. The generated hydroxyl radical would abstract hydrogen from the substrates in the aqueous solutions and the resulting radicals would react further with hydroxyl radicals to form hydroxyl and keto compounds. In the case of β-Ala (1), the α-hydrogen of (1) was eliminated, and the resulting radical (2) react with hydroxyl radical to form isoserine (i-Ser) (3). The i-Ser was further oxidized by a hydroxyl radical to form aminopyruvic acid (5). The compound (5) could easily be oxidized radically to form the final product, glycine (Gly) (6).

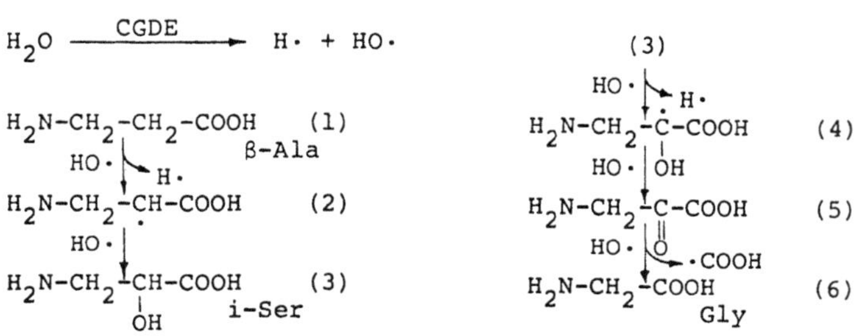

In order to confirm the hypothesis on the degradation of β- and γ-amino acids by CGDE, the time course of the degradation of β-Ala was studied (Fig 1a, 1b) [12]. The peak after glycine in a diagram of amino acid analysis was confirmed as i-Ser by comparison with the authentic amino acid. From the diagram of the time course of β-Ala degradation, i-Ser could be the primary oxidized product, and the amount of i-Ser decreased steadily and the amount of Gly increased depending on the reaction time. The

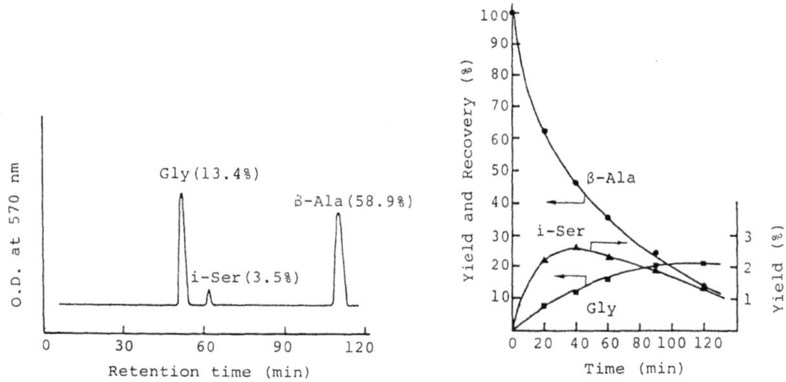

A, Amino acid analysis of the reaction mixture (40 min)

B, Time course of oxidation

Fig. 1 Oxidation of β-Ala by CGDE

time course of the CGDE of β-Ala supports the hypothesis of the oxidative stepwise degradation of β- and γ-amino acids. Fig. 2 shows a diagram of the amino acid analysis of the degradation products of β-NH_2-BA by CGDE [12]. Amino malonic acid (AMA), β-hydroxyaspartic acid (β-OH-Asp), aspartic acid (Asp), serine (Ser), Gly, alanine (Ala), β-amino-α-hydroxybutyric acid (β-NH_2-α-OH-BA) and α-amino-γ-hydroxybutyric acid (β-NH_2-γ-OH-BA) were confirmed and the formation of these amino acids could be explained as shown in the following scheme.

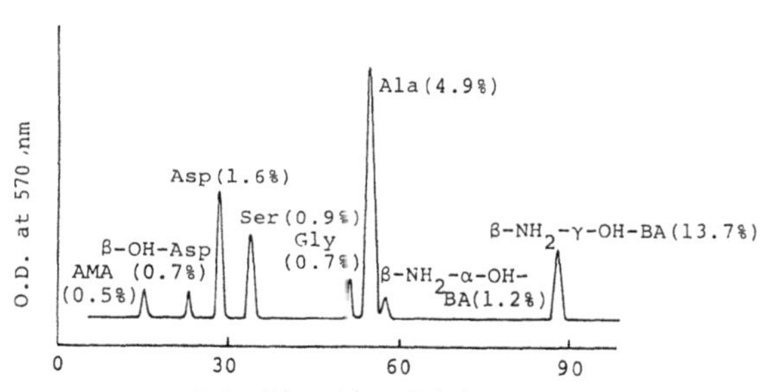

Fig. 2 Oxidation of β-NH_2-BA by CGDE (40 min)

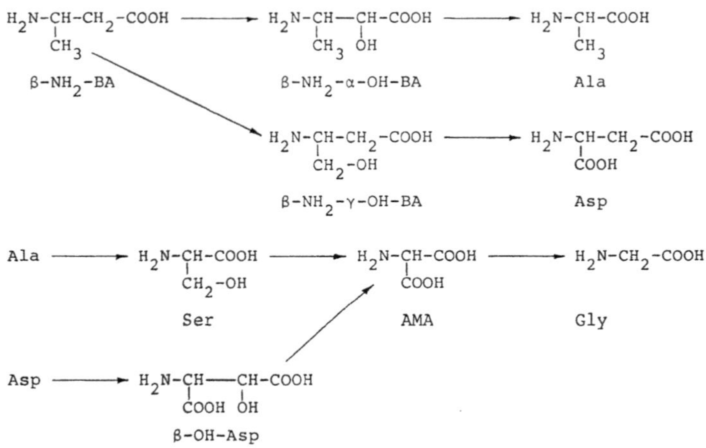

Fig. 3 shows a amino acid analysis of the degradation products of γ-amino butyric acid (γ-NH_2-BA) by CGDE. Gly, i-Ser, α-amino α-hydroxybutyric acid (γ-NH_2-α-OH-BA), γ-amino-β-hydroxybutyric acid (γ-NH_2-β-OH-BA) and β-Ala were identified in the reaction mixture. The formation of these amino acids by degradation of γ-NH_2-BA could also be explained by the following scheme.

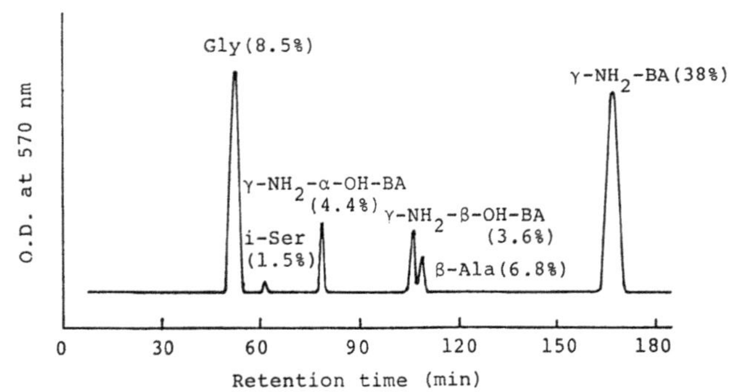

Fig. 3 Oxidation of γ-NH_2-BA by CGDE (40 min)

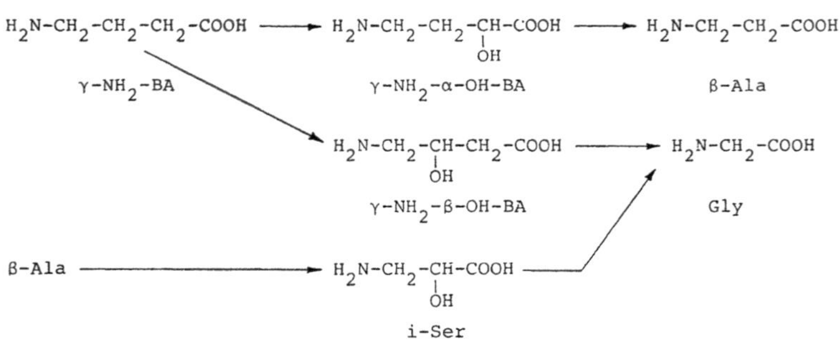

The summerized results indicate that the apparent methylene group elimination of β- and γ-amino acids by CGDE is based on on the oxidative degradation by the hydroxyl radical produced by CGDE. The isolated intermediates show that the substrates were oxidized stepwise to form α-hydroxy acid, α-keto acid and then carboxylic acid. Similarly, oxidation reactions of hydroxy amino acids, Amino alcohols and also aliphatic amines were carried out. Some representative reactions are shown in the following.

2.2 Oxidative degradation of hydroxy amino acids

Fig. 4a shows a time course of oxidation of i-Ser. In this reaction, the yield of Gly, which is a sole amino acid product, reaches up to fifty percent. Fig. 4b shows a time course of CGDE of Ser. The amino acid products are AMA and Gly. Ser was oxidized first to AMA, then AMA gave Gly by decarboxylation. The time course shows that AMA is the primary product of the

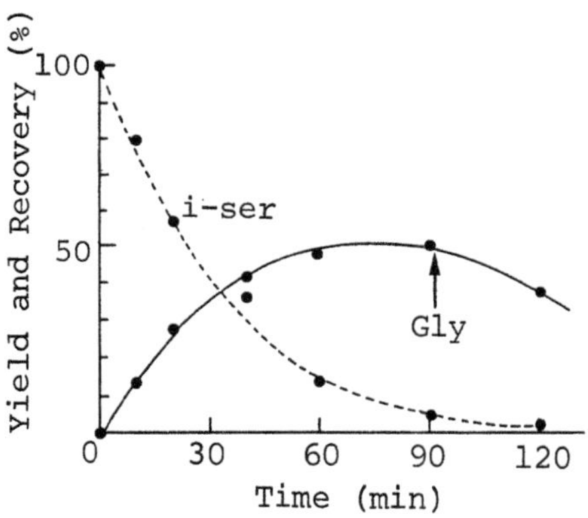

Fig. 4a Oxidation of i-Ser

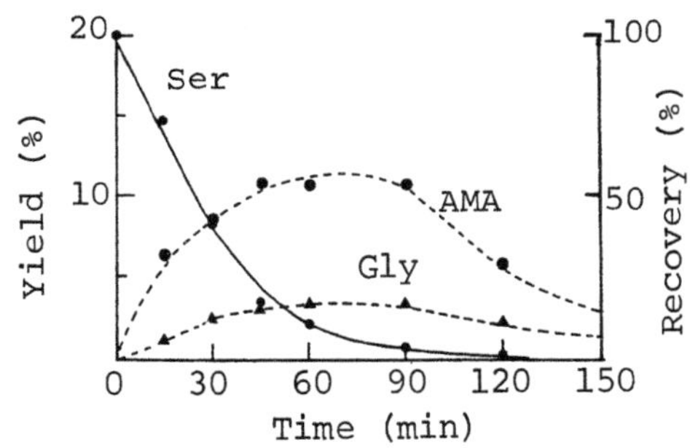

Fig. 4b Oxidation of Ser

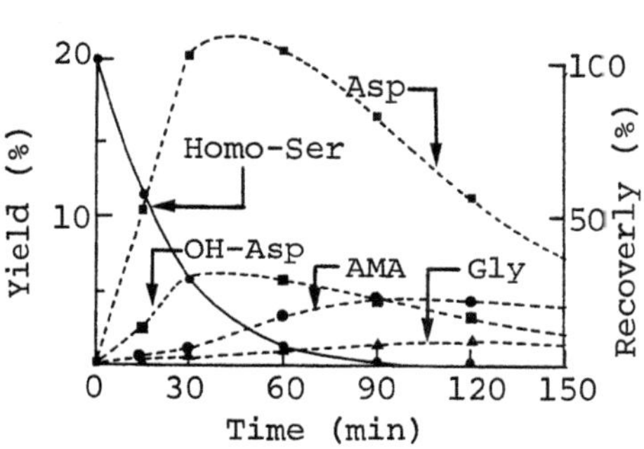

Fig. 5a Oxidation of Homo-Ser

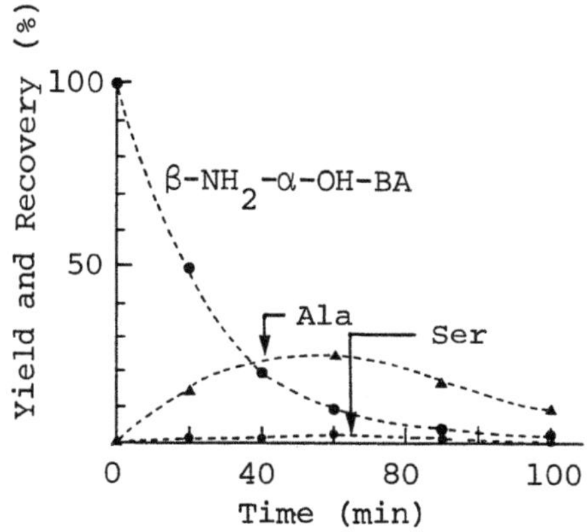

Fig. 5b Oxidation of β-NH_2-α-OH-BA

oxidative dgradation. Fig. 5a shows a time course of oxidation of homoserine (Homo-Ser). The primarily oxidized product is Asp. The yield of Asp reachs up to 20%. The Asp was oxidized to hydroxy aspartic acid (OH-Asp), which was further oxidized to AMA and then gave Gly. The time couse clearly shows a successive oxidative degradation process. Fig. 5b shows a time course of β-amino-α-hydroxybutyric acid (β-NH_2-α-OH-BA). The yield of Ala reachs up to 25%. Small amount of Ser which derived from Ala was observed. Summerized results of the oxidation of several hydroxy amino acids by CGDE are show in Table 1.

Table 1 Oxidation of hydroxy amino acids by CGDE

Substrate	Products (1hr) Yield (%)	Recovery (%)	Maximum Yield (%)
CH_2-CH-COOH NH_2 OH	Gly(48)	13	Gly(51)
CH_3-CH—CH-COOH NH_2 OH	Ala(24), Ser(2.5)	9.5	Ala(24)
CH_2-C(CH_3)COOH NH_2 OH	Gly(22)	30	Gly(27)
CH_2-CH-CH_2-COOH NH_2 OH	Gly(24)	24	Gly(31)
CH_2-CH_2-CH-COOH NH_2 OH	β-Ala(17), Gly(16) i-Ser(2.5)	13	Gly(23)
CH_2-CH-COOH OH NH_2	AMA(10), Gly(3.0)	11	AMA(11)
CH_2-CH_2-CH-COOH OH NH_2	Asp(20), AMA(3.1) OH-Asp(5.4)	6.1	Asp(20)
CH_3-CH-CH-COOH OH NH_2	AMA(4.1), Gly(2.9)	9.4	AMA(4.1)

2.3 Oxidative degradation of aminoalcohols

Fig. 6a shows a time course of oxidation of β-amino ethanol. The yield of Gly reachs up to 34% when the aqueous solution is acidic (pH 2.7). However, the yield of Gly decreases steadily depending on the increase of the pH value. The yield of Gly down to a few % at pH 12. This indicates that the β-position from the charged nitrogen atom in the substrate is prefered for radical formation under acidic conditions. The results obtained in the oxidation of amino alcohols are similar to those obtained in the orientation of carboxylation of aliphatic amines [13]. Fig. 6b is a time course of oxidation of γ-aminopropanol. The primary product is β-Ala and the secondary product i-Ser is formed by oxidation of β-Ala. Gly could be formed from i-Ser and also be derived from direct oxidation of β-carbon of γ-aminopropanol. Fig. 6c shows a time course of oxidation of β-aminoisopropanol. The figure shows again that the yield of Gly is considerably higher (>30%) in acidic conditions (pH 3) than that (~3%) in basic conditions (pH 12).

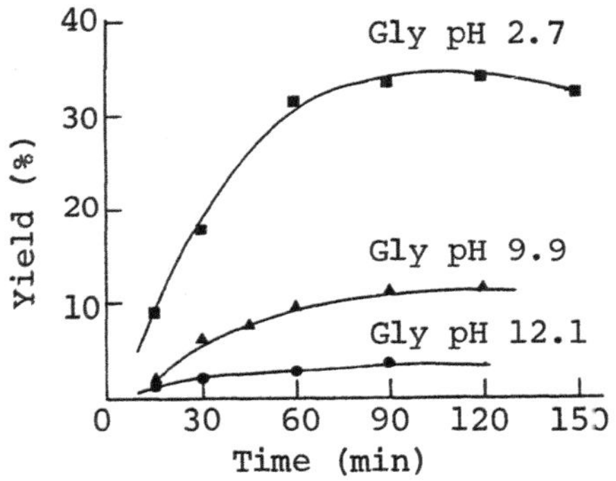

Fig. 6a Oxidation of β-amino ethanol

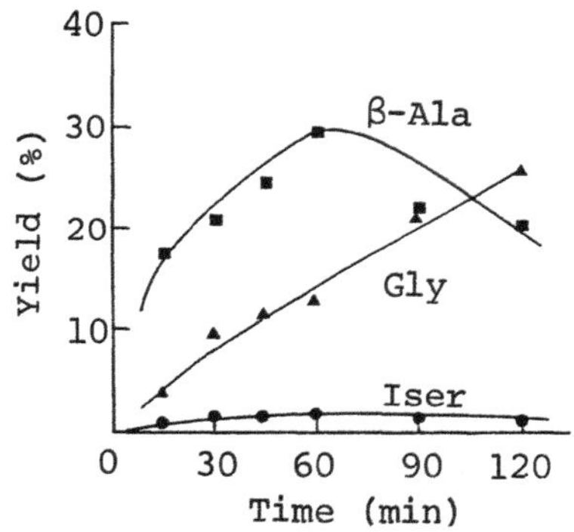

Fig. 6b Oxidation of γ-aminopropanol

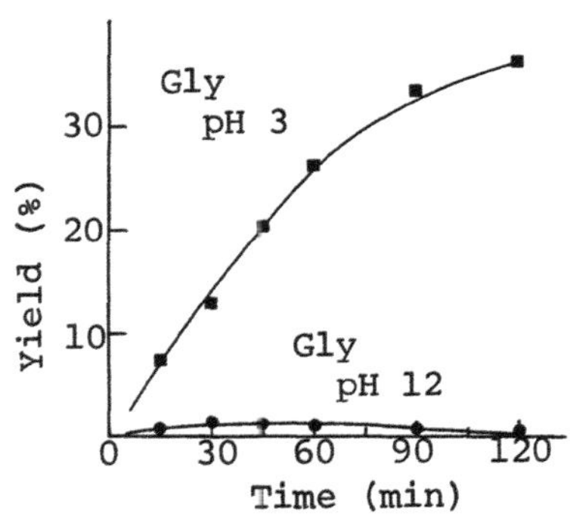

Fig. 6c Oxidation of β-aminoisopropanol

2.4 Oxidation of aliphatic amines by CGDE

Aliphatic amines could be oxidized easily to form amino acids. Fig. 7 shows a time course of oxidation of ethylamine. In acidic conditions (pH 3) the yield of Gly reachs up to 20%, however the yield of Gly in higher pH (pH 12) is low (~1%). This also indicates the selectivity of radical formations in the substrate depending on the pH value of the reaction mixture [12].

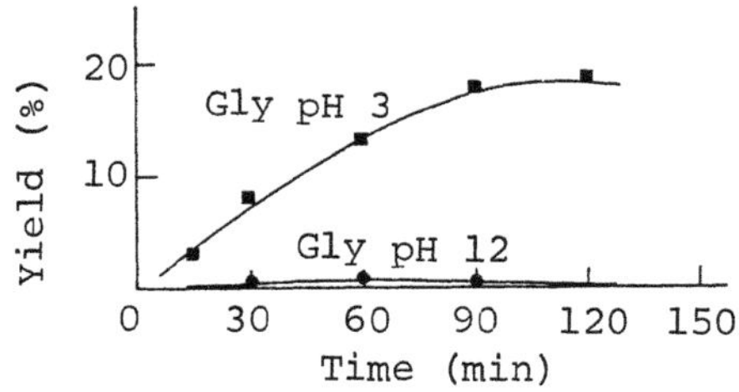

Fig. 7 Oxidation of ethylamine

In the present study, oxidations or oxidative degradations of β- and γ-amino acids, hydroxy amino acids, amino alcohols and aliphatic amines by CGDE are investigated. The CGDE in an aqueous solution is a new effective reaction and is an extreamly strong and clean oxidation process. It is possible to assume that such a process together with the other unique properties of the CGDE [3-13] could play an important role in the chemical evolutionary processes in the primitive hydrosphere to form various bio-organic compounds.

REFERENCES

1) Hickling, A. and Ingram, M.D., J. Electroanal. Chem., 8, 65 (1964).
2) Hickling, A., "Modern Aspects of Electro Chemistry", Bockris, J.O. and Conway, B.E., ed. vol 6, Plenum Press, New York (1971) p. 329.
3) Harada, K. and Iwasaki, T., Nature, 250, 426 (1974).
4) Harada, K. and Iwasaki, T., Chem. Letters, 185 (1975).
5) Harada, K. and Suzuki, S., Naturwiss., 64, 484 (1977).
6) Harada, K. and Suzuki, S., Nature, 266, 275 (1977).
7) Harada, K., Suzuki, S. and Ishida, H., Experientia, 34, 17 (1978).
8) Harada, K., Suzuki, S., Ishida, H., Matsuyama, M. and Tamura M., "Origin of Life", Noda, H., ed. Center for Academic Publication Japan, Tokyo (1978) p. 141.
9) Harada, K., Suzuki, S. and Ishida, H., BioSyztems, 10, 247 (1978).
10) Harada, K. and Terasawa, J., Naturwiss., 65, 259 (1978).
11) Suzuki, S., Tamura, M., Terasawa, J. and Harada, K., Bioorg. Chem., 7, 111 (1978).
12) Harada, K., Chem. Letters, 441 (1980).
13) Terasawa, J. and Harada, K., Chem. Letters, 73 (1980).

GENESIS OF AMINO ACIDS IN THE PRIMEVAL SEA: FORMATION OF AMINO ACIDS FROM SUGARS AND AMMONIA IN A MODIFIED SEA MEDIUM

Hiroshi Yanagawa, Yukiko Kobayashi and Fujio Egami

Mitsubishi-Kasei Institute of Life Sciences,
11 Minamiooya, Machida, Tokyo 194, Japan

Abstract: Sugars such as glycolaldehyde, glyceraldehyde, erythrose, ribose, and glucose were heated with ammonia in a modified sea medium. Amino acids such as glycine, alanine, serine, threonine, aspartic acid, and glutamic acid were obtained from the reaction mixture. This provides the basis for a model of the origin of amino acids in the primeval sea.

It is widely accepted that the origin of life on the Earth took place in the primeval sea more than three billion years ago and the essential components of a living organism must have been already formed abiogenetically and accumulated in the sea by that time. Thus, the fundamental characteristics of organisms were acquired under primeval sea conditions.

Egami has recently found that a close correlation exists between the concentration of transition elements in contemporary sea water and their biological behavior and considers that transition elements, relatively abundant in sea water, such as molybdenum, iron, and zinc must have played important roles in the chemical evolution in the primeval sea (1). On the basis of this idea, we have been conducting studies on the formation of biomolecules in a modified sea medium. This modified sea medium was designed to simulate, experimentally, the chemical evolution in the primeval sea. The concentration of sodium chloride is lower and the concentration of the six transition elements, iron, molybdenum, zinc, copper, cobalt, and manganese is 1,000-100,000 times greater than that of present day sea water. In the course of this research, we recently found that amino acids such as glycine, alanine, β-alanine, serine, threonine, aspartic acid,

Y. Wolman (ed.), Origin of Life, 181–187.

and glutamic acid were produced from formaldehyde and hydroxylamine in a modified sea medium (2,3). From this experimental result and the fact that certain sugar-like substances such as formose (4) are easily produced from formaldehyde under prebiotic conditions (5,6), we came to consider that amino acids may also be formed from sugars in a modified sea medium. In a series of experiments on chemical evolution in the primeval sea, we have first attempted to synthesize amino acids from reactions of various sugars such as triose, tetrose, pentose, hexose, and heptose with hydroxylamine in modified sea medium.

When formaldehyde, glycolaldehyde, DL-glyceraldehyde, dihydroxyacetone, D-erythrose, D-ribose, L-arabinose, D-fructose, D-galactose, D-glucose, D-mannose, and D-sedoheptulose were heated with hydroxylamine in a modified sea medium at 105°C for a week, glycine, alanine, serine, threonine, aspartic acid, and

TABLE 1

Formation of Glycine and Alanine from Sugars and Hydroxylamine in a Modified Sea Medium

Sugars	Amounts of products(μmol/ml)[2]	
	Glycine	Alanine
Formaldehyde	0.72	trace
Glycolaldehyde	3.91	0.47
DL-Glyceraldehyde	1.00	5.64
Dihydroxyacetone	trace	3.89
D-Erythrose	0.97	trace
D-Ribose	0.69	1.08
L-Arabinose	1.07	0.87
D-Fructose	0.19	0.05
D-Galactose	0.87	0.63
D-Glucose	3.40	0.16
D-Mannose	0.30	0.16
D-Sedoheptulose	0.24	trace
Sucrose	0.29	0.18
Cellulose	trace	0
Starch	0.11	trace

[1]The modified sea medium contained 0.01 M (each) $MgSO_4$, $CaCl_2$, and K_2HPO_4, and 0.1 mM (each) $Fe(NO_3)_3$, Na_2MoO_4, $ZnCl_2$, $Cu(NO_3)_2$, $CoCl_2$, and $MnCl_2$. The pH of the medium was adjusted to 6.0 with KOH. Five ml of the reaction mixture containing 0.5 M sugar and 0.25 M hydroxylamine sulfate, and the modified sea medium were put into a pyrex tube, degassed, sealed, and kept at 105°C for a week in a Dry-Block DB-3H(M & S Instrument Co).

[2]Analysis of amino acids was performed after 6 N HCl hydrolysis on a Hitachi KLA-5 amino acid analyzer. Glycine and alanine were further identified by the gas chromatography-mass spectrometry of their trimethylsilyl derivatives.

glutamic acid were obtained. Of these amino acids, both glycine and alanine were obtained in good yield (Table 1). Among the sugars mentioned above, glucose and glyceraldehyde gave the best yields of glycine and alanine, respectively. Glycine and alanine were also produced from sucrose and starch but only a trace amount of glycine was obtained from cellulose. Glycine was obtained in higher yields from formaldehyde, glycolaldehyde, erythrose, glucose, and sedoheptulose. On the other hand, alanine was obtained in higher yields from DL-glyceraldehyde and dihydroxyacetone.

Moreover, when different C_2-compounds such as glycolaldehyde, glyoxal, glycolic acid, and glyoxylic acid were allowed to react with hydroxylamine in a modified sea medium, glycine was obtained mainly (Table 2). The oxidation of alcohol and formyl groups of the C_2-compound to the carboxyl group appears to be a rate-determining step since the yield of glycine increased in the order of glycolaldehyde, glyoxal, and glyoxylic acid.

TABLE 2
Formation of Glycine and Alanine from C_2-compounds and Hydroxylamine in a Modified Sea Medium.[1]

C_2-compounds		Amounts of products (μmol/ml)	
		Glycine	Alanine
Glycolaldehyde	CHO–CH_2OH	3.91	0.47
Glyoxal	CHO–CHO	4.64	0.60
Glycolic acid	COOH–CH_2OH	1.11	trace
Glyoxylic acid	CHO–COOH	22.64	0.84
Methylglyoxal	CH_3–C=O–CHO	0.08	4.33

[1]Reaction mixture (5 ml) of each 0.5 M C_2-compound and 0.25 M hydroxylamine sulfate in a modified sea medium were put into a pyrex tube, degassed, sealed, and maintained at 105°C for a week in a Dry-Block. Other experimental details are as described in Table 1.

Oró has proposed a possible mechanism for the formation of glycine from formaldehyde and hydroxylamine (7). First, oxime is formed from formaldehyde and hydroxylamine, and is then converted into hydrogen cyanide. In the next step, glycinenitrile was produced from formaldehyde, hydrogen cyanide, and ammonia by Strecker's reaction and finally glycine was formed. In our system, glycine might also have been formed from formaldehyde and hydroxylamine by Strecker's reaction and in the case of a C_2-compound such as glycolaldehyde, if it follows the same reaction pathway, alanine must be obtained. However, contrary to our expectations, alanine was not obtained from glycolaldehyde and hydroxylamine but glycine was obtained mainly. Thus we can not explain the formation of glycine from glycolaldehyde and hydroxylamine by Strecker's reaction. We assume a reaction pathway of the formation of glycine and alanine from sugars instead of Strecker's reaction; that is, a C_2-compound such as

TABLE 3
Formation of Glycine and Alanine from Sugars and Ammonia in a Modified Sea Medium

Sugars	Amounts of products(μmol/ml)	
	Glycine	Alanine
NH_4OH System		
Formaldehyde	0.07	0
Glycolaldehyde	0.69	0.47
DL-Glyceraldehyde	0.63	1.17
D-Erythrose	0.49	0.37
D-Ribose	2.02	2.03
D-Glucose	1.36	1.48
$(NH_4)_2SO_4$ System		
Formaldehyde	0	0
Glycolaldehyde	0.23	0.07
DL-Glyceraldehyde	0.12	0.67
D-Erythrose	0.19	0.12
D-Ribose	0.42	0.19
D-Glucose	0.23	0.27

[1]The reaction mixture (20 ml, pH 12) in an ammonium hydroxide system contained 0.5 M sugar, 4 M ammonium hydroxide, and modified sea medium. The reaction mixture (10 ml) in an ammonium sulfate system contained 0.5 M sugar, 0.25 M ammonium sulfate, 0.5 g calcium carbonate, and modified sea medium and the pH was maintained at about 8 during the reaction. Each reaction mixture was put into an Erlenmeyer flask equipped with a condenser and kept at 105°C for a week in a silicon-oil bath. Other experimental details are as described in Table 1.

glycolaldehyde directly reacts with ammonia to form glycine, and C_3-compounds such as glyceraldehyde and dihydroxyacetone also react with ammonia to form alanine. Ammonia was observed to be easily formed from hydroxylamine in our system. In order to substantiate this hypothesis, we next tried to synthesize amino acids from sugars and ammonia in a modified sea medium. As shown in Table 3, glycine and alanine were produced mainly from reactions of glycolaldehyde, glyceraldehyde, erythrose, ribose, and glucose with ammonia at 105°C in a modified sea medium (pH 12). However, it was possible to obtain a trace amount of glycine alone from formaldehyde and ammonia. Moreover, when

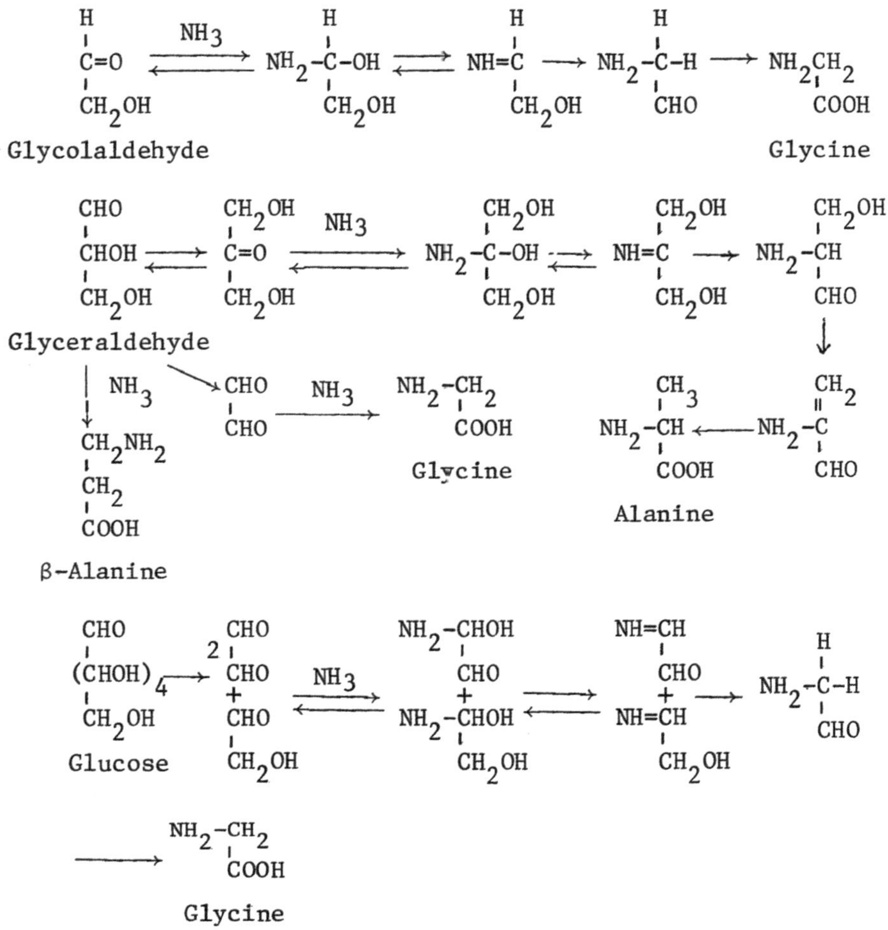

Fig. 1 A possible mechanism for the formation of amino acids from the reaction of glycolaldehyde, glyceraldehyde, and glucose with ammonia in a modified sea medium.

glycolaldehyde, glyceraldehyde, erythrose, ribose, and glucose were heated with ammonium sulfate at 105°C in a modified sea medium (pH 8), glycine and alanine were also produced mainly, but no amino acid was obtained from formaldehyde. Our hypothesis concerning the formation of glycine and alanine from sugars and ammonia is thus supported by these experimental results. On the basis of these results, we propose a possible mechanism for the formation of glycine and alanine from sugars and ammonia (Fig. 1). Glycolaldehyde reacts with ammonia to give an amino alcohol, followed by dehydration, resulting in an imine. The imine is isomerized to an amino aldehyde and finally oxidized to glycine. Glyceraldehyde is easily epimerized to dihydroxyacetone. Dihydroxyacetone first reacts with ammonia to form an imine by a pathway similar to that described above. The imine is further converted into an amino aldehyde, followed by dehydration, resulting in alanine. If glyceraldehyde is partially degraded to a C_2-compound such as glyoxal, it may be converted into glycine by a similar pathway. Erythrose, ribose, and glucose are degraded to C_2- and/or C_3-compounds such as glycolaldehyde and glyceraldehyde, which are then converted into glycine and alanine, respectively. Molybdenum, in a modified sea medium, showed a stimulatory effect on the degradation of sugars. Moreover, this reaction mechanism was supported by the finding that alanine is preferentially produced from methylglyoxal (Table 2).

Finally, we describe a possible scheme for the formation of amino acids from sugars and ammonia in the primeval sea. At the initial stage of chemical evolution, formaldehyde must have been formed from carbon monoxide (8), carbon dioxide (9), and water by UV irradiation and accumulated in the primeval sea. It is well known that formaldehyde reacts by the aldol condensation to form its oligomers, collectively designated as formose (4). Ammonia must have been formed from nitrogen and water by UV irradiation and accumulated in the primeval sea (10). The resulting C_2-compounds such as glycolaldehyde in formose reacted with ammonia to form glycine. The resulting C_3-compounds such as glyceraldehyde reacted with ammonia to form alanine. Moreover, glycine must have been modified with formose, forming various amino acids, which were then converted into proteins in the primeval sea (11).

REFERENCES

1. Egami, F., J. Mol. Evol. 4, 113 (1974)
2. Hatanaka, H. and Egami, F., Bull. Chem. Soc. Japan 50, 1147 (1977)
3. Kamaluddin, Yanagawa, H. and Egami, F., J. Biochem. 85, 1503 (1979)
4. Butlerow, A., Annalen 120, 295 (1861)
5. Gabel, N.W. and Ponnamperuma, C., Nature 216, 453 (1967)

6. Reid, C. and Orgel, L.E., Nature 216, 455 (1967)
7. Oró, J., Kimball, A., Fritx, R. and Master, F., Arch. Biochem. Biophys. 85, 115 (1959)
8. Bar-num, A. and Hartman, H., Origins of Life 9, 93 (1978)
9. Groth, W. and Suess, H., Naturwiss. 26, 77 (1938)
10. Getoff, N., Nature 210, 940 (1966)
11. Yanagawa, H., Kobayashi, Y. and Egami, F., J. Biochem. 87, 855 (1980)

PORPHYRIN-LIKE COMPOUNDS GENESIS UNDER SIMULATED ABIOTIC CONDITIONS

C.I.Simionescu, B.C.Simionescu, M.Leancă, C.Ananiescu and R.Mora

Polytechnic Institute, 6600 Jassy, Romania

Investigations into the synthesis of porphyrins, carried out according to "the cold theory" of the appearance and evolution of the first living forms, showed that porphyrin-like pigments are formed in abiotic simulated conditions. Parallel studies on the classical organic synthesis of porphyrins indicated that pyrrole and formaldehyde were involved in their abiogenesis.

The ready formation of these compounds - and especially that of their metallic chelates - in abiotic conditions is significant, suggesting the possibility of their appearance and intervention during the early stage of chemical evolution.

In the general context of the problem concerning the origin of life, attempts to elucidate the determinant stages involving the self-assembling and the evolution of the first living forms implied many investigations into the abiogenic formation of biologically important compounds.

In some previously published papers we presented certain aspects of an original conception ("the cold theory") on the synthesis of living matter precursors and the emergence of life (1,2). Experimental data, obtained according to the mentioned theory, confirmed the starting theoretical hypotheses, and abiotically synthesized simple organic compounds (3), amino acids (4), purinic and pirimidinic bases (5), sugars (6),

Y. Wolman (ed.), Origin of Life, 189–196.

hydrocarbons (3), fatty acids (7), nucleosides (8), polymeric structures (peptide-like, lipid-like, polysaccharide-like) (1,9) and porphyrin-like pigments (10) were identified in the reaction products.

Although the existence of porphyrins is not essential for the emergence of the first living systems, their presence and especially the presence of their metallic chelates is considered to be necessary for the appearance of photosynthesis, with important implications as to the transition of the Earth primitive reductive atmosphere to an oxidative one. Initially, they could perform effective catalyses of certain reactions; their incorporation into the early organisms determined more and more effective participation, as sensitizers, of photoreactions to obtain chemical energy. The processes providing the useful energy for organisms developed in time, evolving into contemporary bioenergetic processes. Nowadays, porphyrins are involved in two essential processes of life, with major bioenergetical functions - photosynthesis and cellular respiration (11).

In his classical experiment concerning the prebiotical synthesis of certain organic compounds, Miller mentioned the appearance of some red substances with absorption peaks in the 350 - 540 nm range which were not taken as porphyrins due to the lack of fluorescence. Later on, Hodgson and Ponnamperuma (12) identified porphyrin-like pigments in experiments simulating the primaeval reductive atmosphere. The easy formation of porphyrins was also proved by their synthesis from abiotic synthetizable precursors, with the starting reactants benzaldehyde and pyrrole (13) or formaldehyde and pyrrole (14).

The experiments we performed were carried out under flow conditions, using an apparatus described elsewhere (15). Different gaseous mixtures - CH_4 (2 - - 7 Torr), NH_3 (2 - 7 Torr), H_2O (0.03 - 0.005 ml/h) simulating the primaeval atmosphere, were exposed to electrical discharges generated by two external electrodes coupled to a high frequency generator (electrical tension between the electrodes - 12 kV, discharge frequency - 1.66 MHz, power - 40 W). The reaction products were collected in a vessel cooled by liquid nitrogen. In some experiments, water soluble salts of metallic cations ($Zn(CH_3COO)_2$, $Cu(CH_3COO)_2 \cdot H_2O$) were added in the collecting vessel. The reaction products were gradually heated up to the room temperature, reduced to dryness, extracted with benzene - methanol 9/1 v/v, and the soluble part was transferred to n-heptane and chromatographed on silica gel. Chromatographic fractions were obtained with n-heptane, n-heptane - benzene 1/1 v/v, benzene, benzene - methanol 99/1, 98/2, 95/5, 90/10 v/v and methanol. In order to satisfy the analytical criteria previously established for the identification of compounds as porphyrins (15,16), the eluates were examined by absorption and fluorescence spectrophotometry, metal complexing, demetallation, remetallation and solvent partition.

A second set of experiments was performed starting from pyrrole (Py) and formaldehyde (FA) which were condensed at room temperature, in aqueous solutions, in presence or absence of metallic salts (0.4 ml Py, 1.2 ml FA 40% in vol., in water, in 100 ml distilled water, with or without 40 mg metal). In some cases a peridotit rock (containing 9.31% Fe_2O_3, 2.90% FeO, 33.17% MgO, etc.) was introduced instead of the metallic salt.

Taking into account that C_4 - C_{16} monocarboxylic acids were identified in the reaction products yielded by abiotic syntheses (7) and due to their possible role as components of micellar systems, some experiments were performed with Py, FA, peridotit and potassium tetradecanoate introduced in concentrations amounting to twice critical micelle concentration. The formation of tetrapyrrolic compounds was followed by absorption spectrophotometry.

Both before and after chromatography, the organic extracts of the discharge pigments obtained in presence or absence of metallic salts were examined for the absorption features characteristic of porphyrins. Before chromatography no specific spectral features were detected; following the column chromatography, absorption signals in the specific region (380 - 420 nm) were observed for different eluates (Table 1). The absorption spectra of abiotically synthesized porphyrins and zinc-porphyrins are presented in Figure 1.

Chromatographic behaviour of the discharge pigments led to the conclusion that the bulk of porphyrins appears in benzene - methanol 99/1 v/v eluate. The presence of water soluble salts of metallic cations considerably enhanced the free-base porphyrin yields, determining, at the same time, the formation in a prevalent amount of the corresponding metalloporphyrin.

TABLE 1
Characteristic Absorption Features of Different Eluates.

eluate		absorption (nm)
n-heptane		380
n-heptane - benzene 1/1 v/v		380
benzene		390, 435
benzene - methanol	99/1 v/v	390, 505, 545, 590, 615 (390*, 505*, 570*)
	98/2 v/v	390, 435
	95/5 v/v	435 (sometimes)
	90/10 v/v	-
methanol		-

*characteristic absorption features of zinc-porphyrin complex

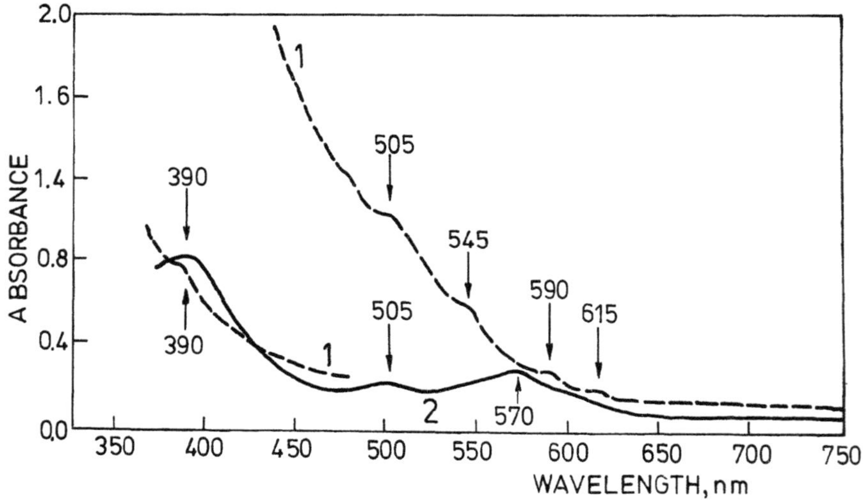

Figure 1. Absorption spectra of abiotically synthesized porphyrins (1) and of zinc-porphyrin complex (2).

Fluorescence spectra of different eluates are presented in Figure 2. The excitation and the emission peak variations, in terms of chromatographic fractions and pigment distributions in the eluates, indicated the formation of a mixture of probably very different substituted porphyrin-like compounds. A typical excitation spectrum of zinc-porphyrins is presented in Figure 3.

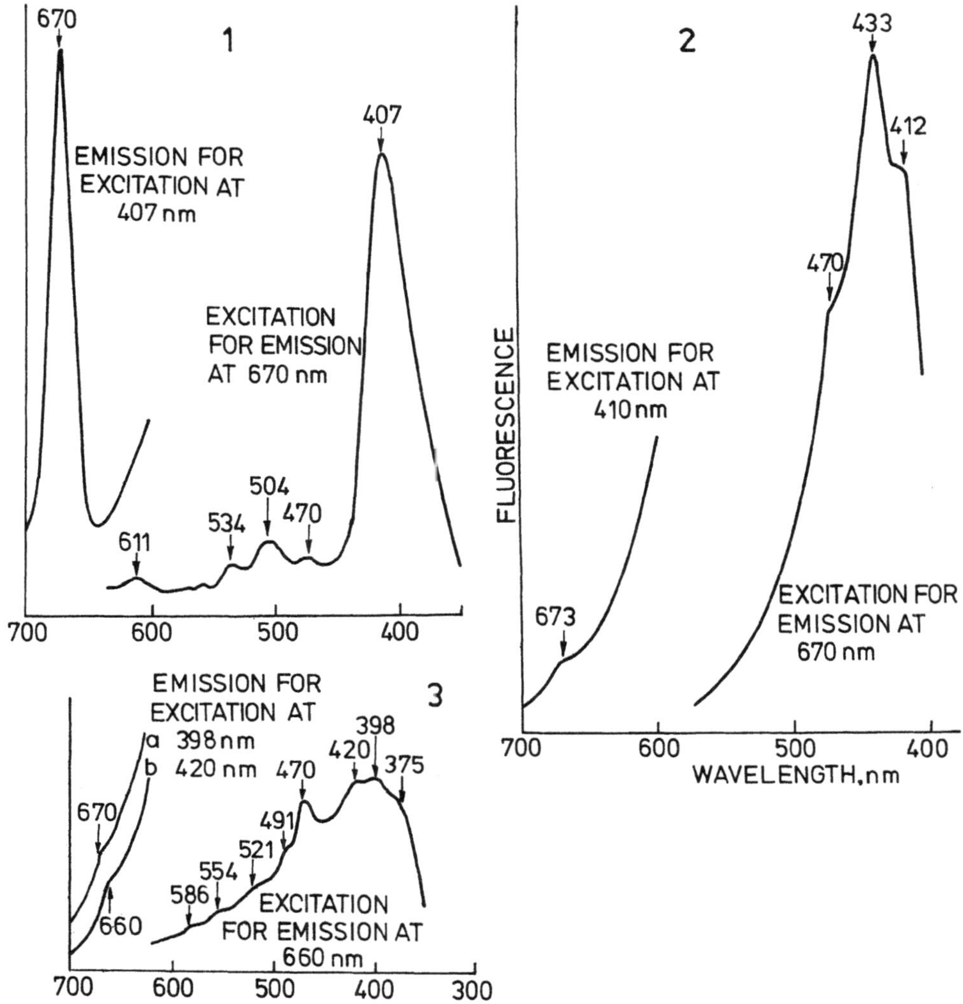

Figure 2. Fluorescence emission and excitation spectra for (1) n-heptane, (2) n-heptane - benzene and (3) benzene eluate.

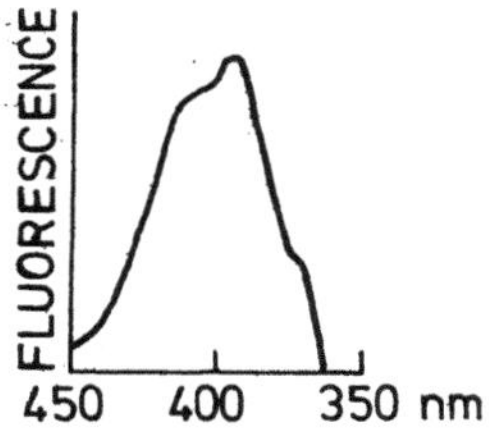

Figure 3. Typical excitation spectrum of zinc-porphyrins generated in plasma reactions.

The free pigments were reacted with zinc and copper acetates, in glacial acetic acid, in order to examine the complexing with diamagnetic and paramagnetic metals. As expected, the complexes with zinc proved to be fluorescent, while those with copper were non-fluorescent (Figure 4, curves 1 - 3).

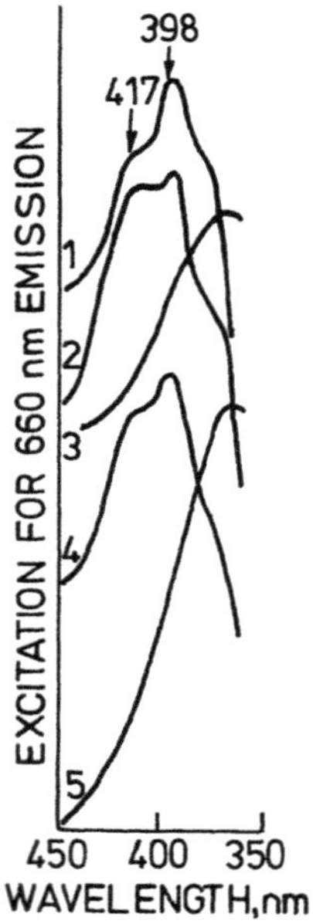

Figure 4. Excitation spectra before (1) and after complexing with zinc (2) and copper (3), after demetallation of copper complex (4) and remetallation with copper (5).

The copper complexes were demetallated with methanesulfonic acid (MSA). The incremental addition of MSA in the spectrofluorimetric cell containing the copper complex led to a gradual appearance of the Soret excitation peak; the spectral characteristics of free-base porphyrins, recorded before metallation (Figure 4, curve 1) and after demetallation (Figure 4, curve 4) appeared to be similar. After a new remetallation with copper acetate (Figure 4, curve 5) the expected loss of fluorescence occurred.

The acid extractibility of the abiotically synthesized pigments was tested by partition between organic solvents and 6N hydrochloric acid. They were readily extracted with aqueous acid solutions and then returned to the organic phase by buffering with sodium acetate.

The experiments performed with Py and FA proved the easy condensing of these intermediates in porphyrinic rings. The metallic cations strongly accelerate the process, leading to the corresponding metalloporphyrins; the presence of the peridotit rock also accelerates the formation of pigments, but in a smaller extent. The existence of both peridotit rock and potassium tetradecanoate in the reaction milieu increased the rate of the reaction as compared with the case when only peridotit was present.

A step by step spectral observation of the progress of the reaction between Py and FA showed the appearance of two absorption bands, at 505 and 435 nm, before the appearance of the Soret band (Figure 5, curve 1). These two peaks diminish in time for the benefit of the absorption bands of porphyrins. The complete separation of tetrapyrrolic compounds obtained both in presence or in absence of metallic salts from the reaction milieu (Figure 5, curve 2) showed that new amounts of porphyrins were formed, the process being preceded by a new diminishing of the 505 and 435 nm bands. It thus appears that intermediate compounds are yielded from Py and FA and that porphyrins arise in time from these compounds.

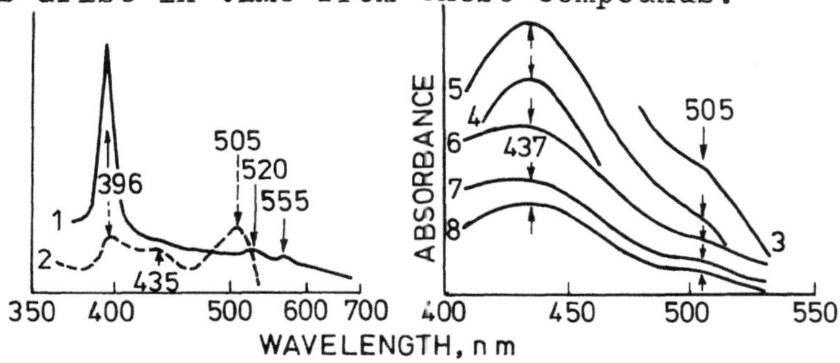

Figure 5. Typical spectrum obtained during the formation of porphyrins from Py and FA with/without metallic salts (1); Cu-porphyrins separated from Py - - FA - Cu acetate synthesis (2); (3) - (8) - organic eluates of discharge pigments: (3) benzene - methanol 99/1 v/v, and (4) benzene, no metal in the synthesis; (5) benzene - methanol 98/2 v/v, Fe in the synthesis; (6) benzene, Mg in the synthesis; (7) benzene, Zn in the synthesis; (8) benzene, Cu in the synthesis.

The reaction between Py and FA was followed in a ^{1}H-NMR spectrometer sample tube, in D_2O. The ^{1}H-NMR spectrum of Py consists in two main signals at 6.1 ppm (protons 3 and 4) and 6.7 ppm (protons 2 and 5). During the reaction, a decrease of the 6.7 ppm signal, assigned to the substitution of 2 and 5 protons, and an appearance of a signal at 5.3 ppm was observed. This signal is assigned to the protons in the structures (a) and (b) which both exist in porphyrins.

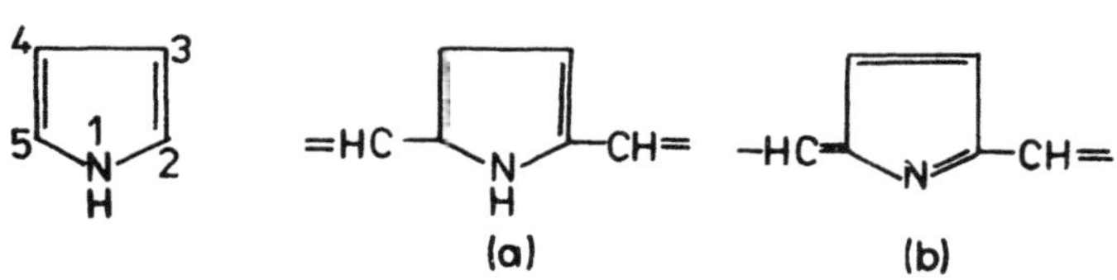

Many chromatographic eluates of the discharge pigments showed absorption shoulders at 435 and/or 505 nm (Figure 5, curves 3 - 8). These absorption could be due to precursors of porphyrins; but, since they also appear in experiments performed with Py and FA, one can consider that abiotic porphyrins are yielded from the same components, which were identified in the discharge products.

REFERENCES

1. Simionescu C.I., Deneş F. and Negulescu I., J. Polym.Sci., Polym.Symp., 64, 281 (1978).
2. Simionescu C.I., Mora R. and Simionescu B.C., Bioelectrochem.Bioenerg., 5, 1 (1978).
3. Simionescu C.I., Dumitriu S., Bulacovschi V., Popa V.I. and Simionescu B.C., Rev.Roum.Chim., 23, 89 (1978).
4. Simionescu C.I., Mora R., Olaru N. and Ioanid E., Compt.rend., Série C, 282, 679 (1976).
5. Simionescu C.I., Lixandru T., Gorea C. and Gorduza V., Compt.rend., Série C, 280, 683 (1975).
6. Simionescu C.I., Dumitriu S., Bulacovschi V. and Popa V.I., Cell.Chem.Technol., 12, 585 (1978).
7. Simionescu B.C., Leancă M., Mora R., Mancaş D. and Simionescu C.I., Rev.Roum.Chim., 24, 1521 (1979).
8. Simionescu C.I., Lixandru T., Gorduza V. and Gorea C., Origins of Life, 8, 237 (1977).
9. Simionescu C.I., Totolin M. and Deneş F., Biosystems, 8, 153 (1976).
10. Simionescu C.I., Simionescu B.C., Mora R. and Leancă M., Origins of Life, 9, 103 (1978).
11. Pullman B., Exobiology, C.Ponnamperuma (Ed)., North-Holland Publ.Co., Amsterdam - London, 1972, 136.
12. Hodgson G.W. and Ponnamperuma C., Proc.Nat.Acad. Sci., 59, 22 (1968).
13. Szutka A., Hazel J.F. and McNabb W.M., Radiation Res., 10, 597 (1959).
14. Szutka A., Nature, 202, 1231 (1964).
15. Simionescu C.I., Simionescu B.C., Mora R., Leancă M. and Ioanid E., Compt.rend., Série C, 284, 743 (1977)
16. Kvenvolden K.A. and Hodgson G.W., Geochim.Cosmochim.Acta, 33, 1195 (1969).

THE ROLE OF ANALYTICAL PROCEDURES IN THE FORMATION OF BIOCHEMICALS FROM EXPERIMENTS SIMULATING THE CHEMICAL EVOLUTION OF PRIMEVAL EARTH

F. STOETZEL, A. SHIMOYAMA, C. PONNAMPERUMA and R. BUVET
Laboratoire d'Energétique Electrochimique et Biochimique
Université Paris Val de Marne
Avenue du Général de Gaulle
F94010 CRETEIL Cédex (France)

Laboratory of Chemical Evolution
Department of Chemistry
University of Maryland
COLLEGE PARK, Maryland (USA)

ABSTRACT

The organic products formed during and at the end of CERES experiments simulating globally the Chemical Evolution of Reactions under Energy Supply on primeval Earth of reducing atmospheres in the presence of condensed water were separated and analysed by rapid ion exchange chromatography without or with previous hydrolysis.

Only small amounts of aminoacids were detected even after more than 100 hours of evolution in solutions heated at 65°C. These quantities of aminoacids were substantially increased by preliminary long hydrolyses with hot concentrated acid. These results confirm that in previously reported experiments, biochemicals were formed mainly during the drastic analytical procedures which followed the simulation experiment proper of the primitive chemical evolution on earth. During these experiments only soluble precursors of these biochemicals were present.

A special run of CERES program (Chemical Evolution of Reactions in primitive aqueous solutions under Energy Supply (1)) was devoted to the study of the influence of analytical procedures on the results obtained with regard to the quantities of aminoacids produced in experiments simulating the primitive evolution of a hypothetically hydrogenated earth periphery.

0.25 mole of CH_4 in the presence of 0.25 mole of NH_3 and of

Y. Wolman (ed.), Origin of Life, 197–199.

0.5 l of 0.1 M NaCl solution were treated in a 10.2 l flask during a 270 hour experiment involving 100 hours of energy supply by spark discharge and 220 hours of heating at 65°C. The hydrogen produced was continuously and slowly evacuated in part through a palladium membrane (2).

Four different treatments of solution samples were performed prior to aminoacids ion exchange chromatography :

a) Simple lyophilisation after neutralisation by acid addition till pH 7 ;
b) Acidification by 10^{-1} N HCl for 1 hour at room temperature followed by neutralisation at pH 7 and lyophilisation ;
c) Acidification by 1 N HCl for 1 hour at room temperature followed by the same neutralisation and lyophilisation ;
d) Hydrolysis by 6 N HCl at 70°C for 12 hours followed by the same neutralisation and lyophylisation.

The resulting quantities of aminoacids were determined by ion- exchange chromatography without involving pH > 2.5 and using a Durrum aminoacid analyser (3).

The results obtained from samples of the solution taken at the end of the experiments are given in table I. Comparable results regarding the role of pretreatment were obtained from solution samples taken in the course of the experiment and from the brown deposit taken from the reactor walls at the end of the experiment.

TABLE I

		Analytical Procedure			
		a	b	c	d
	Asp	47	45	58	228
	Ser	63	67	86	2.5
µ moles	Glu				1.7
produced	Gly	111	109	145	402
	Ala	40	34	46	171
	βAla	17	16		
Total C (at.g.) incorporated		770	749	918	2245
Yield % of C incorporated		0.15	0.15	0.18	0.44

These results show that :

- Notable differences between results are observed according to whether more or less drastic acid treatments are carried out before the aminoacid chromatography.

- Measured aspartic acid, glycine and alanine quantities are slightly increased by 1 hour room temperature treatment in 1 N HCl and sharply increased by 12 hours, 70°C, 6 N HCl hydrolysis.
- Detected serine quantities are comparable to the corresponding quantities of these aminoacids if the sample is not submitted to hydrolysis by hot and concentrated HCl. But, as is typically the case for this aminoacid, they vanish if such a hydrolysis is performed.
- Only smaller amounts of glutamic acid and βalanine are detected.

With respect to classically published results in connection with the formation of biochemicals from experiments globally simulating the primitive evolution of a hypothetically hydrogenated earth periphery these results and previously obtained similar ones (3) show that :

-Serine is in fact produced in comparable amounts to other biologically important simple aminoacids.

-The larger part of the detected amount of aminoacids is in fact produced as such by concentrated and hot acid hydrolysis. These aminoacids are effectively present in the solutions produced by the experiment mainly in the form of precursors such as e.g. nitriles or polymers such as peptides.

-However, part of these aminoacids are present as such in the solutions obtained from the experiment, at least after acidification at pH 2.5.

-Since 1N HCl, room temperature, 1 hour treatment suffices to begin to increase the detected amounts of aminoacids, at least a part of the non-free aminoacids is probably present in the obtained solutions in the form of non-peptidic precursors.

REFERENCES

(1) BUVET R. and STOETZEL F.
Chemical Evolution and Energetics of Reactions in Aqueous Solutions on the Primitive Earth.
Origins of Life (7) 93.107 (1976)

(2) BUVET R. and STOETZEL F.
Energetics of Abiogenic Chemical Systems
Biosystems 7, 2-14 (1975)

(3) FLORES J.J. and PONNAMPERUMA C.
J. Molec. Evolution 2, 1, (1972)

REDUCTION OF THIONUCLEOSIDES: A PREBIOTIC PATHWAY TO DEOXYRIBONUCLEOSIDES

A.D. Patel, W.H. Schrier, M.A. Hrncir and J.J. Nagyvary

Texas A&M University, Dept. Biochemistry,
College Station, Texas 77843, U.S.A.

The biosynthesis of 2'-deoxyribonucleosides is known to take place via the reduction of ribonucleotides, and the Principle of Continuity suggests to search for a similar pathway operating under prebiotic conditions. The facility by which the C-S bond can be reduced in comparison to the C-O would favor 2'-thio-2'-deoxyribonucleosides as the most likely precursors. Previous work from our laboratory has presented a good case for the prebiotic existence of 2'-thio-2'-deoxycytidine which is not less plausible than the present argument for ribocytidine. The thioanalog is obtained from anhydro-araC with several sulfur-containing nucleophiles. The problem of reducing a C-S bond to C-H under prebiotic conditions has not yet received any attention. Besides the traditional Raney nickel treatment, we have now explored such plausible reductions as a) inorganic reduction by Fe(II), b) organic reduction by HCN, and c) photochemical reduction by UV light. All three methods proved to be suitable in producing deoxycytidine in 1 to 8% yield. This may be the best supported model of prebiotic synthesis of a deoxyribonucleoside.

Past research on the origin of life centered on the origins and interrelationships of proteins (proteinoids) and ribonucleotides. Although the role of DNA in the emergence of autonomous living organisms in generally acknowledged, no mechanism has been put forth with much conviction as a major prebiotic pathway for the synthesis of deoxyribonucleotides. Possible prebiotic models for the formation of deoxyribose (1) and the fusion of deoxyribose with purine bases (2) were reported without experimental details. The instability of deoxyribose and the low yield of its

Y. Wolman (ed.), Origin of Life, 201–207.

condensation to β-nucleosides are the unattractive feactures of this model which is also unrelated to the biological pathway. Here we present what we believe to be a plausible model for the abiogenesis of 2'-thio-2'-deoxyribonucleosides which could give rise to a novel nucleic acid analog beside being a precursor of deoxyribonucleosides.

Our proposed concept is demonstrated in the formation of a ribonucleoside analog, 2'-thio-2'-deoxycytidine which can be reduced to deoxycytidine. Since the introduction of the 2'-SH group in the erythro configuration requires an inversion, anhydronucleosides are suitable starting materials (3). We made use of anhydro-araC (4) which was formed in good yield in a simple prebiotic process (5). Our strategy was to introduce a neighboring group containing sulfur, a 3'-O-phosphorothioate, which was capable of rearranging into the 2'-S-Cyd-2':3'P. The organic chemistry pertaining to this reaction sequence, depicted in Scheme 1, has been recently reported from this laboratory (6,7).

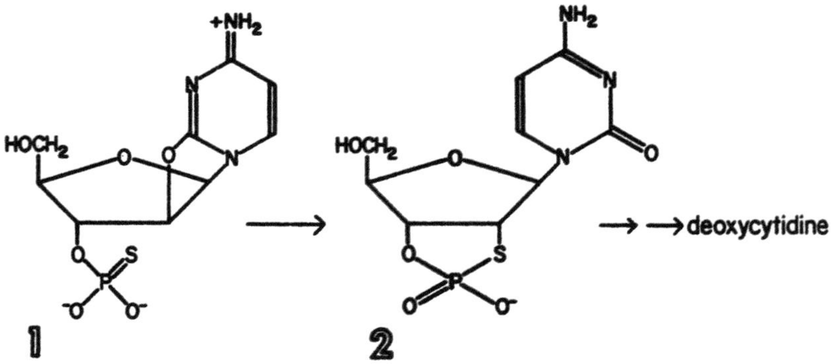

Scheme 1

We have previously proposed (8,9) that polythiophosphate could have played some role in prebiotic evolution. The above synthetic scheme has now been transposed into a non-anhydrous condition compatible with the hypothetical primordial scenario. Our model is not limited to thiophosphates; other thiolated nucleophiles such as thiocarbonate and thiourea may fulfill the same role. Carbon disulfide, a precursor of several potential nucleophiles, was among the photoproducts obtained under similated Jovian conditions (10). We have now shown that the reaction of carbon disulfide with anhydro-araC also provides a pathway to the formation of 2'-thio-2'deoxycytidine which can be reduced in numerous ways to 2'-deoxycytidine.

1. The synthesis of 2'-thio-2'-deoxycytidine with dithiophosphate. Successful thiophosphorylation was achieved not only in the solid state but also in dilute solution according to the conditions given in Scheme 2. Just as in the organic synthesis (6,7), the primary products of thiophosphorylation in water were the isomeric anhydro-araC-3'- and 5'-PS but their ratio (3':5') changed to 10:7 from 5:1 observed in dimethylformamide. The efficiency of the primary thiophosphorylation (acetic acid fraction in Scheme 2) was clearly the determining factor for the overall yield of 2'-S-nucleotide formation. The results of thirty experiments carried out in the solid state depended strongly on the speed of evaporating the solutions and the thickness of the resulting films. The prolongation of the reaction time beyond 24 hr had no beneficial effect on the primary thiophosphorylation. The best yield was around 45% with a pH optimum at 5.5. The corresponding yields obtained in dilute water solution were significantly lower (0.8 to 1.2%), but are comparable with other prebiotic phosphorylations (11). The rearrangement of anhydro-araC 3'-PS to 2'-S-Cyd-2':3'-P (1→2, Scheme 1) is faster in pyridine but it goes to completion in water as well. However, hydrolysis of the phosphate ester bond in both the starting material and the product has to be contended with.

The nucleoside, 2'-thio-2'-deoxycytidine, was obtained as a minor product in the acid hydrolysis of the cyclic phosphorothioate; otherwise, it was obtained by the enzymatic hydrolysis. The structure of this compound was established by its mass spectrum values given in Table 1.

2. 2'-Thio-2'-deoxycytidine formation in CS_2. The starting material, O^2:2'-anhydro-1-β-D-arabinosyl cytosine tosylate (1 g, ca 2.6 mmole) was dissolved in a mixture of pyridine (15 ml), triethylamine (5 ml), and carbon disulfide (5 ml). The reaction mixture was heated at 60°C for 24 hr. By this time 85% of cationic starting material disappeared and an anionic xanthate derivative was formed, which itself was not stable. The thiocarbonate groups could only be removed by 50% NH_4OH (60°C, 4hr) hydrolysis or by strong acid treatment, both of which generated substantial quantities (>50%) of cytosine. Complete separation of the title compound from the various sulfur derivatives required chromatography on DEAE cellulose (0.05 M Et_3NH-HCO_3), silica

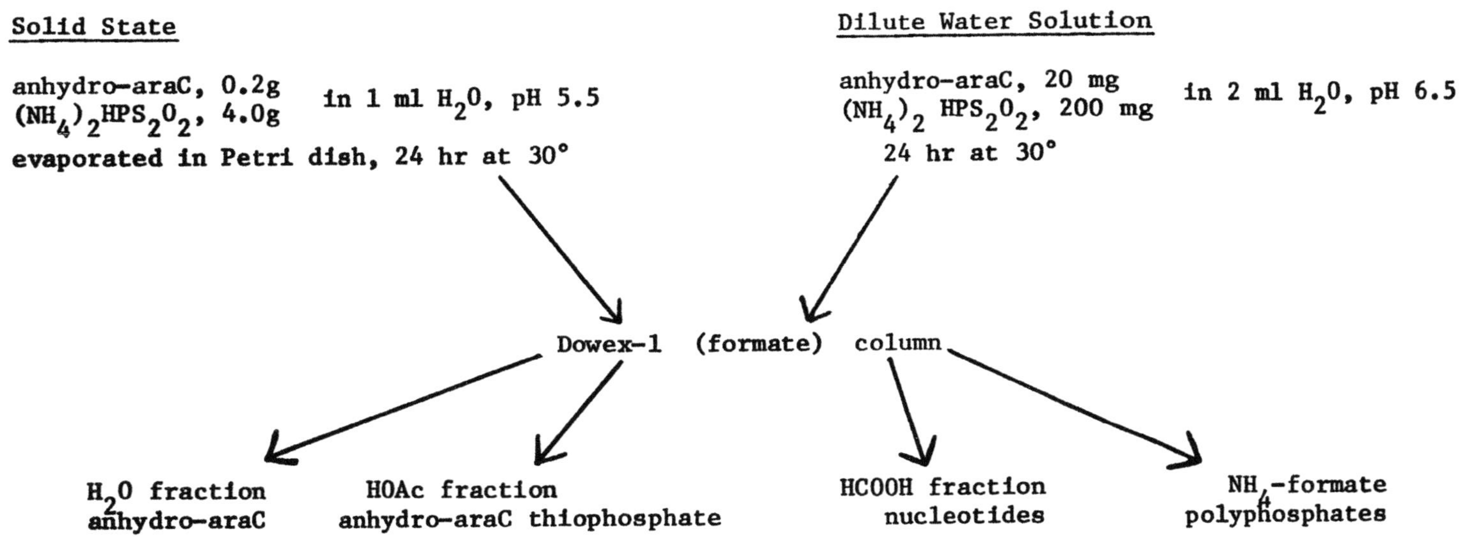

Experimental Conditions for thiophosphorylation

Scheme 2

gel (1-butanol-H_2O 6:1) and Sephadex G-25, adding each time 0.1 ml β-mercaptoethanol to the sample. Yield 1,400 $O.D._{270}$ units, 0.15 mmole (6%) of chromatographically pure 2'-thio-2'-deoxycytidine which gave satisfactory elemental analysis values. TLC in n-butanol:ethanol:1 M NH_4OOCCH_3(30:15:5), R_f = 0.32; on standing, disulfide R_f = 0.14. The structural proof includes NMR (^1H, ^{13}C) spectra and mass spectrum.

TABLE 1.

Mass Spectral Values of Nucleosides. Trimethylsilyl nucleosides were run on a CEC 21-110 instrument at 70 eV and probe temperature of 40°. Underivatised nucleosides were analysed by ^{252}Cf plasma desorption technic (M=bis-2'-S-dCyd disulfide).

mass	ion
1055.0	$(2M + Na)^+$
517.2	$(M + H)^+$
258.1	$(2'\text{-}S\text{-dCyd})^+$
587	$(2'\text{-}S\text{-dCyd-4H} + 4\ Me_3Si)^+$
149.1	$(\text{thioribose} - OH)^+$
228.4	$(\text{dCyd} + H)^+$
300	$(\text{dCyd} + Me_3Si + H)^+$
188	$(C_5H_7O_3 + Me_3Si)^+$

3. The reduction of 2'thioanalogs.

a) Reduction with Fe^{II}. Kaolin (100 mg) and 0.1 ml concd NH_4OH were added to a solution of $FeSO_4$ (100 mg) in 5 ml water. The precipitate was washed to neutrality by centrigugation. The 2'-S-dCyd (100 $O.D._{272}$ units) was incubated with the precipitate in 1 ml water at 50° for 2 hr with constant stirring in a sealed tube under nitrogen. The reaction mixture was filtered and absorbed on a Dowex -50 column (H^+, 1 x 10cm). The products were eluted with 1% NH_4OH and analyzed in several thin layer chromatography systems. The yield of deoxycytidine varied in the range of 3 to 10%; the major product was always cytosine. Unambiguous identification on deoxycytidine was accomplished by mass spectra (Table 1). The spectrum of the Me_3Si derivative which was recorded on a CEC 21-110 instrument contained signals at 300 (dCyd + Me_3Si + H) and at 188 (a dihydrofuran derivative) mass per charge ratios which are a characteristic of deoxycytidine (12). The ^{252}Cf plasma desorption mass spectrum (13) of the underivatized synthetic product gave a signal at 228.4 corresponding to $(M+1)^+$.

b) Photochemical reduction of 2'-SdCyd. A solution of 100 $O.D._{272}$ units of 2'-S-dCyd in 100 ml of methanol-tetrahydrofuran-toluene (1:1:1), containing 5 μl of 2-mercaptoethanol was irradiated with a 450 watt mercury vapor lamp in a quartz equipment under cooling at 25°. After one day the solvent was removed under reduced pressure, and the residue was absorbed on Dowex 50 (H^+) resin as described above. The fraction obtained with dil. NH_4OH (26 $O.D._{272}$ units) was further purified by silica gel TLC in butanol -H_2O (85:15). The deoxycytidine thus obtained (16% yield) was identical with the authentic specimen in several chromatography solvents on TLC and high pressure liquid chromatography (μ-Bondapak C_{18}, methanol-water (9:1) and methanol-$CHCl_3$ (1:1). Qualitative test for deoxyribose was also positive.

Of these two methods of reduction, the photolysis is particularly valuable for the reasons of better yield and obvious prebiotic relevance. The mechanism of thiol photolysis has also been explored by Pryor and Olsen (14). The main event is probably the cleavage of the S-H, the generation of H_2S and R•, and termination by forming RH.

The limiting factor in improving the efficiency of deoxyribonucleoside formation is the lability of the glycosidic bond. The cytosine is being displaced under the formation of an 1,2-episulfide. This decomposition was particularly strong when the reduction was attempted with HCN at 60°. The thiol group was reduced to deoxyribose but only traces of deoxycytidine were found.

The attractive feature of this proposed prebiotic pathway for deoxyribonucleoside synthesis is its similarity to the biosynthetic pathway in the sense that it is based on the reduction of a 2' functionality, i.e. it is in harmony with the Principle of Continuity. Even though the photolytic reduction reported here employs organic solvents, the reaction also takes place in water albeit in lower yield. Further elaboration of details for the other three nucleobases would fill an important gap in the framework of chemical evalution.

ACKNOWLEDGEMENT

This research was supported by a grant from NASA. We thank Drs. R.D. Macfarlane, C.J. McNeal and R.D. Grigsby for the mass spectra, and Dr. Stanley Miller for valuable suggestions.

REFERENCES

1. Oro, J., The Origins of Prebiological System, Academic Press, Inc., New York, (1965) p. 137.
2. Fuller, W.D., Sanchez, R.A., and Orgel, L.E., J. Mol. Evol. 1, 249 (1972).
3. Imazawa, M., Ueda, T., and Ukita, T., Chem. Pharm. Bull. 23, 604 (1975).
4. Abbreviations: anhydro-araC hydrochloride, 0^2:2'-anhydro-1-β-D-arabinosylcytosine hydrochloride; anhydro-araC-3'-PS, 0^2:2'-anhydro-1-β-D-arabinosylcytosine 3'-0-phosphorothioate;2'-S-Cyd-2':3'-P, 2'-thio-2'-deoxycytidine 2':3'-cyclic phosphorothioate; Me_3Si, trimethyl silyl.
5. Sanchez, R.E., and Orgel, L.E., J. Mol. Biol. 47, 431 (1970).
6. Bradbury, E., and Nagyvary, J., Nucleic Acids Res. 3, 2437 (1976).
7. Dunaway-Mariano, D., Tetrahedron 32, 2991 (1976).
8. Slabaugh, M.R., Harvey, A.J., and Nagyvary, J., J. Mol. Evol. 3, 317 (1974).
9. Nagyvary, J., and Provenzale, R.G., Proc. Natl. Acad. Sci. 67, 8 A (1970).
10. Khare, B.N., Sagan, C., Bandurski, E.L., and Nagy, B. Science 199, 1199 (1978).
11. Lohrmann, R., and Orgel, L.E., Science 161, 64 (1968).
12. McCloskey, J.A., in Basic Principles in Nucleic Acid Chemistry, ed. P.O.P. Ts'o, Academic Press, New York, 1974, Vol. 1, p. 209.
13. MacFarlane R.D., and Torgerson, D.F., Science 191, 920 (1976).
14. Pryor, W.A. and Olsen, E.G., J. Amer. Chem. Soc. 100, 2852 (1978).

FACTORS INFLUENCING THE RATE OF NON-ENZYMATIC ACTIVATION OF CARBOXYLIC AND AMINO ACIDS BY ATP

Dail W. Mullins, Jr. and James C. Lacey, Jr.

Laboratory of Molecular Biology
University of Alabama in Birmingham
University Station
Birmingham, AL 35294 USA

The rate of non-enzymatic activation of carboxylic and amino acids by ATP has been found to vary as a function of several factors. These include, (a) the pK_a of the carboxyl group, (b) pH, (c) temperature, (d) salt concentration, (e) the divalent metal cation, (f) the molar ratio of metal cation to ATP, and (g) the nature of the amino acid. The rates of activation of a select group of hydrophobic amino acids (phe, leu, val, ile and met) were found to differ under a variety of conditions, although phe was always activated most rapidly. This amino acid, moreover, was always found to have the greatest affinity for various adenine derivatives as determined by several methods. We conclude that the rate of non-enzymatic activation of amino acids by ATP is at least partially a function of the affinity of the amino acid for this nucleotide.

For the past several years, our research efforts have been directed toward an understanding of the origin of the genetic code and the process of protein biosynthesis, two aspects of the living state which we believe arose simultaneously in evolution, and which have subsequently evolved in concert (for a review, see (1)). As an important part of this program, we have systematically investigated the chemistry of aminoacyl transfer reactions, which are necessarily the component reactions of the process of protein biosynthesis. This work has included studies of the spontaneous transfer of aminoacyl groups from adenylate anhydrides to imidazolides (2), and subsequently to the 2'-OH groups along the backbone of polyribonucleotides (3). The spontaneous polymerization of amino acids esterified in this manner has also been observed (4). More recently, we have been

Y. Wolman (ed.), Origin of Life, 209–215.

investigating the non-enzymatic formation of adenylate anhydrides (activation) using ATP, a divalent metal cation, and a variety of carboxylic and amino acids (5).

Lowenstein (6) and Lowenstein and Schatz (7) were the first to study the chemical activation of carboxylic acids by ATP, and showed that both acetate and glycine could be activated, provided that a divalent metal cation was present. Be^{++}, Ni^{++}, and Ca^{++} ions were found to be the most effective in this regard. These workers used a technique developed by Lipmann and Tuttle (8) to quantitate the degree of activation, trapping the activated species with hydroxylamine to form the hydroxamate, which then yields a brown color (ε_{max} = 495 nm) upon complexation with Fe^{+++} ions. Using a similar technique, Ryan and Fox (9) showed that ATP was by far the most effective nucleotide for catalyzing the activation of glycine, and Fox and his coworkers (10) demonstrated the formation of phenylalanine peptides from ATP, phenylalanine and lysine-rich thermal proteinoid microspheres containing polyadenylic acid.

Based on the results of his early studies, Lowenstein (6) had concluded that ionized carboxyl groups were most readily activated by ATP, although more recent work by Paecht-Horowitz and Katchalsky (11) has suggested just the opposite; that it is the protonated form of the carboxyl group which participates in the activation reaction. Our own studies with regard to this question (5) have indicated that it is the protonated form of the carboxyl group which is most readily activated by ATP, since the extent of activation of a series of carboxylic acids at pH 5 shows a direct correlation with their pK_a's. These data are shown in Fig. 1. Phenylalanine, α-aminoisobutyric acid, and nicotinic acid were chosen because of the commercial availability of their hydroxamates. Although the amino acid hydroxamates all

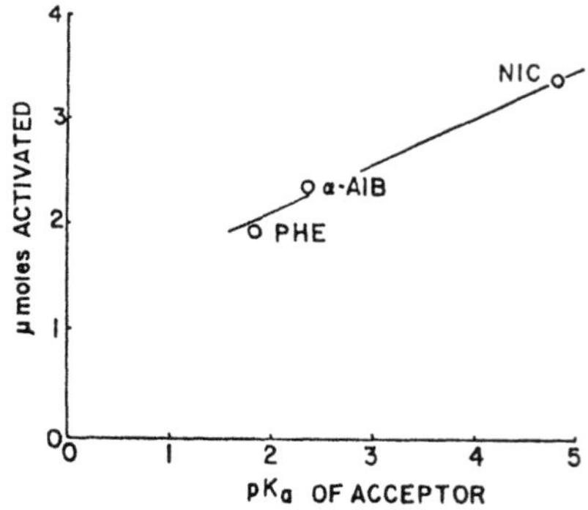

Fig. 1. Activation of phenylalanine (PHE) α-aminoisobutyric acid (α-AIB) and nicotinic acid (NIC) by ATP at room temperature, pH 5.0, as a function of the pK_a of the activated species. Each essay contained, in 3.0 ml at pH 5.0, 300 μmol ATP, 300 μmol $BeSO_4$, 750 μmol hydroxylamine and 600 μmol acceptor. The tubes were incubated at room temperature for 144 h, and 1.0 ml aliquots were withdrawn and assayed for hydroxamate color according to the procedure of Lipmann and Tuttle (8).

have similar extinction coefficients in the assay of Lipmann and Tuttle (8), this is not the case with the hydroxamic acid derivatives of other organic carboxylic acids, and it was necessary to know the precise extent of color formation for these compounds. Although the data of Fig. 1 do not prove such a contention, they are consistent with the idea that it is the protonated form of a carboxylic acid which is activated by ATP, since at a given pH, the higher the pK_a, the greater the amount of protonated form present.

Because N-blocked amino acids, including the C-terminal residue of oligopeptides, typically have higher pK_a's than do their unmodified forms, we suspected from the above data that such derivatives might exhibit enhanced activation by ATP. Although the spontaneous hydroxylaminolysis of N-acetyl and N-formyl groups from amino acids prevented us from studying this question with these modified forms, we were able to look at the relative rates of activation of glycine and a series of glycine peptides, since these are not so readily decomposed by hydroxylamine. As shown in Fig. 2, the addition of successive glycine residues to the amino terminus of an initial glycine has the effect of raising the pK_a of this carboxyl group, with the result that a corresponding increase in the rate of activation is observed. The values reported are corrected for hydroxamate color due simply to some hydroxylaminolysis of the peptides, as determined in control experiments without ATP.

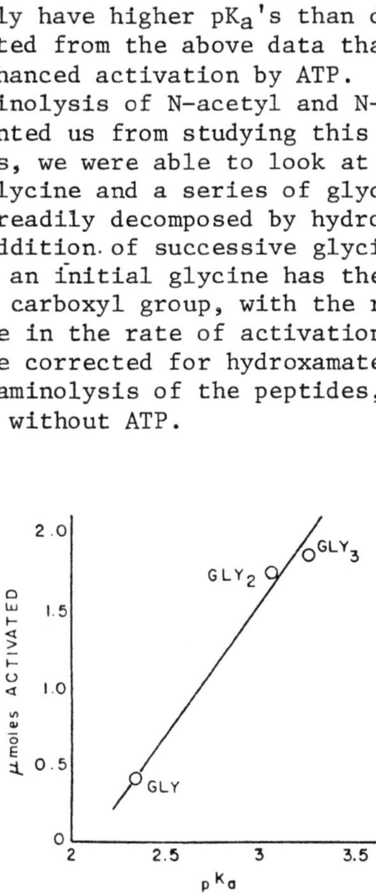

Fig. 2. Activation of glycine and glycine peptides by ATP at room temperature, pH 5.0, as a function of the pK_a of the activated species. Each essay contained, in 2.5 ml at pH 5.0, 250 μmol ATP, 250 μmol $BeSO_4$, 375 μmol glycine or glycine peptide, and 625 μmol NH_2OH. HCl. The solutions were withdrawn and assayed for the amount of hydroxamate present according to the method of Lipmann and Tuttle (8). Parallel samples without ATP were also incubated in order to establish the amount of hydroxamate color formed as a result of hydroxylaminolysis, and these values were used to correct the above readings. Tetraglycine was not soluble at 0.15M, pH 5.0, and so results with this peptide were not included in the above data.

If it is the protonated form of the carboxyl group which is activated, then the rate of activation should be expected to decrease as the pH is raised. Lowenstein et al. (7) had previously shown that the pH optimum for the activation of acetate was 5.2, and that the extent of activation declined on either side of this value. We obtained a similar result with glycine, although we suspect that the observed decline in activation at pH values below 5 is due to the decreased effectiveness of hydroxylamine as a trapping agent below this pH. The sharp decrease in the activation rate above pH 5, however, is no doubt a real effect, since hydroxylamine becomes more effective at higher pH levels. These results are consistent with the idea that it is the protonated form which is activated, since the amount of protonated form decreases as the pH is raised above the pK_a.

Our early activation studies, including those described above, were all carried out at room temperature (23°C), using Be^{++} as a metal ion catalyst. We later learned, however, that the rate of activation could be greatly increased by elevated temperatures, and most of our later studies were carried out at 50°C. Fig. 3 shows the effect of temperature on the non-enzymatic activation of phenylalanine at pH 5, using Be^{++} as the divalent metal cation. While it is perhaps not surprising that increased temperatures should have this effect, it is rather remarkable that ATP is so stable under these conditions, and fully capable of maintaining the activation reaction even after 90 hours at 50°C.

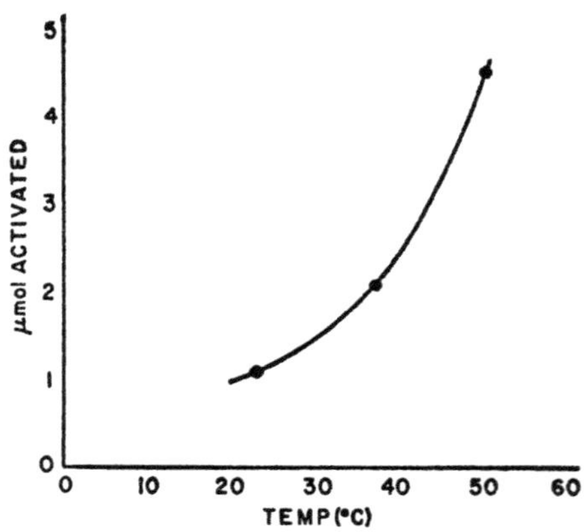

Fig. 3. Effect of temperature on the activation of phe by ATP and Be^{++}, pH 5. Each assay contained in 3.0 ml, 300 µmol ATP, 300 µmol Be^{++}, 750 µmol hydroxylamine and 510 µmol hydroxylamine and 510 µmol phe. The solutions were incubated for 90 h at the various temperatures indicated, at which time 1.0 ml aliquots were withdrawn and assayed for hydroxamate formation according to the method of Lipmann and Tuttle (8).

Because Mg^{++} is the divalent metal ion employed in the amino acid activation reactions of contemporary biological systems, we preferred to use this cation in our own studies, rather than Be^{++}. The activation rate with Mg^{++} at 23°C, however, is exceedingly slow, and its use at this temperature proved impractical. At 50°C, however, the rates of activation with Mg^{++} and Be^{++} are nearly comparable, so that in all of our later studies we were able to use Mg^{++} as the metal ion catalyst. Fig. 4 shows that the rate of activation of phe at 50°C and pH 5 increases with increasing Mg^{++}, essentially plateauing at a Mg^{++}/ATP ratio of 2. This result suggests that a complex between ATP and Mg^{++}, and having a stoichiometry of 2 Mg^{++}:1 ATP, is the most effective form.

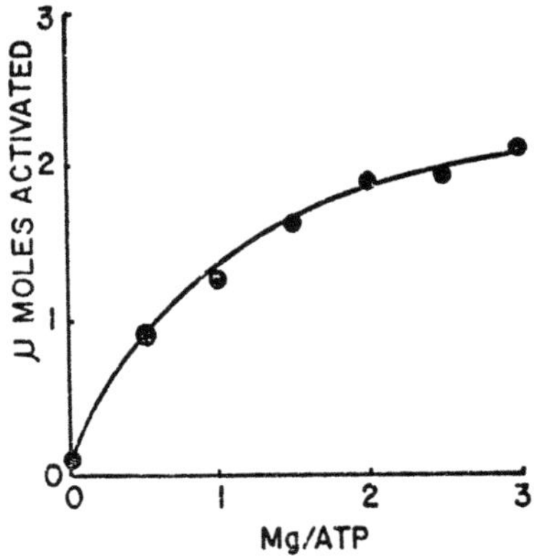

Fig. 4. Effect of varying Mg^{++}/ATP on the activation of phe at 50° C, pH 5.0. Each solution contained, in 3.0 ml, 300µmol ATP, 1.2 µmol hydroxylamine, 450 µmol phe and either 0, 300, 600 or 900 µmol Mg^{++}, as indicated. After incubation at 50° C for 43 h, 1.0 ml aliquots of each solution were withdrawn and assayed for hydroxamate color according to the method of Lipmann and Tuttle (8).

Fig. 5 shows the results of a study of the effect of various divalent metal cations on the activation of a series of hydrophobic amino acids by ATP at 50°C. Included in the study are Mg^{++}, added both as $MgCl_2$ and as a component of synthetic seasalt, Be^{++} and Zn^{++}. All of the amino acids employed are from the first column of the genetic anticode (Table 1 in the article by Lacey and Mullins, this volume), and all have adenylic acid as the central, most important letter of their anticodon. The data are remarkable for several reasons. First of all, as mentioned previously, the catalytic abilities of the various metal ions are roughly comparable at 50°C, even though at 23°C, Mg^{++} and Zn^{++} are considerably less effective than Be^{++}. Second, it should be noted that the order of decreasing rates of activation of these various hydrophobic amino acids (phe > leu > val > ile > met) often parallels the ordering of decreasing relative hydrophobicities of their anticodon dinucleotides (AA > GA > CA > UA) (12), with phenylalanine invariably being activated most rapidly. The more rapid activation of phenylalanine is particularly surprising in view of the fact that it has the lowest pK_a of all the amino acids used. Notice also that the overall selectivity of phenylalanine is generally enhanced by Mg^{++}, the biological ion of choice in these reactions.

The data in Fig. 5 also give some insight into the non-enzymatic activation reaction under more geologically relevant conditions, using synthetic seasalt at both 2x and 4x normal concentration. Although we did not know what effect the high NaCl concentrations would have on the activation reactions, it was anticipated that the rates of activation might be higher in 4x seasalt, since at this concentration the Mg^{++}/ATP ratio is at an optimum of 2 (see Fig. 4). It was also thought that

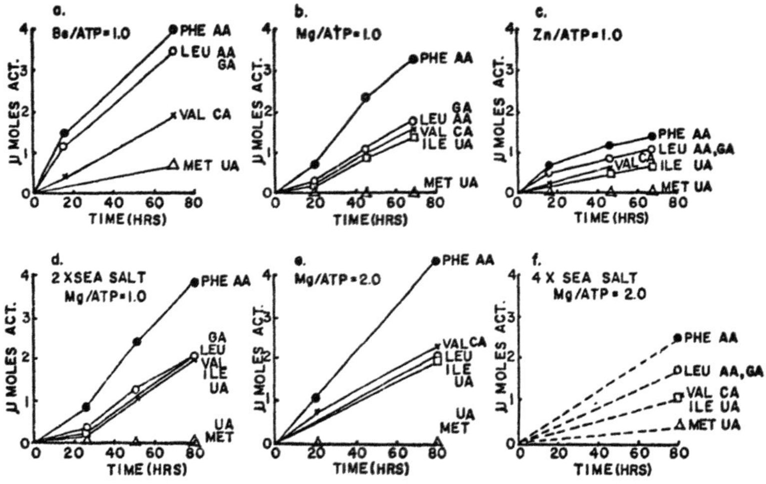

Fig. 5. μMoles of various hydrophobic amino acids (0.15M) activated in an ATP (0.1M), NH_2OH (0.4M) solution at pH 5.0 and 50° C as a function of time. Three divalent cations, Be^{++} (5a), Mg^{++} (5b) and Zn^{++} (5c) were tried at a 1/1 metal/ATP ratio. Seasalt was then used (5d) at sufficient concentration (2x normal) so that the Mg^{++} in the seasalt furnished a Mg^{++}/ATP ratio of 1/1. A Mg^{++}/ATP ratio of 2 was tried (5e) and then 4x seasalt (5f), a concentration which gave a 2/1 Mg^{++}/ATP ratio. The dinucleotide anticodons (3′-5′ direction) for the amino acids are shown, omitting the nucleotide in the "wobble" position, which is degenerate, in most cases. A(TP) is the most hydrophobic of the nucleotides, and the decreasing order of hydrophobicity of the anticodons is AA > GA > CA > UA, as shown in the first column of the genetic anticode (Table 1 in the article by Lacey and Mullins, this volume). Leu has anticodons AA and GA.

higher salt levels might conceivably "force" the hydrophobic interactions between the amino acids and the adenine ring of ATP, and thus possibly enhance activation. Surprisingly, however, it was found that the presence of 4x seasalt markedly inhibited the activation reactions, although at 2x seasalt there was little or no effect (see Fig. 5, b,d,e and f). The reason for these results is not readily apparent, but perhaps the salt influences the ionization of the molecules involved. We have already shown that ionization of the amino acid is important.

In an attempt to understand the physico-chemical basis for the selectivities of activation observed in these experiments, we carried out a series of studies designed to determine the degree of selective affinity between each of these amino acids and the adenine ring system (either as adenosine or ATP). These results, as well as other data from the literature, are reviewed in Table 2 of the article by Lacey and Mullins (this volume), and all indicate a clear preference of phenylalanine for adenine derivatives. We have tentatively concluded from these studies that the rate of chemical activation of an amino acid by ATP is a direct function of the affinity of the amino acid for this nucleotide. As other data here show, affinity is not the sole factor, since the rate of activation of amino acids is not always proportional to their affinity for ATP. Nevertheless, the fact that real differences have been observed in this system indicates that selective affinities between amino acids and nucleotides were important during prebiotic chemical evolution, and that the innate differential rates of activation may have played a major role in the origin of protein synthesis and genetic coding.

ACKNOWLEDGMENT We gratefully acknowledge the continued support of the National Aeronautics and Space Administration (NGR 01-010-001).

REFERENCES

1. Lacey, J.C., Jr. and Weber, A.L., Precamb. Res. 5: 1 (1977).
2. Lacey, J.C., Jr. and White, W.E., Jr., Biochem. Biophys. Res. Commun. 47: 565 (1972).
3. White, W.E., Jr., Lacey, J.C., Jr. and Weber, A.L., Biochem. Biophys. Res. Commun. 51: 283 (1973).
4. Lacey, J.C., Jr., Weber, A.L. and White, W.E., Jr., Origins of Life 6: 273 (1975).
5. Mullins, D.W., Jr. and Lacey, J.C., Jr., J. Mol. Evol. (in press).
6. Lowenstein, J.M., Biochim. Biophys. Acta 28: 206 (1958).
7. Lowenstein, J.M. and Schatz, M.N., J. Biol. Chem. 236: 305 (1961).
8. Lipmann, F. and Tuttle, L.C., J. Biol. Chem. 159: 21 (1945).
9. Ryan, J.W. and Fox, S.W., BioSystems 5: 115 (1973).
10. Fox, S.W., Jungck, J.R. and Nakashima, T., Origins of Life 5: 227 (1974).
11. Paecht-Horowitz, M. and Katchalsky, A., J. Mol. Evol. 2: 9 (1973).
12. Weber, A.L. and Lacey, J.C., Jr., J. Mol. Evol. 11: 199 (1978).

HCN OLIGOMERIZATION - ISOLATION AND PRELIMINARY CHARACTERIZATION OF A NEW PRECURSOR OF ADENINE.

Andries B. Voet and Alan W. Schwartz.

Department of Exobiology, Faculty of Science, The University, Nijmegen, The Netherlands.

Fractionation of the H_2O-soluble HCN oligomers was performed by means of elution from active charcoal, chromatography on Sephadex G-15 and ion-exclusion HPLC. In contrast to previously published studies, at least eight different fractions were shown to posess different spectral properties after Sephadex G-15 column chromatography and to yield different sets of products upon hydrolysis. One of these fractions, which gave adenine and hypoxanthine after hydrolysis with 6 *M* HCl, was purified further by means of cation exclusion HPLC. In this way, a precursor of adenine (and hypoxanthine) has been isolated and is undergoing structural characterization. A tentative structure is suggested and a new reaction scheme leading to the synthesis of adenine from *cis*-diaminomaleonitrile without isomerization is proposed.

Oró and Kimball (1) demonstrated in 1961 a possible route for the synthesis of adenine from HCN, with the trimer aminomalononitrile as intermediate. This proposal was indirectly supported by Ferris and Orgel (2) in 1965 while Sanchez *et al.*(3) and Ferris and Orgel (4) in 1966 extended these earlier studies and demonstrated that the most probable precursor was the tetramer diaminomaleonitrile (DAMN). Subsequently several groups have identified other products, including pyrimidines, amino acids, urea, oxalic acid and several hydantoins (5-10).

Aqueous HCN solutions of at least 0.1 *M* at pH=9.2 will undergo selfcondensation, to form a black precipitate called azulmic acid and a water soluble fraction (11,12). There are several ways to fractionate the soluble products. Ferris and coworkers (13) have applied ion exchange chromatography on

Y. Wolman (ed.), Origin of Life, 217–223.

Dowex, to separate the oligomeric products into neutral, acidic and basic fractions. Hydrolysis of the amphoteric and acidic fractions are reported to produce purines and pyrimidines. Another method which has been applied is fractionation on a Sephadex column. Matthews and Moser (10,14) utilized this method in a search for peptides, which they propose may be produced directly form HCN. This assumption, however, has been disputed by Ferris and coworkers (13,15), who were unable to find evidence for peptide linkages. As part of this work it was suggested that Sephadex is not efficient enough to separate the complex mixture of HCN oligomers. We have found, however, that under suitable conditions and particulary after a prefractionation on charcoal (which separates such compounds as ammonium formate and oxalic acid from heterocyclic structures), the products can be separated by Sephadex G-15 to yield a number of structurally distinct fractions. By further purification by means of HPLC, we were able to isolate a new precursor for the purine adenine.

Most of the previous research in this area has concentrated on identification of the products formed from HCN oligomerization after hydrolysis. Mechanistic studies have been performed largely on model structures or by starting with a presumed intermediate of HCN oligomerization (9, 19,20). An understanding of the mechanisms involved in this oligomerization, however, should ideally be based on direct identification of the precursors of biologically important molecules. The present communication is a preliminary report describing the isolation and some properties of a precursor of adenine, isolated directly from the HCN oligomerization products. The structure of the precursor also suggests a reaction scheme for the formation of purines from HCN which differs in several important respects from previously proposed schemes.

To 27g (40ml) freshly distilled HCN in 900 ml H_2O, was added sufficient conc. NH_4OH to adjust the *p*H to 9.2. The solution was made up to 1 L and allowed to stand at least six months at room temperature. After six months the soluble fraction was separated from the solid azulmic acid by filtration. Fractionation and purification of the soluble HCN oligomers were performed with different analytical techniques. A first purification was performed on active charcoal. The HCN oligomers were adjusted to *p*H 7 with formic acid and then applied to a charcoal column (the charcoal was prepared as described by van der Velden and Schwartz (16)). This column was eluated with H_2O, 1*M* HCl followed by H_2O, and finally with concentrated formic acid. Evaporation of the formic acid fraction gave 1.5 to 4.5 g of a brown, shiny material depending on the oligomerization time of the HCN solution. The product was applied to a Sephadex G-15 column for a second fractionation (fig 1). The effluents were monitored by U.V. transmission at 254 nm.

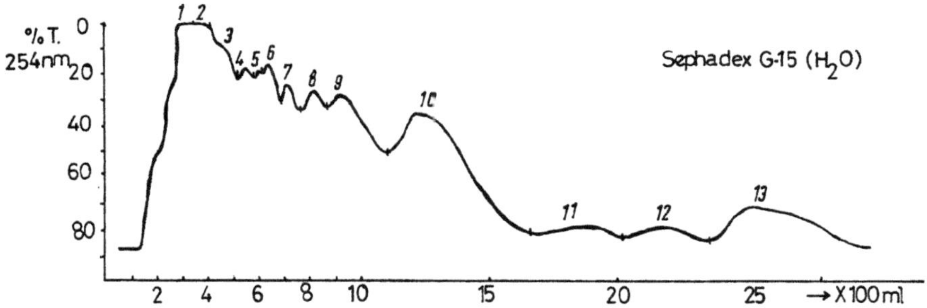

Fig.1.

Fractionation of the charcoal-purified products on Sephadex G-15 with H_2O as eluate.

The eluate was divided into thirteen fractions which were each evaporated to dryness. Each fraction was applied to an Aminex A25 and A6 HPLC column in the same manner as described in a previous paper (16).

Most of the latter fractions show remarkable differences in U.V. absorbing compounds. For example, fraction 6 and 10. (fig 2).

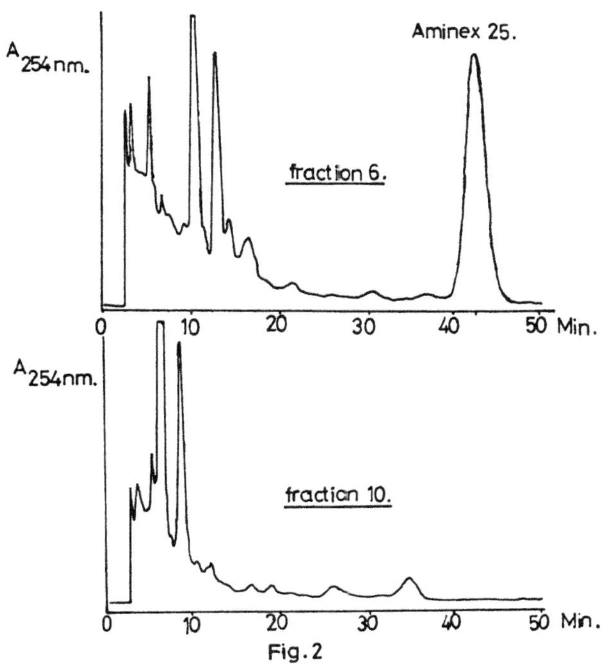

Fig.2

After hydrolysis of the fractions with 6 *M* HCl for 20h at 110°C, they were analyzed by A 25 and A6 HPLC. 4,5 - dihydroxypyrimidine was one of the compounds detected after hydrolysis of fraction 4. Although adenine was detectable after hydroloysis in more than one fraction, fraction 13 proved to be of special interest since a single precursor peak was observed for adenine which, after hydrolysis yielded adenine (and a smaller amount of hypoxanthine) in a relatively high yield in relation to other products in that fraction (fig 3). Fraction 13 was therefore further purified by HPLC on Aminex 25 to yield the adenine precursor, which was characterized by U.V., I.R., N.M.R. and mass spectrometry.

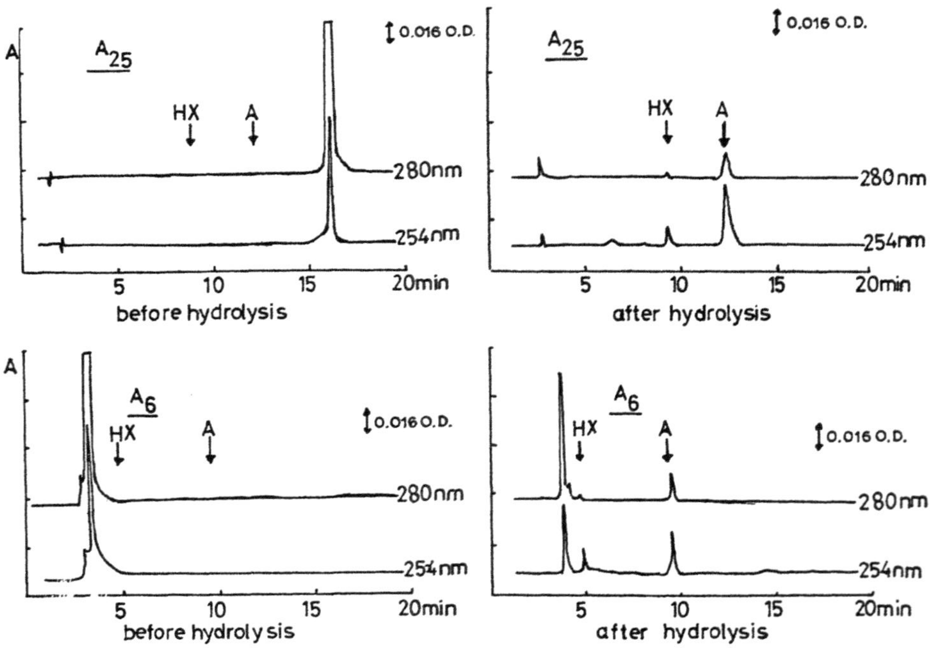

Fig.3

The U.V. spectrum showed maxima at 222,290 nm in H_2O, 222,295 nm in 0.1 *M* NH_3 and 209,282 nm in 0.1 *M* HCl. A broad band in the I.R. at 3100 cm-1 combined with a peak at 1672 cm-1 suggested the presence of an amide functional group. Proton exchange data by means of ^1H-NMR indicated two NH_2 groups and one C -H bond. Low resolution mass spectrometry gave $^m/e$ values of 178, 161, 135, 108, 81, 54, 53. The last five values correspond to those obtained from adenine. By means of high resolution ms, a mass of 178.0596 was determined for the molecular ion, indicating a molecular formula of $C_6H_6N_6O$, for which we tentatively suggest, in connection with the other spectral data, a 2 or 8 carbamoyl-6-aminopurine (fig 4). (The N-H bond at $N_{7(9)}$ is only marginally detectable in such purines).

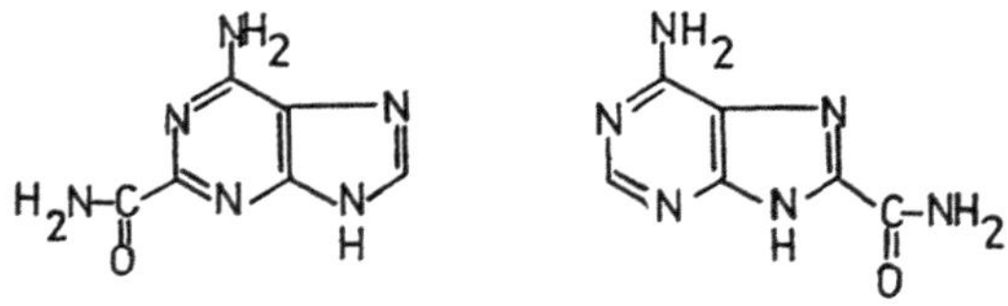

Fig. 4.

Althrough the structure of the precursor $C_6 H_6 N_6O$ still must be confirmed, it is perhaps not premature to suggest the following reaction scheme for its formation starting from HCN: Scheme I (shown is the formation of adenine via 2-carbamoyl-6-aminopurine. The analogous pathway to 8-carbamoyl-6-aminopurine is not illustrated).

SCHEME I.

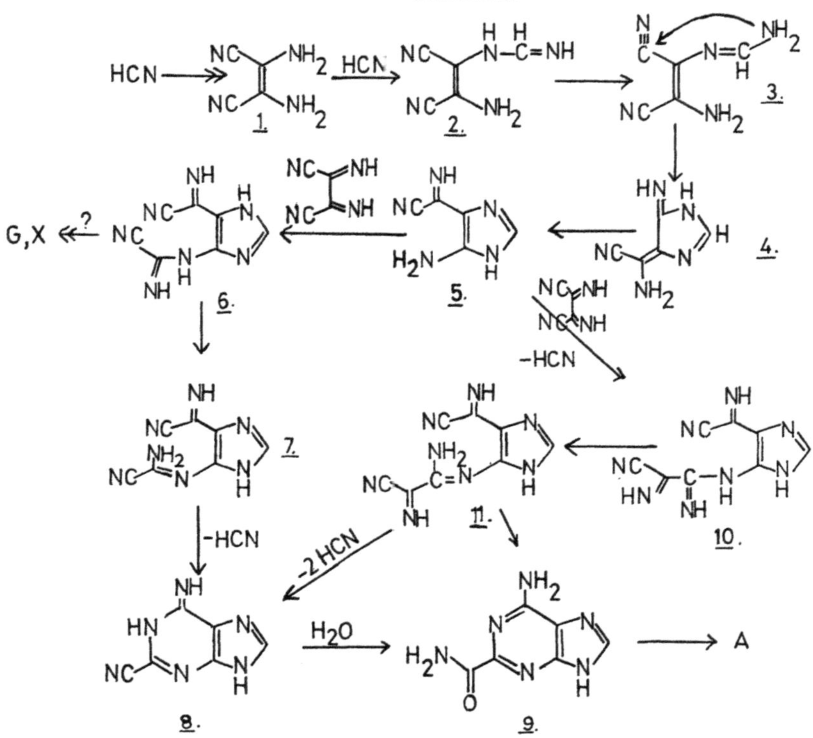

The reaction scheme leading from *cis*-diaminomaleonitrile 1 to 5. has already been suggested for non aqueous conditions by Shuman *et al* (17).It is important to point out that adenine has previously been thought to be formed via AICN or AICAI by reaction of the HCN trimer with formamidine (21,22) or by reaction of the trans-tetramer with formamidine (9). However, the predominant isomer of diaminomaleonitrile formed by non-photocatalyzed oligomerization of HCN is known to be *cis* rather than *trans* (9). Furthermore, formamidine is only formed in concentrated ammoniacal solutions (8). It has previously been pointed out by Ferris and Edelson (18) that an oxidation step is necessary to explain the production of several of the products of the HCN oligomerization. Diiminosuccinonitrile is the precursor of urea, one of the major products of the oligomerization and is formed by the oxidation of diaminomaleonitrile (19).This oxidation can apparently proceed in the absence of oxygen by means of internal redox reactions (18). We suggest that condensation of 5. with diiminosuccinonitrile provides the most likely pathway for the formation of the adenine precursor in fraction 13 and also provides a possible route to guanine and xanthine by hydrolysis of structure 6. The mechanism suggested by Ferris *et al* (8) involving as yet unidentified, higher molecular weight oligomers formed from *trans*-diaminomaleonitrile, may be valid for some of the other fractions which we have observed also produce adenine upon hydrolysis. These fractions, as well as the further details of the pathway suggested in Scheme I and the structures of precursors to other purines and pyrimidines are currently under investigation.

REFERENCES.

(1) Orõ, J., and Kimball, A.P., *Arch. Biochem. Biophys.* 94, 217 (1961)
(2) Ferris, J.P., and Orgel, L.E., *J. Amer. Chem. Soc.* 87, 4976 (1965)
(3) Sanchez, R.A., Ferris, J.P., and Orgel, L.E., *Science*, 153, 3731 (1966)
(4) Ferris, J.P., and Orgel, L.E., *J. Amer. Chem. Soc.* 88, 1074 (1966)
(5) Orõ, J., and Kamat, J.S., *Nature* 190, 442 (1961)
(6) Ferris, J.P., Wos, J.D., and Lobo, A.P., *J. Mol. Evol.* 3, 311 (1974)
(7) Ferris, J.P., Joshi, P.C., and Lawless, J.G., *BioSystems* 9, 81 (1977)
(8) Ferris, J.P., Joshi, P.C., Edelson, E.H., and Lawless, J.G., *J. Mol. Evol.* 11.293 (1978)
(9) Sanchez, R.A., Ferris, J.P. and Orgel, L.E., *J. Mol. Biol.* 30, 223 (1967)
(10) Matthews, C.N. and Moser, R.E., *Nature* 215, 1230 (1967)
(11) Boulay, P.; *J. Pharmac. Chim* (2) 16, 180 (1830);
Ann. Chim. Physique (2) 43, 282 (1830)
(12) Völker, T. *Angew. Chem.* 72, 379 (1960)
(13) Ferris, J.P., Donner, D.B. and Lobo, A.P., *J. Mol. Biol.*74, 4, 499 (1973)
(14) Matthews, C.N., *Science* 203, 1136 (1979)

(15) Ferris, J.P., *Science* 203, 1135 (1979)
(16) van der Velden, W. and Schwartz, A.W., *Geochim. Cosmochim. Acta* 41, 961 (1977)
(17) Shuman, F.R., Shearin, W.E. and Tull, R.J., *J. Org. Chem.* 44, 25, 4532 (1979)
(18) Ferris, J.P. and Edelson, E.H., *J. Org. Chem.* 43, 21, 3989 (1978)
(19) Ferris, J.P. and Ryan, T.J., *J. Org. Chem.* 38, 19, 3302 (1973)
(20) Ferris, J.P., Donner, D.B. and Lotz, W., *J. Amer.Chem.Soc.* 94, 6986 (1972)
(21) Orő, J. and Kimball, A.P., *Arch. Biochem. Biophys.* 96, 293 (1962)
(22) Orő, J., *Nature* 191, 1193 (1961).

CYANAMIDE MEDIATED SYNTHESES OF LEU, ALA, AND PHE PEPTIDES UNDER PLAUSIBLE PRIMITIVE EARTH CONDITIONS

J. R. Hawker, Jr. and J. Oró

Departments of Biophysical Sciences and Chemistry
University of Houston, Houston, Texas 77004

ABSTRACT. Using the model of a primitive earth evaporating pond, peptides were formed in yields of up to 21%, 30% and 35%, respectively for alanine, phenylalanine, and leucine when aqueous solutions of the appropriate amino acid were evaporated and heated for 24 hrs at 80°C with cyanamide and one or more of the following components: adenosine 5'-triphosphate, 4-amino-5 imidazole carboxamide, or $MgCl_2$. The greatest peptide yield was achieved at about pH 3. But peptide formation was demonstrated for a system of leu, cyanamide and $MgCl_2$ adjusted to pH 7 with NH_4OH. The incorporation of radioactive amino acids into products was found to be increased in the presence of $MgCl_2$ at all pH's studied, which ranged from 3 to 7. The major side product in all reactions appears to be an amino acid-cyanamide adduct. The peptides were analyzed and identified by thin layer chromatography, acid hydrolysis, and enzymatic degradation.

Cyanamide has been proposed as a plausible condensing agent for polymerization of amino acids under primitive earth conditions (1,2). Cyanamide has been used under mild aqueous conditions to achieve peptide synthesis (3,4) but with low yields. It has been found to be more effective under evaporating conditions at moderate temperatures (5,6) in promoting peptide synthesis. It has also been used in a variety of other possible prebiotic condensation reactions, such as the synthesis of deoxythymine oligonucleotides (7), the synthesis of acylglycerols (8), and the phosphorylation of nucleosides with apatite (9).

Investigations in this laboratory have used cyanamide as a condensing agent because it was probably being made continuously

Y. Wolman (ed.), Origin of Life, 225–232.

in the early atmosphere (1,20). In addition, hydrolysis of cyanamide is relatively slow; an aqueous solution of cyanamide (pH 6.8) was unchanged after five days of heating at 65°C (4). Thus the model of an evaporating pond could have served as the means by which cyanamide participated in prebiotic condensation reactions under evaporating conditions.

Previous work in this laboratory (3,4) has shown that a system involving 4-amino-5-imidazole carboxamide (AICA), adenosine 5'-triphosphate (ATP), amino acid and cyanamide will condense amino acids to peptides. This was also observed if ATP was omitted. In the present work, other variations of this system, as well as other amino acids were investigated. The divalent metal ion Mg^{2+} was studied in some of the reactions for its possible enhancing effect on condensation reactions. It has been suggested to be a possible inorganic catalyst in prebiotic conditions, such as the formation of phosphoramidates from ATP and amino acid (10). Also, inorganic ions may have been relatively abundant in primitive waters. Mg^{2+} has been shown to catalyze transphosphorylation reactions and amino acyl adenylate formation from amino acid and ATP (11). Moreover, peptide formation has been reported from a system of amino acid, nucleoside 5'-triphosphate, imidazole and $MgCl_2$ (10,12,13). Finally, peptide synthesis in the presence of $MgCl_2$ was studied at a neutral pH.

The following chemicals and materials were used in this investigation: Cyanamide (Sigma, m.p. 42°C), 4-amino-5-imidazole carboxamide hydrochloride (Sigma), adenosine 5-triphosphate disodium salt (Sigma), L-leucine (Mann), L-leucyl-L-leucine (Sigma) L-leucyl-L-leucyl-L-leucine (Sigma), (^{14}C)-L-leucine (Amersham), (^{14}C)-L-phenylalanine (Amersham), L-phenylalanine (Sigma), L-phenyl-alanyl-L-phenylalanine (Sigma), L-phenylalanyl-L-phenylalanyl-L-phenylalanine (Sigma), L-alanine (Sigma), L-alanyl-L-alanine (Sigma), L-alanyl-L-alanyl-L-alanine (Sigma), (^{14}C)-L-alanine (Amersham), magnesium chloride (Matheson, Coleman & Bell) carboxypeptidase A (Sigma), and ACS Counting Scintillant (Amersham). Silica gel 60 EM TLC glass plates were obtained from Curtex Matheson Scientific.

Typical reaction mixtures consisted of aqueous solutions (.6-1 ml) of amino acid (0.075 mmoles), cyanamide (0.25 mmoles), and $MgCl_2$ (0.055 mmoles). Some reactions had 4-amino-5-imidazole carboxamide (AICA) (0.25 mmoles) and/or ATP (0.055 mmoles) added to the reaction mixture. Still others had $MgCl_2$ omitted. Reaction mixtures containing labeled amino acids were prepared by adding ≃1 μc (^{14}C)-L-phenylalanine (513 mc/mmole), 0.2 μc (^{14}C)-L-leucine, or 2.5 μc (^{14}C)-L-alanine to 0.61 ml of the appropriate reaction mixture in order to quantitate yields. After mixing, the solutions were heated at 80°C for 24 hrs under evaporating conditions in open glass vials. After heating, the reactions were dissolved in water or 0.1 N HCl and the reaction products analyzed. The initial pH of most reactions was 3 to 4. Reactions containing only amino acid and cyanamide (±$MgCl_2$) had a pH of 5-6. Some reactions were adjusted to pH 3 with 0.1 N HCl or 1 M CH_3COOH and other reactions were brought to pH 7 with 1 M NH_4OH as a part of a pH study. The final pH of different reactions, after heating and dissolving in water, was also measured. For reactions with an initial pH of 3-4, the final pH ranged from 4.5-6. For reactions with an initial pH of 5-6, the final pH remained the same. For reactions that were brought to pH 7, the final pH ranged from 7-8.5.

Thin layer chromatography was performed on silica gel G glass plates using solvent system I; n-butanol:acetic acid:water (4:1:1, v/v/v) for one dimensional separation. For two dimensional separation of products, solvent system II: chloroform: methanol:17% NH_3 (2:2:1, v/v/v), first dimension; and solvent system III: phenol:water (75:25 w/w/), second dimension, were used. Amino acids and peptides were detected with a Cd-ninhydrin reagent (14). Some plates were also sprayed with a chlorine/ o-toluidine reagent (15) which hydrolyzes peptide bonds. Still other plates were sprayed with the Sakaguchi reagent (16,17) to detect arginine-like guanidino groups which may have formed from the reaction of cyanamide and amino acid.

To determine the yield of products, reactions containing (^{14}C) amino acid were spotted and developed on TLC plates then scanned with a Berthold LB 2760 Radiochromatogram Scanner to locate the position of radioactive products. Some plates were then sprayed with the Cd-ninhydrin reagent to compare their radiochromatogram peaks with their visualized spots. For other plates the radioactive areas were scraped off, suspended in ACS Counting Scintillant (Amersham) and counted on a Packard Model 3380 Liquid Scintillation Spectrometer to quantitate yields.

Peptides were identified by cochromatography of synthesized peptides with amino acid and peptide standards, acid hydrolysis, and enzymatic degradation of suspected peptides with Carboxypeptidase A according to the procedure of Nooner et al., (6).

Peptides in yields of up to 35% have been obtained for leu, up to 30% for phe, and up to 21% for ala. Leu and phe gave di-, tri- and even tetrapeptides in detectable quantities. Ala gave only dipeptides in detectable quantities. No peptide synthesis occurred if cyanamide was omitted. Table 1 contains the R_f values of leu, ala and phe peptides.

TABLE 1 R_f Values of Peptides

compound (a)	Solvent System I	Solvent System II
leu	.359	-
$(leu)_2$	.537	-
$(leu)_3$	.591	-
$(leu)_4$	.627	-
phe	.393	-
$(phe)_2$	.555	-
$(phe)_3$	.600	-
$(phe)_4$	.659	-
ala	-	.76
$(ala)_2$	-	.89

(a) The products from the leu, phe, and ala reactions were chromatographed on Silica Gel 60 precoated plates from EM reagents with a layer of thickness of 0.25 mm. Small discrete spots representing unreacted amino acid and peptides were visualized after spraying the plate with the Cd-ninhydrin or chlorine/o-toluidine reagent.

$MgCl_2$ appeared to have an enhancing effect on peptide synthesis and on the incorporation of amino acid into product in some of the reactions of both leu and phe. $MgCl_2$ was not tested in any of the ala reactions. Yields of various leu, phe, and ala reactions are shown in Table 2. For both leu and phe, in a complete system of amino acid, ATP, AICA, and cyanamide, addition of $MgCl_2$ had little effect on peptide yield. The same was true if ATP was omitted. But for both of these amino acids, in reactions containing only amino acid, ATP, and cyanamide or amino acid and cyanamide only, $MgCl_2$ had a definite effect on peptide yield as shown in Table 2. For leucine in both of these reactions, peptide synthesis only occurs if $MgCl_2$ is present.

The optimum initial pH for peptide synthesis was <4 in agreement with earlier studies (3,4,6), although, as mentioned earlier the final pH of reactions with an initial pH of 3-4 ranged from 4.5-6. For three of the reaction systems: 1) amino acid, ATP, AICA, cyanamide, 2) amino acid, AICA, cyanamide, and 3) amino acid ATP, cyanamide, no peptide synthesis occurred if the initial pH was above 4, whether $MgCl_2$ was present or not.

TABLE 2 Yields of Leu, Phe, and Ala Products

				% Yield of Peptides				
Reaction (1)	pH (2)	AA (3)	% Adduct	di-	tri-	tetra-	Total	% Rec. (4)
A_1	3	21	36	10	9	-	19	76
* A_2	3	15	49	14	4	-	18	82
B_1	4	38	24	-	-	-	-	62
* B_2	3	4	41	13	6	10	29	74
C_1	5	56	9	-	-	-	-	6
* C_2	5	27	30	22	-	-	22	79
* C_3	3	3	41	17	9	9	35	79
* C_4	7	23	33	11	7	-	18	74
D_1	3	25	32	12	4	3	19	76
* D_2	3	10	38	6	4	5	15	63
E_1	4	14	30	3	5	5	13	57
* E_2	4	13	35	8	8	14	30	78
F_1	6	51	24	-	-	-	-	75
* F_2	6	40	25	8	-	-	8	73
G_1	4	21	32	22	-	-	22	75

Reactions were heated at 80°C for 24 hrs

(1) Reaction Components
A = leu, cyanamide, ATP, AICA
B = leu, cyanamide, ATP
C = leu, cyanamide
D = phe, cyanamide, ATP, AICA
E = phe, cyanamide, ATP
F = phe, cyanamide
G = ala, cyanamide, ATP, AICA

(2) Initial pH rounded to nearest integer
(3) Per cent of free amino acid remaining at end of reaction
(4) Per cent recovery of amino acid plus amino acid products

*$MgCl_2$ added to this reaction

But in a system of leu, cyanamide and $MgCl_2$ adjusted to pH 7 with 0.1M NH_4OH di- and tripeptides with a total yield of about 18% were formed (See Table 2). Peptide formation was not observed if $MgCl_2$ was omitted or if the system was adjusted to pH 7 with NaOH. The final pH of this reaction, after heating and dissolving in water, was about 8.5. Thus this reaction did proceed at a neutral to slightly basic pH. Calculations of Bada and Miller (18) estimate there was a 0.001-0.01M concentration of ammonium ions dissolved in primitive oceans. Thus this kind of reaction may have occurred during evaporation of primitive earth ponds and tidepools. The necessity of $MgCl_2$ for this reaction indicates the role inorganic ions of primeval waters may have had in prebiotic reactions.

The major side product in all reactions studied (in yields up to 48%) appears to be an amino acid-cyanamide adduct. From a study of radiochromatograms and TLC's it was found to migrate just below the dipeptide (for leu and phe). It was not visualized right away as were other amino acids and peptides. It was usually only visualized after sitting several hours or overnight. Since our ninhydrin reagent contains some acetic acid, we suspected that the amino group of the amino acid was blocked and then hydrolyzed slowly after ninhydrin spraying (from the acid present) thus allowing the ninhydrin reaction to proceed. It is possible that the cyanamide had reacted with the amino group of the amino acid generating the following guanidino derivative:

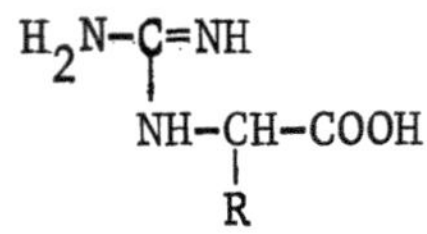

This possibility is in line with a report in the literature by Halmann (4) who studied cyanamide condensation reactions of glycine. He reported that the major product in his reaction was glycocyamine which has the same structure as the presumed compound above. Thus we tried the Sakaguchi reagent (16,17), which detects the guanidino group of arginine, in hopes that it would also be positive with the above product. This reagent also detects urea, the hydrolysis product of cyanamide, by giving a different color reaction from that of arginine. Results so far have shown the color reaction of arginine at the position of the presumed amino acid-cyanamide adduct with the color reaction of urea at the tail of this spot. Other evidence in support of this structure is that this compound was not hydrolyzed by treatment with Carboxypeptidase A. Also, acid hydrolysis of this compound yielded the amino acid and urea.

As for the role of AICA and ATP in these reactions, a previous paper (6) has suggested that the role of AICA was to maintain an acidic environment to allow cyanamide mediated peptide synthesis to proceed and that the role of ATP was to increase the viscosity of the reaction mixture. Initially it was thought that ATP might be involved in nucleotide-amino acid interactions which would have facilitated peptide synthesis. But the present results do not provide any firm answers to this question. The yields were not appreciably different between a system of amino acid, cyanamide, and $MgCl_2$ and a system of amino acid, cyanamide, and $MgCl_2$ and ATP, although it has been noticed that the incorporation of amino acid into product appears to be greatly enhanced in the presence of ATP and $MgCl_2$. An example is the system of leu, cyanamide and ATP (see Table 2, reaction B). If no $MgCl_2$ is present, the total yield of product was about 24%. But if $MgCl_2$ is added, the percentage of product formed was about 69%.

It is possible that any differences in yield may have been due in part to differences in pH and in part to the presence of ATP or that the effect of ATP was pH dependent. Indeed, peptide synthesis was demonstrated at pH 7 with only a system of leu, cyanamide and $MgCl_2$, but if ATP was added to this system at pH 7 no peptide synthesis occurred. Though perhaps nucleotide-amino acid adducts, e.g., amino acyl adenylates or 2'(3') amino acid esters of ATP, could be formed during the course of the reaction, no evidence for this has been found yet. Products of this type might have significance in the origin of the genetic code and protein biosynthesis.

The first step in the activation of cyanamide is presumed to be the protonation of cyanamide to yield a carbodiimide (19):

$$H_2C{\equiv}N \xrightarrow{H^+} H_2N-\overset{+}{C}{=}NH \qquad \text{protonated carbodiimide}$$

The rest of the mechanism postulated is as follows: Cyanamide protonation is followed by nucleophilic attack by the carboxyl group of an amino acid; then condensation of this activated amino acid with another amino acid to yield a dipeptide and urea:

$$H_2N-\overset{+}{C}{=}NH + O-\overset{O}{\overset{\|}{C}}-\underset{R}{CH}-\overset{+}{NH_3} \longrightarrow H_2N-\underset{+}{\overset{NH_2}{C}}-O-\overset{O}{\overset{\|}{C}}-\underset{R}{CH}-\overset{+}{NH_3} \qquad \text{activated amino acid}$$

$$\xrightarrow{NH_2CHRCOOH} NH_2-\underset{R}{C}-CO-NH-\underset{R}{CH}-COOH + NH_2CONH_2$$

$$\text{dipeptide} \qquad\qquad \text{urea}$$

Alternatively the cyanamide can react with the amino group of the amino acid giving rise to the guanidino derivative.

$$H_2N-\overset{+}{C}{=}NH + H_2N\underset{R}{C}HCOOH \longrightarrow H_2N-\underset{NH=\underset{R}{C}H-COOH}{C}{=}NH$$

ACKNOWLEDGMENTS. This work was supported in part by a grant from the National Aeronautics and Space Administration, NGR-44-005-002. We also wish to thank Ms. K. Rewers for her technical assistance.

REFERENCES

1. Oró, J. Ann. N.Y. Acad. Sci. 108: 464 (1963).
2. Steinman, G., Lemmon, R. M., Calvin, M. Proc. Natl Acad. Sci. (U.S.) 52:27 (1964).
3. Ponnamperuma, C., Peterson, E., Science 147: 1572 (1965).
4. Halmann, M. Arch. Biochem. Biophys. 128:808 (1968).
5. Sherwood, E., Nooner, D. W., Eichberg, J., Epps, D. E., Oró, J. Origin of Life, Proceedings of the Second ISSOL Meeting and the Fifth ICOL Meeting, Ed. H. Noda, pp. 105-111, Tokyo. Japan (1978).

6. Nooner, D. W., Sherwood, E., More, M.A., Oró, J. J. Mol. Evol. 10:211 (1977).
7. Sherwood, E., Oró, J., J. Mol. Evol. 10:193 (1977).
8. Eichberg, J. E., Sherwood, E., Epps, D. E., Oró, J., J. Mol. Evol. 10:221 (1977).
9. Schwartz, A. W. Biochim. Biophys. Acta 218:477 (1972).
10. Lohrmann, R., Orgel, L.E. Nature 244:418 (1973).
11. Lowenstein, J. M. Biochim. Biophys. Acta 28: 206 (1958).
12. Weber, A. L., Caroon, J. M., Warden, J. T., Lemmon, R. M., Calvin, M. Biosystems 8:277 (1977).
13. Sawai, H., Lohrmann, R., Orgel, L.E. J. Mol. Evol. 6: 165 (1975).
14. Heilmann, J., Barollier, J., Watzke, E. Z. Physiol. Chem. 309:219 (1957).
15. Nitecki, D. E., Goodman, J. W. Biochemistry 5:665 (1966).
16. Rydon, H. N., Smith, P. W. G. Nature 169:922 (1952).
17. Sakaguchi, S. Biochem. J. (Japan) 5:25 (1925).
18. Bada, J. L., Miller, S. L. Science 159:423 (1968).
19. Hulshof, J., Ponnamperuma, C. Origins of Life 7:197 (1976).
20. Lohrmann, R., J. Mol. Evol. 1:263 (1972).

TEMPLATE-DIRECTED SYNTHESIS OF OLIGOGUANYLIC ACIDS—METAL ION CATALYSIS

P. K. Bridson, H. Fakhrai, R. Lohrmann,
L. E. Orgel and M. van Roode

The Salk Institute for Biological Studies
San Diego, California 92138

The condensation of guanosine 5'-phosphorimidazolide on a poly(C) template is subject to regiospecific metal-ion catalysis. In the presence of the Pb^{2+} ion, predominantly [2'-5']-linked oligomers up to the 40-mer are formed. The Zn^{2+} ion catalyzes the formation of predominantly [3'-5']-linked oligomers up to the 35-mer. The Sn^{2+}, Sb^{3+} and Bi^{3+} ions also catalyze the formation of [2'-5']-linked products, but less efficiently than the Pb^{2+} ion.
The fidelity of synthesis is high in the Zn^{2+}-catalyzed reaction —A, U, and C are incorporated about 1/2% as often as G. The discrimination is only about 10 to 1 with Pb^{2+}.
The prebiotic significance of our results is discussed briefly.

Abbreviations: Im, imidazole; ImpA, adenosine 5'-phosphorimidazolide; ImpG, guanosine 5'-phosphorimidazolide; Gp, guanosine 2'(3')-phosphate; pG, guanosine 5'-phosphate; pGp, 5'-phosphoguanosine 2'(3')-phosphate; $(Gp)_n$ (n = 2, 3...) oligomers of pG; $G(pG)_n$ (n = 1, 2...), oligoguanylates terminated by a free 5'-OH group. poly(C), polycytidylic acid; poly(G), polyguanylic acid; numbers given as superscripts between a nucleoside and a phosphate indicate the type of internucleotide linkage, e.g. G^2pG is a guanylyl-[2'→5']-guanosine.

Y. Wolman (ed.), Origin of Life, 233–239.

We have published a series of papers in which we describe the templat directed condensation reactions of adenosine derivatives in considerable detail. We have shown that poly(U) directs the synthesis of oligoadenylic acids at least up to the octamer from a suitable adenosine derivative, the 5'-phosphorimidazolide, ImpA—in the absence of poly(U) only the dimer an traces of the trimer could be detected. The product formed in this templat reaction is predominantly [2'-5']-linked, unlike the nucleic acids which are exclusively [3'-5']-linked (1, 2). Recently, we have also shown that the Pb^{2} ion catalyzes the formation of the longer oligomers, and increases the proportion of the natural [3'-5']-linkage in the products (3).

The corresponding reactions of the 5'-phosphorimidazolide of guanosin (ImpG) could not be studied in comparable detail, owing to the difficulty of separating oligo G's in paper chromatographic systems. The development of an RPC-5 column (4) that enables us to resolve oligomers at least up to the 40-mer, and even allows separation of linkage isomers, in some cases, has enabled us to extend our work considerably. Here we report on the eff of Zn^{2+} and Pb^{2+} on the efficiency and stereoselectivity of the template-directed oligomerization of ImpG.

Polycytidylic acid (5) and unlabelled and [^{14}C]-labelled guanosine 5'-phosphorimidazolide (ImpG) were prepared by modifications of published procedures (6). $(Gp)_n$ markers were prepared by partial alkaline hydrolys of poly(G). Paper chromatography was carried out on Whatman 3MM pape using n-propanol-conc. ammonia-H_2O (55:10:35) as solvent. High performance liquid chromatography (Waters) was carried out on an RPC-5 colum The column was eluted with a linear gradient (0-0.06M) of $NaClO_4$ at pH 12 Enzyme degradations with pancreatic RNase and ribonuclease T_1 were carried out by modifications of published procedures (7).

All template-directed reactions were carried out at 0°C as described previously (2, 3) (see caption to Figure for reactants and concentrations). Pancreatic ribonuclease was used to degrade the poly(C) prior to analysis of oligo(G) products.

The traces in Fig. 1(a) and 1(b) are representative of the HPLC elutio profiles of the products obtained from the self-condensation of ImpG in a 2, 6-lutidine buffer, in the presence of the Pb^{2+} ion (a), or the Zn^{2+} ion (b).

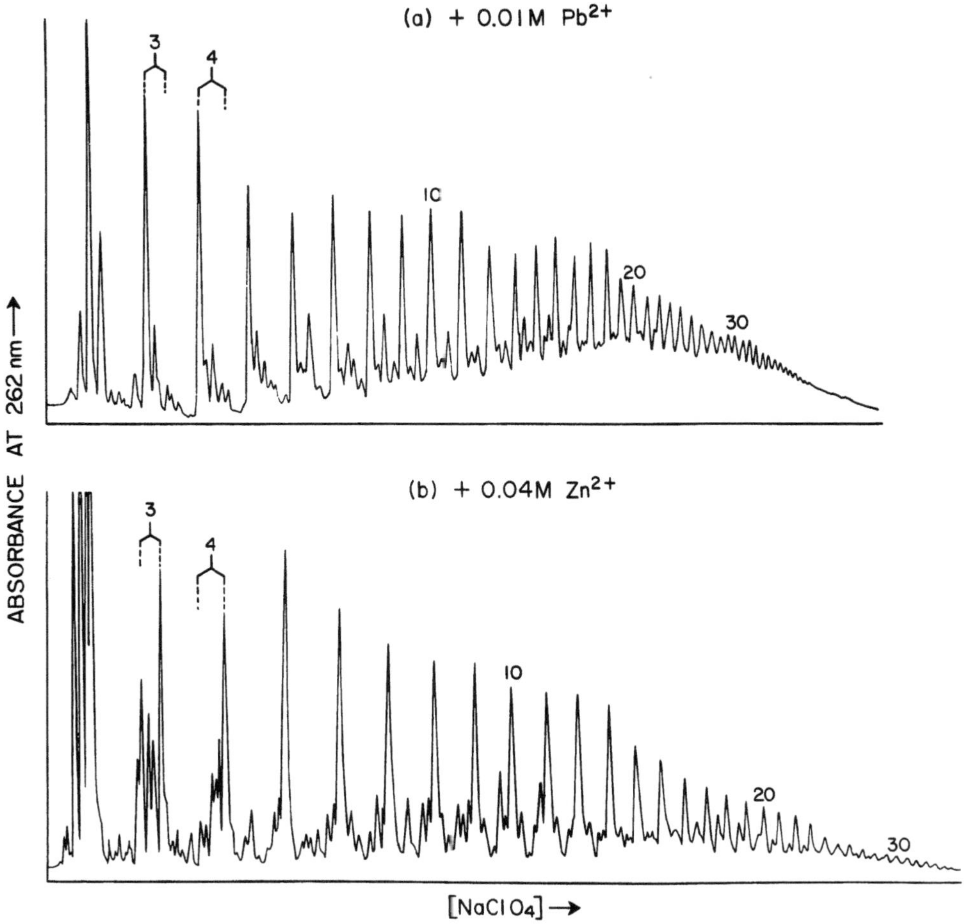

FIGURE I. Elution Profiles of Products from the Template-Directed Self-Condensation of ImpG in the Presence of 0.01M Pb^{2+} (a) or 0.04M Zn^{2+} (b). The positions of the major 10-, 20- and 30-mer peaks are indicated. The positions of the all [2'-5']- and all [3'-5']-linked isomers of $(pG)_3$ and $(pG)_4$ are also indicated. Reaction conditions: 0.02M $ImpG^*$ (0.25 mCi/mmole), 0.04M poly(C), 0.4M $NaNO_3$, 0.5M $Mg(NO_3)_2$, 0.4M 2,6-lutidine buffer, pH 7.0. After 12 days at 0°C, excess EDTA was added, and the pH adjusted to 7.9 with Tris. Pancreatic RNase (0.25 mg/μmole poly(C)) was added, and the solution incubated at 37°C for 8 hr. Material on chromatogram (a) accounts for 86.2% of the products, and on (b) 74.6%. The remaining product, pG, eluted with the voic volume (not shown).

We first identified a variety of short oligomers in our products, including G^2pG, G^3pG, pG^2pG, pG^3pG and a mixture of $(pG)_3$ isomers, by co-chromatography with authentic markers. In the Zn^{2+}-catalyzed reaction, we then assumed that the regularly spaced major peaks were the oligomers $pG^3(pG)_n$. We confirmed this assumption by showing that for all values of n up to 35 the $(pG)_n$ oligomers had retention times close to those of the corresponding enzymatically-prepared [3'-5']-linked $(Gp)_n$ oligomers. We als showed that treatment of the reaction mixture with ribonuclease T_1 destroy the material responsible for the major peaks and yielded only monomer anc small amounts of very short oligomers as hydrolysis products.

The analysis of the oligomers obtained with Pb^{2+} was more difficult. Alkaline hydrolysis of the total material on the origins of paper chromatograms yielded only G, Gp and pGp establishing that we were dealing with a family of [2'-5']- and [3'-5']-linked oligonucleotides. We isolated material from the main peaks and from the clusters of subsidiary peaks which accompany the main peaks. The oligomers responsible for the main peaks were completely resistant to ribonuclease T_1 and, therefore, entirely [2'-5']-linked. The material giving the minor peaks was degraded to shorter oligomers, showing that they contained [3'-5']-links.

In control reactions in which the template poly(C) was omitted, only small yields of dimer and trimer were obtained when the Zn^{2+} ion was present. Oligomers up to the 8-mer were obtained with Pb^{2+}, with the higher oligomers in low yield.

Recently, we have begun to study the dependence of the length of the products of template-directed reactions on the length of the oligo(C) templates. The results are unexpected. In the Pb^{2+} catalyzed reaction short oligo(C)'s facilitate the formation of much longer oligo G products. Preliminary observations suggest that the same is true in the Zn^{2+} catalyzed reaction.

We have also measured the fidelity of the poly(C)-directed self-condensation of ImpG. The reactions were carried out as described above, but with the inclusion of an amount of ImpU, ImpC or ImpA equal to the inpu of ImpG. In each case we prepared two reaction mixtures, one with [^{14}C]-labeled ImpG and the "wrong" nucleotide unlabeled, and one with the labels reversed. A comparison of the radioactivity incorporated into oligomers in the two reactions allows one to determine the ratio of G to the "wrong" nucleotide that is incorporated.

The results of this experiment are very different for Pb^{2+} and Zn^{2+}. The "wrong" base (U, C or A) is incorporated about 10% of the time when the Pb^{2+} ion is used as catalyst. However, with Zn^{2+} as catalyst, the "wrong" base is incorporated less than 1/2% of the time. A fidelity of 200 approaches within about an order of magnitude that achieved by RNA polymerases.

In another recent series of experiments we have surveyed a large number of metal ions as catalysts for the poly(C)-directed oligomerization of G. Our results may be summarized as follows:—

(1) The Sn^{2+}, Sb^{3+} and Bi^{3+} ions have some activity in promoting the formation of [2'-5']-linked oligomers. None is nearly as effective as Pb^{2+}.

(2) No metal ions other than Zn^{2+} promote the formation of [3'-5']-linked oligomers.

(3) The Mg^{2+} in our buffers cannot be replaced by Ca^{2+}, Sr^{2+} or Ba^{2+}. Mn^{2+} can replace Mg^{2+} up to about 50%.

(4) In a preliminary experiment we found that alternating poly(Arg Leu) can replace Mg^{2+} in the Zn^{2+}-catalyzed condensation.

In this section we discuss briefly the relevance of our results to theories of the origins of life. Discussion of other aspect of our work will be published together with our detailed experimental results.

The replication of nucleic acids is the central molecular process on which genetic continuity is based. It has often been postulated that a similar template-directed reaction played an important role in the early evolution of life on the Earth. However, it has proved difficult to achieve such a reaction efficiently without the help of enzymes. Thus we are faced with the classical "chicken and egg" problem—no replication without enzymes but no enzymes without replication (and the genetic code!)

The demonstration that a reaction similar to nucleic-acid replication can proceed without the help of complex polypeptide catalysts would overcome the difficulty. If a very simple system could be shown to support molecule replication, it would be much easier to understand how life evolved on the Earth.

Our results are a step in this direction. The smallest biologically important nucleic acids, the t-RNA's, are 70-100 bases long. Thus the demonstration that a nucleic-acid like molecule some tens of bases long can direct the synthesis of its complement is a suitable goal for work on this aspect of the origins of life. We are encouraged that such simple catalysts as the Pb^{2+} and Zn^{2+} ions can catalyze the formation of oligo G's in this size range on oligo(C) or poly(C) templates.

Our results, however, are only a first step towards molecular replication. We know that the incorporation of pyrimidines in non-enzymatic template-directed reactions is much more difficult than the incorporation of purines (1,2). Furthermore, the replication of heteropolymers cannot proceed through triple-helices, while our reaction may well involve a 2 poly(C) ImpG helix. Thus our substrates and reaction conditions will probably have to be modified substantially before we are able to synthesize the complement of an arbitrary oligonucleotide polymer.

The regiospecificity of the template-directed condensation of ImpG or poly(C) is intriguing. In the absence of a template, the 2'-OH of a nucleotid is about ten times as good a nucleophile against phosphoramidazolides as th 3'-OH (6). Thus it is not surprising that the Pb^{2+} ion should catalyze the formation of [2'-5']-linked oligomers. However, the formation of largely [3'-5']-linked oligomers when the Zn^{2+} ion is used as catalyst was not anticipated.

All DNA and RNA polymerases are metallo-enzymes containing zinc. This may reflect the evolution of the enzyme from a much simpler Zn^{2+}-containing catalyst. The fact that Zn^{2+} alone of the common metal ions promotes the formation of [3'-5']-linked oligomers under our condition is certainly suggestive. However, we will have to accumulate a great deal of further experience with similar systems, before we can draw any firm conclusion. Are there other conditions under which Co^{2+}, for example, shows the same regioselectivity as Zn^{2+} does in our system?

It is not likely that 5'-phosphorimidazolides were the true prebiotic substrates of nucleic-acid replication. However, it is not out of the questic that some related phosphorimidazolide or phosphoramidate played that role phosphate and nucleotides are activated as N-derivatives of lysine or histidine in several enzymic reactions. Alternatively, the correct substrate may be nucleoside 5'-diphosphates or triphosphates. We are currently investigating the metal-ion catalysis of template-directed reactions of ATP and GTP.

ACKNOWLEDGEMENT The work described in this paper was supported by NASA Grant No. NGR 05-067 and NIH Grant No. GM 13435.

REFERENCES

(1) Orgel, L. E., Lohrmann, R., Acc. Chem. Res. 7, 368 (1974).

(2) Lohrmann, R., Orgel, L. E., J. Mol. Evol. 12, 237 (1979).

(3) Sleeper, H. L., Lohrmann, R., Orgel, L. E., J. Mol. Evol., 13, 203 (1979).

(4) Pearson, R. L., Weiss, J. F., Kelmers, A. D., Biochim. Biophys. Acta 228, 770 (1971).

(5) Steiner, R. F., Beers, R. F., Jr., J. Polym. Sci. 30, 17 (1958).

(6) Lohrmann, R., Orgel, L. E., Tetrahedron 34, 853 (1978).

(7) Sulston, J., Lohrmann, L., Orgel, L. E., Miles, H. T., Proc. Nat. Acad. Sci. U.S. 60, 409 (1968).

CHEMICAL EVOLUTION OF MODEL SYSTEMS OF PRIMEVAL EARTH PERIPHERY AND THERMAL POLYMERISATION OF AQUEOUS SOLUTIONS OF CYANIDES.

F. STOETZEL and R. BUVET

Laboratoire d'Energétique Electrochimique et Biochimique
Université PARIS VAL DE MARNE
F 94010 CRETEIL Cédex (France)

ABSTRACT

The Chemical Evolution of Reactions under Energy Supply (CERES) in systems possibly simulating the primitive earth periphery was studied as it concerns the composition of the gas phase and redox and spectroscopic properties of aqueous solutions.

For comparison, the evolution of solutions of cyanide in similar conditions has been studied.

This comparison shows that the spectroscopic and chemical evolution of aqueous solutions in CERES systems, and more generally of any system simulating the primeval evolution of earth periphery from methane and ammonia or nitrogen, cannot be simply accounted for on the basis of formation, polymerisation and transformation of cyanides.

For several years and until she was obliged to stop her work before her death, Francine STOETZEL and I have developed under the name of CERES program, which means Chemical Evolution of REactions in primitive Solutions or Chemical Evolution of Reactions under Energy Supplies, a series of experiments aimed at exploring the physiological and evolutionary properties of systems simulating several kinds of possible primitive earth peripheries, starting from their equilibrium state and placing them under continuous supplies of absorbable energy (1).

Most frequently we started from methane at 0.5 to 0.7 atmosphere in a 10 l flask, representing 0.2 to 0.3 moles, in the presence of aqueous solutions containing only simple inorganic compounds and particularly ammonia, heated at 65°C during 20 hours per day ; and we supplied energy in the system from spark

Y. Wolman (ed.), Origin of Life, 241–246.

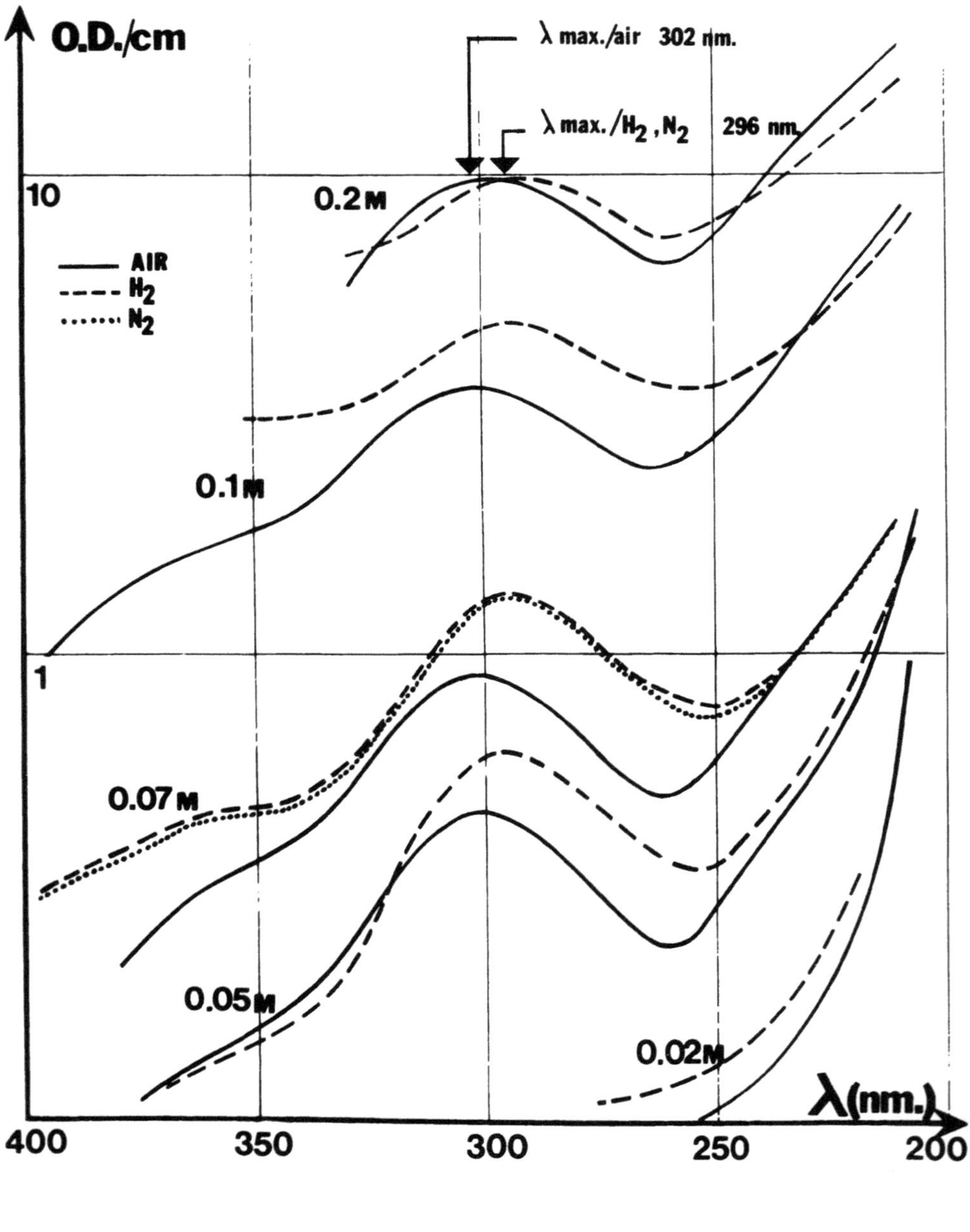

Figure 1 :
Spectra of 0.02 to 0.2 M cyanide aqueous solutions heated in closed vessels at 65° ± 0.2°C in ammonia buffer (0.2 M NH_4Cl + 0.2 M NH_3)
under hydrogen, air or nitrogen (for 0.07 M cyanide concentrations only).

discharges in the gas phase, which was initially made of methane, ammonia and water, during 10 hours per day. Samples of gaseous and aqueous phases were taken periodically and we determined their physicochemical properties by analytical tests modifiying as less as possible the solutions composition.

During the last runs we did before Francine STOETZEL stopped her work we observed, by using a palladium gauge (2) for measuring separately the partial pressure of hydrogen formed by the spark discharge, that every day when the solution was heated during 10 hours after the discharge was stopped, the partial pressure of hydrogen slowly decreased. Which probably meant that some kind of hydrogenation of unsaturated compounds formed in the aqueous phase occurred and could represent a functional precursor of some electron exchange from the atmosphere to the solution i.e. in the opposite direction to the contemporary one.

Consequently, we decided to examine if any similar phenomenon could also occur with simpler solutions obtained from the chemical evolution of KCN alone, taken in the same range of low concentrations as it is produced in CERES experiments, and in ammonia buffer also at similar concentrations as involved in CERES systems.

To this end, we simply placed diluted KCN 0.02 to 0.2 M solutions in 0.2 M NH_3 + 0.2 M NH_4Cl ammonia buffer, stirred under hydrogen and heated at 65 $\pm$ 0.2 °C for 24 hours in Warburg manometers.

We never observed in these systems any absorption of hydrogen, which makes a first difference with complete CERES experiments from spark excitation of gas phase.

But moreover most interesting results were obtained when comparing the ultraviolet spectra of the obtained solutions with either the spectra of the same solutions samely heated under air or with the spectra of the solutions obtained in CERES experiments.

The ultraviolet spectra of cyanide solutions at different concentrations after 24 hours of heating under air, nitrogen and hydrogen are represented by figure 1. The presence of the caracteristic absorption peak at 296 nm shows that the the diaminomaleodinitrile, which is the well known tetrameric first detectable product obtained by polymerization of cyanides (3,4,5) is formed at 65°C even from cyanide concentrations as low as 0.05 M.

The control experiment performed under nitrogen at 0.07 M confirms that hydrogen does not react with polymerized cyanide solutions, since spectra obtained either under hydrogen or under nitrogen are exactly the same.

However, it appears that the presence of oxygen from the air modifies very noticeably the spectroscopic evolution of diluted cyanide solutions, at least in this range of concentrations since the differences tend to vanish when the concentra-

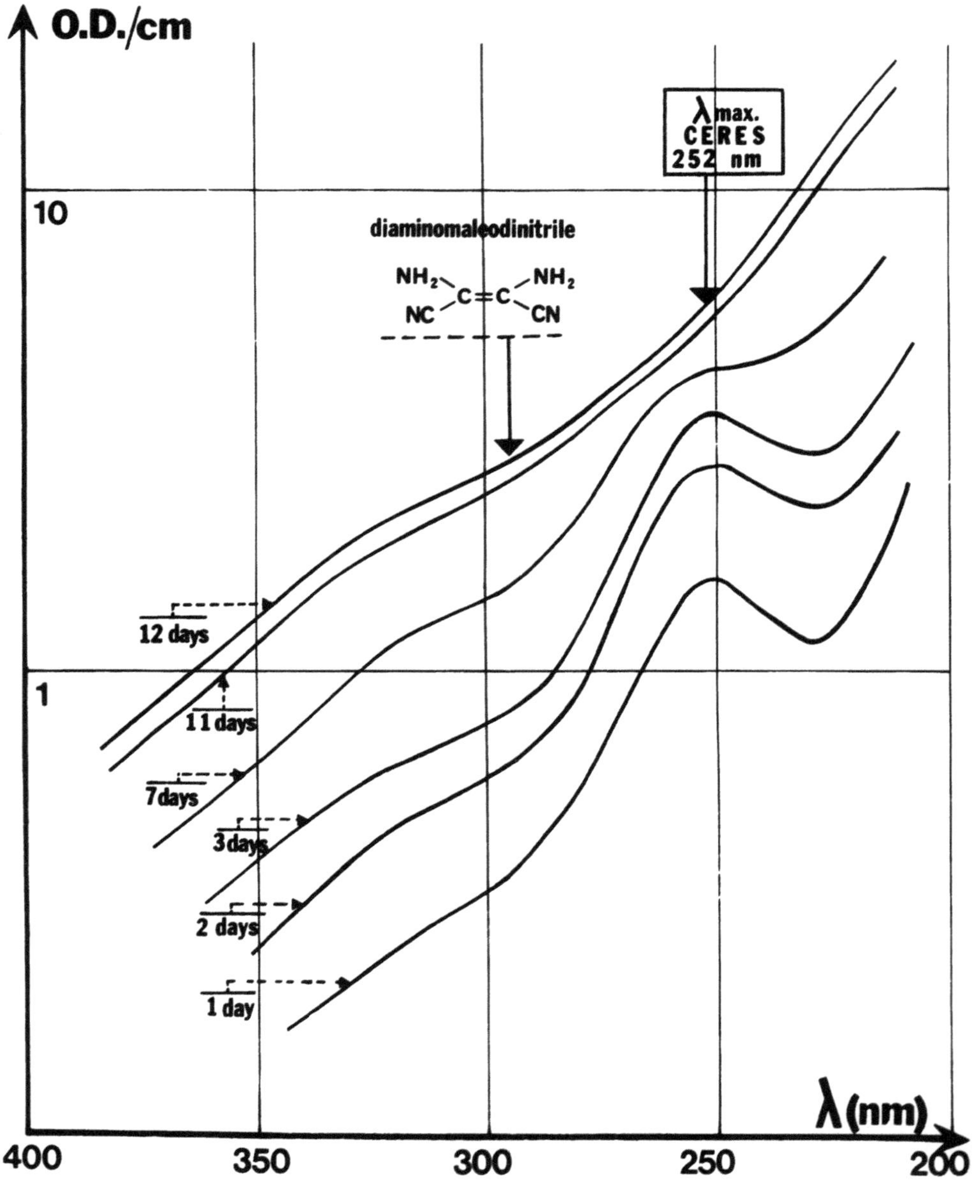

Figure 2 :

Spectra of the aqueous solutions obtained from a CERES run starting with 10.2 l of CH_4 at 0.5 atm (0.21 mole) treated by spark discharge for 10 h per day in the presence of 0.5 l of 0.42 M NH_3 and 0.1 M NaCl.

tion is increased. Unfortunately we had not the opportunity of performing experiments in Warburg manometers under air.

But, finally, most remarquable as it regards the primitive formation of biochemicals is the result of the comparison with spectra obtained in typical CERES experiments which are shown on figure 2.

This comparison clearly shows that the absorption peak observed at 252 nm in CERES experiments and which could certainly also be observed in all other global simulations of terrestrial primitive evolution cannot absolutely be accounted for by any of the spectroscopic features of solutions of cyanide polymers.

In fact, we also observed as usual in our heated solutions of cyanides, after some hours, even at these low concentrations, the formation of the well known black precipitate of cyanide high polymers, though in all CERES runs we have always observed that the solutions remained clear even after 200 hours of heating, which makes a third difference between both kinds of evolutions.

From these results we can conclude the following points :

- the cyanide tetramer, diaminomaleodinitrile, which is often considered as an important precursor in the formation of biochemicals, is in fact not detectable at noticeable concentrations in the solutions obtained from the evolution of complex mixtures of atmospheric precursors formed by gas phase excitation of reduced atmospheres.
- Another compound formed in these complex mixtures, which is neither simply ammonia nor hydrogen, completely modifies the evolution of cyanide solutions from the classically known polymerisation of cyanides which passes through the formation of the tetramer.

At present we can but consider that the responsible of the observed shift from the normal polymerization of cyanides should be one of the other products formed in the gas phase under the effect of the spark discharge. Acetylen and formaldehyde are the most acceptable candidates for such a purpose although they are known for being produced by spark discharges in smaller quantities than hydrogen cyanide (6)

ACKNOWLEDGMENTS

This paper has been written in memory of Francine STOETZEL who contributed so much by her work and human qualities to the success of the Third International Conference on the Origin of Life, Pont à Mousson (1971) and to the development of researches on the Origin of Life in the Laboratory of Biochemical Energetics successively in Paris and Créteil.

REFERENCES

(1) BUVET R. and STOETZEL F.
Chemical Evolution and Energetics of Reactions in Aqueous Solutions on the Primitive Earth.
Origins of Life (7), 93-107 (1976)

(2) BUVET R. and STOETZEL F.
Energetics of Abiogenic Chemical Systems
Biosystems 7, 2-14 (1975)

(3) ROBERTSON P.S. and VAUGHAN J.
J.Am.Chem.Soc.,80, 2691 (1958)

(4) TOUPANCE G., SEBBAN G. and R. BUVET
Etapes initiales de la polymérisation de l'acide cyanhydrique et synthèses prébiologiques.
J.Chim.Phys,67, 1870 (1970)

(5) SANCHEZ R.A., FERRIS J.P. and ORGEL L.E.
J.Mol.Biol,30, 223 (1967)

(6) RAULIN F. and TOUPANCE G.
The role of sulphur in chemical evolution
J.Molec. Evol., 9, 329-38 (1977)

THE POLYMERIZATION PRODUCTS OF α-AMINOPROPIONITRILE. THE COMPONENT SEPARATION USING CATION-EXCHANGE RESIN.

S. Morimoto, K. Kawashiro, H. Yoshida, and K. Sugiura.

Faculty of Engineering, Tokushima University,
Minamijosanjima 2, Tokushima 770, Japan.

By the batch separation on Amberlite CG-50, the total products were separated into the three parts, the acidic and neutral (A), the weakly basic (B), and the basic (C) part, using water, aq. acetic acid solution, and aq. ammonia solution as the eluent, respectively. The three parts were fractionated through the column of Sephadex G-10 and the most part of the polymeric product was obtained in the basic part (C). The amino acid analyzer chromatograms of the each gel-filtration fraction showed the peptides , peptide amides, and peptide nitriles in the respective parts, and the related imino-compounds in the three parts.

The purpose of the present work is to separate the polymerization product of α-aminopropionitrile (APN) according to the different chemical properties of the components, using the cation -exchange resin, Amberlite CG-50. The polymerization product (APN -polymer) used was obtained from the mixture of APN and α,α'-iminodipropionitrile (IDPN), after the reaction at the temperature near 0°C during 8 years.[1)]

The batch method of ion-exchange separation is shown in Figure 1. After an aq. solution of APN polymer was equilibrated with Amberlite CG-50 in a flask, the resulting solution was filtrated and rinsed with water. The filtrate and rinse were lyophilized to yield Section A, the acidic and neutral component. The elutes

Y. Wolman (ed.), Origin of Life, 247–254.

with aq. 2% acetic acid and aq. 5% ammonia were also lyophilized to yield Section B, the weakly basic component and Section C, the basic component, respectively. The batch separation yields are shown in Table 1. The excess of the summation may be due to some reaction such as hydrolysis.

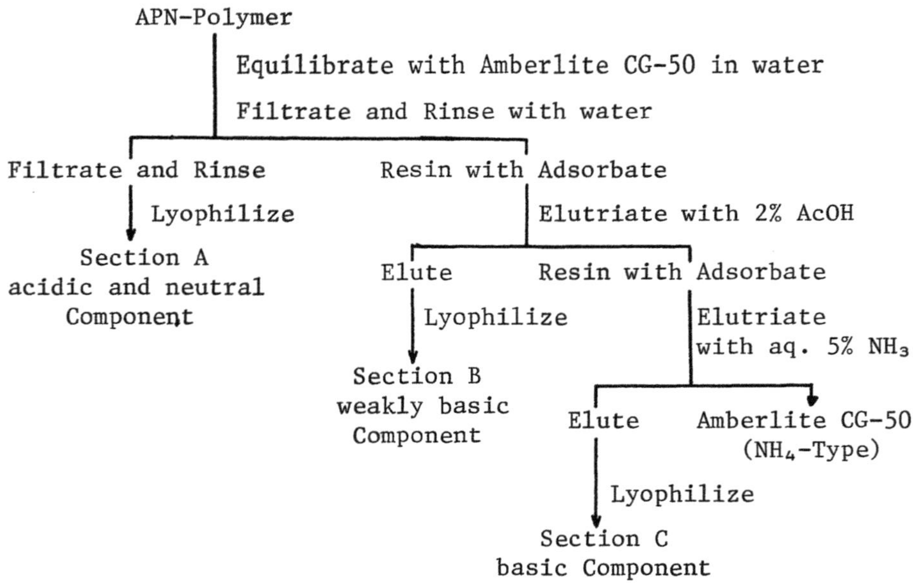

FIGURE 1. The batch separation of APN-Polymer using cation-exchange resin.

TABLE 1.

The Batch Separation Yields of APN-Polymer using Cation-Exchange Resin.

Section A : Elutriated with water
" B : " " aq. 2% AcOH
" C : " " aq. 5% NH_3
Amberlite CG-50 : 300 ml

Ex. No.	APN-Polymer gr	Section A gr (%)	Section B gr (%)	Section C gr (%)	Total gr (%)
IR-6	14.97	2.70	8.08	6.04	16.82
		(18.0)	(54.0)	(40.3)	(112.3)
IR-7	14.94	1.69	7.56	6.80	16.05
		(11.3)	(50.6)	(45.5)	(107.4)
IR-8	15.00	3.22	6.73	6.89	16.84
		(21.5)	(44.8)	(45.9)	(112.2)

The Section A and B of the batch ion-exchange separation were fractionated through the column of Amberlite CG-50 and the ninhydrin color yield (O. D.) of each fraction was observed to give the chromatograms shown in Figure 2. The Section A contained the small amount of the weakly basic component (eluted with aq. 1 % AcOH). The Section B contained only the trace amount of the acidic and neutral component (eluted with water).

For the purpose of gel-filtration according to molecular sizes, the three Section, A, B, and C were fractionated through the column of Sephadex G-10 and each fraction obtained was lyophilized to yield the solid product and give the chromatograms shown in Figure 3. The sum of the each fractional yields was divided into two parts, the Polymer and Oligomer Section, and are shown in Table 2, being calculated on the basis of the total re-

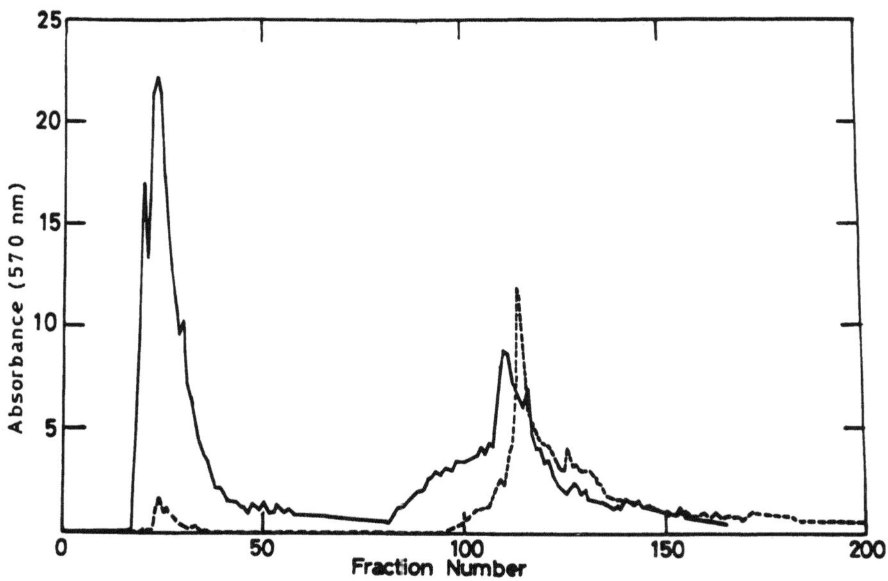

FIGURE 2. The ion-exchange column chromatogram of each Section of batch ion-exchange separation.
Amberlite CG-50, 3.0 x 100 cm : Eluent, water and aq. 1% AcOH
Flow rate, 18-20 ml/cm^2·hr : Each fraction, 30 ml
——————— Section A (acidic and neutral component)
------- Section B (weakly basic component)

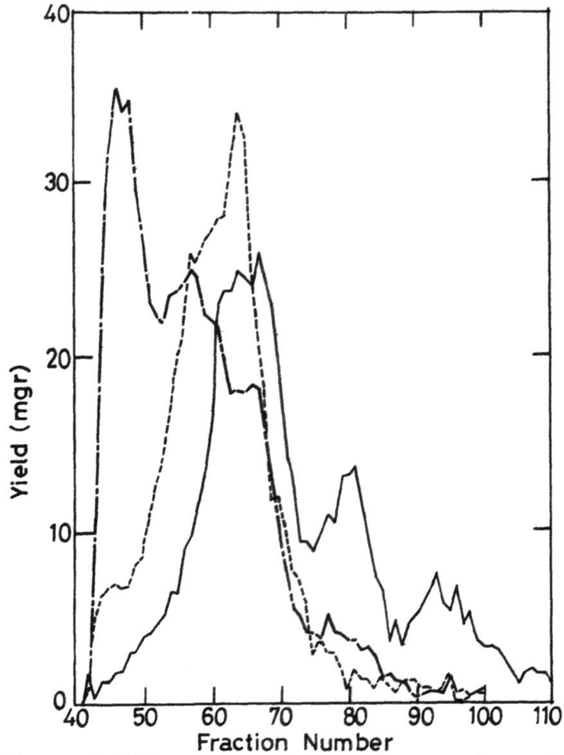

FIGURE 3. The gel-filtration chromatogram of each section of the batch ion-exchange separation.
Sephadex G-10, 2.0 x 100 cm : Eluent, aq. 2% AcOH
Flow rate, 98 ml/hr : Each fraction, 3 ml
——————— Section A (acidic and neutral component)
- - - - - - - Section B (weakly basic component)
—·—·—·—· Section C (basic component)

TABLE 2

The Gel-Filtration Yields of Each Section of the Batch Ion-Exchange Separation (IR-8).

	Polymer Section[b)]	Oligomer Section[b)]
Section A	4.7%(0.9%)	95.3%(18.2%)
Section B	14.2 (5.7)	85.8 (34.3)
Section C	39.6 (16.2)	60.4 (24.7)

a) The batch ion-exchange yields were calculated as 19.1(Section A), 40.1(Section B), and 40.9%(Section C), on the basis of the total recovered weight(100%).

b) The Polymer and Oligomer Section include Fraction 41-52 and 53-110,respectively. The over-all yields of ion-exchange and gel-filtration are shown in the parentheses.

covered weight(100%). The over-all yields of ion-exchange and gel-filtration in the parentheses had their highest value 34.3% at the Oligomer Section of Section B and the highest of Polymer Section was 16.2% at the Section C.

The solid product obtained by lyophilizing each fraction of the gel-filtration was dissolved in the small amount (0.5 ml) of water to make the concentrated fractional solution and an aliquot (0.1 ml) of the solution was charged into an amino acid analyzer to give the chromatogram as shown in **Figure** 4-7.

α,α'-Iminodipropionitrile IDPN, α,α'-imino-propionamide-propionic acid I(PAM,PAC), tetraalanine $(Ala)_4$, alanine Ala, trialanine $(Ala)_3$, and dialanine $(Ala)_2$ were observed in the Section A as shown in Figure 4.

Hexaalannylamide $(Ala)_6NH_2$, pentaalanylamide $(Ala)_5NH_2$, tetraalanylamide $(Ala)_4NH_2$, trialanylamide $(Ala)_3NH_2$, dialanylamide $(Ala)_2NH_2$, α-aminopropionitrile APN, α,α'-iminodipropionamide IDPAM, ammonia NH_3, and alanylamide $AlaNH_2$ were observed in the Section B as shown in Figure 5.

α,α'-Iminodipropionimide IDPIM, dialanyl-α-aminopropionitrile $(Ala)_2APN$, NH_3, and $AlaNH_2$ were observed in Section C as shown in Figure 6 (Fraction 62,65,69). As the fraction number decreased, the number of peaks observed decreased as shown in Figure 7. The Section B and C included the small and large amount of the reddish brown product which was eluted with aq. 1N NaOH, but not with the buffer (pH 5.28).

The elution volumes of the peptide, peptide amides, and the peptide nitriles were observed, using the amino acid analyzer (Aminex A-4, Buffer pH 3.25).[2] The relationships between their elution volumes (Ve) and molecular weights (M) were obtained as shown in Figure 8, indicating the combination of the two separation mechanisms, ion-exchange and gel-filtration.

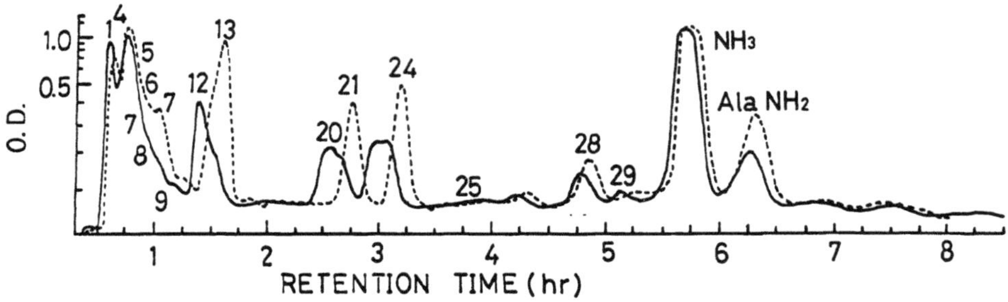

FIGURE 4. The amino acid analyzer chromatogram of each gel-filtration fraction(concentrated) of the neutral and acidic component(IR 8A). Buffer pH 3.25 only

——— Fraction 60 - - - - - Fraction 61

6 IDPN, 7 I(PAM,PAC), 12,14 $(Ala)_4$, 13 Ala, 19,20,22,23 $(Ala)_3$ 21,24 $(Ala)_2$

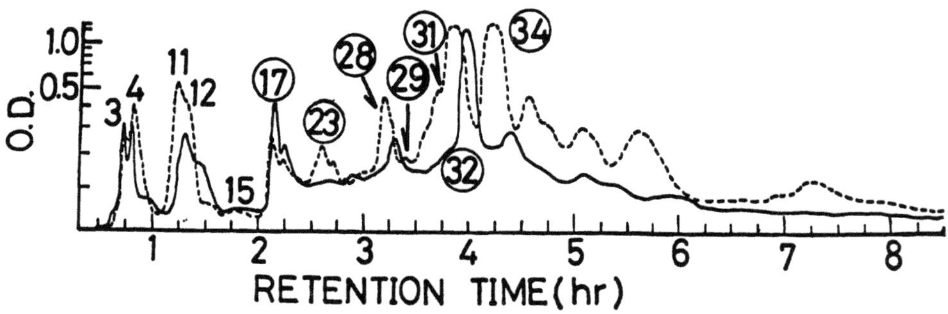

FIGURE 5. The amino acid analyzer chromatogram of each gel-filtration fraction(concentrated) of weakly basic component(IR 8B). Buffer change(pH 3.25→5.28) 0.5 hr

——— Fraction 58 - - - - - -Fraction 61

12 $(Ala)_6NH_2$, 15 $(Ala)_5NH_2$, ⑰,⑱ $(Ala)_4NH_2$, ㉓,㉕ or ㉖ $(Ala)_3NH_2$, ㉘,㉚ $(Ala)_2NH_2$, ㉙ APN, ㉛ IDPAM, ㉜ NH_3, ㉞ $AlaNH_2$

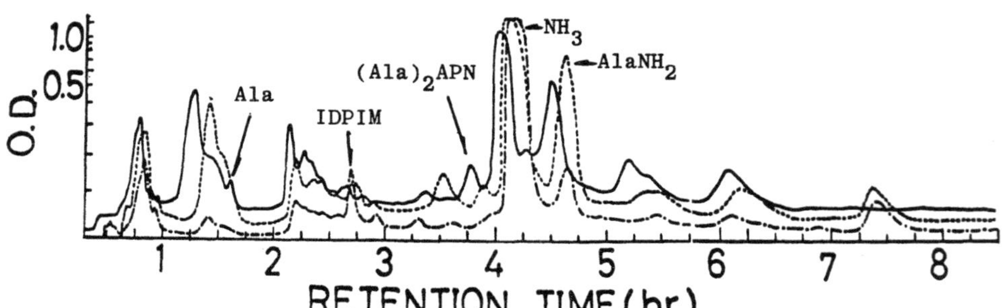

FIGURE 6. The amino acid analyzer chromatogram of each gel-filtration fraction(concentrated) of basic component (IR 8C). Buffer change(pH 3.25→5.28) 0.5 hr

——— Fraction 62 - - - - - -Fraction 65

—·—·—·— Fraction 69

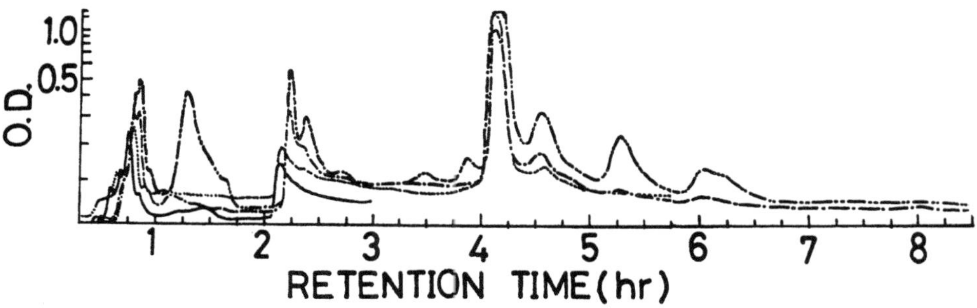

FIGURE 7. The amino acid analyzer chromatogram of each gel-filtration fraction(concentrated) of basic component (IR 8C). Buffer change(pH 3.25→5.28) 0.5 hr

——— Fraction 46 ------ Fraction 49
-·-·-·- Fraction 56 -··-··- Fraction 60

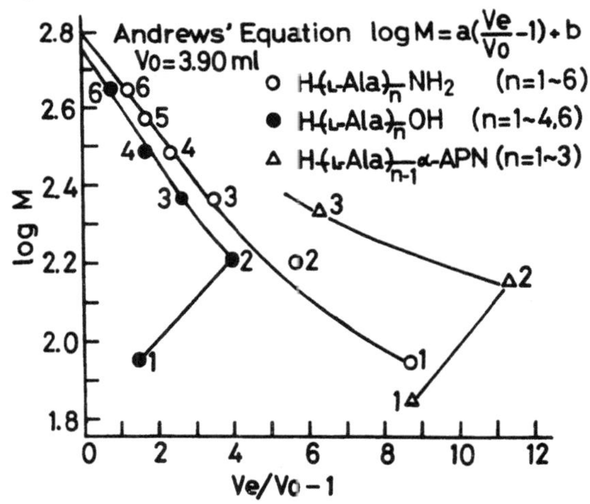

FIGURE 8. The elution volumes of peptides, peptide amides, and peptide nitriles on the amino acid analyzer chromatogram. Buffer pH 3.25 only.

The possible mechanism of the formation of these products is as follows. The peaks emerging earlier than α,α'-iminodipropionitrile (peak No. 6) on the chromatograms of the Section A (Figure 4) suggest the formation of imino peptides, although α,α'-iminodipropionic acid can not be observed, because this is negative to the ninhydrin reaction.

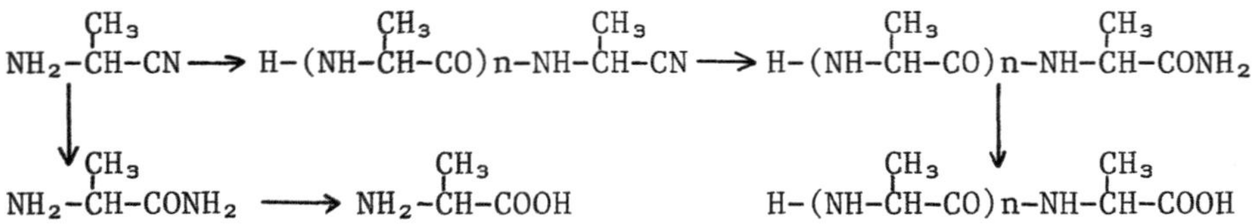

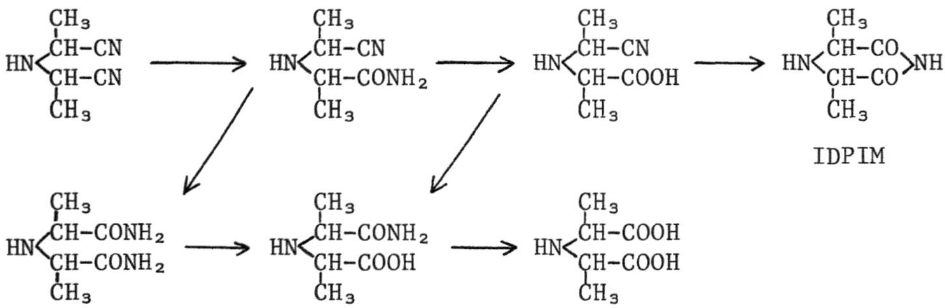

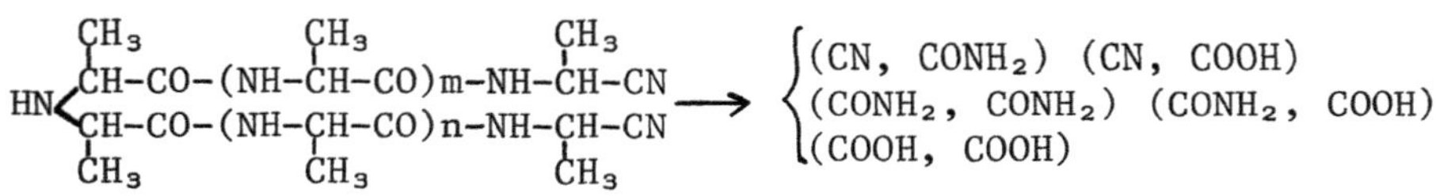

From these results it was suggested that the polymeric products obtained in the Polymer Section of the Section C may have some other structural units as well as the alanyl residues in their molecules. The mass spectrometry will be useful not only to confirm the present tentative identification but also to investigate the unidentified products. The ion-exchange separation of the present work may make it easier than otherwise.

REFERENCES

1) Morimoto, S., Kawashiro, K., and Yoshida, H., Origins of Life, 8, 341 (1977).

2) Kawashiro, K., Morimoto, S., Yoshida, H., and Sugiura, K., Origins of Life, 8, 347 (1977).

THE COLD THEORY OF THE ORIGINS OF LIFE. ASSUMPTIONS FOR THE THEORY AND EXPERIMENTAL EVIDENCE

C.I.Simionescu and F.Dénes

"Petru Poni" Institute of Macromolecular Chemistry, 6600 Jassy, ROMANIA

A possible evolution model is suggested on the base of experimental results obtained in low temperature conditions and by using cold-plasma as an energy source.

The investigations carried out in the last three decades, under simulated prebiological Earth conditions have proved the possibility of obtaining a great number of biologically important molecules. Three main theories have been developed, suggesting possible ways for the appearance of the protobiopolymers: the "Thermal Theory" /1/, the "Adsorption Theory" /2/ and the "Theory of Dehydrating Agents" /3,4/. These theories are not in acord, however, with the "natural" development of the macromolecular compounds required by the early living systems. Almost all of the reactions carried out in simulated conditions involve manmade or natural (extracted) "biomonomers" representing in this way a break of continuity of chemical evolution. More than that, these reaction systems and theories are focused on obtaining and elaborating the corresponding theoretical support for a certain class of biopolymers (polypeptides, nucleic acids, etc.). Matthews and co-workers claimed also that heteropolypeptides are formed directly from hydrogen cyanide /5/. A hybrid theory was also developed, which requires both hypohydrous conditions and adsorption phenomena /6,7/. These theories present, at the same time, a number of weak points in their explanations:
-If one would admit the existence of a prebiotic "hot"

Y. Wolman (ed.), Origin of Life, 255–262.

globe,a very humid atmosphere would characterize the Earth,where polycondensation processes would not develop.

-How can one explain that the higher limits of the temperatures did not arise above the destruction temperatures of the protobiopolymers?

-How could the protobiological polymeric structures have "survived" in high-energy radiation conditions on hypohydrous surfaces?

-How can the stability of the dehydrating agents in the primitive oceans be explained while an "unlimited" number of reactions could take place between the large number of co-existing compounds,the "biomonomers" and the dehydrating agents respectively?

-How can the MONOMER $\rightleftharpoons$ POLYMER + H_2O equilibrium's shifting to the right side of the equation in dilute aqueous solutions be explained?

Beside these theories which certainly present very important experimental findings and theoretical explanations regarding the formation of protobiopolymers and the first living systems,our investigations suggest a different possible (new) way for the appearance of protobiopolymers and the protocell (Cold Theory). The main statements which support the Cold Theory are as follows:

1.The fourth state of the matter in the vicinity of our planet determines important changes in the structure of the gas molecules,which are components of the atmosphere,through electrical phenomena (ionosphere).

2.The atmospheric pressure decreases exponentially with height,above the ground level. It is known also, that the temperature in the atmosphere decreases with height until a minimum level (close to -60°C) is reached,then the temperature inversion occurs (at about 50,000 meters the temperature is still below 0°C /8/.

3.Cold plasma polymerization processes show that the adsorption phenomena play a crucial rôle in the processes which lead to macromolecular compounds. These phenomena are much more intense at low temperatures. Large low temperature-spots could existed on the surface of the abiotic Earth or they could be present in the atmosphere as ice particles or ice-coated grains of various cosmic materials.

4.There are no doubts that the atmosphere of the Earth was at first reducing,but later oxidizing. Between the reducing and the oxidizing stage there must have been a transitional period.

5.All of the energies available on the primitive Earth with potential action on the gaseous phase components of the primeval atmosphere, present an important characteristic: their action on gaseous mixtures determines the generation of a large number of active species mono- and multifunctional free radicals especially. It follows that the absolute amount of energy on the Earth is what was important for the molecular evolution, rather than its quality.

6. The "survival" of the protobiological compounds could have been possible by their trapping in the upper layer ice surfaces (frozen oceans).

7. The low temperatures could create the anhydrous conditions in the "polycondensation reactions" (trapping processes of the free radicals and their recombination in the warming up periods).

The low temperature environments on the primeval Earth could have the following consequences:

-The decrease of the atmospheric pressure value by partial condensation of ammonia and water determined the practically free penetration of the short wave solar radiations through the methane rich atmosphere.

-The crystalline structure of the ice and the possible alternative melting-freezing processes could concentrate the organic matter through the phenomenon of expelling the "impurities" during the recrystallization and could create matricial effects.

-The ice-trapped protobiological compounds ensured the "instant" high concentration values during the heating climate periods.

The recombination of the active species do not follow the common chemical thermodynamics, resulting in a built up of protobiopolymers in aqueous environments by recombination processes. All of these statements suggest the possibility that the macromolecular precursors of living matter have been generated through the accumulation and the combination of the active species (generated in the gas-phase) on ice-covered surfaces, including the frozen oceans.

Consequently our experiments were carried out in the apparatus shown in figure 1, which is presented as a laboratory model for the conditions of the primeval Earth/9/. Depending on the relative ratio of the feed composition, polypeptide-like and polysaccharide-like structures become predominant /9,10/. Polypeptide-like /9/ and polysaccharide-like materials /10/ have been preferentially formed when ammonia and water were the main components of the feed composition succesively. The main polymer structures are

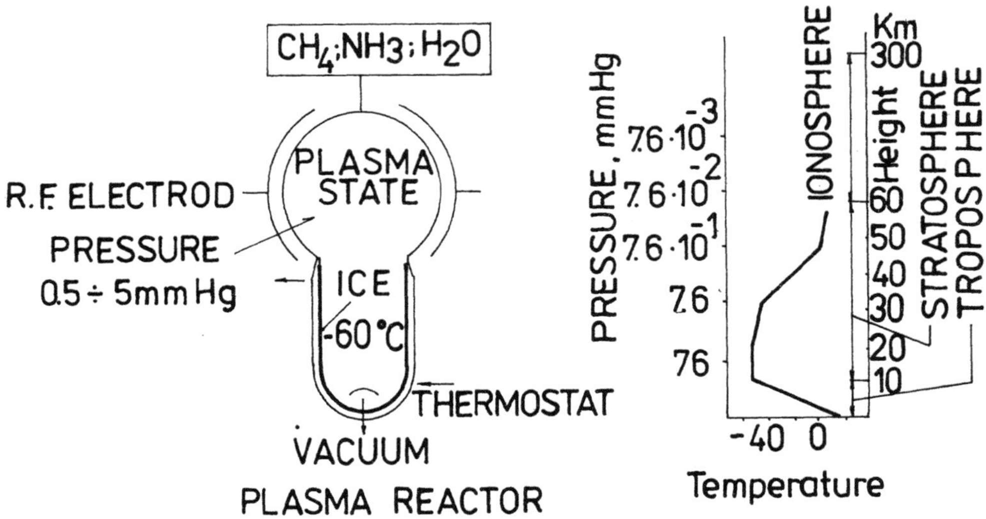

Fig.1-Laboratory model for the conditions of the primeval Earth

accompanied by linear and cyclic polymers and copolymers of hydrogen cyanide and nitriles /9/. The components identified subsequent to the hydrolysis of the raw material were amino acids, sugars, purines, pyrimidines and unknown neutral and basic ninhydrin positive derivatives /9,10/. The very low concentration of free amino acids, sugars and bases in the raw material should also be mentioned. It should be pointed out that these protobiopolymers are as a matter of fact mixtures of copolymers /9/. The evidence which prove the polypeptide-like material existence in the raw reaction product are catalogued in table 1. Table 2 shows the nitrogen content of the fractionated raw material. Analytical data prove the presence of peptide linkages for the nitrogen-containing fractions and for the raw material. A black insoluble material was also identified on the top of the column /9,10/. Table 3 summarizes the experimental data which prove the polysaccharide-like structures formation in the nitrogen-free fractions. It is noteworthy that similar experimental conditions led to tetrapyrollic (porphyrine-like) structures also /11/. The comparative analysis of the raw material obtained at various feed compositions, i.e. various ratios of methane, ammonia and water, show that the ammount of (hydrophobic) hydrocarbon-like components increased in the reaction milieu with the increase of the methane concentration in the initial gaseous mixture. The data which prove the hydrocarbon-like components formation are summarized in table 4.

TABLE 1
The Evidence which Prove the Polypeptide-like Structures Existence in the Raw Reaction Product

The components resulted subsequent the hydrolysis of raw material: glycine,alanine,glutamic acid,aspartic acid and trace serine.

The molecular weight (GPC): tens of thousands;limited heterogeneity.

The degree of transformation of the raw material into amino acids by acid hydrolysis: 30-70 %.

Peptide-like material content in the raw reaction product: 0.36 - 0.65 mg/mg (Lowry method).

Structural analysis of the lyophylised raw product: the presence of the absorptions for peptide bond (IR).

Structural analysis of the hydrolysed and lyophylised raw material: the absence of vibrations peculiar to peptide bond and the presence of $-NH_3^+$ absorption.

Elemental analysis of the raw material: N-33 %, C-45 %, H-7 %, O-5 %.

The relative ratio of the polymeric and low molecular weight material in the raw reaction product:predominant peptide-like compounds,low quantity of black polymer and very low quantity of free amino acids.

TABLE 2
Elemental Analysis of Raw and Fractionated Products Obtained with a High Ratio of $H_2O/CH_4,NH_3$

Sample	N(%)	C(%)	H(%)	O(%)
Raw material	32.33	36.44	5.95	25.28
Fraction 1	0.00	42.12	6.70	51.18
Fraction 2	10.60	30.70	5.31	53.39
Fraction 3	29.10	36.71	5.64	28.59
Fraction 4	14.40	31.00	5.47	49.13
Fraction 5	16.50	30.50	5.60	47.40
Fraction 6	5.03	24.07	4.80	66.10
Fraction 7	0.82	30.90	5.15	63.13
Fraction 8	0.00	Insufficient amount of sample		

TABLE 3
The Evidence which Prove the Polysaccharide-like Structures Presence in the Nitrogen-free Fractions

The polymeric nature of the raw material: the molecular weight (GPC)-tens of thousands.
Elemental analysis of the fractions 1,7 and 8 (see table 2).
Structural analysis of lyophylised nitrogen-free fractions: absorptions (IR) peculiar to mono- and polysaccharide-like compounds.
Monosaccharide composition (TLC,GC) of nitrogen-free fractions (hydrolyzates): galactose,xylose.

TABLE 4
The Evidence which Prove the Lipid-like Structures Presence in the Raw Material

The "macromolecular" nature of fraction extracted with chloroform: the molecular weight (GPC,MS)-200-500.
Structural analysis (IR,NMR,MS) of the chloroform extracted and lyophylised fraction: predominant =CH-, $-CH_2-$ and $-CH_3$ composition (saturated linear and branched hydrocarbon chains and/or saturated condensed cycles.
Elemental analysis of the raw material: N-14 %, C-53 %, H-6 %, O-27 %.
Hydrophobic nature: solubility in chloroform, heptane, methyl cellosolve, etc.

The "protobiopolymers" from the raw material, especially the lipid-like structures, selfassemble in uniform sized (10-50 μm) very stable microspheres. The properties of the cold-conditions generated microspheres are as follows: limited size-heterogeneity, thermal stability, resistance in large pH range (6-9 pH), stability in radiant energy conditions, electrical conductivity (semiconductive properties), membrane-like properties (membrane potential, selective retention of biologically active compounds), supramolecular order in the wall-membrane structure (polarized light image).

The low temperature surface spots on the abiotic Earth have played a very important rôle in the appearance of the early "biopolymers". The importance of low temperatures has also been suggested even in the sixties /12,13/, however negligeable amount of experimental data was presented in order to support the suggestions.

The protobiopolymers have appeared simultaneously and they accumulated differentially, determined by the composition change of the primeval atmosphere. Low temperatures favorized certainly the existence of a methane rich atmosphere which determined the priority of the hydrophobic (lipid-like) structures accumulation.

The formation of the protobiopolymers has started, probably, in the warming up periods of the climate by recombination mechanisms, during the ice-melting up processes.

The structural and conformational variety of protobiopolymers assured their "adaptation" to the environment by the first natural selection processes on molecular level.

The main reaction products have a polymeric nature. The negligeable amounts of free "biomonomers" allow us to suggest that the protobiopolymers have preceded the small molecular weight compounds in their chronological appearance. The later cathegory of substances have been generated mainly by hydrolyzation and decomposition mechanisms, probably.

The protobiopolymer mixtures and those rich in hydrophobic structures especially, selfassemble in stable, uniform sized individual structures (microspheres).

The wall-membrane structure of the microspheres (protomembrane) presents membrane-like properties.

The special electrical properties of the protocell's wall and the presence of the tetrapyrollic structures suggest the possibility of development, through these membrane-like systems, the early reversible transducers of various energies (light, redox, electrochemical energies, etc.).

These findings allow us to propose the following (new) chemical evolution model:

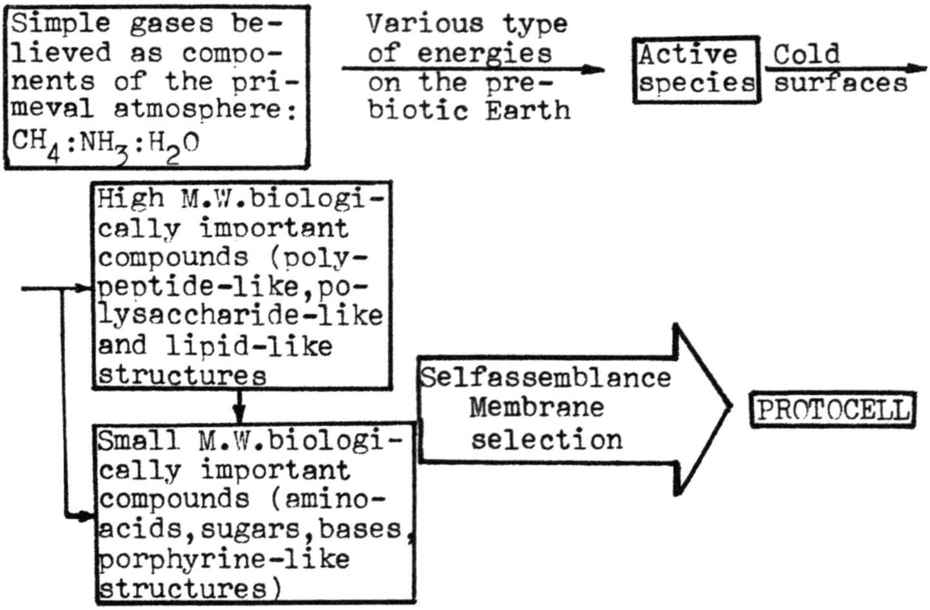

REFERENCES

1) Fox,S.W. and Dose,K.,Molecular Evolution and the Origins of Life,Second Edition,W.H.Freeman and Co.,San Francisco,1976,p.139.

2) Katchalsky,A. and Ailam,G.,Biochim.Biophys.Acta., 140,1 (1967).

3) Ponnamperuma,C.and Peterson,E.,Science,147,1572(1965)

4) Ferris,J.P.,Goldstein,G. and Beaulieu,D.J.,J.Am. Chem.Soc.,92,6598 (1970).

5) Matthews,C.N.and Moser,R.E.,Nature,215,1230 (1967).

6) Akabori,S.,Kagaku,25,54 (1955).

7) Akabori,S.,Okawa,K.and Sato,M.,Bull.Chem.Soc.Japan, 29 (5),608 (1956).

8) Gordon,R.B., Physics of the Earth,Rinehard and Winston Inc.,New York,1972,p.11.

9) Simionescu,C.I.,Dénes,F.and Negulescu,I.,J.Polymer Sci.,Polymer Symposia,64,281 (1978).

10) Simionescu,C.I.,Totolin,M.and Dénes,F.,BioSystems, 8,153 (1976).

11) Simionescu,C.I.,Simionescu,B.C.and Mora,R.,Bioelectrochemistry and Bioenergetics,5,1 (1978).

12) Sanchez,R.,Ferris,J.and Orgel,L.E.,Science,153,72 (1966).

13) Kliss,R.M. and Matthews,C.N.,Proc.N.A.S.,48,1300 (1962).

THE APPEARANCE OF PROTOBIOPOLYMERS AND PROTOMEMBRANES IN ACCORDANCE WITH THE "COLD THEORY" OF THE ORIGINS OF LIFE

C.I.Simionescu, F.Dénes and M.Totolin

"Petru Poni" Institute of Macromolecular Chemistry, 6600 Jassy, ROMANIA

Starting from simple gas mixtures (CH_4,NH_3,H_2O) suggested as being models of the primeval åtmosphere for certain time periods,we succeded in synthesizing functional and structural protobiopolymers. Methane rich atmosphere led to lipid-like hydrophobic products which selfassemble in stable uniform sized microspheres,carrying membrane-like properties.

Taking into account the assumptions of the "Cold Theory" /1/,in previous experiments /2,3/ we proved the formation of protobiopolymers starting from mixtures of methane,ammonia and water vapor,in cold plasma state (RF glow discharge),over a low temperature (-60°C) ice surface. Polypeptide-like and polysaccharide-like materials have been preferentially formed when ammonia and water respectively were the main components of the feed composition,while a high methane content led to the appearance of lipid-like (hydrocarbon-type) structures /4/.

In all of the cases the final raw reaction products and their fractions (GPC,extractions) have been investigated. Spectral analysis (IR,NMR,MS),chromatography determinations (PC,TLC,GC),gel permeation measurements,elemental analysis,electron microscopy, scanning electron microscopy and optical microscopy prove the formation of those structures and their polymeric nature.

The molecules of lipid-like compounds are forced

Y. Wolman (ed.), Origin of Life, 263–269.

by their structures to undergo specific orientations with respect to the aqueous medium and to form suitable organized entities. It is possible that such molecules could have been of importance in promoting the initiation and development of the protobiopolymer's organization through selfassembly in well defined individual structures (micelles,monolayers,bilayers,protomembranes). These early "membranes" determined during the chemical evolution,through selective retention and encapsulation processes,the qualitative change of the unorganized biologically important molecular mixture in a superior organized matter.

In previous experiments /5-7/ neutral lipids, phospholipids and membrane-like structures have been synthesized in various experimental conditions.

Our experiments carried out by using an initial mixture with high methane content /4/ led to raw materials which after extraction with chloroform (72 hours) and evaporation of the chloroform fraction,resulted in yellow solid product. Structural analysis performed on the saturated linear,branched and saturated cyclic solids show that their chemical nature is close to hydrocarbon-type derivatives. The predominant functional groups of the compounds appear as $=CH-$, $-CH_2-$ and $-CH_3$ beside the very small quantity of $-CH_2-CO-$ groups. Molecular weight measurements indicate a range between 200 and 500. Elementary analysis and amino acid determinations were carried out on the lyophylized crude final reaction products also. Glycine and alanine have been identified beside the other ninhydrine positive compounds. Pyrollic structures have also been identified in the raw products of the experiments carried out in similar conditions /8/. The protobiopolymers from the raw materials,especially the lipid-like structures selfassemble in uniform sized (10-50 µm) very stable microspheres (Fig.1 and 2). The entities survive in harsh thermal (they can be heated up to 80°C) and radiant (200 W mercury lamp placed at 100 mm from the sample) energy conditions. They disaggregate partially under the action of energy impacts and reassemble in the absence of them. The microspheres are very stable,at the same time,in a large pH range (6-9 pH). Electrical measurements show that the lyophylized lipid-like raw material presents semiconductive properties. The specific resistivity plotted versus the temperature indicates a linear variation /9/. Membrane-like properties were recorded also. Optical microscopy performed on

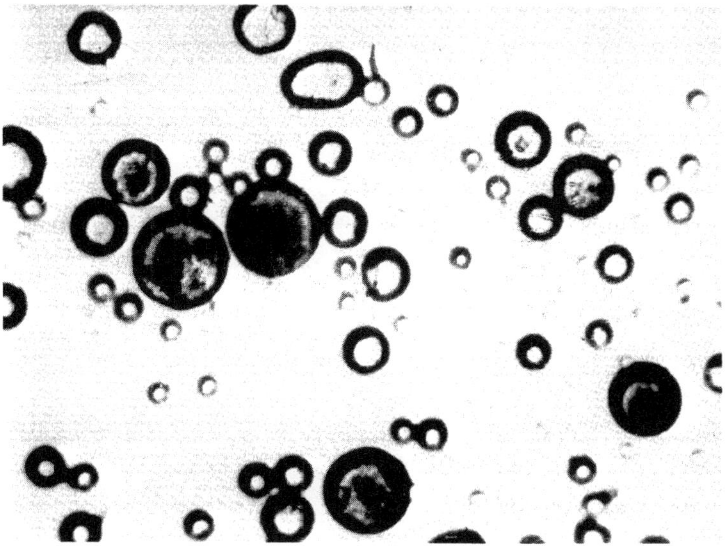

Fig.1-Photomicrograph of self-assembled protobiopolymers (microspheres obtained from methane rich feed mixture). Magnification X 600

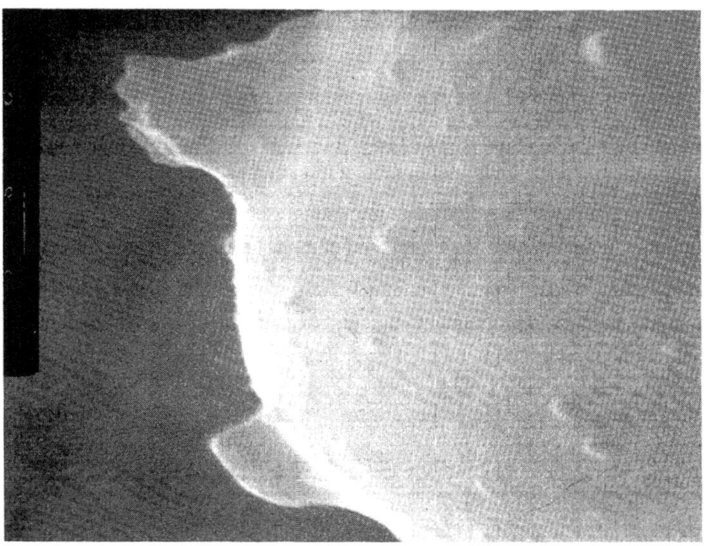

Fig.2-Scanning electron microscopy image of the microspheres. Magnification X20,000

the microspheres in polarized light conditions (crossed Nicol prism) shows the existence of a supramolecular order (Fig.3).

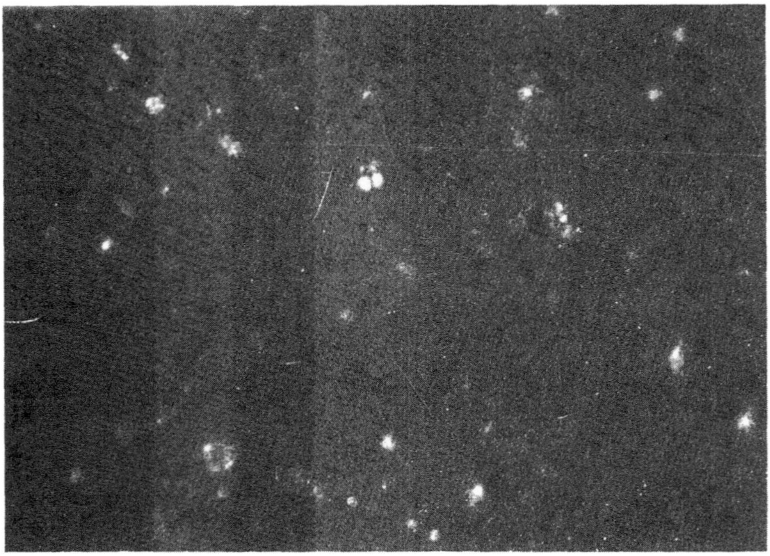

Fig.3-Crossed Nicol prism image (polarized light) of self-assembled protobiopolymers. Magnification X 120

Double layers were evidenced in the wall-membrane of the microspheres by electron microscopy (Fig.4).

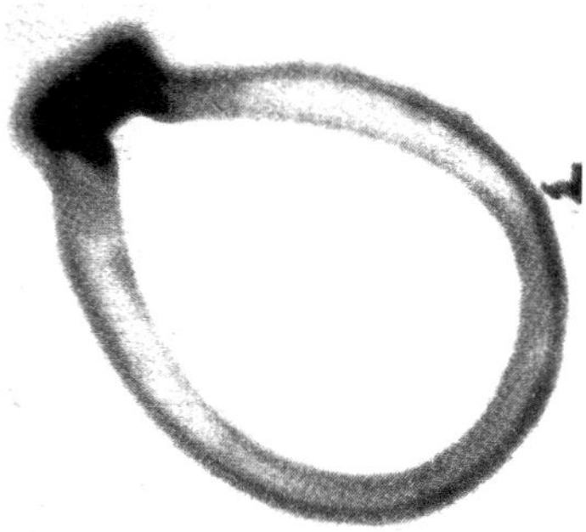

Fig.4-Electron micrograph of a section of osmium tetroxide-stained microsphere. Magnification X 8600

The boundary potential is around 30 mV (Fig.5). The potential has been recorded between the two phases, inside and outside the wall of the microspheres (raw reaction product) by conventional techniques, by introducing aproximately 2 μm diameter glass microelectrodes filled with 3M KCl (5-15 MOhm) in the microspheres.

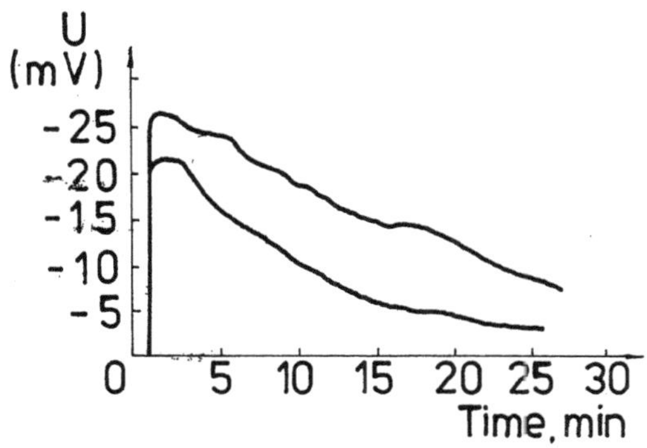

Fig.5-Boundary potential of microspheres

It is noteworthy that the microspheres can selectively retain biologically active compounds (e.g.,trypsine) /10/.

The fact that the microspheres synthesized in cold plasma conditions present membrane-like and special electrical properties and they are accompanied by pigments (e.g.,pyrollic structures) raises the possibility that such entities might also be able to promote synthetic reactions utilizing radiation energy. The ultimate source of the energy for all of the living systems is the Sun; the compounds that store and transfer such energy in contemporary organisms are predominantly ATP. It has been proposed at the same time /11,12/ that inorganic poly (pyro) phosphates preceded ATP in evolution. Reasons for inferring that the first photosynthesis were not mediated by clorophyll have been presented also /13,14/. The intensity of UV radiation is believed to have been greater on the surface of the primitive Earth because of the absence of the ozon shield.

Consequently in some of our experiments microsphere assisted photophosphorylation were investigated starting from KH_2PO_4. In comparison to the previous experiments of the lipid-like structures formation, instead of water,50 ml 5 % KH_2PO_4 aqueous solution was deposited as ice on the cylindrical part of the reactor /1/. On the end of the reaction (after 6 hours) the liquid phase raw material was irradiated with a 500 W mercury lamp at a distance of 200 mm, for 48 hours.

The inorganic polyphosphate was characterized by

paper chromatography (molybdate test, R_f values)(Fig.6) and by GPC (R_f values, UV detection at 280 nm)(Fig.7). The analytical data clearly show the formation of polyphosphate.

Fig.6-Paper chromatography (eluent: 1 M, sodium formate, 3.4pH)

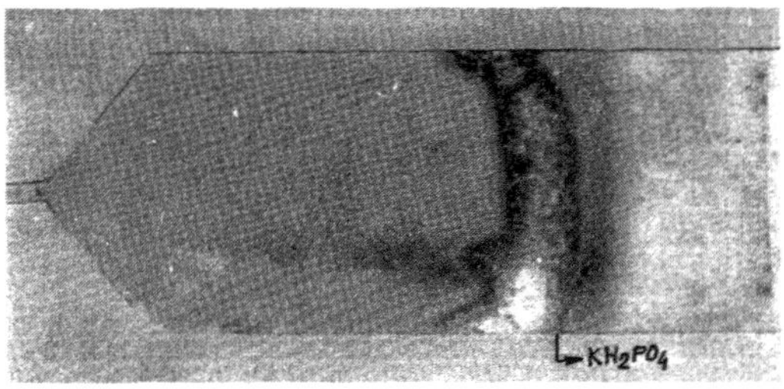

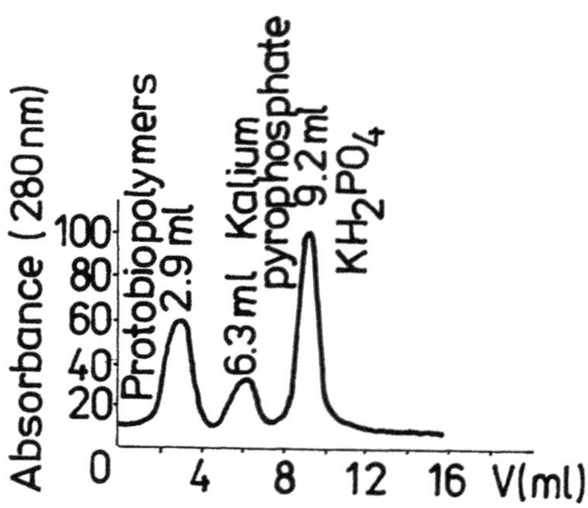

Fig.7-GPC diagram with UV detection. Column packing-Sephadex G 10; column dimensions: height 330 mm, Ø 8 mm; eluant: distilled water; V_o=2.9 ml.

This experiments bring additional data for the fact that the inorganic polyphosphate could play a very important rôle in the early "bioenergetical" processes.

REFERENCES

(1) Simionescu, C.I. and Dénes, F., Formation of Protobiopolymers-The Cold Theory, Abstract, The Second ISSOL

Meeting and the Fifth International Conference on the Origin of Life, Kyoto, Japan, April 5-10, 1977.
2) Simionescu, C.I., Dénes, F. and Negulescu, I., J.Polym. Sci., Polymer Symposia, 64, 281 (1978).
3) Simionescu, C.I., Totolin, M. and Dénes, F., BioSystems, 8, 153 (1976).
4) Simionescu, C.I., Dénes, F. and Totolin, M., BioSystems (in press).
5) Hargreaves, W.R., Mulvihill, S.J. and Deamer, D.W., Nature, 266, 78 (1977).
6) Seleznev, S.A., Fedorov, L.M., Mikhailov, A.J., Borisova, A.M. and Kuzina, S.I., Zh.Evol.Biokhim., 12 (5), 411 (1976).
7) Fox, S.W. and Dose, K., Molecular Evolution and the Origins of Life, Second Edition, W.H.Freeman and Co., San Francisco, 1976, p.139.
8) Simionescu, C.I., Simionescu, B.C. and Mora, R., Bioelectrochemistry and Bioenergetics, 5, 1 (1978).
9) Simionescu, C.I. and Dénes, F., Cellulose Chemistry and Technology (in press).
10) Simionescu, C.I., Dénes, F. and Brandsch, R., J.Polym. Sci., Polymer Symposia, C 1, 66 (1979).
11) Baltscheffsky, H., in The Origin of Life and Evolutionary Biochemistry, Dose, K., Fox, S.W., Deborin, G.A. and Pavlovskaya, T.E., Eds., Plenum Press, New York, 1974, p.9.
12) Kulaev, I.S., in The Origin of Life and Evolutionary Biochemistry, Dose, K., Fox, S.W., Deborin, G.A. and Pavlovskaya, T.E., Eds., Plenum Press, New York, 1974, p.271.
13) Evstigneev, V.B., Origins of Life, 6, 435 (1975).
14) Krasnovsky, A.A., in The Origin of Life and Evolutionary Biochemistry, Dose, K., Fox, S.W., Deborin, G.A. and Pavlovskaya, T.E., Eds., Plenum Press, New York, 1974, p.233.

TERRESTRIAL EVOLUTION OF POLYMERIZATION OF AMINO ACIDS: HEAT TO ATP

S. W. Fox and T. Nakashima

Institute for Molecular and Cellular Evolution
University of Miami, Coral Gables, FL 33134, U.S.A.

Sets of amino acids containing sufficient trifunctional monomer are thermally polymerized at temperatures such as 65°; the amino acids order themselves. Various polymers have diverse catalytic activities. The polymers aggregate, in aqueous solution, to cell-like structures having those activities plus emergent properties, e.g. proliferatability. Polyamino acids containing sufficient lysine catalyze conversion of free amino acids, by ATP, to small peptides and a high molecular weight fraction. The lysinerich proteinoid is active in solution, within suspensions of cell-like particles, or in other particles composed of lysine-rich proteinoid and homopolyribonucleotide. Selectivities are observed. An archaic polyamino acid prelude to coded protein synthesis is indicated.

Experiments indicate that formation of a kind of protein could have occurred most easily on the primitive Earth by the warming under hypohydrous conditions of sets of amino acids containing sufficient trifunctional amino acid (1). A principal question has been that of how such synthesis could have evolved to a cellular synthesis of peptides energized by ATP instead of heat. A cellular synthesis would have had to occur in or near aqueous solution. The evolutionary link has been found to be a cell-like structure aggregated from thermal proteinoid and containing a lysinerich proteinoid within the aggregated micro-structure.

Self-instructing of amino acids can be traced through this sequence of events. Heating of glutamic acid, glycine, and tyrosine, for example, yields a polymer in which the major

Y. Wolman (ed.), Origin of Life, 271–276.

component is an equimolar compound of pyroglutamyltyrosylglycine and pyroglutamylglycytyrosine (Table 1).

TABLE 1

Tyrosine-Containing Tripeptides Expected from Random Polymerization of Glutamic Acid, Glycine, and Tyrosine and Those Found.

Peptides Expected Randomly		Peptides Found
Gly-α-glu-tyr	α-Glu-α-glu-tyr	
Gly-γ-glu-tyr	α-Glu-γ-glu-tyr	
Gly-gly-tyr	γ-Glu-α-glu-tyr	
Gly-tyr-glu	γ-Glu-γ-glu-tyr	
Gly-tyr-gly	< Glu-α-glu-tyr	
Gly-tyr-tyr	< Glu-γ-glu-tyr	
Tyr-α-glu-glu	α-Glu-gly-tyr	
Tyr-γ-glu-glu	γ-Glu-gly-tyr	
Tyr-α-glu-gly	< Glu-gly-tyr	< Glu-gly-tyr
Tyr-γ-glu-gly	α-Glu-tyr-glu	
Tyr-α-glu-tyr	γ-Glu-tyr-glu	
Tyr-γ-glu-tyr	< Glu-tyr-glu	
Tyr-gly-glu	α-Glu-tyr-gly	
Tyr-gly-gly	γ-Glu-tyr-gly	
Tyr-gly-tyr	< Glu-tyr-gly	< Glu-tyr-gly
Tyr-tyr-glu	α-Glu-tyr-tyr	
Tyr-tyr-gly	γ-Glu-tyr-tyr	
Tyr-tyr-tyr	< Glu-tyr-tyr	

< = pyro

The dominant fraction in this thermal polymer is that containing these two tripeptides (2). For tyrosine-containing tripeptides alone, thirtysix would result if the syntheses were statistically random; Table 1 lists them.

The amount found is 19.2 times the amount of indicated tripeptide complex expected from calculations based on the statistically random assumption. (This is a quantitative evaluation of nonrandomness.)

Although the evaluation of nonrandomness through amino acid sequences in peptides is new (2), a number of laboratories have accumulated data yielding the same qualitative inference of nonrandomness (3).

Most of the thermal copolyamino acids aggregate to cell-like structures on contact with water. While lysinerich proteinoid itself aggregates under conditions of high salt concentration (4), such products have less stability than those from acidic proteinoid. Mixtures of acidic and lysinerich proteinoids yield relatively stable, but dynamic, microspheres. One of the dynamic properties of lysinerich proteinoid itself is catalysis of peptide formation from amino acids and ATP (5). This property is incorporated into the microspheres composed of acidic and lysine-rich proteinoids.

Since lysinerich and acidic proteinoids are themselves made by heat, and since the mixed microspheres derived therefrom catalyze peptide formation, the evolution of terrestrial peptide bond synthesis from geothermal copolymerization of amino acids to synthesis of peptides in protocells formed from those same thermal polymers is modelled. In the course of this evolution, the peptide bond synthesis evolves from activation by heat to activation by ATP, and from hypohydrous conditions to hydrous media, as within a cell. The evolutionary link is thus a lysinerich proteinoid.

Peptides have been formed from glycine, lysine, phenylalanine, or proline in the presence of lysinerich proteinoid in aqueous solution (5); the proteinoid catalysis is evidently applicable to all amino acids.

As indicated earlier (6), lysinerich proteinoid can be complexed with homopolyribonucleotides much as can acidic proteinoid. Such phase-separated particles containing poly A (polyadenylic acid) and lysinerich proteinoid were shown to catalyze the synthesis of peptides of phenylalanine (6). Conditions have since been identified for formation of oligopeptides from each kind of amino acid in solution and each kind of homopolyribonucleotide in the phase-separated particles.

An example of the selectivity observed with individual amino acids when the polynucleotide is held constant is seen in Fig. 1. The yields of Fig. 1 are for oligopeptides. Most of the incorporated amino acid is however found in a fraction at the origin on the chromatogram. In this experiment, peptides formed most easily in the descending order of gly>phe>pro>lys; this rank is subject to the conditions of the experiment. The rank is also determined by the identity of the polynucleotide in the phase-separated particle, as suggested in other experiments (7).

Fig. 2 shows an individual chromatogram of the product in formation of glycine peptides from glycine and ATP in a suspension of phase-separated particles composed of lysinerich proteinoid and poly A. The fraction at the origin is clearly evident.

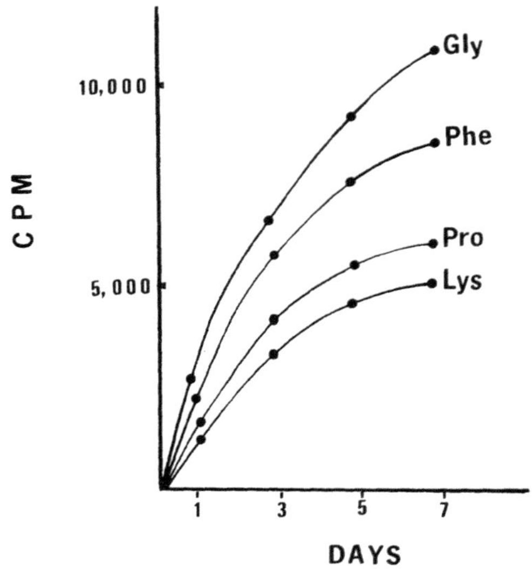

Fig. 1. Time-course curves for formation of oligopeptides of glycine, phenylalanine, proline, or lysine in suspensions of poly A-lysinerich proteinoid. Temperature was 25°. Concentrations of lysinerich proteinoid was 6.0 mg/ml, that of each amino acid was 0.3 mg/ml, and poly A was 3.0 mg/ml.

Fig. 2. Diglycine formed from glycine and ATP as catalyzed by lysinerich proteinoid complexed with polyadenylic acid.

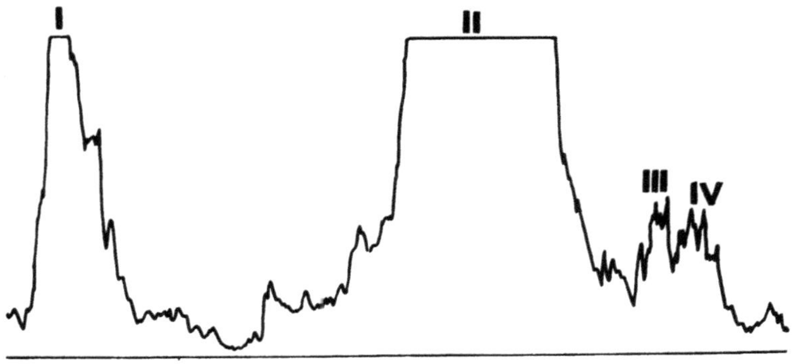

Fig. 3. I is high molecular weight fraction incorporating phenylalanine. II is phenylalanine. III is phenylalanylphenylalanine. IV is phenylalanylphenylalanylphenylalanine.

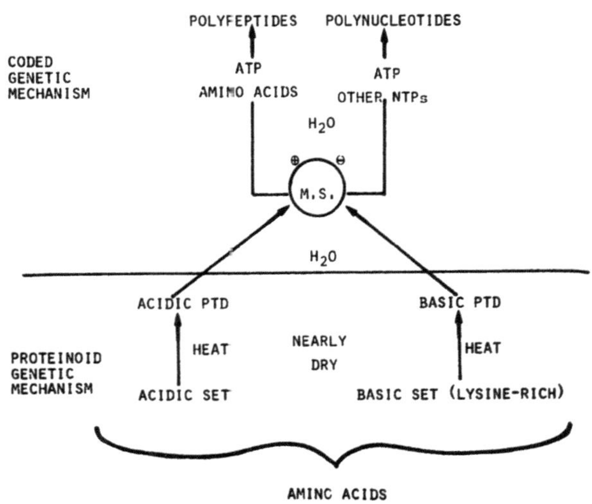

Fig. 4. Origin of a coded genetic mechanism from a "self-ordered" proteinoid genetic mechanism.

When the amino acid is phenylalanine instead of glycine, Fig. 3 (6) resulted.

When two amino acids, such as glycine and phenylalanine, are added to suspensions of nucleoproteinoid microparticles, mixed dipeptides result. The dipeptide is predominantly glycylphenylalanine when poly A is within the particle but predominantly phenylalanylglycine when the polynucleotide is poly C or poly U.

Evolution of the steric ordering phenomena is suggested as originally rooted in a self-instructing phase during polymerization The polymers consequently possess specificities because of the steric interactions of the amino acid sidechains. Such sidechain stereochemistry principally determines the specificity of the whole copolyamino acid molecule.

The origin of the coding system from a self-ordered proteinoid mechanism (8,9) is thus visualized (Fig. 4), as based in selective interactions of the type demonstrated for polynucleotides and polyamino acids (10).

ACKNOWLEDGEMENT. This work was aided by Grant No. NGR 10-007-008 of the National Aeronautics and Space Administration.

REFERENCES

1. Fox, S. W. and Dose, K., Molecular Evolution and the Origin of Life, revised ed., Dekker, New York, 1977, p. 153.
2. Nakashima, T., Jungck, J. R., Fox, S. W., Lederer, E., and Das, B. C., Intl. J. Quantum Chem. QBS4, 65 (1977).
3. Ref. 1, p. 158.
4. Rohlfing, D. L., Origins Life 6, 203 (1975).
5. Nakashima, T. and Fox, S. W., J. Mol. Evol., in press (1980).
6. Fox, S. W., Jungck, J. R., and Nakashima, T., Origins Life 5, 227 (1974).
7. Nakashima, T. and Fox, S. W., Proc. Natl. Acad. Sci. U.S. 69, 106 (1972).
8. Calvin, M., Chemical Evolution, Oxford University Press, 1969, p. 155ff.
9. Dillon, L. S., The Genetic Mechanism and the Origin of Life, Plenum Press, New York, 1978, p. 209.
10. Fox, S. W., Molec. Cell. Biochem. 3, 129 (1974).

POTENTIOMETRIC TITRATION BEHAVIOR OF POLYAMINO ACIDS PREPARED BY THERMAL POLYCONDENSATION

Etsuo Kokufuta* and Kaoru Harada**

Institute of Applied Biochemistry* and Department of Chemistry**, The University of Tsukuba, Sakura-mura, Niihari-gun, Ibaraki, 305, Japan

Abstract: Potentiometric titrations of thermally prepared polyamino acids were carried out at different ionic strengths. The intrinsic dissociation constants of carboxyl groups or amino groups in the polymers were evaluated by the curves of the apparent dissociation constant against the degree of dissociation and the Henderson-Hasselbalch plots. The ratio of α- and ω-linkages of amino acid residues was also evaluated by the analysis of the titration data.

1. INTRODUCTION

Polyamino acid can be easily prepared by heating suitable mixtures of α-amino acids. These polymers can range in amino acid composition from homopolymers to polymers (proteinoids) which contain amino acid residues common to protein and have many properties of protein (1,2). The thermally prepared polymers are regared as models for abiotic protein since the polymerization conditions belived to have been prevalent on the primitive earth. Several studies on the chemical and biological properties of the thermal polyamino acids have been reported (2). However, little attention has been paid to the physicochemical properties.

In the previous studies (3-6), we demonstrated that the analyses of potentiometric titration data give significant informations about the dissociation properties of the thermal polyamino acids, which are related to the α- and ω-linkages of amino acid residues having carboxyl or amino groups. In this paper, the potentiometric titration behavior was also investigated for thermally prepared polyaspartic acid (PAA), copolymer of glutamic acid and

Y. Wolman (ed.), Origin of Life, 277–284.

alanine (CGA), polylysine (PL), and copolymer of lysine and alanine (CLA). The results obtained are discussed in terms of possible peptide chain preparation by the thermal polycondensation.

2. MATERIALS AND METHODS

The polyamino acids (PAA, CGA, PL, and CLA) used in this studies were prepared by thermal polycondensation of free L or DL amino acids (1,3-6). The polymers were purified by dialysis, followed by lyophilization. The molecular weight of polymers obtained was sufficiently high for the studies of the titrations. The amino acid residues were found to be racemized mostly when L-amino acids were used as the starting materials for the polycondensation.

The potentiometric titrations were carried out at 25±0.01°C by using Toadenpa Automatic Titration Apparatus. The sample solutions at different polymer concentrations were titrated in a nitrogen atmosphere with 0.1N NaOH or HCl. The ionic strength of titration system was adjusted with NaCl. Each system was titrated at least twice, and only differences betewen duplicates of 0.04 pH units or less were tolerated.

3. RESULTS AND DISCUSSION

3.1 Polyaspartic acids

The potentiometric titration curves of PAA prepared from DL-aspartic acid agreed with those from L-aspartic acid within experimental error. As shown in Fig. 1, the titration curve of

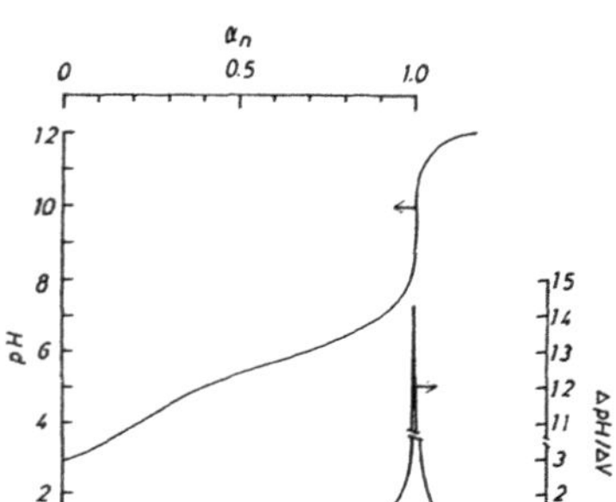

Fig. 1. Potentiometric titration curve of PAA prepared from DL-aspartic acid (ionic strength = 0). The differential titration curve ($\Delta pH/\Delta V \sim V$ curve) was obtained by the graphical differentiation of the titration curve ($pH \sim V$ curve). The degree of neutralization (α_n) was expressed in the ratio of V to V_{end} (V_{end}: titrant volume at the end point).

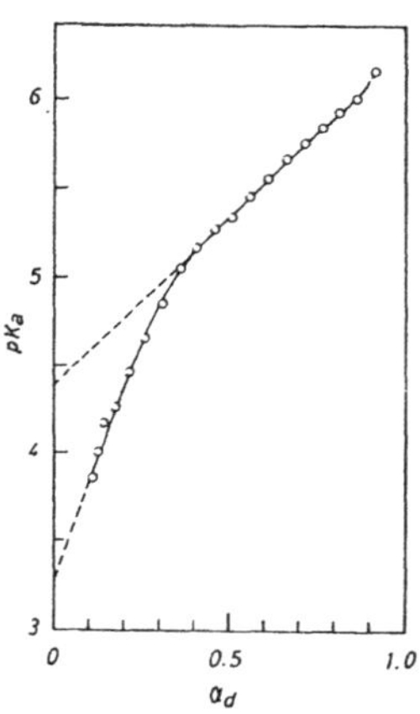

Fig. 2. Dependence of pKa on α_d at the ionic strength = 0. The apparent dissociation constant (pKa) and the degree of dissociation (α_d) calculated from the titration curve given in Fig. 1 by Eqs. (1) and (2), respectively.

PAA shows an inflection point at the degree of neutralization (α_n) of 0.29, whereas this inflection point was not observed in the titration curves obtained at the polymer concentration below about 10^{-3} mol/l and at ionic strength above 0.5. From these results, it could be assumed that the α_n region below and above the inflection point is assigned to the dissociation processes of α- and β-carboxyl groups in PAA, respectively. In order to confirm this assumption, the intrinsic dissociation constants (pKo) of α- and β-carboxyl groups were evaluated by two different approaches.

In the first method, the pKc values were estimated by graphical extrapolation of the curve of the apparent dissociation constant (pKo) against the degree of dissociation (α_d). The values of pKa and α_d were calculated by the following equations:

$$pK_a = pH + \log (1 - \alpha_d)/\alpha_d \qquad (1)$$

and

$$\alpha_d = \alpha_n + (C_{H^+} - C_{OH^-})/C_p \qquad (2)$$

where C_{H^+} and C_{OH^-} represent the molarity of protons and hydroxyl ions, and C_p is the polymer concentration in mol/l. Fig. 2 shows the curve of pKa vs. α_d obtained from the titration curve in Fig. 1. The pko values of α- and β-carboxyl groups were estimated by extrapolation of pKo from the α_d region below 0.3 and above 0.4, respectively. The results obtained are: 3.26 for α-COOH and 4.83 for β-COOH.

On the other hand, the pKo values of α- and β-carboxyl groups were also evaluated by the Henderson-Hasselbalch equation;

$$pH = pK + n \log (1 - \alpha_d)/\alpha_d \qquad (3)$$

where pK represents the average dissociation constant, and n is the empirical constant which denotes the magnitude of the interaction between protons and polyion as the deviation of n from unity. When the contribution of the electrostatic potential of polyions is fully eliminated by the addition of a neutral salt, the dissociation behavior of polymeric acids approaches to that of monomeric acids so that the pKo value can be estimated by (7):

$$pK_0 = pK + 0.434(e^2\kappa/3DkT) \qquad (4)$$

Here, the term $0.434(e^2\kappa/3DkT)$ (e, electronic charge; κ, Debye-Hückel parameter; D, the dielectric constant of the solvent; k, Boltzman constant; T, absolute temperature) represents the correction term related to the free energy consumed for building up the ionic atmosphere. The results obtained by these relations are summarized in Table I. The pKo values estimated from the pK values under the condition of n=1 are: 3.23 for α-COOH and 4.32 for β-COOH.

The pKo values obtained by Henderson-Hasselbalch method are equal to those by the extrapolation method within experimental error. The average values by both analytical method are: 3.25 for α-COOH and 4.35 for β-COOH. The pKo value of α-carboxyl groups is

TABLE I

pK and n values at Different Ionic Strengths Obtained by Henderson -Hasselbalch Equation and the pKo Values Calculated by Eq.(4)

Ionic strength	pK	n	$\frac{0.434e^2\kappa}{3DkT}$	pK_O	
0.1	3.12	1.00	0.108	3.23	Dissociation process of α-carboxyl group ($\alpha_n < 0.29$)
0.5	—	—	0.240	—	
1.0	—	—	0.340	—	
0.1	4.61	1.22	0.108	4.78	Dissociation process of β-carboxyl group ($0.29 < \alpha_n < 1$)
0.5	4.14	1.06	0.240	4.38	
1.0	3.98	1.00	0.340	4.32	

comparable to pK_1 of N-acetylaspartic acid (pK_1 3.37 and pK_2 4.91) which is not affected by the charged $^+NH_3$- group because of the N-acethylated amino group. The pKo value of β-carboxyl group is comparable to the pK_2 of glycylaspartic acid (pK_1 2.81, pK_2 4.45 and pK_3 8.60) of which the pK_2 could not be affected by primarily dissociated carboxyl group because the $-COO^-$ group could be form a Zwitter ion with the charged $^+NH_3$- group. On the basis of the results mentioned above, it can be indicated that the α- and β-carboxyl groups in PAA dissociate in the α_n region below and above 0.29, respectively. This signifies that the α- and β-linked aspartyl residues in PAA are formed by alkaline hydrolysis of thermally prepared anhydropolyaspartic acid (APPA), and the ratio of the α- and β-linkages is about 7:3.

In order to confirm further the results mentioned above, the ratio of the α- and β-linkages was investigated by following analytical method (4). When we consider a system containing 2 species of weakly monobasic acids, the concentration of each acid (C_1 and C_2) can be expressed by following equations;

$$\left.\begin{aligned} Y &= C_1 + C_2X \\ Y &= (1/K_1)(K_1 + C_{H^+})\alpha_d^t C_t \\ X &= K_2(K_1 + C_{H^+})/K_1(K_2 + C_{H^+}) \end{aligned}\right\} \quad (5)$$

if the corrections for the activity coefficients are negligible. Here, K_1 and K_2 represent the dissociation constants of acids, α_d^t is the sum of the degree of dissociation for each acid, and C_t is the total acid concentration. Eq.(5) is considered to be the formal equivalent of Speakman's equation (8) which is used to determine the dissociation constants of weakly dibasic acids such as adipic acid. Although Eq.(5) was derived from the model of monomeric acid, it is possible to analyze the titration data of PAA obtained at high ionic strength, because the dissociation behavior of PAA approaches to those of monomeric acids at ionic strength of 1.0 (see Table I). The corrections for the activity coefficients are considered to be negligible within experimental error. Thus, the values of C_1 and C_2 are estimated by means of the slope and the intercept on the Y-axis of a straight line which are obtained by plotting Y against X. When the values of K_1 (5.62×10^{-4}) and K_2 (4.47×10^{-5}) are obtained from the averaged pKo values of α- and β-carboxyl groups by the relation of pK= -logK, C_1 and C_2 represent the concentrations of α- and β-carboxyl groups, respectively. The results obtained are: $C_1=2.31 \times 10^{-4}$ mol/l and $C_2=5.91 \times 10^{-4}$ mol/l. Moreover, the sum (8.92×10^{-4} mol/l) of C_1 and C_2 agrees with the total concentration (8.92×10^{-4} mol/l) of carboxyl groups in the titration system, which was determined by an amino acid analysis after quantitative hydrolysis of PAA with HCl. Therefore, the analytical results of the titration data by Eq.(5) also reveals that the ratio of α- and β-linked aspartyl residues in PAA is about 7:3.

On the other hand, the analytical method by Eq.(5) was also applied to PAA prepared by thermal polycondensation of asparagine ($AspNH_2$) and isoasparagine ($IsoAspNH_2$) in the salt solution. The K_1 and K_2 in Eq.(5) were used the same values as described above. The analytical results show: 75±6% α-linkage for PAA from $AspNH_2$ in 1N NaCl (pH 7); 40±5% α-linkage for PAA from $AspNH_2$ in H_2O (pH 4); 85±8% α-linkage for PAA from $IsoAspNH_2$ in 1N NaCl (pH 7). These results signify that the mechanism of the formation of PAA from $AspNH_2$ and $IsoAspNH_2$ is not a simple intermolecular transamidation, but the reaction could involve another mechanism to interconvert the α- and β-linkages during the aqueous thermal condensation. The 5-membered imide was hydrolyzed easily at neutral pH to form α- and β-aspartyl residues. If this is the case, the ratio of α- and β-linkages of the polymer might be

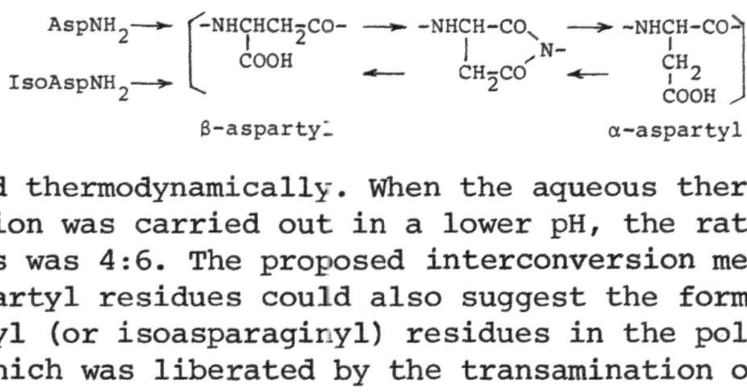

determined thermodynamically. When the aqueous thermal polycondensation was carried out in a lower pH, the ratio of α- and β-linkages was 4:6. The proposed interconversion mechanism of α- and β-aspartyl residues could also suggest the formation of asparaginyl (or isoasparaginyl) residues in the polymer with ammonia which was liberated by the transamination or by the hydrolysis of the starting material. The aqueous thermal polycondensation of $AspNH_2$ and $IsoAspNH_2$ is not only interesting as a possible abiotic origin of polypeptide on the primitive earth, but also interesting as an origin of α-aspartyl residue in the protoprotein.

3.2 Copolymer of Glutamic Acid and Alanine

The titration curves of CGA at different ionic strengths showed a gradual change in pH with increase of α_n, followed by a rapid inflection at $\alpha_n=1$. The curves of pKa <u>vs.</u> α_d obtained by Eqs.(1) and (2) showed that the pKa values are almost independent of α_d in the α_d range below 0.6-0.7. The n values estimated by Eq.(3)

were close to unity at the ionic strength above 0.1. These titration results indicate that the CGA behaves like monomeric acid because the density of carboxyl groups in the copolymer is small due to the copolymerization of alanine.

The carboxyl groups in CGA were characterized by the pKo values of 3.93-3.98, when the titration data were analyzed by Eqs.(3) and (4). On the other hand, the pKo value of CGA was also evaluated from the value $[(pKa)_{\alpha\to o}]$ obtained by extrapolating pKa to $\alpha_d \to o$:

$$pK_0 = (pK_a)_{\alpha\to 0} + 0.434\,(e^2\,\kappa/3DkT) \qquad (6)$$

The results obtained are pKo=3.96-3.98, as calculated from the $(pKo)_{\alpha\to o}$ values at ionic strengths of 0.1, 0.5 and 1.0. These results are in fair agreement with those estimated by Eqs.(3) and (4).

The pKo values of CGA are different from those (4.69-4.78) of poly-α-L-glutamic acid (9) and are similar to those (3.70-3.94) of γ-linked polyglutamic acids such as bacterial poly-γ-D-glutamic acid (10). Thus, it is assumed that the glutamyl residues in the copolymer are γ-linked. In order to confirm this assumption, the titration data at ionic strength of 1.0 was analyzed by Eq.(5). The values of K_1 ($1.58\text{X}10^{-4}$) and K_2 ($1.86\text{X}\ 10^{-5}$) were obtained from the average pKo values of poly-α-L-glutamic acid and poly-γ-D-glutamic acid, respectively. As shown in Fig.4, Y is independent of X. This indicates that the glutamyl residues in CGA are mostly or entirely γ-linked.

On the basis of the results described above, the propagation of the peptide bond could be explained as shown below. Glutamic acid was converted to pyrrolidone-2-carboxylic acid. The carboxylic acid of the lactam could react with the amino groups of the other amino acid to form peptide bonds. The other possible peptide chain propagation is that the amino group of the other amino acid reacts with the lactam ring to form a γ-glutamyl linkage by a transamidation reaction. The titration results indicate that no α-linked glutamyl residues are present in the copolymer. Therefore, these results support the view that the peptide chain propagates mainly by the transamidation mechanism.

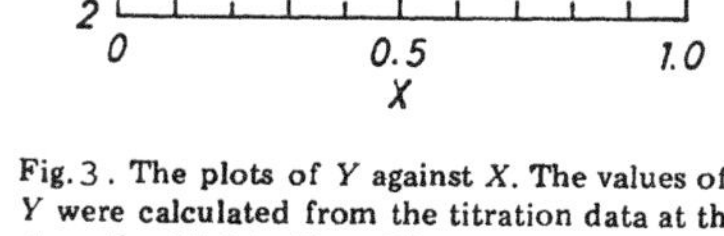

Fig.3. The plots of Y against X. The values of X and Y were calculated from the titration data at the ionic strength of 1.0 by Eqn. (5).

3.3 Polylysine and Copolymer of Lysine with Alanine

The curves of pKa vs. α_d for PL and CLA showed monotonous changes, whereas that for poly-L-lysine (9) prepared by the Leuchs method shows a characteristic pattern due to the helixcoil transition. The monotonous behavior could be due to the facts that the constituent amino acid of PL and CLA are almost racemized. On the other hand, the titratable amino groups in PL and CLA are smaller than the total amino group contents estimated by elemental analysis. The n values for PL and CLA increased with increase of ionic strenth and were not close to unity even at the ionic strength of 1.0. Moreover, no remarkable difference between n values for PL and CLA was observed, although the density of amino groups in CLA are smaller than that in PL. These results can be explained by the view that both PL and CLA behave as a branced random-chain polyelectrolyte.

The pKo values of the titratable amino groups in PL and CLA were evaluated by the analytical methods of Eqs.(3) and (4) and of Eq.(6). The results by the former method give the pKo values of 7.20 -7.40 for PL and 7.30-7.45 for CLA, and the others also give those of 7.29-7.56 for PL and 7.64-7.65 for CLA. The pKo values obtained are evidently different from that (10.11-10.25) of poly-α-L-lysine (9), indicating that the titratable groups in PL and CLA are mostly α-amino groups. The ratio of α- and ε-amino groups were investigated by the same analytical method as used for PAA and CGA, although the form of analytical equation for polybases is different from that of Eq.(5). The X and Y in Eq.(5) for polybases are modified as follows (6);

$$X = \frac{K_1 + C_{H^+}}{K_2 + C_{H^+}}, \quad Y = \frac{(K_1 + C_{H^+})(1 - \alpha_d^t)C_t}{C_{H^+}} \qquad (7)$$

where K_1 and K_2 represent the dissociation constants of protons from ε- and α-ammonium groups, respectively. In the cases of PL and CLA, the titration data were analyzed by some values of K_1 and K_2 selected at the condition where the plots of Y vs. X were best fit to a linear relationship, since the appropriate K values were not obtained from the pKo values of model compounds. The fitness was evaluated by means of correlation coefficient (r) obtained by the following equation;

$$r = \frac{\frac{1}{N}\sum_{i=1}^{N}(X_i - \bar{X})(Y_i - \bar{Y})}{\sigma_X \sigma_Y} \qquad (8)$$

where $\bar{X}$ and $\bar{Y}$ are average values of X_i and Y_i, respectively, σ_X and σ_Y are the standard deviations of X_i and Y_i, respectively, and N is the data number. The calculation of X, Y, and r at various K_1 and K_2 were performed on an electronic computer (TOSPAC 5600) using the FORTRAN program. The results obtained show in Table II. It is found that about 95% of titratable amino groups in PL and CLA is α-amino groups. These results suggested that the lysyl residues in the thermal polymers of lysine is composed mainly or entirely of ε-linked residues.

TABLE II

The Values of K_1 and K_2 Chosen to Give Maximum γ at Various Ionic Strengths (μ) and the Values of C_1 and C_2 at γ_{max}[a]

Polymer	μ	pK_1	pK_2	r_{max}	$C_1 \times 10^3$ (M)	$C_2 \times 10^3$ (M)	$C_2/(C_1 + C_2)$	$C_t \times 10^3$ (M)
PL	0.1	10.5(10.4)	7.2(7.1)	0.997	0.10	2.33	0.96	2.48
	0.5	10.8(10.6)	7.5(7.3)	0.991	0.18	2.34	0.93	2.57
	1.0	10.9(10.6)	7.6(7.3)	0.990	0.15	2.38	0.94	2.49
Copoly-(lys, ala)	0.1	10.7(10.6)	7.4(7.3)	0.992	0.15	2.02	0.93	2.11
	0.5	10.9(10.7)	7.6(7.4)	0.991	0.11	2.02	0.95	2.09
	1.0	11.0(10.7)	7.7(7.4)	0.991	0.13	1.98	0.94	2.15

[a] The K value was expressed as the negative logarithmic form by the relation of $pK = -\log K$. The values in parentheses are pK values corrected by the term of $0.434(e^2\kappa/3DkT)$ in Eq.(4) or (6). Total concentration (C_t) of amino groups in the titration system was obtained by potentiometric titration curve.

The fact mentioned avobe might indicate the following propagation of the lysyl residue in the polycondensation process. Lysine was first converted to a 7-membered α-amino lactam. The amino group of the other amino acid reacted with the lactom to form ε-lysyl residue by a transamination reaction, and the transamination processes were repeated to form polymer containing both ε-lysyl residues and branching points.

REFERENCES

1) Fox, S.W. and Harada, K., "Laboratory Manual of Analytical Methods of Protein Chemistry", Alexander, p.127 and Lundgren, H.P., eds., vol.4, Pergamon Press, Elmsford, N.Y.(1966),p.129.
2) Fox, S.W. and Dose, K., "Molecular Evolution and the Origin of Life", Marcel Dekker, N.Y.(1977),p.138.
3) Kokufuta, E., Suzuki, S., and Harada, K., BioSystems, 9, 211 (1977).
4) Kokufuta, E., Terada, T., Suzuki, S., and Harada, K., BioSystems, 10, 299 (1978).
5) Harada, K., Matsuyama, M., and Kokufuta, E., Polym. Bull., 1, 177 (1978).
6) Kokufuta, E., Terada, T., Tamura, M., Suzuki, S., and Harada, K., Arch. Biochem. Biophys., 196, 23 (1979).
7) Harris, F.E. and Rice, S.A., J. Polym. Sci., 15, 151 (1959).
8) Speakman, J.C., J. Chem. Soc., 855 (1940).
9) Ciferri, A., Puett, D., Rajagh, L., and Hermans, J., Biopolymers, 6, 1091 (1968).
10) Waley, S.G., J. Chem. Soc., 617 (1855).

ON THE POLYCONDENSATION OF AMINO ACID ADENYLATES ON MONTMORILLONITES

Frederick R. Eirich

Polytechnic Institute of New York

ABSTRACT

Supplementing earlier findings by A. Katchalsky et al., and current work of M. Paecht-Horowitz, a number of studies on the mechanisms of adsorption underlying the catalysis mechanism show, at least, 3 types of adsorption, i.e. two ion exchange mechanisms on clay faces and edges, and a general, possibly ion-pair mediated, polar adsorption. X-ray data on dry clay-peptide sandwiches show clearly adsorption in well defined layers, while analysis of rates of polymerization, of blocking experiments of clay agglomeration, and of the effects of polypeptide preadsorption show that the clay faces and edges are involved in the enhancement of the polycondensation which also requires that the clay platelet stacks can expand by interlayer adsorption during the reaction.

Prebiotic chemical evolution has been widely accepted as a fact.[1] Our knowledge has advanced to the point where questions into the spontaneous formation of quite complicated molecules and systems are being asked and worked on, such as the formation of polypeptides, polysugars, and polynucleotides. In the absence of genetic codes, of replication and metabolism, i.e. before life as we know it became established, one must look into very different modes of self-synthesis for the prebiotic precursors of the

Y. Wolman (ed.), Origin of Life, 285–293.

macromolecules that we know constitute the basis of life. A large number of different pathways have been described in the literature.[2] Below we wish to report on the continuation of a series of studies that were originated by A. Katchalsky and have been continued since by the undersigned with the assistance of Dr. M. Paecht-Horowitz as co-investigator and E. Katzir-Katchalski as consultant.[3,4,5]

The original observations showed that amino acids (AA) activated by nucleotides, in particular by adenylmonophosphates (AMP), e.g. alanyladenylate (Alad), but also most other amino acid adenylates (AAA) which in aqueous solution at 20°C and pH >7 will mostly hydrolize, will polycondense nearly completely to peptides under liberation of adenylic acid under the same conditions, when finally divided bentonite type clays, especially sodium montmorillonite (NaMM, or MM) is added. Characteristically, the molecular weight distribution of the oligomers and higher peptides was found to be continuous but to show discrete degrees of polymerization (DP), usually multiples of 2,3 and 4.

Since then we have established further that all AAA's except lysine and histidine will react, that is i.e. polymerize, as well as co-polymerize to form sequences of short blocks of AA's, and further that the ionic composition of the bentonites of the exchangeable and of the octahedral ions does affect the DP distribution, while not changing the nature of the reaction, i.e. maintaining its speed, degree of completion and discrete DP distribution. Blocking experiments with reagents which occupy selectively either the faces or the edges of the clay platelets show by the resulting reaction interference that both, faces and edges, are involved in the mechanism by which the clay promotes the polycondensation.[6] Other details of the structure of the clays which were found to be important are summarized in previous work,[4] but the most critical requirement is that the individual clay platelets are capable of full dispersion and reaggregation. Multivalent ions or organic molecules which cement the platelets in tight stacks whose interlamellar spaces cannot be readily expanded are not conducive to assisting polycondensation.

The necessary adsorption studies were carried out and revealed the following features:[7,8]

Amino acids and oligomers adsorb on Na$^-$ MM, at pH <7, at first, only by a regular ion exchange mechanism whereby the terminal - NH_3^+ groups displace equivalent Na^+ ions until the full cation exchange capacity (CEC) is employed. Adsorption of at least an order of magnitude less is found at pH >7, when the carboxylate ions become adsorbed on the clay platelet edges. At pH's below 8, higher oligopeptide concentrations lead to amounts adsorbed above the CEC plateau, but it is only for the AA's with more than one basic function, e.g. lysine, that one or more higher plateaus are reached. For polypeptide adsorption, the CEC plateau is barely indicated and the about 8X higher adsorption plateau for the dense surface occupation of clay lying AA residues is reached directly; small multiples (1.5 to 3X) are also possible. Nucleic acids, e.g. AMP, are very little adsorbed, i.e. according to the anion EC capacity of the clay platelets edges.

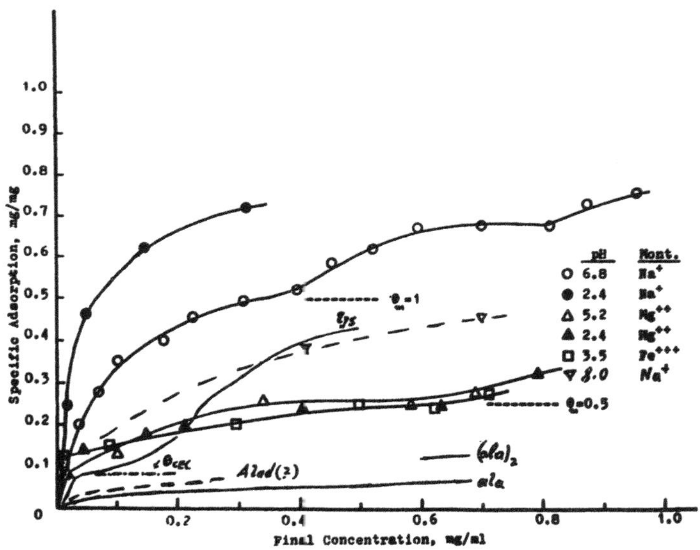

Fig. 1 Isotherms for the Adsorption of Poly-D, L-Alanine on Sodium-, Magnesium-, and Ferric Montmorillonite at 25°C

Since there are no methods for obtaining the thicknesses of the layers of the adsorbed oligo-or polypeptides on individual clay platelets while in dilute dispersion, we must rely on the evidence from measurements on clay samples dried at the various stages of peptide adsorption. X-ray data show that the peptides are intercalated between parallel clay platelets, that the interplatelet distances increase with the amount adsorbed while showing distinct steps at multiples of 1/2 of the surface occupation density of a full flat monolayer (Θ), i.e. at Θ/2, Θ and 3Θ/2 (7).

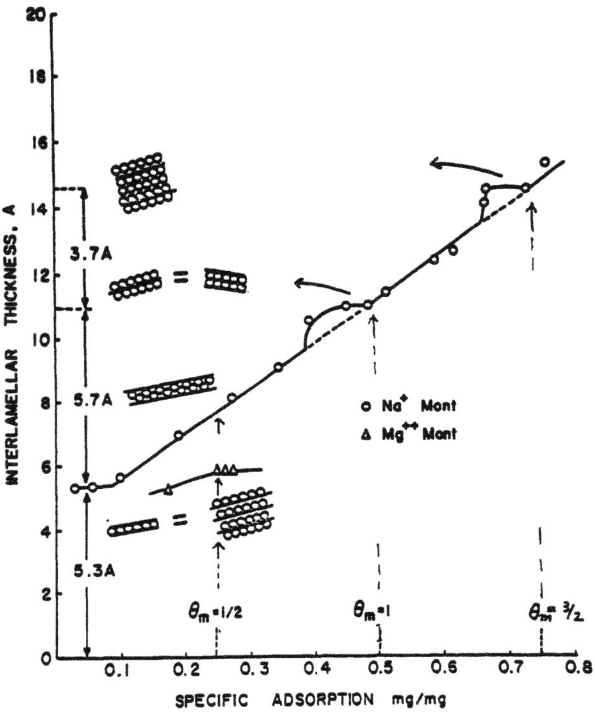

Fig. 2 Relation between the amount of Poly-D, L-Alanine on Sodium-, and Magnesium-Montmorillonite and the resulting Interlamellar Thicknesses, at pH = 3.

Fig. 2 shows the manner of interlameller space increases, whereby it is of note that spacing for Θ/2 is that for a single monolayer of oligomers between 2 platelets, or for the same layer of hydrated divalent ions; the platelet distances for Θ=1 correspond to 2 monolayers per platelets, i.e. one on either side, and so forth. Interestingly, at Θ=1/2, the precipitated and/or lightly dried clay-peptide sandwiches are very poorly redispersible in water, while at Θ=1 they are readily so.

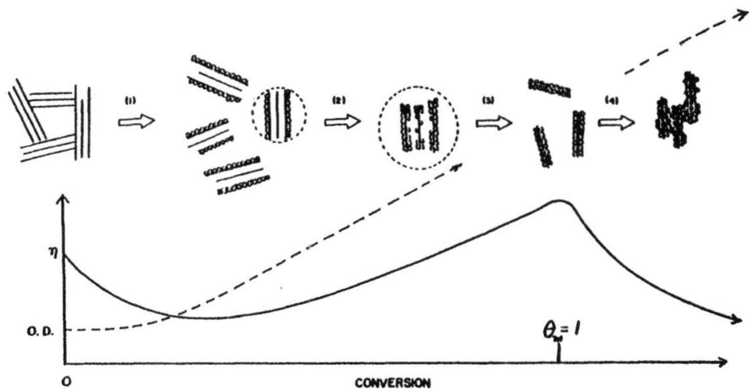

-39-

Fig. 3 The course of optical density (O.D.) and of specific viscosity of a Montmorillonite dispersion during polycondensation of alanine adenylate, forming oligo alanine peptides; on top stages of Montmorillonite disaggregation and reaggregation during the reaction.

Fig. 3 shows likely types of the modes of stacking at Θ=1/2, 1, 3/2 and so on.

Computer integrated Fourier IR analyses show absorption differences at OH and H bands supporting the presence of CEC type adsorption as well as of a more general one by way of surface H-bonding. There are therefore at least 3 distinct adsorption mechanisms for our peptides, nucleotides, AA's and oligomers, namely cationic and anionic up to EC on clay faces or edges, a physical adsorption over most of the clay platelet faces, and an adsorption of individual molecules by a simultaneous combination of these mechanisms e.g. a pincerlike 2-point adsorption of Alad on edges and faces.

A critical factor for the course of the polycondensation processes and for our understanding them, was found to be the degree of platelet aggregation. When, at the beginning, the monomeric Alad is added to the fully dispersed clay, the reaction is hardly different from that of adding the monomer to water at pH 7.5. That is, hydrolysis is the preponderant event and single platelets do not offer protection for the anhydride bonds. Only when some oligomers are produced, are adsorbed and starting stacking of the clay platelets, will increasingly more oligomer and polymer be formed at a gently self accelerating rate. When more monomer is added than can be quickly adsorbed on the clay stacks, this monomer undergoes mostly hydrolysis. Thus, for best results, the monomer must be added in successive portions of no greater quantities than can be quickly adsorved, whereby the addition intervals are determined by the time for the completion of the reaction of the previously added aliquot, as shown by the stopping of the pH drop (due to nucleic acid liberation) as seen on the pH-stat.(2)

Thus the locus, where the anhydride bond is protected against OH-ion attack but remains accessible to the-NH_2 displacement reaction, must be enhanced by the stacking, i.e. must be either the interplatelet layers, or the aggregation of edges. Since the latter are not effective when they are produced by aggregation via cations or organic basis, both of which tie the plates at about 5.5 Å distances, it must be the expandable intercalation space, or an expandable system of ribs of platelet edges, which is catalytically affected. Most likely, it is the concave nooks of unevenly overlapping platelets in expandable stacks which present the condensation sites, schematically shown on Fig. 4.(4) At these sites the anhydride groups are seen to be harbored by the more acidic interplatelet moiety, a situation not true at the edges. These anhydride bridges may then be attacked by neighboring terminal amino groups which fluctuate between the protonated and the unprotonated state in which latter they can attack the anhydride bonds and displace the nucleic acid.

This hypothesis is supported by the observation that pre-adsorption on the clay of "dead" polypeptides, i.e. without the activating terminal nucleotide group, enhances polycondensation greatly. Firstly, preadsorbed polymer prestacks the platelets prior to the monomer addition, and allows the immediate start of the polycondensation reaction without the lag period. Secondly, the DP's of the products of this polycondensation

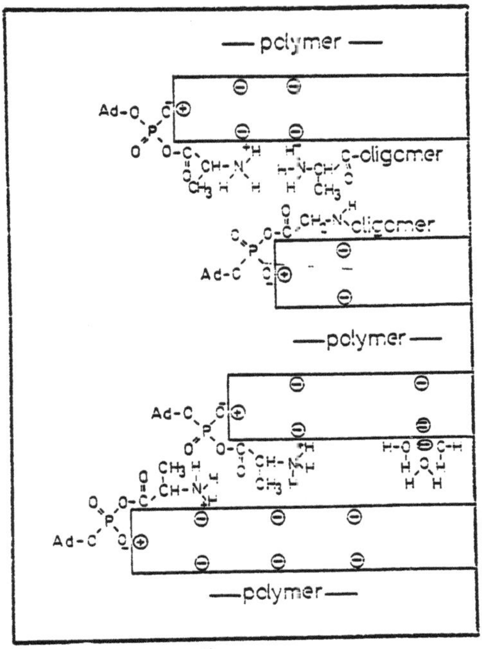

Fig. 4 Suggested Geometry for Adenylate Condensation at Edges and Interlayers of Montmorillonite Stacks.

undergo a peculiar cycle, as one gradually increases the preadsorbed amounts of peptides from Θ to more than 2x the full monolayer, in that they drop first precipitously from average DP's of approximately 20 to values of about 6, when the preadsorbed polymer covers 1/2 of the full clay surface, then rise again to DP's of approximately 40 when Θ approaches 1, drop again to DP$\sim$12 for Θ=3/2, with less clear periodicities thereafter.[4,9] Thus, whenever the clay is stacked by the preadsorbed polymer to the rather firm single monolayer intercalation, polycondensation proceeds much less efficiently to lower DP's, similar as if the clay platelets were held by single layers of hydrated divalent cations; but if enough polypeptide is preadsorbed to fit 2 molecular layers between the platelets, the polycondensation is greatly enhanced. Clearly, expansion of the clay stacks is an essential prerequisite for effective polycondensation.

The state of clay aggregation thus plays an important role and must do so as it changes during the course of the polycondensation reaction. Adding successively more dead polymer to a fully dispersed clay, or developing the peptides during the course of the reaction, does produce, though, two different ocurses of the increase in aggregation as followed by optical density measurements. For the addition of D-L-polyalanine, the optical density increases at first linearly, then at a rather constant, and further at a declining rate towards a point at which flocculation occurs. During the course of the condensation reaction without preadsorption, the O.D. increases at first very little but then at a gently increasing rate, reflecting the adsorption of gradually increasing DP's.[9] With preadsorbed polyalanine the relatively high DP's of the polypeptide added are not easily displaced by the early, lower, DP oligomers formed by reaction so that the tight monolayer sandwiches stay unexpanded and hinder the course of polymerization, whereas during the polymerization without polyalanine whatever single layer clay sandwiches have been formed can be expanded by the increasing amounts of increasingly higher DP's, with their power to displace, or add to, the earlier lower ones.[4]

On our spectral DP distribution we still have only guesses. The distribution may be due to an accumulation of preferred stack sizes, themselves given by geometrically determined adsorption amounts (as for instance in micelle formation). Another possible cause for it could be the manner of stepwise monomer addition which, after every reaction interval, finds decreasing numbers of living species to add on. This will account for the formation of distinct DP's, but whether a mathematical model can be set up which replicates the distributions observed will have to be left to a computer study. The possibility that the molecular weight distribution might be due to the formation of oligopeptide rings on the clay surface which then add to one another to form multiples of a preferred number of peptides in the rings, has been argued at greater length in a previous report[4], but so far tests, e.g. of charged ion fluorescence emission spectra, to prove this have failed.

To sum up this part of our ongoing study, our data add up to a very plausible mechanism for the enhancement of AAad polycondensation on MM, but for a conclusive confirmation more details have to be ascertained.

REFERENCES

(1) M. Calvin, Angew. Chem. Internat. Edn., 13, 121 (1974); C. Ponnamperuma, "The Origins of Life" E. P. Dutton, New York, 1972.

(2) H. Matsubeia, T. Yamancho, Evolution of Proteins, Japan Scient. Soc. Press, Tokyo, 1978.

(3) M. Paecht-Horowitz, I. Berger, A. Katchalsky, Nature, 228, 636 (1970).

(4) F.R. Eirich, Polytechnic Ann. Reports to NASA, Grant No.

(5) M. Paecht-Horowitz, Origins of Life, 5, 173(1974).

(6) M. Paecht-Horowitz, Biosystems 9, 93 (1977).

(7) Sh.H. Hsu, Ph.D. Dissertation, Polytechnic Inst. of New York, 1977.

(8) A. Banin and M. Paecht-Horowitz, to be published.

(9) M. Paecht-Horowitz, in press.

POLYMERIZATION OF SERINE GUANYLATE IN THE PRESENCE OF MONTMORILLONITE

Mella Paecht-Horowitz

Dept. of Soil and Water Sciences, The Hebrew University, Rehovot, Israel

ABSTRACT

Serine guanylate was prepared and its polymerization studied in the presence of montmorillonite and in its absence. In water, without clay, serine guanylate polymerizes in the same way as does serine adenylate. In the presence of montmorillonite, serine guanylate polymerizes to a lesser extent and produces also lower degrees of polymerization than does serine adenylate. It is postulated that the reason for this difference in behaviour might lie in the fact that guanylic acid is much more acidic than adenylic acid; hence would bind much more strongly to the edges of montmorillonite and thus, by blocking these sites, would inhibit the catalytic activity of the clay.

1. INTRODUCTION

In our search for possible models for prebiotic peptide synthesis, we have found previously that adenylates of amino acids polymerize in the presence of certain clay suspensions differently from the way they polymerize in water in the absence of clay, under otherwise the same conditions (Paecht-Horowitz, 1976).

As a further step we wanted to see whether this different polymerization (higher degrees of polymerization, discrete spectra of degrees of polymerization, block copolymerization) is typical of adenylates of amino acids only, or whether mixed anhydrides between amino acids and other nucleotides would give the same results as adenylates.

As serine adenylate produced relatively high degrees of

Y. Wolman (ed.), Origin of Life, 295–299.

polymerization in the presence of montmorillonite (2) and gave also in its presence high yields of copolymerization with other amino acids (3), we prepared serine guanylate and studied its polymerization in the presence of montmorillonite and compared it to the polymerization of serine adenylate under the same conditions.

2. EXPERIMENTAL

Serine guanylate was prepared according to the method of Moldave et al. (1) for alanine adenylate, with slight modifications necessary because of the high insolubility of guanylic acid in pyridine. 1 mMole guanylic acid and 1 mMole N-carbobenzoxy-serine-O-benzyl ester were dissolved in 10 ml of a mixture of pyridine water 3/7. To this solution, 8 gr. DCC, dissolved in 16 ml. pyridine, were added, and the whole mixture stirred for 2 hrs. at room temperature. The precipitate formed was separated by suction filtration, washed three times with 2 ml. pyridine (30%) each, and the solution left in a separating funnel, at room temperature for several minutes, until it separated into two layers. The lower layer was introduced into 200 ml. cold acetone and left overnight in a freezer. The next day the precipitate formed was separated by centrifugation, washed twice with 100 ml of cold acetone each, then the precipitate was extracted twice with 25 ml. of methyl cellosolve at room temperature. The combined methyl cellosolve extractions were introduced into 250 ml ether and left overnight in a freezer. Next day the precipitate of N-carbobenzoxy-O-benzyl-serine guanylate was centrifuged off, washed with ether, and dried in a dessicator over P_2O_5 and paraffin wax platelets. The yields were very low, 5-10%. Polymerization experiments were carried out as described previously (3).

3. RESULTS AND DISCUSSION

In water, without any clay, serine guanylate behaves exactly like alanine or serine adenylate. Their polymerization results are given in Table I.

In the presence of montmorillonite serine guanylate polymerizes in a way similar to this of serine adenylate, but to a lesser extent. Also the degrees of polymerization obtained are lower with serine guanylate than with serine adenylate. Figure 1 shows the results obtained from the two substances under the same conditions.

Preliminary copolymerization experiments show that copolymerization too takes place to a lesser extent with serine guanylate than with serine adenylate. It seems to us that the reason for the different behaviour of serine adenylate lies in the following: Montmorillonite particles adsorb cations on their interspatial faces and anions on their edges (7). Guanylic acid,

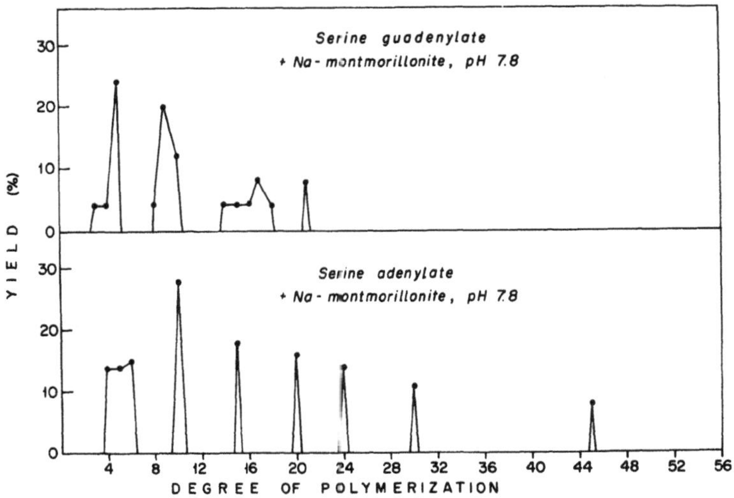

Fig. 1. Polymerization results of serine adenylate and serine guanylate in the presence of Na-montmorillonite (0.5 g/l) at room temperature, when the monomer was added in very small portions to the suspension and the pH was kept constant at pH 7.8

being a stronger acid than adenylic acid(pK's adenylic acid: 3.80; 6.15; pK's guanylic acid: 2.20; 5.75), is probably bound much stronger to the edges of the clay than is adenylic acid and hence is removed with much more difficulty from there than is the latter. Yet we have found previously that for polymerization purposes, the amino acid nucleotide has to be attached to both the inter-spatial faces and the edges of the clay particle, and if one of the sites is not available, polymerization will stop (5).

We have here probably the same phenomenon we encountered with the adenylates of the basic amino acids. The basic amino acids are so strongly adsorbed on the interspatial faces of montmorillonite, that they desorb by drastic means only; hence there is practically no polymerization of basic amino acids in the presence of montmorillonite. With guanylic acid apparently the same thing happens; only with opposite charges and the blocked sites in this case are not the faces but the edges. We are now starting some adsorption experiments of guanylic acid on montmorillonite in order to verify this hypothesis. Should our assumption prove right, the possibilities to envisage are the following:

a) that for the polymerization of serine from serine guanylate montmorillonite is not a suitable support, but a more acidic clay mineral might be more indicated as in the case of

Table I

Degrees of polymerization obtained in water at pH 7.8, in the absence of clay, from serine guanylate and serine adenylate

D.P.	Serine adenylate %	Serine guadenylate %
1	19	16
2	14	14
3	9	10
4	7	9
5	7	8
6	7	7
7	6	7
8	5	7
9	4	7
10	4	5
11	4	5
12	4	5

lysine adenylate, hectorite — more basic than montmorillonite — gives better polymerization results than the latter (6); b) that from the point of view of prebiotic synthesis, guanylic acid and may be other nucleotides too, have evolved at a much later stage than adenylic acid, and that, if indeed formation of primitive peptides took place by montmorillonite clay catalysis, they were formed from adenylates mostly; c) that on the contrary, montmorillonite having been the catalyst, and guanylates and other nucleotides of amino acid being present, the fact that the latter did not behave in the same way as the adenylates of amino acids provided some sort of selection, necessary for evolution, and might have influenced the sequences of amino acids in the sequences of amino acids in the prebiotic peptides; d) that we have to take into account that at the very beginning amino acids were not activated by nucleotides, but in some other way.

ACKNOWLEDGEMENT

This work was supported by Grant No. NGR 33-006-070 from NASA

REFERENCES

1. Moldave, K., Castelfranco, P., and Meister, A., 1959. J. Biol. Chem. 234, p. 841.

2. Paecht-Horowitz, M., 1974. The Origin of Life and Evolutionary Biochemistry, eds. K. Dose, S.W. Fox, G.A. Deborin, and T.E. Pavloskaya. Plenum Publishing Corp., New-York. pp. 373-385.

3. Paecht-Horowitz, M., 1974. Origins of Life 5, p. 173.

4. Paecht-Horowitz, M. Origins of Life 7, p. 369.

5. Paecht-Horowitz, M., 1977. BioSystems 9, p. 93.

6. Paecht-Horowitz, M., 1978. J. Mol. Evol. 11, p. 101.

7. Van Olphen, H., 1963. An Introduction to Clay Colloid Chemistry (Interscience Publishers, John Wiley and Sons, New York - London - Sydney), Chapter 8, p. 165.

QUANTUM MECHANICAL CONFORMATIONAL ANALYSIS OF AMINOACYLADENYLATES: IMPLICATION IN THE ORIGIN OF LIFE.

Henri BROCH, Daniel CABROL, Dane VASILESCU

Laboratoire de Biophysique - Université de Nice -
Parc Valrose - 06034 Nice Cedex - France.

The knowledge of the conformational possibilities of aminoacyladenylates can contribute to the understanding of two differents processes correlated with the origin of life :

- the activation phase of aminoacids during the protein biosynthesis, with intervention of tRNA synthetase
- the prebiotic peptide synthesis using montmorillonite as catalyst.

We discuss and speculate these two pathways with regard to minimal energy conformations computed with quantum PCILO method.

It is now well etablished that aminoacyladenylates play a central part concerning the protein synthesis, in two different ways :

1) as a normal intermediate in the aminoacylation of tRNA with the contribution of the enzyme tRNA synthetase (1-3)
2) in prebiotic peptide synthesis without enzyme intervention (4-7).

In previous papers we have pointed out the general problem set by the conformational aspect of these important biomolecules (8-13). Two sub-systems may be distinguished in an aminoacyladenylate (see fig. 1) :

- the 5'AMP stem
- the aminoacylphosphate stem with a specific residue.

With the example of Tyr-5'AMP (fig. 1) we define the rotational parameters along the single bonds of the molecule :
χ (C1'-N9) ; ψ (C4'-C5') ; ϕ (C5'-O5') ; ω (O5'-P5) ; ω' (P5-O3) ; ϕ' (O3-C') ; ψ (C'-Cα) ; ϕ (Cα-N) ; Θ (Cα-Cβ) ; Θ (Cβ-Cγ) ; etc ...

The conformational energy is calculated using the quantum

Y. Wolman (ed.), Origin of Life, 301–308.

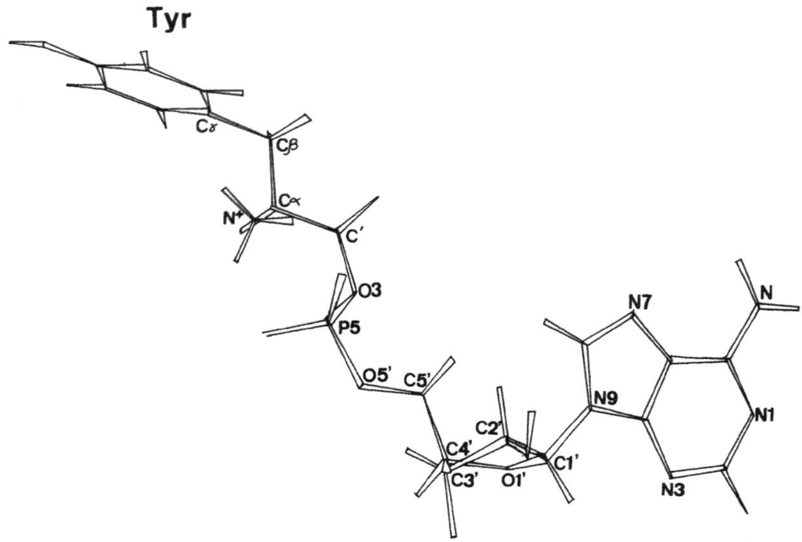

Fig. 1 : Perspective view of tyrosyladenylate

PCILO method (Perturbative Configuration Interaction using Localized Orbitals) (14) ; then this energy is minimized with regard to all rotational parameters.

Generally the puckering of the ribose in 5'AMP is C3'endo and the position of adenine is anti (χ = 180° ± 90°) ; the prefered values of ϕ (C5'-O5') and ψ (C4'-C5') are ≃ 180° and ≃ 60° (position noted gg) respectively. Concerning ψ (C4'-C5') other possibilities are observable : ≃ 180°(gt) and ≃ 300°(tg).

All studied aminoacyladenylates are in the ionized form (a negative charge distributed on the two free oxygens of the phosphate and a positive charge on the ammonium group).

In each aminoacyladenylate the most important fact is the existence of a double seven-sided ring (P5-O3 - C'-Cα - N^+-H1···O1 and P5-O3 - C'-Cα - N^+-H2···O2) in the minimal energy conformation of glycylphosphate stem, corresponding to a strong electrostatic interaction between the oxygens O1 and O2 of the phosphate group and the hydrogens H1 and H2 of the NH_3^+ group (see fig. 2a). This folding corresponds to the values of angular parameters : ω' = 190° ; ϕ' = 320° ; ψ (Cα-C') = 50° and ϕ (Cα-N) = 30° (or 150° or 270°).

When glycyladenylate is dissolved in water, we have simulated the hydration of the glycylphosphate stem by introducing six water molecules around the ammonium and phosphate hydrophilic groups and considering the set as a supermolecule. The global energetic minimisation leads to a slight unfolding of the seven sided ring (see fig. 2b) ; this conformation is one of the most folded we could obtain, taking the steric constraint of the water molecules

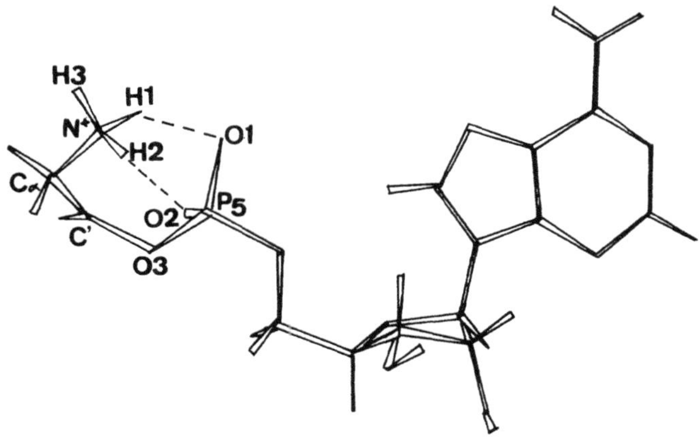

Fig. 2a : The seven membered rings in glycylphosphate stem P5-O3-C'-Cα-N+-H1···O1 and P5-O3-C'-Cα-N+-H2···O2

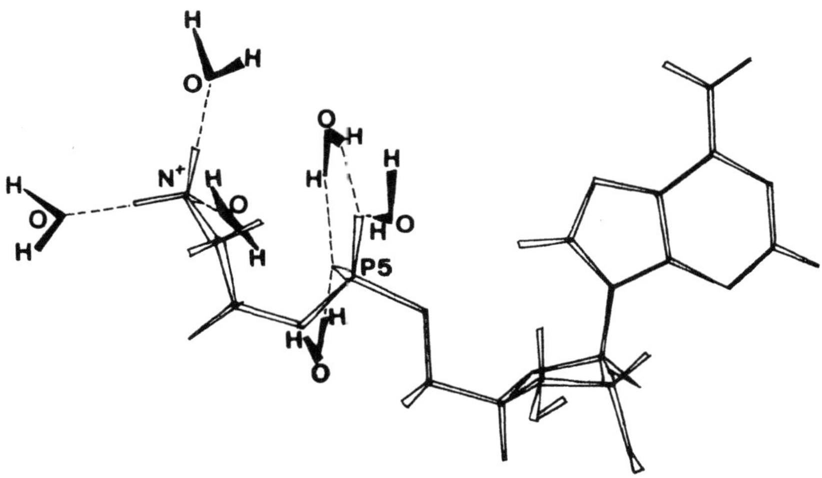

Fig. 2b : Hydration scheme of glycylphosphate stem

into account (12).

We have investigated the conformations of 9 aminoacyladenylates concerning the residues : Asp, Cys, Gly, His, Met, Phe, Pro, Ser and Val, with the 5'AMP stem in configuration : C3'endo -gg-anti (details are given in ref. 10 and 11).

In this work, we present the special case of Tyr-5'AMP, in relation with recent experimental investigations concerning the binding of an analogue of Tyrosyladenylate - i.e. Tyrosinyladenylate - to crystalline Tyr-tRNA synthetase (15) and we discuss some aspects related to the prebiotic peptide synthesis mechanism of clay catalyzed polymerisation of aminoacyladenylates.

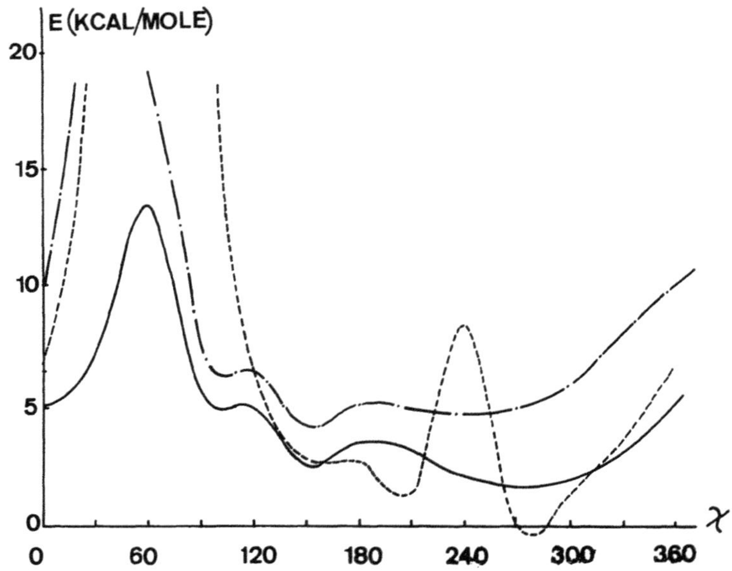

Fig. 3 : PCILO energy curves E(χ) for 5'AMP (ribose in C3'exo)
ψ rotamers : ---- gg ; -·-·-· gt ; —— tg

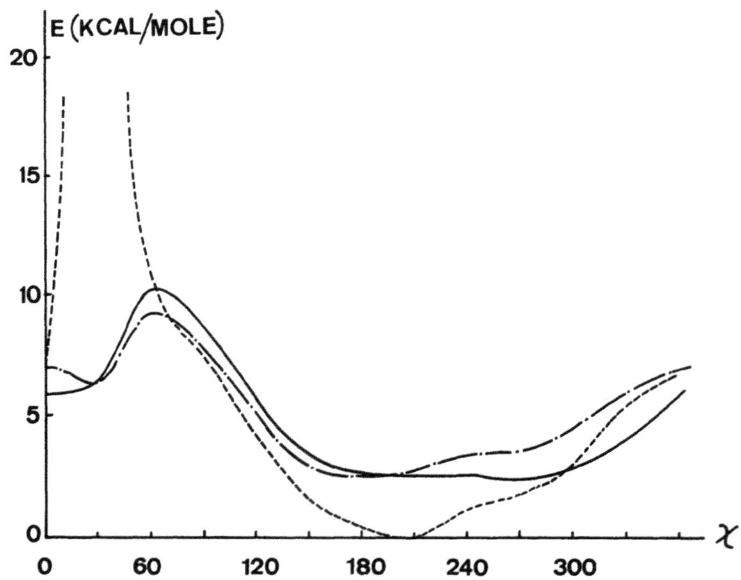

Fig. 4 : PCILO energy curves E(χ) for 5'AMP (ribose in C3'endo)
ψ rotamers : ---- gg ; -·-·-· gt ; —— tg

In the case of Tyr-5'AMP bind to Tyr-tRNA synthetase, Monteilhet and Blow have found an uncommon conformation consisting in a C3'exo puckering of ribose and a tg position for the exocyclic bond.

In order to determine the specific influence of the C3'exo puckering on the other conformational parameters, we have studied 5'AMP with the two typical ribose puckerings. Figures 3 and 4 show the E(χ) energetic curves for the three gg, gt and tg rotamers. A major difference between the results obtained for C3'exo (fig.3) and C3'endo (fig. 4) ribose puckerings is the marked preference for the tg rotamer on the gt one observed in the C3'exo case ; we had already shown in a previous work (16) this implication of the sugar puckering on the rotation round the exocyclic bond.

For ψ (C4'-C5') = 60° rotational barriers centered on χ = 30° and 60° result from the interaction of N3 and H(C2) of the adenine with the oxygens of the phosphate group ; the weak one observed only for C3'exo, centered on χ = 240°, originates from interaction of H(C8) of the adenine with O5'.

Concerning the rotation around the glycosydic bond, the C3'exo puckering case shows a quasi-syn orientation of adenine for the gg and tg rotamers and an anti orientation for the gt one, while for the C3'endo puckering the base is always in an anti orientation, with a greater flexibility.

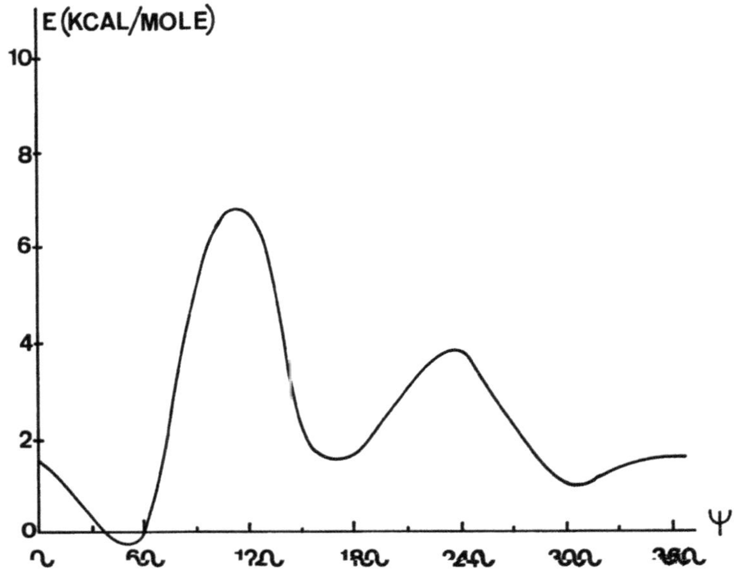

Fig. 5 : PCILO energy curve E(ψ(C4'-C5')) for Tyr-5'AMP ribose in C3'exo and χ =200° ; φ =180° ; ω =330°; ω'=190° φ'=320°; ψC'-Cα=50°; φN-Cα=30°; θCα-Cβ=190°; θCβ-Cγ = 90°

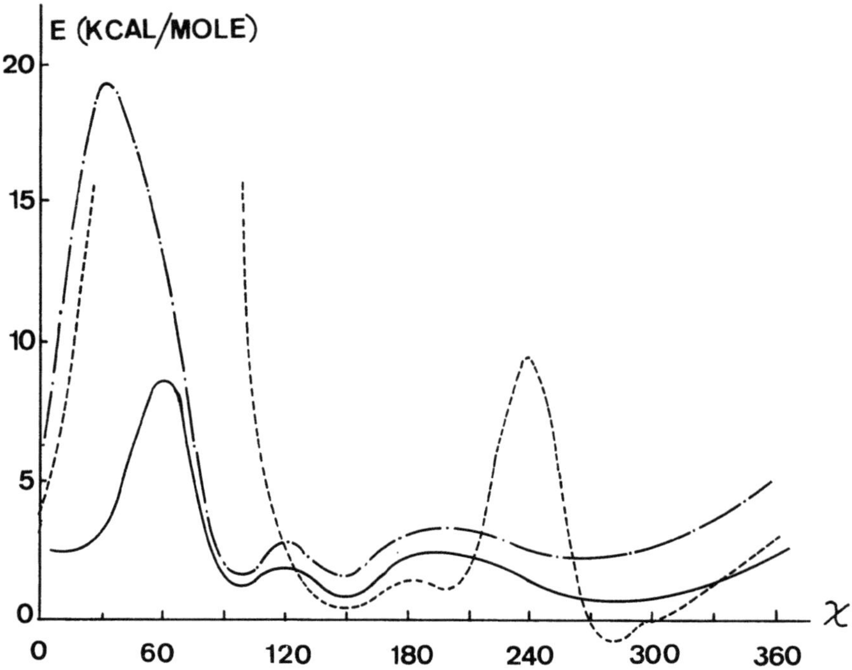

Fig. 6 : PCILO energy curve E(χ) for Tyr-5'AMP (ribose in C3'exo)
ψ rotamers : ---- gg ; -·-·-· gt ; —— tg

In the case of tyrosyladenylate, figure 5 shows energetic variation with respect to rotation round the exocyclic bond ; this curve has been obtained with the base in anti orientation, glycyl-phosphate stem in its double seven side ring form and remainder of the molecule in the minimal energy conformation determined for Phe-AMP (11). Classification of the three expected rotamers is inchanged and transconformation gg → gt requires much more energy than the gg → tg one. This result confirms the previously noted implication.

Another consequence of the C3'exo puckering is the shift of the ω (O5'-P5) minimum from 330° to 30° when ψ (C4'-C5') takes the 180° or 300° value. Figure 6 shows the E(χ) energetic curves obtained for the three gg, gt and tg rotamers ; the respective positions of the minima reinforce the previous observations on the preference of the tg rotamer against the gt one. Adenine is in a quasi-syn orientation for ψ = 60° and in an anti orientation for ψ = 300° or 180°. But we have to note that for the tg rotamer, the adenine rotation round the glycosydic bond is easier. This indicates the influence of the introduction of the aminoacid residue in the 5'AMP molecule.

The Tyr-AMP molecule in its in vacuo minimal energy conformation for ψ = 300° is represented on figure 7 in the right exins-

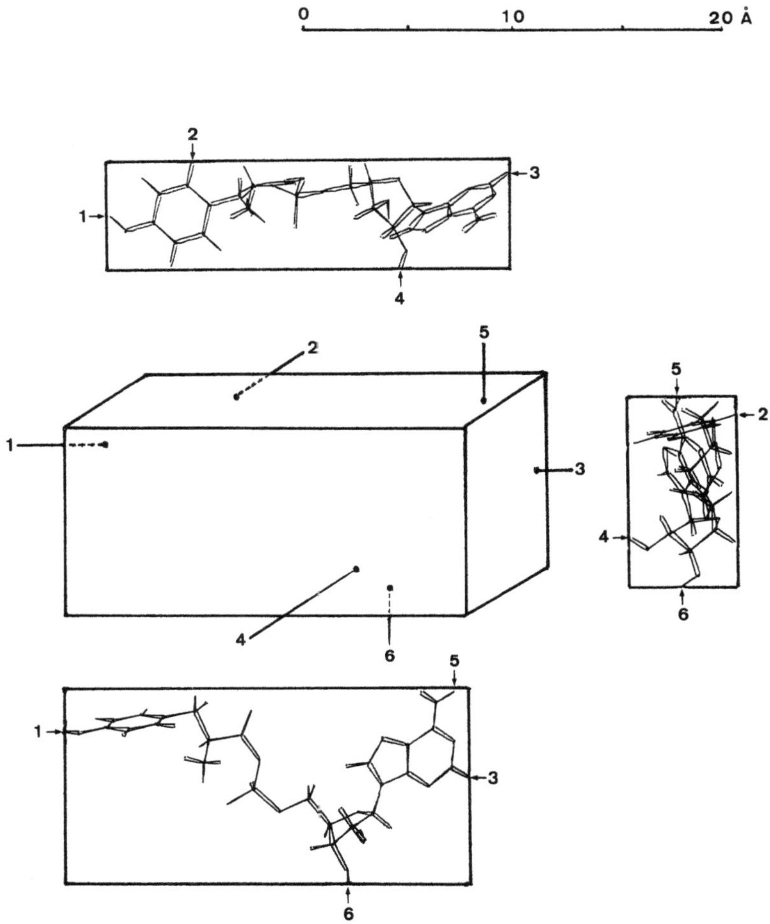

Fig. 7 : Tyrosyladenylate in its in vacuo minimal energy conformation (for ribose in C3'exo and ψ (C4'-C5') = 300°)

crit parallepiped of minimal volume.

This steric hindrance is in agreement with the fact that the molecule cannot enter enterely in the 10 Å deep cleft of the catalytic site of the tyrosyl-tRNA synthetase. The aminoacyl group being the specific part of the molecule, the adenine must lie out. The flexibility shown by the adenine, favored by the C3'exo puckering and the ψ orientation, in agreement with assumptions developped in (15), should facilitate its adjustement on neighboring polypeptide.

Another interesting aspect of these conformational considerations can be underligned concerning the mecanism of montmorillonite catalysis of polymerization. In neutral or acidic conditions, the ammonium group is engaged in the "double seven sided membered ring" with the phosphate group. In absence of clay, this stable structure should forbid the polymerization of aminoacyladenylates. In presence of clay, the catalysis by montmorillonite clays demonstrated by the works of Paecht-Horowitz and Katchalsky could begin by the opening of this bridged structure due the compensation of the ammonium and phosphate groups by the charged sites of the clay.

At high pH, the amino group remains uncharged and the bridge cannot be closed giving an increased flexibility to the aminoacyl-phosphate part of the aminoacyladenylate. This could facilitate the polymerization in absence of clay.

ACKNOWLEDGEMENT. This work was carried out with help from CNRS and DRET.

REFERENCES :

1 - BERG P., J. Am. Chem. Soc. 77, 3163 (1955)
2 - LAGERKVIST U., ÅKESSON B. and BRÄNDÉN R., J. Biol. Chem. 252, 1002 (1977)
3 - MULVEY R.S. and FERSHT A., Biochemistry 17, 5591 (1978)
4 - BUCHANAN J.M. in "The Origin of Prebiological Systems". FOX S. Edit., Academic Press - New York (1965) p. 101
5 - PAECHT-HOROWITZ M., Origins of Life 7, 369 (1976)
6 - PAECHT-HOROWITZ M. and LAHAV N., J. Mol. Evol. 10, 73 (1977)
7 - PAECHT-HOROWITZ M., Bio Systems 9, 93 (1977)
8 - VASILESCU D., BROCH H., CORNILLON R. and LESPINASSE J.N., J. Theor. Biol. 57, 395 (1976)
9 - BROCH H., CORNILLON R., LESPINASSE J.N. and VASILESCU D., J. Theor. Biol. 57, 407 (1976)
10 - VASILESCU D., BROCH H., CORNILLON R. and LESPINASSE J.J., Studia Biophys. 66, 167 (1977)
11 - BROCH H., VASILESCU D. and PERONNEAU H., Studia Biophys. 68, 107 (1978)
12 - BROCH H. and VASILESCU D., Studia Biophys. 69, 21 (1978)
13 - BROCH H. and VASILESCU D. in "Biomolecular Structure, Conformation, Function and Evolution" Proceedings of the Madras Int. Symp. Pergamon Press, New York (1979) p. 329
14 - CLAVERIE P., DAUDEY J.P., DINER S., GIESSNER-PRETTRE C., GILBERT M., LANGLET J., MALRIEU J.P., PINCELLI U. and PULLMAN B., QCPE 10, 220 (1972)
15 - MONTEILHET C. and BLOW D.M., J. Mol. Biol. 122, 407 (1978)
16 - BROCH H., CORNILLON R., LESPINASSE J.N. and VASILESCU D., Biopolymers 14, 695 (1975).

ENVIRONMENTAL CONDITIONS FOR THE FORMATION OF MARIGRANULES AND KINETIC STUDIES ON THE FORMATION

Hiroshi Yanagawa and Fujio Egami

Mitsubishi-Kasei Institute of Life Sciences,
11 Minamiooya, Machida, Tokyo 194, Japan

Abstract: We found that higly organized particles, "marigranules", were produced from glycine and acidic, basic, and aromatic amino acids in a modified sea medium. The marigranules had membrane-like structures and were packed with polymers having molecular weights of 1,800, 6,800, 15,000, and 82,000 daltons. It is considered that these marigranules represent a new type of protocell-like structures that were present on primitive earth. Here we described the environmental conditions for the formation of marigranules and present kinetic studies made on them. Among amino acids, tryptophan was an essential amino acid for the formation of marigranules and among the metal components of modified sea medium, cupric ion was absolutely necessary for the formation of marigranules. The form of marigranules was dependent on temperature and time. The time course for the formation of these marigranules was oscillatory.

Egami (one of the present authors) recently found that a close correlation exists between the concentrations of different chemical elements in contemporary sea water and their biological behavior and he suspects that transition elements relatively abundant in sea water such as molybdenum, iron, and zinc must played an important role in the course of chemical evolution in the primeval sea because the inorganic composition of the primeval sea was essentially similar to that of the sea today (1). Based upon this idea, we have attempted an experimental approach to the chemical evolution that occurred in the primeval sea. The modified sea medium has been designed to simulate experimentally the chemical evolution in the primeval sea, that is, to accelerate the reaction in the laboratory. Thus, this modified sea medium contains 1,000-100,000 times more of the following

Y. Wolman (ed.), Origin of Life, 309–312.

six transition elements, iron, molybdenum, zinc, copper, cobalt, and manganese and a lower concentration of sodium chloride than does present day sea water. In the course of the research, we recently found that highly organized particles were produced in a higher yield from glycine and acidic, basic, and aromatic amino acids in the modified sea medium (2-4). We have called the organized particles produced in the modified sea medium "marigranules" because they are granules produced under conditions similar to the marine environment. The marigranules (C, 58.22; H, 3.76; N, 14,23; ash 7.62%), which are highly organized particles of 0.3 to 2.5 μm in diameter, were ununiformly packed with KOH-soluble polymers and also had membrane-like structures. Their infrared spectrum suggested the presence of peptide bonds. The interior structure of marigranules was solubilized by treatment with KOH solution. The KOH-solubilized component was characterized by gel filtration and polyacrylamide gel electrophoresis. The component consisted of polymers having molecular weights of 1,800, 6,800, 15,000, and 82,000 daltons, and 34% of the total nitrogen was shown to be that of primary amino groups following treatment with elastase. This suggests that the polymers had at least one peptide bond per three amino acid residues. Thus, it appears that the interior of marigranules consists of polymers having an elastin-like structure. The surface structure of the marigranules was solubilized with SDS and Triton X-100. The SDS-solubilized components were 3,400, 350, and 150 daltons in size.

We also found that organized particles were phase-separated from a concentrated aqueous solution of freeze-dried powder prepared from the reaction mixture mentioned above, and we designated these particles as "marisomes" (5). The marisomes (C, 63.67; H. 6.71; N, 10.56; ash, 6.54%) had a spherical structure from 2 to 3 μm in diameter. They were found to be empty particles with a soft envelope, by scanning electron microscopic observation. The marisomes could be completely solubilized by SDS, ethanol, and Triton X-100. The solubilized marisomes were phase-separated again by dialysis against water. Marisomes were also solubilized in boiling water and were phase-separated again on cooling. Thus the system was reversible. Marisomes consisted mainly of lipophilic polymers having molecular weights of approx. 3,400 daltons. The molecular size and lipophilic properties of the marisomes were quite similar to those of the surface structure of the marigranule.

Taking into consideration the characteristics of both of these particles, we were led to the working hypothesis that a marisome may be the precursor of a marigranule; that is, the marisome was first phase-separated from the reaction mixture and then amino acids and their oligomers were absorbed into the marisome, in which they were polymerized to give elastin-like polymers with cross-links. Thus, both the organized particles, marisomes and marigranules, are good models representing the course of evolution of protocells.

In this paper, we describe the environmental condition for the formation of marigranules and the kinetic studies carried out on them. Marigranules were produced from a reaction of glycine and acidic (L-Glu and L-Asp), basic (L-His, L-Arg, and L-Lys), and aromatic (L-Phe, L-Trp, and L-Tyr) amino acids at 105°C for 8 weeks under a N_2 atmosphere in a modified sea medium enriched with six essential transition elements (Mo^{6+}, Zn^{2+}, Fe^{3+}, Cu^{2+}, Mn^{2+}, and Co^{2+}). We examined in detail the environmental conditions (temperature, composition of the modified sea medium, composition of the amino acids, the presence of oxygen, and time for the formation of marigranules).

Marigranules were not produced below 27°C; however, they could be produced at 37°C or above (50°C, 60°C, 75°C, 90°C, and 105°C). The size of the marigranules increased with increasing temperature. The diameters of marigranules produced at varying temperatures are, 0.05-0.2 μm at 37°C, 0.05-0.45 μm at 50°C, 0.06-0.8 μm at 60°C, 0.2-1.0 μm at 75°C, 0.3-1.0 μm at 90°C, and 0.3-1.0 μm at 105°C. Marigranules produced at 60°C or below (37°C and 50°C) were much softer than those produced at 105°C, and which were packed with polymers having molecular weights of 110,000, 45,000, 25,000, and 1,000. The influence of metal ions on the formation of marigranules was examined by the single omission method. The single omission of Mg^{2+}, Ca^{2+}, $PO_4{}^{3-}$, Fe^{3+}, Mo^{6+}, Co^{2+}, and Mn^{2+} had no effect on the formation of marigranules. However, the single omission of Cu^{2+} and Zn^{2+} decreased significantly the formation of marigranules. Especially, when Cu^{2+} was omitted from the modified sea medium, no normal spherical structure of marigranules was formed at all but the abnormal cubic structure was formed. Therefore, Cu^{2+} appears to be an essential element for the formation of marigranules. When 100-fold Mo^{6+} (10 mM) at a concentration the same as that of standard modified sea medium, was added to the reaction system, very large particles (4 μm in diameter) were produced. The optimum composition of amino acids for the formation of marigranules was a mixture of glycine and acidic (L-Glu and L-Asp), basic (L-His, L-Arg, and L-Lys), and aromatic (L-Phe, L-Trp, and L-Tyr). In particular, aromatic amino acids were required for the formation of marigranules. Among aromatic amino acids, tryptophan was essential. When tryptophan was omitted from the reaction mixture, the spherical structure of marigranules was not formed but an abnormal cubic structure was formed. The presence of oxygen in the reaction system affected the formation of marigranules. The marigranules produced under partially aerobic condition were larger in number and size than those under completely anaerobic condition. Thus, the presence of a small amount of oxygen was preferable to the formation of marigranules. The time course of the formation of marigranules was examined in detail. It was found that the formation of marigranules was oscillatory in number and in size. This oscillation may result from natural selection among marigranules.

REFERENCES

1. Egami, F., J. Mol. Evol. 4, 113 (1974)
2. Yanagawa, H. and Egami, F. in Origin of Life: Proc. 5th Intern. Conf. Origin of Life (Noda, H., ed.) pp. 385, Center for Acad. Publ. Japan, Tokyo, 1978
3. Yanagawa, H. and Egami, F., Proc. Japan Acad. 54, 10 (1978)
4. Yanagawa, H., Kobayashi, Y. and Egami, F., J. Biochem. 87, 855 (1980)
5. Yanagawa, H. and Egami, F. Proc. Japan Acad., 54, 331 (1978)

MEMBRANE LIPIDS AND THE ORIGIN OF LIFE

J. Oró, G. Holzer, M. Rao
University of Houston, Houston, TX

T. G. Tornabene
Colorado State University, Fort Collins, CO

ABSTRACT. There are two major types of amphiphilic lipids found in the plasma membranes of all living cells. The acylglycerol ester derivatives of ordinary bacteria and eukaryotes, and the isopranyl ether derivatives of Archaebacteria. Due to the importance of membranes in the origin of life and to the limited work done in this field we have undertaken experiments on the prebiotic synthesis of the building blocks of membrane lipids of prokaryotes and eukaryotes, and we have studied the less known membrane lipid components of Archaebacteria. Concerning the former, we have been successful in synthesizing fatty acids, glycerol, glycerophosphate, mono-, di- and tripalmitoyl glycerol esters and the corresponding phosphatidic acids. Glycerol has been obtained by reduction of the products from the formose reaction. Furthermore, under the prebictic conditions of an evaporating pond model and using cyanamide as a condensing agent we have synthesized glycerol phosphate and all the other complex lipids. These amphiphilic molecules have the ability to self-assemble into stable bilayered microvesicles which show structure and functions similar to cell membrane lipid bilayers. Concerning the Archaebacteria (Methanogens, Halophiles, Sulfolobus and Thermoplasma) we have analyzed their membrane lipids and found them to be derivatives of diphytanyl glycerol diethers or di-C_{40}-isopranyl diglycerol tetraethers. Such a unique membrane ether lipid composition is ideally suited for stability to the high temperature and other extreme environmental conditions of archaen times. This and other evidence suggests that the extant Archaebacteria evolved from a common anaerobic ancestral line when the primitive Earth was evolving from anoxidic to oxidic conditions at about 3.7 Aeons ago.

Y. Wolman (ed.), Origin of Life, 313–322.

During the past years the major emphasis of most investigators in prebiotic chemistry has been directed toward the synthesis of biomonomers and polymers under primitive Earth conditions. These efforts have left little doubt that most of the important components in biological system could have been formed from simple molecules such as H_2, N_2 (NH_3), H_2S, CH_4, CO_2 (CO) and H_2O as a result of various kinds of energy and mineral effects. Fig. I summarizes the abiotic reaction sequences leading to the formation of the major biomolecules as we know them today (see Ref. 1 for a review). We have now reached a crucial state in chemical evolution, where the available macromolecular components could associate to a system with an enhanced chance of survival and the capacity to evolve to more complex structures, e.g. protocells. Although the actual mechanisms leading to such structures are not yet understood, they must have evolved within the confines of a protected microenvironment, similar to those provided by the contemporary cell membranes. The importance of simple membranes as components of protocells has been emphasized by several authors (1-4).

Clusters of organic macromolcules which had the potential of self-replication were most certainly not like modern cells. However they had to be in a thermodynamically stable state, thus the first percursor of cells must have been confined within a boundary or a protomembrane. Oparin was one of the first investigators to suggest the formation of membranes around macromolecules in a process called coacervation (5). Coacervation is the spontaneous separation of a polymer containing aqueous phase into two aqueous phases, one with a high polymer concentration, the other one with a relatively low concentration. Such phenomenon could have occurred in the primordial organic broth, leading possibly to the formation of small droplets, rich in macromolecules. Such coacervate droplets could interact with the surrounding environment, absorb more compounds that can be incorporated into its structure and finally after reaching a certain size would break up into smaller droplets. In a number of experiments Oparin and his colleagues (6) were able to demonstrate that enzymes enclosed in coacervate droplets can act on substrate molecules which are adsorbed from the medium and products are released into the medium. Obviously the concept of these droplets should be understood as a model because they are formed from complex biological materials, such as polysaccharides, polypeptides or DNA rather than simpler compounds that could be formed under conditions of the primitive Earth.

Another mechanism of self-forming cell like structures, the so called microspheres, was described by Fox and his colleagues (7,8). Microspheres are formed from crosslinked amino acid polymers of proteinoids, and have properties that suggest the presence of a membrane-like surface although they contain no lipids,

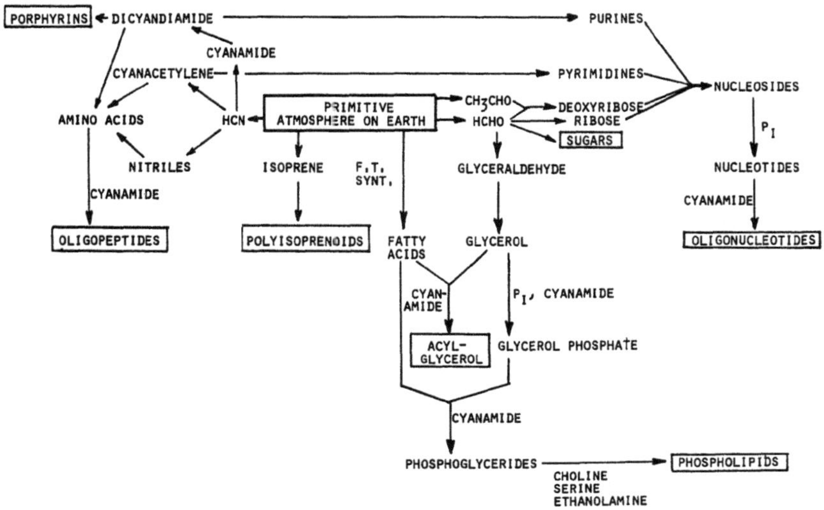

FIGURE 1 : CHEMICAL EVOLUTION OF BIOMONOMERS AND POLYMERS

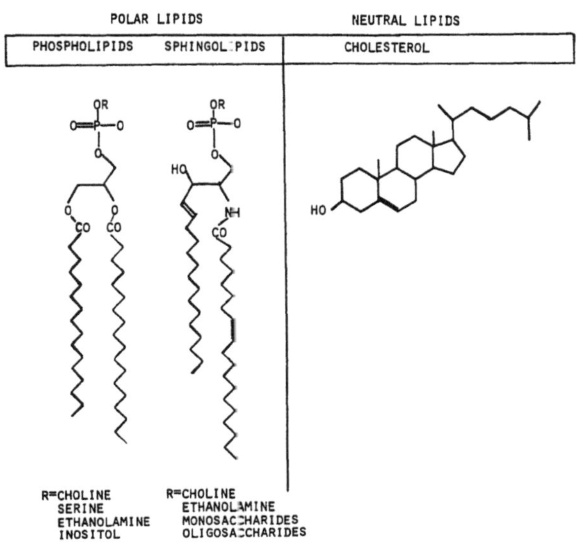

FIGURE 2 : LIPID COMPONENTS OF THE BIOMEMBRANE

and do not form bilayered microvesicles. They can be stimulated to undergo division or association and they are capable of accelerating chemical reactions. It was also shown that microspheres associate selectively with homopolynucleotides and under certain conditions are capable of promoting peptide synthesis(9). However there are some limitations in considering the microspheres as prototypes for primitive membranes (4).

The major components found normally in biomembranes are phospholipids. One of the prominent features of phospholipids is their ability to form spontaneously, bilayered structures referred to as liposomes. These microvesicular structures have a polar outer surface, an intermediate hydrophobic core, which is a liquid-crystalline matrix, and a polar inner surface. The characteristics of cell membranes, which are to some extent also the intrinsic properties of spontaneously formed bilayered liposomes, allow selective interactions with the external aqueous environment, which are essential for the survival of any cell. Some of these functions are to provide a stable medium for controlled synthesis or hydrolysis of macromolecules and to permit the selective entry or extrusion of small molecules, water and ions. There is now extensive evidence that amphiphilic molecules can form many different kinds of liposomes when dispersed in water (4,10,11). Liposomes are described not only as microscopic spherical lipid bilayers with internal aqueous compartments, but also as tubular structures consisting of lipid bilayers or multilayers (4). As vesicles the liposomes are selectively permeable (12), can absorb small molecules and interact with proteins to form lipid-protein conjugates (13) which have enzymatic activity (14).

A number of recent investigations (4,15) have been directed to gain a better understanding of the processes involved in the incorporation of proteins into primitive lipid membranes and the mechanisms of their particular functions, such as enzymatic activity and ion transport.

Before discussing the availability of the structural components which are necessary in the formation of primitive lipid membranes, it might be useful to point out the structure of the biomembrane. The molecular components are shown in Fig. 2. The actual model of a biomembrane, as demonstrated in Fig. 3, is redrawn from Nicolson (16). The principal structure of the membrane is a lipid bilayer with choline phospholipids and glycol-lipids on the outer surface, and serine and ethanolamine phospholipids in the cytoplasmic side. Most proteins are associated with the cytoplasmic side of the bilayer, however some of them extend across it. Those proteins which extend across the membrane have usually a carbohydrate portion which is encased in the outer surface to prevent the protein chain from diffusing into the cyto-

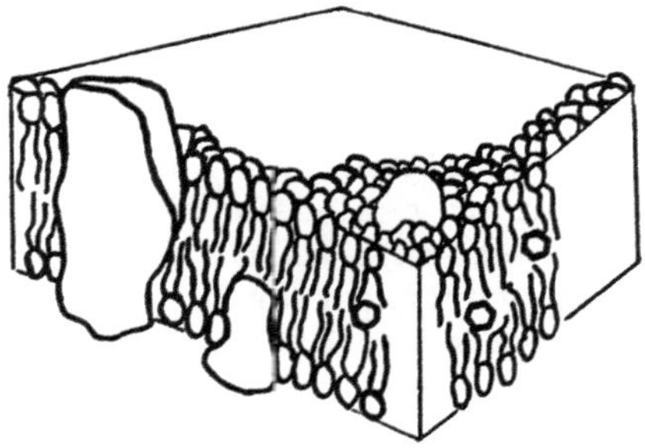

FIGURE 3 : MODEL OF THE BIOMEMBRANE, BILAYER OF PHOSPHOLIPIDS WITH INTERCALATED PROTEINS AND CHOLESTEROL.

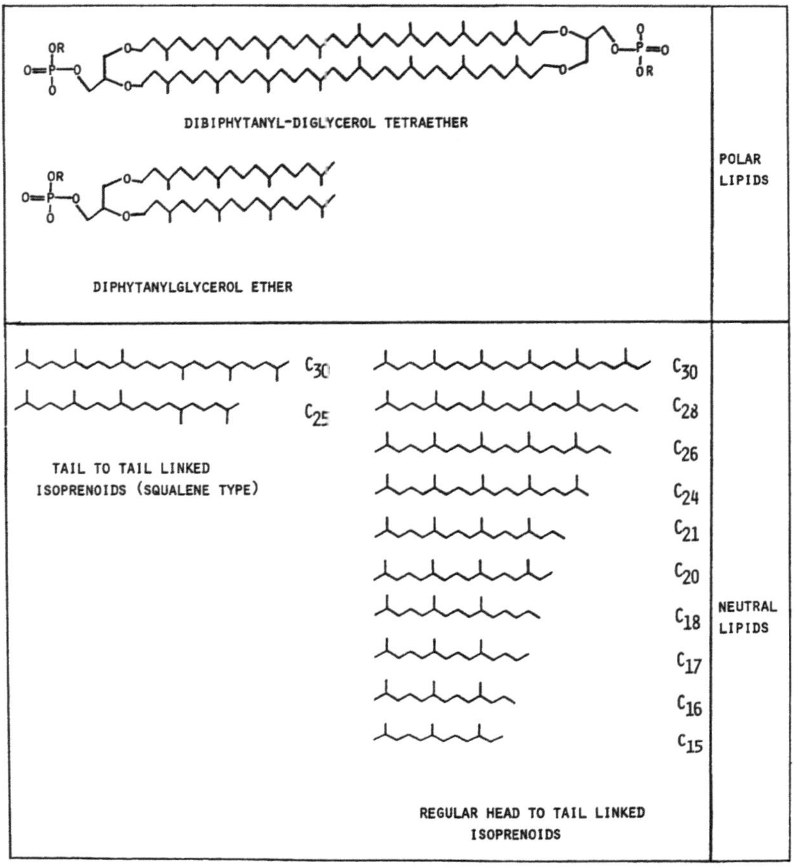

FIGURE 4 : LIPID COMPONENTS OF THE MEMBRANE IN ARCHAEBACTERIA.

plasm. Membrane proteins display specific functions such as enzymatic activity and they serve as carriers for metabolites across the membrane.

Two approaches can be followed in the study of protomembranes and the development of cell-like structures. One is the reconstitution of plasma membrane models from prebiotic precursors and the experimental verification of their properties. An alternative approach is the study of lipid membranes in some of the organisms which are presumed to be more ancient. In the latter case one of the obvious problems is to establish criteria which allow to estimate the "primitivity" of the organisms. Studies in molecular evolution attempt to answer this question by the direct comparison of informational macromolecules of different species of microorganisms. Recently such a comparative analysis of r-RNA sequences of methanogenic bacteria was carried out (17,18). The study showed that these bacteria are distinctly different from the vast majority of prokaryotic bacteria and therefore should be distinguished by a different phylogenetic classification. The organisms are designated as Archaebacteria (19) and include now a series of bacteria from extreme environments such as the extreme halophiles (20) and thermoacidophiles (21,22). Archaebacteria are unique organisms for the most part. Their 16 S r-RNA sequences have practically nothing in common with their counterparts in eukaryotic and prokaryotic cells. The results of r-RNA sequence homologies suggest that Archaebacteria are extremely ancient (18). Furthermore, Archaebacteria do not contain the components typically found in bacterial cell walls, such as muramic acid (23). The methanogens have a unique metabolism based on CO_2 and H_2 as carbon and energy sources (24). But perhaps one of the most unique properties of these organisms is that the membrane of Archaebacteria always consist of glycerol ether lipids of polyisoprenoid alcohols (25) and that their neutral lipids are composed of isoprenoid hydrocarbons (26). Moreover, geochemical studies have shown that a high similarity exists between the isoprenoid content of Archaebacteria and the distribution of these compounds in ancient sediments and petroleum (27). Recently a C_{40} -isoprenoid chain, which is part of the tetraether assembly in Archaebacteria, was isolated from kerogen of an Eocene Formation (28). Although there are a number of questions to be answered before we can fully accept that Archaebacteria are the direct descendants of primitive forms, the unusual lipid composition of these bacteria can provide us with viable information for more appropriate models of protocellular membranes. The molecular components of the polar and neutral lipids in Archaebacteria are shown in Fig. 4. A model of the ether lipid membrane is illustrated in Fig. 5.

The feasibility of any models of protocellular membranes depend on the availability of the molecular components in the

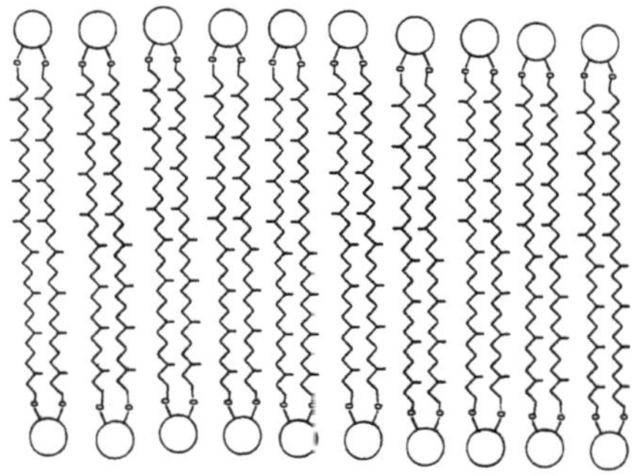

MONOLAYER OF DIBIPHYTANYL GLYCEROL TETRAETHERS

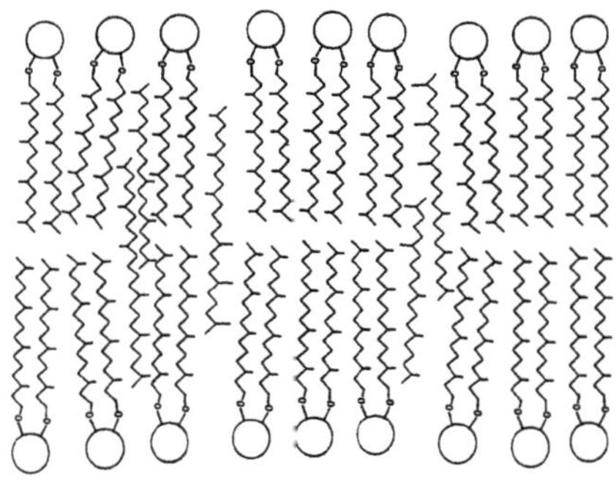

BILAYER OF DIPHYTANYL GLYCEROL ETHERS WITH INTERCALATED SQUALENE AND LOWER HOMOLOGS

FIGURE 5 : MODEL OF THE MEMBRANE IN ARCHAEBACTERIA

CO_2 + H_2 $\xrightarrow{\text{FISCHER TROPSCH}}$ FATTY ACIDS	(29,30)
CH_2O $\longrightarrow$ GLYCERALDEHYDE $\longrightarrow$ GLYCEROL	(31)
GLYCEROL + FATTY ACIDS $\xrightarrow{\text{CYANAMIDE}}$ ACYLGLYCEROLS	(32)
GLYCEROL + PHOSPHATE $\xrightarrow{\text{CYANAMIDE}}$ GLYCEROL PHOSPHATE	(34)
GLYCEROLPHOSPHATE + FATTY ACID $\xrightarrow{\text{CYANAMIDE}}$ PHOSPHATIDIC ACID	(33)
PHOSPHATIDIC ACID + CHOLINE (SERINE) $\xrightarrow{\text{CYANAMIDE}}$ PHOSPHATIDYL CHOLINE (SERINE)	(35)
ISOPRENE $\xrightarrow{\text{UV}}$ MONOTERPENES	(36)
ISOPRENE $\xrightarrow{\gamma\text{-RADIATION}}$ CYCLIC AND ACYCLIC ISOPRENOIDS	(36)

FIGURE 6 : PREBIOTIC SYNTHESIS OF LIPID MEMBRANE COMPONENTS.

primitive Earth environment and the likelihood of their abiotic synthesis. During the past years most of the lipid components of the normal biomembrane have been obtained in prebiotic reaction sequences (Fig. 6). The synthesis of fatty acids from CO and H_2 (29,30) in a Fischer-Tropsch type synthesis and the formation of glycerol from glyceraldehyde which is a condensation product of formaldehyde (31) stimulated the search for systems that were capable of forming complex lipids. In a series of studies we demonstrated that mono-, di- and tripalmitoylglycerols could be obtained in good yield from glycerol and ammonium palmitate using cyanamide as condensing agent (32). When glycerol was replaced by glycerol-3-phosphate, phosphatidic acid was formed (33). Glycerol-3-phosphate was obtained from glycerol and ammonium phosphate using cyanamide or urea as condensing reagent (34).

Recent attempts to attach choline or serine to phosphatidic acid have also been proven successful. Phosphatidylcholine was obtained in good yield from phosphatidic acid and choline under similar conditions as above (35). Thus it is now reasonable to assume that the major lipid components of the cell membrane could have been formed in the environment of the primitive Earth.

Some of our future studies will be designed to assess the probability of obtaining ether lipids--as found in Archaebacteria ---under prebiotic conditions. These ether lipids are ideally suited for stability to extreme pH and high temperatures, conditions that might have existed during certain periods of the primitive Earth.

One of the first steps in the prebiotic assembly of these ethers is the synthesis of the isoprenoid moiety of the lipid molecule. In recent experiments we have obtained cyclic and acyclic isoprenoids by the irradiation of isoprene with UV or λ-radiation (36). It is hoped that further investigations along these lines will provide a better understanding of primitive life forms and their evolution.

ACKNOWLEDGMENTS. This work was supported in part by a grant from the National Aeronautics and Space Administration, NGR-44-005-002 to J.O, and in part by Department of Energy Grant EE-77-5-02-4478 to T.G.T.

REFERENCES

1. Oró, J., Sherwood, E., Eichberg, J. , Epps, D.E., in: Light Transducing Membranes (ed. D. Deamer), Academic Press, New York, 1978, p. 1.

2. Bernal, J. D., in: The Origin of Life, Weidenfeld and Nicolson, London, 1967.

3. Bangham, A.D., Prog. Biophys. Mol. Biol. 18, 29 (1968).

4. Hargreaves, W. R. and Deamer, D., in: Light Transducing Membranes (ed. D. Deamer), Academic Press, New York, 1978, p. 23.

5. Oparin, A.I., "The Origin of Life on Earth", Academic Press, New York, 1957.

6. Oparin, A.I., in "Prebiotic and Biochemical Evolution (Eds. Kimball, A.P. and Oro, J., North Holland, Amsterdam, 1971, p. 2.

7. Fox, S. W. and Dose, K., Molecular Evolution and the Origin of Life, Marcel Dekker, New York, 1977.

8. Fox, S. W. in: Light Transducing Membranes (ed. D. Deamer), Academic Press, New York, 1978.

9. Snyder, W. D. and Fox, S. W., Biosystems 7, 222 (1975).

10. Gebicki, J. M. and Hicks, M., Nature 243, 232 (1973).

11. Gebicki, J. M. and Hicks, M., Chem. Phys. Lipids 16, 142 (1976).

12. Papahadjopoulos, D. and Watkins, J.E., Biochim. Biophys. Acta 135, 639 (1967).

13. Bangham, G. C., Chem. Phys. Lipids 1, 225 (1967

14. Racker, E., J. Biol. Chem. 247, 8198 (1972).

15. Israelachvili, J., in: Light Transducing Membranes (ed. D. Deamer), Academic Press, New York, 1978, p. 91.

16. Nicolson, G. L., Biochim. Biophys. Acta 457, 1 (1976).

17. Fox, G. E., Magrum, L. J., Balch, W.E., Wolfe, R.S. and Woese, C. R., Proc. Natl. Acad. Sci. USA, 74,4537 (1977).

18. Balch, W.E., Magrum, L. J., Fox, G.E., Wolfe,R.S.,Woese, C. R., J. Mol. Evol. 9, 305 (1977).

19. Woese, C. R., Magrum, L. J., Fox, G.E., J. Mol. Evol. 11, 245 (1978).

20. Kates, M., Yengoyan, L.S., Sastry, P. S., Biochim. Biophys. Acta. 98, 252 (1965).

21. Langworthy, T.A., Biochim. Biophys. Acta 487, 37 (1977).

22. Langwortny, T.A., Mayberry, W. R., Smith, P. F., J. Bacteriol. 119, 106 (1974).

23. Kandler, O., and König, H., Arch. Microbiol. 118, 114 (1978).

24. Zeikus, J. G., Bacteriol. Rev. 41, 514 (1977).

25. Tornabene, T.G. and Langworthy, T.A., Science 203 , 51 (1979).

26. Tornabene, T.G., Langworthy, T.A., Holzer, G., Oró, J., J. Mol. Evol. 11, 259 (1978).

27. Holzer, G., Oró, J., Tornabene, T.G., J. Chromatogr. 186, 795 (1979).

28. Michaelis, W. and Albrecht, P., Naturwiss. 66, 420 (1979).

29. Nooner, D.W., Gibert, J. M., Gelpi, E., Oró, J., Geochim. Cosmochim. Acta 40, 915 (1976).

30. Leach, W. W., Nooner, D.W., Oró, J. in: Origin of Life, Center for Academic Publication, Japan Scientific Societies Press, Tokyo, Japan, 1978, p. 113.

31. Lohrmann, R., J. Mol. Evol. 1, 263 (1977).

32. Eichberg, J., Sherwood, E., Epps, D., Oró, J., J. Mol.Evol. 10, 221 (1977).

33. Epps, D.E., Sherwood, E., Eichberg, J., Oró, J., J. Mol. Evol. 11, 279 (1978).

34. Epps, D.E., Nooner, D.W.. Eichberg. J.. Sherwood. E.. Oró, J. J. Mol. Evol. 14, 235 (1979).

35. Rao. M.. Eichberg. J.. Oró, J., unpublished results.

36. Siddiqui, N., Holzer, G.. Oró, J., unpublished results.

RESOLUTION OF UNDERIVATIZED AMINO ACIDS BY HIGH PRESSURE LIQUID CHROMATOGRAPHY, USING CHIRAL ELUANTS

E. Gil-Av*, P.E. Hare**, A. Tishbee* and S. Weinstein*

*Department of Organic Chemistry, Weizmann Institute of Science, Rehovot, Israel.
**Geophysical Laboratory, Carnegie Institution of Washington, Washington, D.C. 20008.

The investigation of a number of important problems in research on the origins of life hinges on the determination of the configuration or enantiomeric composition of optically active substances of biological significance. There are many biological molecules which, in principle, could serve as the asymmetric probes in relevant experiments. However, for a variety of reasons, protein amino acids are, at least thus far, the most suitable class of compounds for such studies.

Well known examples are: the search for optically enriched α-amino acids as evidence for the possible existence of life in space; experiments on the origin of optical activity through preferential degradation of one enantiomer in racemic mixtures of α-amino acids, e.g., by irradiation with longitudinally polarized electrons. Analysis of L to D amino acid ratio as indication for the presence of indigenous amino acids in meteorites, and of possible recent contaminations in fossils. Geochemical dating based on the state of racemization of amino acids in the proteinaceous material in fossils.

The intensified activity in exobiology, triggered by space research, has contributed to the development of a number of gas chromatographic methods of resolution of amino acid derivatives. These procedures are based either on the chromatography of diastereomeric derivatives such as N-TFA-2-butyl esters (TFA = trifluoroacetyl) on ordinary achiral phases (1), or on the use of chiral phases such as N-TFA-L-valyl-L-valine cyclohexyl ester (2) or N-docosanoyl-L-valine-t-butylamide (3) which resolve N-TFA isopropyl esters of amino acids. Both approaches have now become

Y. Wolman (ed.), Origin of Life, 323–330.

routine, and meet with many of the requirements as to sensitivity and efficiency posed by the enantiomeric analysis of protein amino acids (4).

A number of laboratories have worked in the last decade on the resolution of underivatized amino acids by liquid chromatography. In particular, supports, such as polystyrene (5,6) and later polyacrylamide (7) to which ligands such as L-proline were linked covalently and equilibrated with Cu(II) ions, were shown to have high stereoselectivity in interacting with enantiomeric α-amino acids. Chiral recognition is due to the formation of ternary diastereomeric coordination complexes involving the cation present, the ligand covalently linked to the support and the amino acid to be resolved.

These significant findings, first reported over ten years ago (5), have not, however, led to generally useful procedures for the enantiomeric analysis of mixtures of amino acids. On the polystyrene type of resin the rate of diffusion is slow resulting in broad peaks and long analysis times, and as to the polyacrylamide supports, the initial promising publications (7) have not, as yet, been substantiated by further work.

Recently, we have used the principle of chiral recognition through coordination to a metal ion, in conjunction with a different chromatographic mode of operation, for improving the methods of enantiomeric analysis. It was found that by the simple expedient of dissolving a chiral coordination complex, such as $Cu(L\text{-}Pro)_2$, in an aqueous mobile phase convenient and efficient resolution of underivatized amino acids can be obtained using as support either a cation exchange resin (8) or a reversed phase C-18 silica gel (9).

The mode of operation is illustrated in Fig. 1, which is self-explanatory (for experimental details see 8, 9). The eluates are detected by post-column derivatization to fluorescent compounds. Reagents such as o-phthalaldehyde (plus thioethanol) or fluorescamine attack only primary amino groups, so that the background of chiral reagent in the mobile phase (L-proline or other suitable chiral ligands with a secondary amine function) does not interfere.

Table I reveals some of the main findings of the experiments with $Cu(L\text{-}Pro)_2$ as the chiral ligand in the eluant. Before discussing the results it should be pointed out that the degree of the stereoselective effect is measured by the amount of deviation from unity of the resolution factor ($r_{L/D}$), which represents the ratio of the retention times of the peaks corresponding to the two enantiomers (see footnotes to the Table). A value of $r_{L/D} > 1$ means that the L-enantiomer emerges last, and for $r_{L/D} < 1$ the

TABLE I

Resolution Data Of Amino Acids Chromatographed With An Aqueous Mobile Phase Containing Cu(II)-(L-Pro)$_2$.

Amino Acid	Form	Reversed Phase Column[a]		Ion Exchange Column[b]	
		t_R' min	$r_{L/D}$[c]	t_R'	$r_{L/D}$[c]
Alanine	D	0.20	2.92	17.2	1.00
	L	0.58		17.2	
Valine	D	3.80	4.82	17.7	0.92
	L	18.3		16.3	
Isoleucine	D	13.7	4.53	23.7	0.91
	L	62.0		21.5	
Alloisoleucine	D	13.8	4.38	25.5	0.92
	L	60.4		23.5	
Leucine	D	15.9	2.56	28.6	0.99
	L	40.7		28.3	
Methionine	D	9.4	2.41	24.6	0.97
	L	22.7		23.8	
Tyrosine	D	16.3	2.28	39.4	0.78
	L	37.2		30.8	
Phenylalanine	D	69.7	1.94	55.0	0.89
	L	135.0		48.7	
Glutamic acid	D	0.46	1.98	3.3	1.00
	L	0.91		3.3	
Serine	D	-	1.00	13.4	0.86
	L			12.9	
Threonine	D	-	1.00	13.7	0.95
	L			13.0	
Allothreonine	D	0.40	4.50	11.4	0.88
	L	1.83		10.0	
Lysine	D	0.14	2.29		
	L	0.32			
Histidine	D	2.73	0.74		
	L	2.02			
Arginine	D	1.07	1.85		
	L	1.98			

[a] Column, 15x0.46 cm (Supelco LC-18); mobile phase, aqueous solution of cupric acetate (0.008M) and L-proline (0.017M), pH=5.0; flow rate, 30 ml/hr; temp., 25°C; pump pressure, 30 atm. Retention time of cysteic acid is taken as the column void volume (3.75 min).

[b] Column 12x0.2 cm (5μ DC 4a cation exchange resin); mobile phase, aqueous solution of cupric sulphate (0.008M), L-proline (0.016M), in 0.1N sodium acetate buffer, pH=5.5; flow rate 10 ml/hr; temp. 75°C; pump pressure 200 atm. Cysteic acid retention, 1.3 min.

[c] $r_{L/D}$ = resolution factor = ratio of the adjusted retention time of the L-isomer over that of the D-isomer.

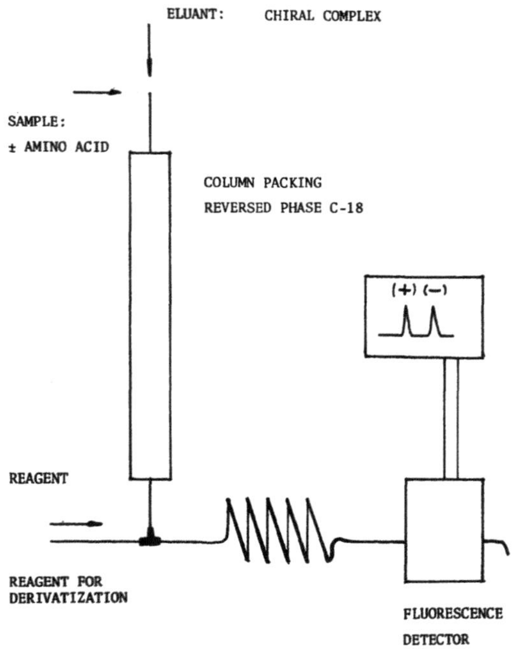

Fig. 1. Scheme of chromatographic set-up.

L-enantiomer elutes first. The order of emergence will depend on the particular amino acid considered and on the mode of operation. The following conclusions can be drawn on the basis of the data in Table I:

Reversed phase colum.-

1) The reversed phase shows, in general, far higher efficiency than the ion exchange resin, with $r_{L/D}$ values in the range of 1.8-4.8. It is to be noted that such high separation factors are hardly ever found for isomeric closely related non-enantiomeric compounds.
2) Under the experimental conditions (notably pH $\sim$5.0-5.5) the L-isomer emerges after the D-isomer, with the exception of histidine, which through its imidazole group can coordinate with the copper in a fashion different from that of the other amino acids.
3) Threonine and aspartic acid could not be resolved under the conditions given in the Table. However, recently Hare (10) has found that at pH=7 the enantiomers could be separated and show inversed order of emergence ($r_{L/D}\sim$0.8 for Asp).
4) Serine could not be resolved under any circumstances tried.

Ion exchange column.-

1) The order of emergence is inverted ($r_{L/D}<1$), as compared with the reversed phase column. This indicates that, in contrast to the above mode, the process of chiral recognition occurs essentially in the mobile phase (8,9).
2) As already mentioned, the ion-exchange resin shows less stereoselectivity than the reversed phase. Alanine, for instance, is not resolved and for leucine only a shoulder was observed. On the other hand, the aromatic and hydroxy amino acids have $r_{L/D}$ values which show a significantly larger deviation from unity than the neutral amino acids. Under the conditions given in the Table, serine and threonine are separated into their enantiomers on the ion exchange column only.
3) The acidic amino acids could not be resolved and the basic amino acids were not eluted from the resin under the experimental conditions.

The method of detection used has the advantage of high sensitivity, and permits readily to detect picomoles of eluate. On the other hand, it is obviously not suitable for eluates containing a secondary instead of a primary amino group. Also cysteine and cystine could not be detected in the effluant and may have been oxidized by the Cu(II) ions.

These results show the great interest of the method for the enantiomeric analysis of amino acids. A noteworthy feature of the approach are the many factors which can be varied and which affect the performance of the system, e.g., the nature of the chiral ligand and the metal ion, the nature and concentration of organic solvents added to the aqueous eluant, temperature, pH, ionic strength, etc. (8,9,11,12).

The influence of these parameters is now being studied systematically. Very recently, it was found that many remaining difficulties mentioned above can be overcome by the use of new chiral ligands (12). An example of such a superior ligand is N,N-di(n-propyl)-L-alanine which, as shown in Fig. 2, permits to resolve readily serine and aspartic acid at pH=5.5, as well as threonine and cysteic acid. Furthermore, with a N,N-dialkylsubstituted ligand in the eluant, detection of proline and similar eluates should be possible by post column oxidation with chlorosuccinimide of the secondary to a primary amine (13), followed by reaction with fluorescamine; the tertiary amine group of the chiral ligand should not interfere.

These results represent a further step towards the development of a convenient, efficient and sensitive routine procedure for the simultaneous qualitative, quantitative and enantiomeric analysis of underivatized protein amino acid mixtures. A very useful version of such a method would consist in the modification of an amino acid analyser for the optional determination of enantiomeric ratios. The coupling of a reversed phase with a classical ion exchange column could also solve some remaining problems of overlap of different amino acid peaks.

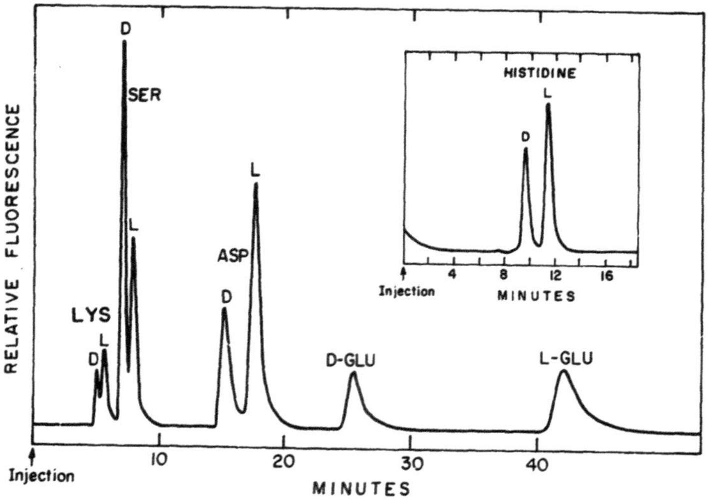

Fig.2. Chromatogram of some α-amino acids on a reversed phase column. Column, 25x0.46cm packed with 5μ Nucleosil C-18; mobile phase, aqueous solution of cupric acetate (0.004M) and L-N,N-di-n-propylalanine (0.010M), pH 5.5, flow rate 40 ml/h; temp.,ambient; pump pressure, 150 atm.

Already in its present stage of development, the new approach has proved to be extremely useful, e.g., in the investigation of the amplification of optical activity by crystallization in the presence of suitable chiral impurities by Addadi et al. (14).

Minute single crystals, precipitated from a supersaturated solution of D,L-aspartic acid in the presence of a small amount of L-alanine, could be readily analysed (14) for their enantiomeric composition, as well as for the presence of adsorbed impurities (as little as 0.03%). Also, the procedures developed are very convenient for establishing (14) the preference for adsorption of one enantiomer of an amino acid by the crystals of another optically active amino acid (e.g., L-Glu/D-Glu on L-Thr), a phenomenon which determines the amplification process.

Another important application is related to geochemical research on fossil proteinaceous material, for which the racemization of aspartic acid is often of particular interest. It can be seen in Fig. 2 that there is a "window" for aspartic acid in the chromatogram, so that the determination of its enantiomeric composition can be carried out particularly well. The analysis of a "real world sample", i.e., the hydrolysate of a high molecular weight fossil protein isolated by S.Weiner (15) is shown in Fig.3.

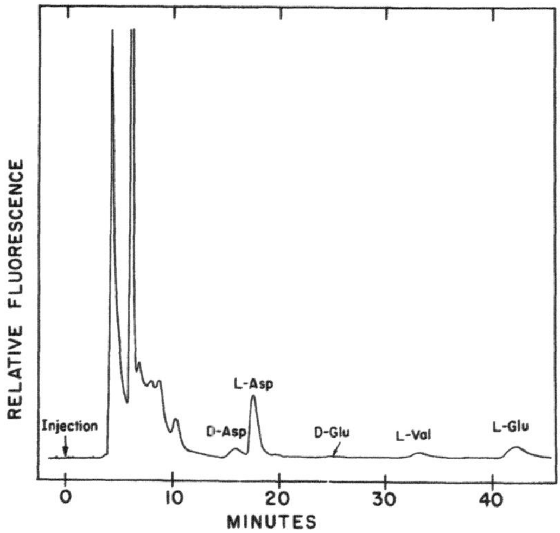

Fig.3. Chromatogram of the hydrolyzate of a high molecular weight fossil protein age >50000 years.

This example demonstrates the characteristic features of the method: No need for derivatization prior to chromatography, high resolution factors and high sensitivity. It should also be recalled that the procedure permits to check readily for non-enantiomeric impurities, which might be hidden under the aspartic acid peaks, by running a second chromatogram with a complex of opposite configuration dissolved in the eluant (8,9).

REFERENCES

(1) Gil-Av, E., Charles-Sigler, R., Fischer, G. and Nurok, D., J. Gas Chromatogr. 4, 51 (1966); Pollock, G.E. and Oyama, V.I., ibid. 4, 126 (1966).
(2) Gil-Av, E., Feibush, B. and Charles-Sigler, R., "Gas Chromatography 1966", ed. A.B. Littlewood, Inst. of Petroleum, London, 1967, p. 238; Feibush, B. and Gil-Av, E., Tetrahedron 26, 1361 (1970).
(3) Charles, R., Beitler, U., Feibush, B. and Gil-Av, E., J. Chromatogr. 112, 121 (1975).
(4) Frank, H., Nicholson, G.J. and Bayer, E., J. Chromatogr. Sci. 15, 174 (1977).
(5) Rogozhin, S. and Davankov, V.A., J. Chem. Soc. D 490 (1971).
(6) Davankov, V.A. and Semechkin, A.V., J. Chromatogr. 141, 313 (1977).
(7) Lefebvre, B.L., Audebert, R. and Quivodron, C., Isr. J. Chem. 15, 69 (1977); and references therein.
(8) Hare, P.E. and Gil-Av, E., Science 204, 1226 (1979).

(9) Gil-Av, E., Tishbee, A. and Hare, P.E., J. Amer. Chem. Soc. 102, 5115 (1980).
(10) Hare, P.E., unpublished data.
(11) Previously, for instance, metal ions such as Zn(II), Co(II) and Hg(II) in conjunction with L-Pro, and ligands such as L-Pro-L-valine coordinated to Cu(II), have been found to show stereoselectivity on a reversed phase column (9); Zn(II) coordinated to L-Asp-L-Phe-O-Me has also been shown to resolve some α-amino acids in the reversed phase mode, Gilon, C., Leshem, R., Tapuchi, Y. and Grushka, E., J. Amer. Chem. Soc. 101, 3403 (1979).
(12) Weinstein, S., manuscript in preparation.
(13) Weigele, M., DeBernado, S. and Leimbruber, W., Biochem. Res. Comm. 50, 359 (1973).
(14) Addadi, L., van Mil, J. and Lahav, M., submitted for publication.
(15) Weiner, S., unpublished data.

STEREOSELECTIVE INTERACTIONS OF SMALL BIOLOGICAL MOLECULES

Young Hwan Kim, A. Tishbee and E. Gil-Av

Department of Organic Chemistry, Weizmann Institute of Science, Rehovot, Israel.

In a preceding communication from this laboratory (1) we have discussed the importance of chromatographic methods for the resolution of optical isomers as tools in research on problems related to the origins of life. Besides its analytical significance, the study of chromatographic separation based on stereoselective interactions between chiral molecules acting, respectively, as selectors and selectands (2), has also considerable interest for the development of concepts on chiral recognition and its mechanism (3,4). The ideas evolved could be relevant to the problem of amplification of optical activity discussed in this section. This is particularly true for systems involving biological molecules.

Recent chromatographic research has shown that the selector-selectand association, required as a condition for bringing about chiral recognition, can be effected through a variety of intermolecular forces, e.g., hydrogen bonding (2,3), charge transfer complexation (5), coordination to metals (6), ionic and dipole interactions, as well as combinations of such forces (7). In the present paper we wish to report on the pursuit of our work on charge transfer complexation and chiral recognition, and its extension to stationary phases consisting of relatively small biological molecules.

Several years ago (5), it was found that R(-)-2-(2,4,5,7-tetranitrofluorylideneaminooxy)propionic acid (TAPA), linked covalently to silica gel or coated on it, can act as an efficient stereoselective stationary phase for the resolution of helicenes, which are ortho-condensed polyaromatic hydrocarbons. The overlap of the aromatic rings in these compounds (see structure of the

Y. Wolman (ed.), Origin of Life, 331–336.

hexahelicene) leads to a chiral molecular shape. Derivatives with 5-14 rings are known and have all been resolved on TAPA columns (5).

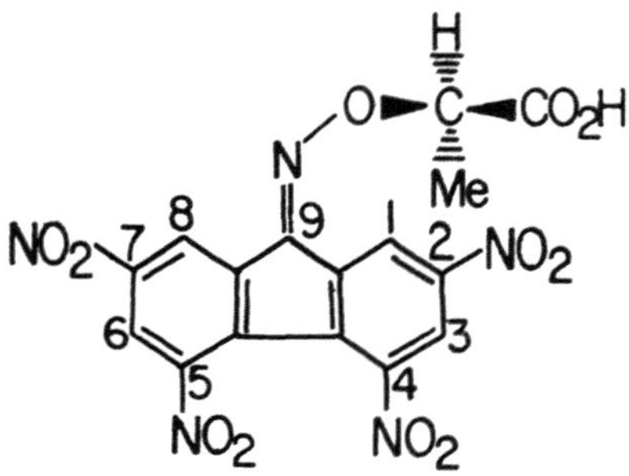

R(—)-TAPA

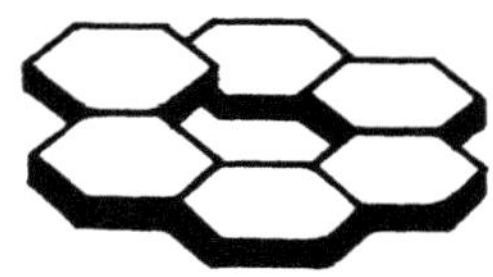

HEXAHELICENE

There exist a large number of biological molecules, such as the nucleotides, the nucleosides and the flavins, which have certain general structural features recalling those of TAPA. Thus, riboflavin (Vitamin B_2) and TAPA have both a tricyclic system, known to form charge transfer complexes with various compounds, particularly aromatic hydrocarbons (8), and a side chain attached to the central ring and containing one (or more) asymmetric center(s):

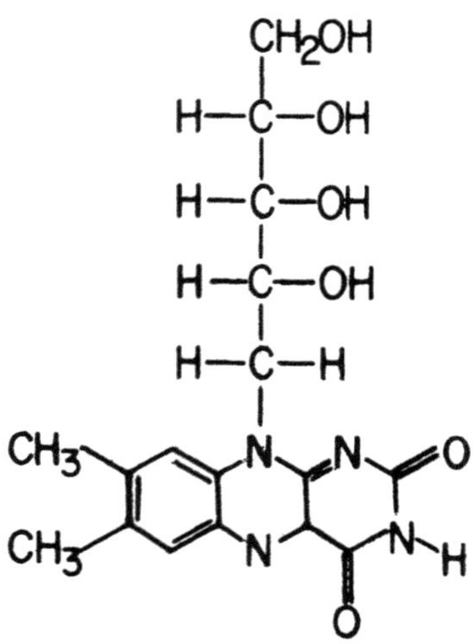

RIBOFLAVIN

By simply coating Vitamin B_2 dissolved in water on a 5μ silica gel (5-8%) a highly efficient column was prepared, on which separation of the enantiomers of [6] to [14]-helicene could be achieved by high pressure liquid chromatography with CH_2Cl_2/n-hexane as the mobile phase (9). The resolution factors (ratio of the retention time of the peak emerging last over that emerging first) increases with the number of rings in the molecule. The values of the resolution factors - 1.05 to 1.13 - are lower (10) than observed on TAPA (5). The order of emergence on riboflavin is P(+) before M(-) (11), i.e., the reverse of that on R(-)-TAPA. The excellent separation of the [13]-helicene isomers on riboflavin is shown in Fig. 1,A [see (9) for details].

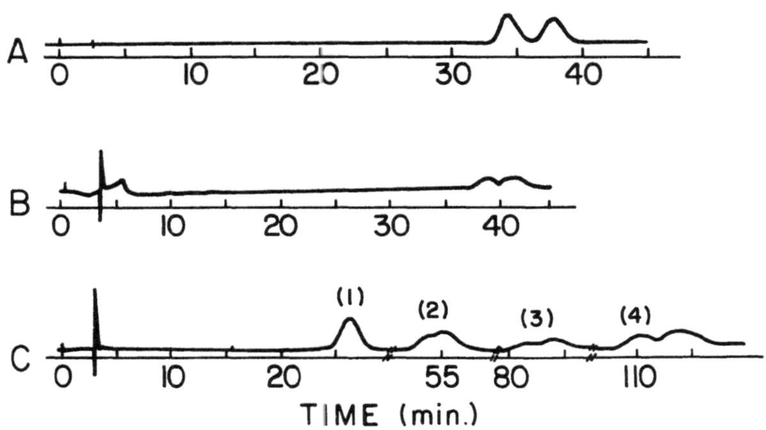

Fig. 1. Resolution of [13]-helicene by HPLC on 5μ silica gel coated with (A) riboflavin, (B) 5'-AMP and (C) 3'-AMP (four recyle steps); column 20x0.46 cm, eluant CH_2Cl_2/n-hexane; detector UV, 254 nm.

Having obtained these encouraging results, we decided to extend our investigations to nucleotides and nucleosides. These compounds, like Vitamin B_2, also contain a ring structure capable of charge transfer complexation (8), and a sugar - ribose - attached to the ring as the chiral handle. Adenine derivatives appeared to be particularly promising, since purines are stronger charge transfer complexing agents than pyrimidines. Adenosine, deoxyadenosine, 3'-adenosine monophosphate (3'-AMP) and 5'-adenosine monophosphate (5'-AMP) were examined for stereoselectivity towards [10] to [14]-helicene under essentially the same experimental conditions as Vitamin B_2 (12).

All these relatively low molecular weight compounds were found to manifest some degree of chiral recognition for the helicenes examined, though, in general, the resolution was considerably less good (10) than on Vitamin B_2. Thus, whereas 5'-AMP showed the highest resolution factors in the series examined (e.g., 1.064 for [13]-helicene with CH_2Cl_2/n-hexane, 1:9; see also Fig. 1,B), the corresponding value for 3'-AMP was <1.03. A peak resolution similar to that obtained for 5'-AMP for [13]-helicene required four recyle steps on 3'-AMP (see Fig. 1,C).

Discussion and conclusions.

As chiral discrimination is a distinctive feature of life, it is natural that attempts to resolve optical isomers by chromatography were started with the use of biological substances, or derivatives thereof, as the stationary phase. By now it is established (3,4) that derivatives of the protein amino acids are excellent solvents for the resolution of various classes of compounds, which are capable of interacting with such phases through hydrogen bonding.

Another class of biological substances employed as stationare phases already in the earliest experiments on chromatographic resolution, are the carbohydrates. For instance, the classical separation of the enantiomers of Troeger's base was carried out on lactose (13), aromatic α-amino acids were resolved on cellulose (14), and N-acyl-α-amino acids (15) and o-hydroxylated ter- and tetraphenyl derivatives (16) on starch. It should be mentioned that for polymeric carbohydrates, preserved in their original shape, resolution can be ascribed largely to the presence of chiral cavities. This has been demonstrated, in particular, for triacetyl cellulose, which can be considered as a chiral molecular sieve (17).

The experiments reported here focus attention on the discrimination of certain optical isomers by biological substances by the combined effect of charge transfer complexation and chiral recognition through a sugar residue. These observations are of considerable interest, as modern carbohydrate chemistry puts much emphasis on the information content of sugars and their function in molecular recognition. On the other hand, the role of charge transfer complexation in biological processes is receiving increasing attention. The pursuit of the present studies appears, therefore, to be of topical interest.

The chromatographic approach to the study of chiral discrimination, as described in this paper, has the advantage of simplicity and of high sensitivity. The example of the resolution of [13]-helicene, by recycling on 3'-AMP, demonstrates that the method permits to detect even very small effects.

By examining, as stationary phases, a systematically modified series of compounds, the influence of structural factors on chiral recognition could be investigated and eventually a model for the mechanism evolved.

The experiments reported refer to the resolution of synthetic molecules, the helicenes, not found in nature. It will be a future objective to extend this work to the resolution of solutes of biological significance. Another objective will be to carry out the experiments with aqueous solutions, instead of organic solvents, as the mobile phase.

One particularly pertinent aspect of such work would be to examine compounds containing simple structural units which might be formed under primitive earth conditions. The demonstration of chiral differentiation for biological, optically active substances by such primeval molecules would permit to propose amplification models for optical activity based on partitioning between two phases, such as an aqueous solution and clay.

Finally it should be mentioned that the present data are relevant to the mechanism of stereoselective mutagenicity of certain metabolites of polyaromatic hydrocarbons (18,19). It has been suggested (18) that this phenomenon could be due to a stereoselective physical interaction of the mutagen with the DNA molecule prior to covalent binding to a base of the strand. The charge transfer complexation with chiral recognition for optically active aromatic hydrocarbons, which we report, supports this view and might provide a model for the interaction occurring with DNA.

REFERENCES

(1) Gil-Av, E., Hare, P.E., Tishbee, A. and Weinstein, S., These Proceedings.
(2) Selector and selectand are cybernetic terms for separating agent and sample input, respectively. They are used to avoid ambiguities or restrictions that might arise from the use of terms such as solvent-solute, stationary phase-mobile phase, etc.; see (5), p. 205, footnote.
(3) Feibush, B., Balan, A., Altman, B. and Gil-Av, E., J. Chem. Soc. Perkin II 1230 (1979).
(4) Weinstein, S., Leiserowitz, L. and Gil-Av, E., J. Amer. Chem. Soc. 102, 2768 (1980).
(5) Mikes, F., Boshart, G. and Gil-Av, E., J. Chromatogr. 122, 205 (1976).
(6) Davankov, V.A. and Semechkin, A.V., ibid. 151, 313 (1977); Gil-Av, E., Tishbee, A. and Hare, P.E., J. Amer. Chem. Soc. 102, 5115 (1980).
(7) Kyba, E.B., Koga, K., Sousa, L.R., Siegel, M.G. and Cram, D.J., ibid. 95, 2696 (1973); and subsequent papers.
(8) Slifkin, M.A., "Charge Transfer Interactions of Biomolecules", Academic Press, London, 1971.

(9) Kim, Y.H., Tishbee, A. and Gil-Av, E., J. Amer. Chem. Soc. 102, in press (1980).
(10) The comparison is based on chromatographic results obtained under conditions optimized, respectively, for good resolution (5,9,12), and gives only an approximate idea of relative stereoselectivity. A better measure would be the difference in the stability constants of complexation for the enantiomeric solutes in an inert solvent.
(11) For the nomenclature of helicenes, see (5), p. 206, footnote.
(12) Kim, Y.H., Tishbee, A. and Gil-Av, E., to be published.
(13) Prelog, V. and Wieland, P., Helv. Chem. Acta 27, 1127 (1944).
(14) Dalgliesh, C.E., J. Chem. Soc. 3940 (1952).
(15) Krebs, H. and Schumacher, W., Chem. Ber. 99, 1341 (1966).
(16) Haynes, P.K., Hess, H. and Musso, H., Ber. 107, 3733 (1974).
(17) Hesse, G. and Hagel, R., Liebigs Ann. Chem. 996 (1976).
(18) Meehan, T. and Straub, K., Nature 277, 410 (1979).
(19) Chang, R.L., Wood, A.W., Levin, W., Mah, H.D., Thakker, D.R., Jerina, D.M. and Conney, A.H., Proc. Natl. Acad. Sci. (USA) 76, 4280 (1979).

POSSIBLE SELECTIVE ADSORPTION OF ENANTIOMERS BY NA-MONTMORRILLONITE

Elaine Friebele, Akira Shimoyama,[1] Cyril Ponnamperuma

Laboratory of Chemical Evolution, Dept. of Chemistry
University of Maryland, College Park, MD. 20742

If clay minerals can preferentially adsorb two enantiomers of an organic molecule, they might provide a mechanism for concentration of homogeneous optical isomers. The conflicting reports concerning the selective adsorption of the L-enantiomer of amino acids by clays prompted us to examine the preferential adsorption of optical isomers of amino acids. Racemic amino acids were incubated with Na-montmorillonite at 100% CEC at three hydrogen ion concentrations. The amino acids used were (D,L) α-alanine, (D,L) α-aminobutyric acid, (D,L) valine, (D,L) norvaline. Quantities of amino acids adsorbed on the clay as well as those non-adsorbed were determined by ion exchange chromatography. Analysis of enantiomers was by gas chromatography. A mass balance was obtained from these experiments, leaving no question that 100% of the amino acids was recovered. Differences in the quantities of D and L enantiomers in any of the fractions were no larger than a few percent. Although a large difference in the adsorption of amino acid enantiomers was not observed, our analysis may indicate a small preferential adsorption by Na-montmorillonite.

The role of clay minerals as surfaces for differential adsorption of organic molecules has been included in many discussions of the origin of life since the idea was first suggested by Bernal (1). If this characteristic of clays could be confirmed, the limited number of functional molecules in

[1]present address: Mining College, Akita University
Akita 010, JAPAN

Y. Wolman (ed.), Origin of Life, 337–346.

biological systems could be explained by selective adsorption of certain monomers on clays during the course of chemical evolution. However, relatively little experimental data is available in support of such a hypothesis. The main basis for preferential adsorption appears to be the different acid-base properties of biochemical monomers. Non-α-amino acids, which have higher isoelectric points, are adsorbed to a greater extent, especially at low pH, than α-amino acids (2,3). Thymine and uracil and their corresponding nucleosides are not adsorbed on montmorillonite, while other nucleic acid bases and their nucleosides and nucleotides are adsorbed (4,5,6,7). Other biological, as well as nonbiological purines and pyrimidines, which are more basic, are more easily adsorbed by cationic exchange, especially at low pH.

It is attractive to consider the selective adsorption of enantiomers by clays as a possible explanation for the origin of optical activity in biological molecules, especially in regard to proteins. If clays preferentially adsorb the L enantiomers of amino acids, they might provide a mechanism for concentration of homogeneous optical isomers prior to polymerization into proteins. However, as optical isomers are physically and chemically identical, and as clays have no known asymmetric centers, there is no theoretical basis for asymmetric adsorption. To test the hypothesis of selective adsorption of enantiomers, we studied the adsorption of racemic amino acids on Na-montmorillonite at different hydrogen ion concentrations, analyzing the distribution of the enantiomers in the adsorbed as well as the nonadsorbed portions of the mixtures, thereby obtaining total quantities of each enantiomer in the experiments as a check for loss or degradation of material.

Preparation of clay. Na-montmorillonite of Wyoming Bentonite was obtained from the Source Clay Minerals Repository of the Clay Minerals Society at the University of Missouri-Columbia. The clay was suspended in deionized-distilled water and fractionated to less than 2 μm size particles. The clay was treated with 30% hydrogen peroxide and 0.2 N HCl to purge the clay of any organic or bacterial contaminants and then dialyzed. Adding sodium chloride and dialyzing again, Na-montmorillonite was obtained.

Adsorption Experiments. 200 mg of the prepared clay and 0.1 meq of the racemic amino acid was used in each experiment. This quantity corresponds to 50% CEC of the clay. Valine and norvaline were used in the same adsorption experiment in equimolar quantities, making the total quantity of amino acids equal to 100% CEC. (D,L) α-alanine was obtained from the Nutritional Biochemical Company. (D,L) α-aminobutyric acid, (D,L) valine, and (D,L) norvaline were from Schwartz-Mann.

After adjustment to pH 3, 7, or 10 with NaOH or HCl and an adsorption period with continuous shaking for 24 hours, the clay-amino acid mixtures were centrifuged. The supernatant contained the amino acids which were not adsorbed on the clay. The clay pellet, which contained the adsorbed amino acids, was washed with distilled water of the same pH as the experiment. After centrifugation, we obtained two fractions: weakly adsorbed amino acids in the supernatant, and strongly adsorbed amino acids, those resistant to washing and remaining on the clay. The latter were removed from the clay extraction with redistilled 1N HCl, then lycphilized to remove the HCl. The quantities of amino acids in each of the three fractions (non-adsorbed, weakly adsorbed, and strongly adsorbed) were analyzed by ion exchange chromatography on a Durrum 500 amino acid analyzer. In most cases a single extraction of the clay with 1N HCl was sufficient to completely remove the amino acids which were strongly adsorbed, but in a few cases, when large amounts of amino acids were strongly adsorbed at pH 3, a second acid extraction was performed and included in the strongly adsorbed fraction.

For enantiomer analysis, N-trifluoroacetyl-isopropyl esters of the amino acids in each fraction were prepared. The derivatized D and L isomers were then separated by gas chromatography using a 1/4" (O.D.) x 114" column packed with 10% N-lauroyl-L-valyl-t-butyl amide on Chromosorb W AW, 100/120 mesh. The samples were run at a helium flow of 16 to 20 ml/min, isothermally at 100° for valine and norvaline, at a 90° for α-alanine and α-aminobutyric acid. Peak areas were determined with an Infotronics electronic digital integrator, model CRS-204. In the few instances in which impurities interfered with accuracy of peak area calculations, peak heights were used to determine D:L ratios.

A first set of adsorption experiments was performed with great care to guard against contamination, although aseptic techniques were not used. Glassware was soaked in 6N HCl overnight and rinsed with deionized-distilled water just before use. Gloves were worn when handling mixtures and solutions. Only deionized distilled, water, collected as it condensed after distillation, was used for the clay suspension for all solutions.

In a second series, the adsorption experiments were repeated under sterile conditions. In these cases, all glassware and solutions were autoclaved, and the clay suspension was refluxed for 30 minutes. The clay-amino acid mixtures were prepared in a laminar flow hood, and general sterile technique was applied to the handling of the solutions thereafter.

Statistical calculations according to the following formula

(8) were carried out to determine if the mean D:L ratios of the standard and sample were equal:

$$X = \frac{\bar{x}_1 - \bar{x}_2}{\left[\sigma^2_{1/n} + \sigma^2_{2/m}\right]^{1/2}}$$

X is a statistic having a standardized normal distribution:

$\bar{x}_1$ = standard mean D:L ratio

$\bar{x}_2$ = sample mean D:L ratio

σ_1 = standard deviation of standard mean
σ_2 = standard deviation of sample mean

n and m are the number of GC analyses of standard and sample, respectively. If X is large, the probability that an X equal to or greater than that value will occur within the normal distribution is very small (< 5%); thus the hypothesis that $x_1 = x_2$ is rejected, and significant difference between x_1 and x_2 is established.

Amino acid analyses of all fractions of the adsorption experiments revealed that full recovery (within $\pm$ 5%) of the amino acids had been obtained. Thus, no error in estimating the quantities of amino acids adsorbed will occur, since the adsorbed fraction was measured directly, and we are confident that no loss of material occurred during the experiments.

To further check the validity of our results, we calculated quantities of D and L enantiomers of the amino acids in each fraction from amino acid quantities and the percentages of enantiomers obtained from GC analysis. Since we began with racemic mixtures, the total quantities of D and L enantiomers obtained by our analyses should be equal, within experimental error. If, as in some experiments, the quantities of D and L enantiomers did not balance within the calculated standard deviation, the data was rejected. All experimental data presented here has undergone this mass balance test.

The relative abundances of D and L enantiomers of the amino acids adsorbed and nonadsorbed are presented in Table 1 and Table 2. Adsorption of the strongly adsorbed fraction is attributed to cation exchange, and the weakly adsorbed fraction, to a weaker interaction, probably hydrogen bonding with the clay interlayer water molecules (3). The existence of a statistically significant difference between the standard and sample D:L ratios is denoted by an asterisk. In all the strongly and weakly adsorbed fractions showing a significant difference, the D:L ratios are less than one, i.e., there is a preference for the L enantiomer, although a very small one. The largest difference that we see, for α-aminobutyric acid at pH 10, corresponds to a 2% greater abundance of the L enantiomer than the D enantiomer. The smallest significant difference, for norvaline at pH 3, is equal to a 0.5% greater quantity of the L enantiomer in the strongly adsorbed fraction.

TABLE 1

Adsorption of (D,L) Amino Acids on Na-Montmorillonite

	α-Alanine				α-Aminobutyric Acid			
Sample	D:L	σ	n		D:L	σ	n	
Standard	1.028	0.029	10	(A)	1.010	0.024	5	(A)
	1.005	0.014	4	(B)				
pH 3								
Adsorbed S	1.022	0.009	5	(A)	1.005	0.002	3	(A)
W	1.018	0.011	5	(A)	0.985	0.012	4	(A)
Nonadsorbed	1.024	0.006	4	(A)	1.014	0.015	4	(A)
pH 7								
Adsorbed S	0.976*	0.028	5	(B)	0.996	0.005	4	(A)
W	1.013	0.028	4	(B)	0.993	0.032	3	(A)
Nonadsorbed	1.013	0.020	6	(B)	1.021	0.038	3	(A)
pH 10								
Adsorbed S	0.953*	0.012	4	(B)	0.925*	0.046	7	(A)
W					0.988*	0.004	3	(A)
Nonadsorbed	1.011	0.006	3	(B)	1.012	0.021	4	(A)

S - Strongly adsorbed; W - Weakly adsorbed
* - Statistically significant difference between standard and sample D:L Ratios
n - Number of GC analyses
(A) - Sterile conditions not used
(B) - Sterile conditions used

TABLE 2

Adsorption of (D,L) Amino Acids on Na-Montmorillonite

Sample	Valine D:L	Valine σ	Valine n		Norvaline D:L	Norvaline σ	Norvaline n	
Standard	1.008	0.022	10	(A)				
	1.011	0.014	10	(B)	1.006	0.031	10	(B)
pH 3								
Adsorbed S	0.977*	0.013	5	(A)	0.986*	0.020	4	(B)
W	0.973*	0.026	3	(A)	1.017	0.011	3	(B)
Nonadsorbed	1.020*	0.005	3	(A)	1.006	0.027	6	(B)
pH 7								
Adsorbed S	0.996	0.013	6	(A)	0.979*	0.016	5	(B)
W	1.031*	0.021	3	(A)	0.995	0.013	4	(B)
Nonadsorbed	1.005	0.011	3	(A)	0.997	0.023	5	(B)
pH 10								
Adsorbed S	1.015	0.021	4	(B)	0.986	0.035	4	(B)
W	1.008	0.021	3	(B)	1.011	0.025	3	(B)
Nonadsorbed	1.010	0.011	3	(B)	1.000	0.012	3	(B)

S - Strongly adsorbed; W - Weakly adsorbed
* - Statistically significant difference between standard and sample D:L Ratios
n - Number of GC analyses
(A) - Sterile conditions not used
(B) - Sterile conditions used

Although replicates of the experiments were performed, the reproducibility of the results has not been thoroughly tested because some of the data sets were rejected as invalid when total measured D and L isomer quantities did not balance. However, we give an example of two replicate experiments of the adsorption of (D,L) valine at pH 3. For comparison, the D:L ratios in experiments A and B are as follows: strongly adsorbed, 0.977 and 0.979, weakly adsorbed, 0.973 and 0.992, and nonadsorbed, 1.020 and 1.001, respectively. A comparison of the acutal quantities of D- and L-valine adsorbed and non-adsorbed is shown in Figure 1. The adsorbed fractions both contain slightly larger quantities of the L enantiomer; the actual quantities of the isomers in each fraction of the replicate experiments are very close. However, the fact that the D:L ratio for the nonadsorbed fraction is lower in

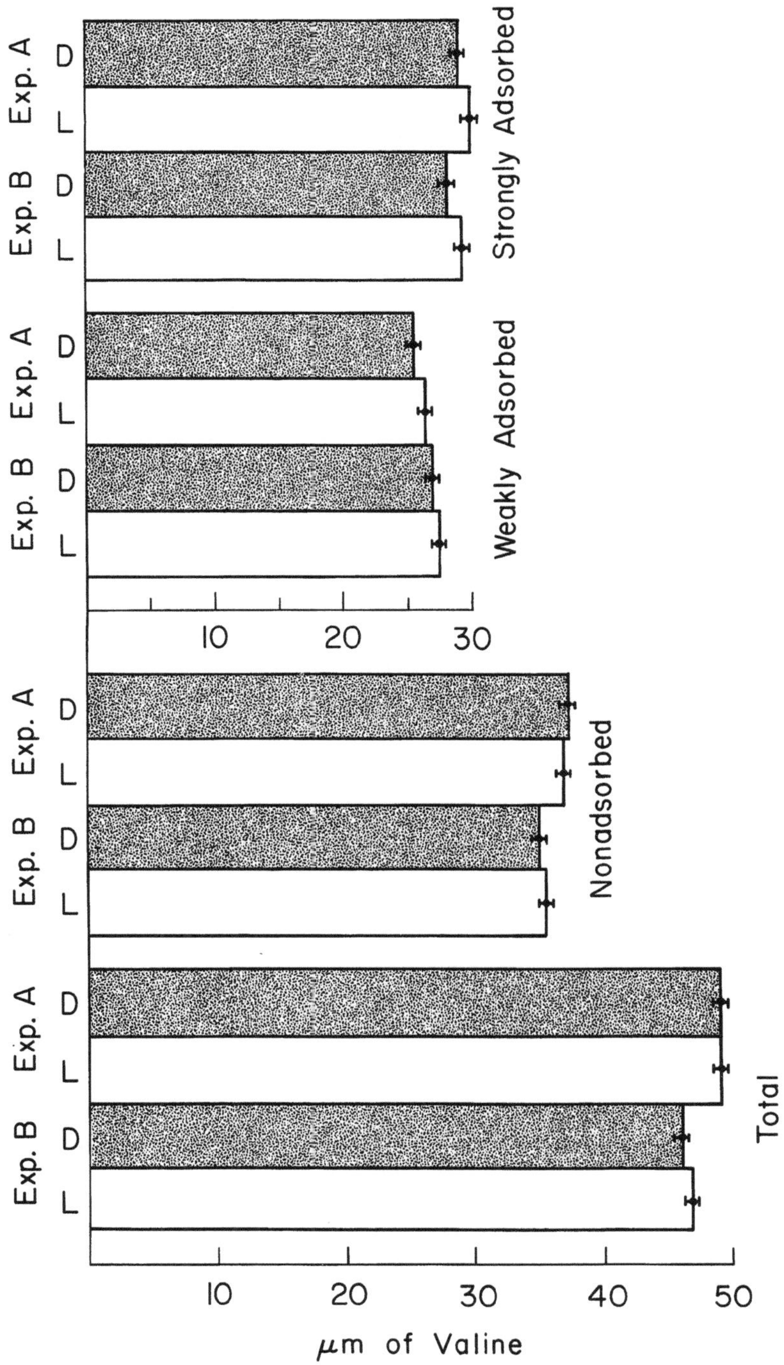

Figure 1. Adsorption of (D,L) Valine on Na-montmorillonite at pH 3, 50% CEC.

experiment B makes the total quantities of D and L enantiomers slightly unbalanced. In this experiment, the total difference between quantities of D and L enantiomers measured in micromoles is about twice the standard deviation, which is a measure of analytical variation.

This study of adsorption of amino acid isomers on Na-montmorillonite shows that there is certainly no large preferential adsorption of either enantiomer by the clay. Other researchers, using indirect methods, have concluded from their observations that a large selection of isomers by the clays occurs (9,10). Jackson measured differences in ultraviolet light adsorption of the supernatants of mixtures of clay with D- and L-phenylalanine (9). However, when the adsorbed fraction, which is often less than 10% of the total amino acids used, is estimated from the nonadsorbed fraction, the error is greatly magnified. Bondy and Harrington (10) studied the inhibition of adsorption of radioactive amino acid isomers by the addition of unlabeled isomers. Their findings left some unanswered questions. For example, the addition of L-leucine inhibited the adsorption of tritiated L-leucine while the addition of D-leucine did not. However, neither unlabeled isomer inhibited adsorption of the tritiated D-leucine. Direct measurement of the quantities of D and L enantiomers adsorbed on the clay and accounting for the total quantity of amino acids seems to be the only foolproof approach to studying preferential adsorption by clays. Our efforts to guard against material loss and degradation, and to successfully balance total quantities of amino acid isomers in the experiments seems to be an improvement on the methods used in earlier studies of this subject.

The differences in D and L enantiomers adsorbed on the clay corresponds to 0.5-2.0% greater quantities of L-amino acids than D-amino acids. Standard deviations of GC analyses generally range from 0.01 to 0.8%, in good agreement with the standard deviations of 0.2-0.4% obtained by Bonner _et al_ (11) in their quantitative evaluation of this method of enantiomer analysis. (The standard deviation for our strongly adsorbed fraction, α-ABA, at pH 10 was an exception at 1.5%). According to the studies of Bonner _et al_, in which known quantities of the enantiomers were used, the mean absolute error of this method is 0.43%. Thus, the differences in D and L isomers that we see in some of the adsorbed fractions are greater than both the standard deviation and the absolute error, and this is evident in the fact that a statistically significant difference between the mean D:L ratios of these samples and their standards has been established.

We plan to continue these studies using high performance

liquid chromatography employing a chiral eluant to separate enantiomers (12). The elimination of pre-column derivatization, and the highly sensitive fluorescence detection of post-column o-phthaldehyde derivatives should reduce analytical variability and either confirm or invalidate our conclusion that the slightly higher quantities of L-amino acids measured in the adsorbed fractions are larger than experimental and analytical variation. More replicate experiments are also planned to test the reproducibility of these results.

What is the basis for a small but significant preferential adsorption of L-amino acids on Na-montmorillonite? As stated by Wellner (13) montmorillonite has no known chirality. But no thorough studies have been done to determine if chiral sites do exist on clays in small quantities. Some type of asymmetry might result from the distribution and arrangement of octahedral vacancies in which Al^{3+} is replaced by Mg^{2+} in the clay structural layers. Thus, an occasional chirality might result from the arrangement and frequency of cation exchange sites on the two facing silica tetrahedral layers. Whether or not hydrogen ion concentration has an effect on the differential adsorption of amino acid enantiomers by the clay is unclear. There is no definite pattern of preferential adsorption occurring at one pH and not at others (Table 1 and 2). It is reasonable to assume that the pH would affect the dissociation of clay interlayer water molecules and the amount of clay layer expansion; of course, the ionic forms of the amino acids is determined by the solution pH.

Although this possible preferential adsorption of L-amino acids is small at 0.5-2%, if it were taking place during chemical evolution over millions of years, the cumulative effect could have been to change the distribution of amino acid enantiomers in the primordial seas so that L isomers were gradually concentrated on clays. More research with other clays, perhaps other cations, and other amino acids under different types of conditions is needed to finally determine the role of clays in the origin of optical activity in biological molecules.

ACKNOWLEDGEMENTS: This work was supported by NASA Grant NGR 21-00-317 and NSF Grant EAR 77-00013. The authors wish to thank J.J. Kemper, University of Maryland for editorial assistance.

REFERENCES

(1) Bernal, J.D. The Physical Basis of Life, London, Routledge and Kegan Paul (1951). pp. 34-35

(2) Sieskind, O., and Wey, R. Compt. Rend. 248:1652 (1959)

(3) Friebele, E., Shimoyoma, A., and Ponnamperuma, C. J. Mol. Evol. in press (1980)

(4) Shaw, J.G. Canad. J. Biochem. 43:829 (1965)

(5) Lailach, G.E., Thompson, T.D., and Brindley, G.W. Clays Clay Min. 16:285 (1968a)
(6) Lailach, G.E., Thompson, T.D., and Brindley, G. W. Clays Clay Min. 16:296 (1968b)
(7) Odom, D.G., Rao, M., Lawless, J.G. and Oro, J. J. Mol. Evol. 12:365 (1979)
(8) Cooper, B.E. Statistics for Experimentalists, Oxford, Permagon Press, Ltd. pp. 87-89 (1969)
(9) Jackson, T.A. Experientia 27:242 (1971)
(10) Bondy, S. and Harrington, M., Science 203:1243 (1979)
(11) Bonner, W.A., Van Dort, M.A., and Flores, J. Anal. Chem., 46:2104 (1974)
(12) Hare, P.E. and Gil-Av, E. Science 204:1226 (1979)
(13) Wellner, D. Science 206:484 (1979)

SPONTANEOUS AND INDUCED GENERATION OF OPTICAL ACTIVITY IN RACEMIC 4,4'-DIMETHYL-1,1'-BINAPHTHYL

Richard E. Pincock, May D. Lu, and Fu-Ning Fung

Department of Chemistry, University of British Columbia
Vancouver, Canada, V6T 1Y6

In contrast to the easy spontaneous generation of optically active samples of 1,1'-binaphthyl by crystallization from a melt, the 4,4'-dimethyl derivative of binaphthyl (DMB) does not undergo a similar transformation to a chiral form. However, the presence of chiral crystals of 4,4'-diaminobinaphthyl stereospecifically induces the resolution of DMB at temperatures where the optically active product is unstable. Although thermodynamic control leads to the stable (optically inactive) racemate form of DMB, kinetic control of nucleation and crystal growth gives rise to eutectic crystals and to optically active samples of DMB.

Steric hindrance prevents a stable coplanar arrangement of the two naphthalene units in 1,1'-binaphthyl (1, X=H) and this compound therefore possesses axial chirality in its preferred molecular conformation. In solution or in a melted state the hindrance is small enough to allow the resulting two enantiomers to interconvert by rotation around the central single bond (*e.g.*, $t_{1/2}$ 15 min at 50°, *ca.* 0.5 sec at 150°) (1). In a fluid phase racemic binaphthyl is therefore produced as the thermodynamically controlled product.

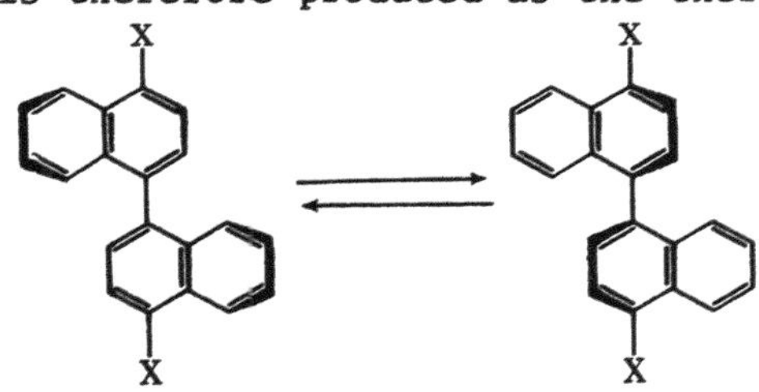

1 X=H ; 2 $X=CH_3$; 3 $X=NH_2$

Y. Wolman (ed.), Origin of Life, 347–353.

However, when a melt of pure binaphthyl (mp 145°) is held in a supercooled state, nucleation and growth of crystalline binaphthyl preferentially gives rise to thermodynamically stable samples which are optically active (2). This provides an apparently unique example of proven spontaneous generation of optical activity, i.e., no external influence of any type need be present in order to produce macroscopic optically active samples (3,4).

Small amounts, even traces, of certain materials eliminate the 50-50 spontaneous distribution of + and - samples; a strong bias towards production of one of the two enantiomers of binaphthyl can be introduced by the presence of a variety of chiral compounds. For example, 2.5% D- or L-mandelic acid prompts the production of an excess of (+) and (-) binaphthyl, respectively, in about 80% of samples; and 5% of these enantiomeric agents virtually assures 100% of (+) and (-) rotations, respectively, in solidified binaphthyl samples (5).

The above mentioned characteristics of spontaneous and induced generation of optically active 1,1'-binaphthyl suggested that various derivatives of binaphthyl would present interesting parallel (or contrasting) results. The 4,4'-dimethyl derivative possesses a minimum of molecular difference from the unsubstituted parent compound (i.e., the same steric arrangements around the interannular bond and little polar effect of the substitutents). An additional factor in its choice for study was that samples of this compound had never been obtained in a resolved form. It therefore offered a test of the general possibility of utilizing spontaneous and induced methods for resolution of configurationally mobile R$\rightleftharpoons$S systems.

EXPERIMENTAL. Preparation of 2, X=CH_3. Racemic 4,4'-dimethyl-1, 1'-binaphthyl (i.e. DMB) was obtained by $CuCl_2$ coupling of the Grignard reagent from 4-bromo-1-methylnaphthalene prepared by the method of Fieser and Seligman (6). In a typical preparation 20g of the bromide, 3.7g of magnesium and 100ml of dry tetrahydrofuran was refluxed overnight. Anhydrous $CuCl_2$, 20g, was added during cooling of the reaction mixture in an ice bath. The reaction was refluxed 3 hrs and worked up by addition of dilute HCl and by extraction into benzene to yield 13.7g of crude solid. Chromatography on alumina with elution by petroleum ether produced colorless DMB, mp 137-142°, 4.9g (37%). Several recrystallizations from acetone gave crystals with mp 142-142.5° (reported mp 146-146.5°) (7). Anal. Calc. for $C_{22}H_{18}$: C, 93.58%; H, 6.42%. Found C, 93.64%; H, 6.58%. This preparation, even with modifications of crystallization solvent, rate of cooling or attempted initiation of crystallization by eutectic seeds (see below), always gave the racemate form of DMB. This form is readily characterized by the I.R. spectrum of the solid suspended in nujol. The racemate has a strong absorption split into peaks at 750 and 758 cm^{-1}, which in the eutectic form is merged into one peak. Another prominent difference is a strong absorption at 826 cm^{-1} for the racemate and 836 cm^{-1} for the eutectic form (with a weak shoulder at 826 cm^{-1}).

Resolution Experiments. Numerous attempts to generate optically active samples of DMB by solidification of a melt (techniques similar to those successful in resolution of binaphthyl) (3,4) gave only optically inactive samples (by polarimetry) and no noticable eutectic crystalline form (by I.R. spectroscopy). Temperatures of 90-130° and nucleation by selective cooling, vibration, or sonic treatment gave only the racemate form of DMB. Addition of *ca.* 5% of either enantiomer of mandelic acid, camphor, alanine, proline, or ammonium α-bromo-camphor-π-sulfonate to DMB and solidification of the supercooled melt gave rotations insignificantly different from zero in all individual samples.

Naphthidine as a Resolving Agent. Naphthidine, **3** X=NH_2, in its optically active eutectic form was obtained by the classical resolution (8) or by the spontaneous resolution method (9). A single crystal of eutectic naphthidine, which could be chosen either from resolved naphthidine or from a sample which was overall racemic (9), was lightly crushed with a spatula and weighed into a set of 1 ml ampules. To these were added weighed amounts of racemate DMB and the ampules sealed in air. The samples were held in an oil bath at 150° for *ca.* 4 min to melt all the DMB while the naphthidine (mp 204°) remained as a solid. (Prolonged heating resulted in complete dissolution of the naphthidine and, finally, only racemate DMB). The samples were transfered to a bath at 98 to 113° and held undisturbed; shaking or sometimes even light contact with the vials gave racemate product. The nucleation and crystallization were slow; samples were held undisturbed normally for 15 hrs and as long as 48 hrs. The samples were then completely dissolved in acetone for polarimetric analysis (2,9). Consistent results, with (+)-naphthidine giving (+)-DMB and (-)-naphtidine giving (-) DMB in excess, were easily obtained (see below for typical rotation results). Optical induction was obtained with as little as 0.1% naphthidine, as much as 10% was also used, but optimum results seemed to be obtained with 1% naphthidine in 50mg size samples. Crystals of eutectic naphthidine produced by thermal treatment at 197-203° (9) seemed to give the highest rotations, possibly because they contain little or no racemate seeds.

Kinetics of Racemization of DMB. Larger amounts of resolved DMB were obtained by combining naphthidine induced optically active samples having the same sign of rotation, dissolving the mixture in acetone, and passing the solution rapidly through a short column of alumina (to remove naphthidine) with elution by petroleum ether. Evaporation of solvent gave colorless, crystalline DMB; 248 mg $[\alpha]_D$ + 99°, 508mg $[\alpha]_D$ - 110°, and 965mg $[\alpha]_D$ + 77° for three different preparations used in kinetic runs reported below. Kinetic studies of racemization in polycrystalline samples (see Figure I) were carried out as previously described (2,9).

Several kinetic comparisons of the rate of thermal racemization of crystals obtained directly from the naphthidine induced resolution were carried out. Several of the largest, well formed crystals of DMB formed in the induced resolution at 103° were hand picked after simply breaking a vial with a hammer. Two single crystals weighing 2.28mg and 2.15mg gave specific rotations of +114° and +104°, respectively. Similar crystals, 2.67, 2.84 and 2.61 mg, individually sealed in 1 ml vials and heated at 108°, the contents dissolved in acetone, gave $[\alpha]_D$ + 67 after 24 hrs, + 63 after 48 hrs, and + 34 after 72 hrs, respectively. Assuming the initial crystal had $[\alpha]_D$ of +110°, these rotations correspond to half-lives of 33 hrs, 60 hrs, and 42 hrs, respectively. For comparison, the kinetics of racemization of similar, but polycrystalline samples was made. About 200mg of combined selected single crystals was lightly crushed with a spatula. Division of this into *ca.* 15mg samples sealed in 1 ml vials ($[\alpha]_D$ initial was + 104°) and heating at 103° gave a somewhat scattered but roughly consistent set of decreasing rotations with a half-life of 4 hrs.

RESULTS AND DISCUSSION. A great many attempts to crystallize DMB from a melt into optically active samples were completely unsuccessful. Whereas with similar procedures binaphthyl readily and spontaneously generates optically active samples and rarely gives zero rotations (3), its dimethyl derivative refused to resolve into a different crystal form and always gave rotations of zero. The form returned is a racemate, *i.e.*, composed of alternating layers of R and S molecules (10) and no signs of a possible eutectic form, with separate crystals containing R (or S) molecules, was evident in any crystallization of DMB.

Similarly, and again in contrast with induced resolutions of binaphthyl (5), the crystallization of supercooled DMB in the presence of a variety of chiral materials failed, with one exception given below, to generate optically active samples. The existence of a relatively stable eutectic form, which is one of the requirements for an easy spontaneous or induced resolution, is possible with binaphthyl but apparently not with its dimethyl derivative. The contrast of results with these two compounds, with only a minor molecular structural difference between them, points out the highly specific and critical nature of the solid state. The ability to preferentially grow chiral crystals (*i.e.*, with binaphthyl) or grow racemate crystals (*i.e.*, with DMB) is greatly affected by the slight structural change between these compounds, which from the viewpoint of solution reactions, would be minimal. Relative crystal energies are, as yet, unpredictable and production of desired crystal forms is essentially impossible to control.

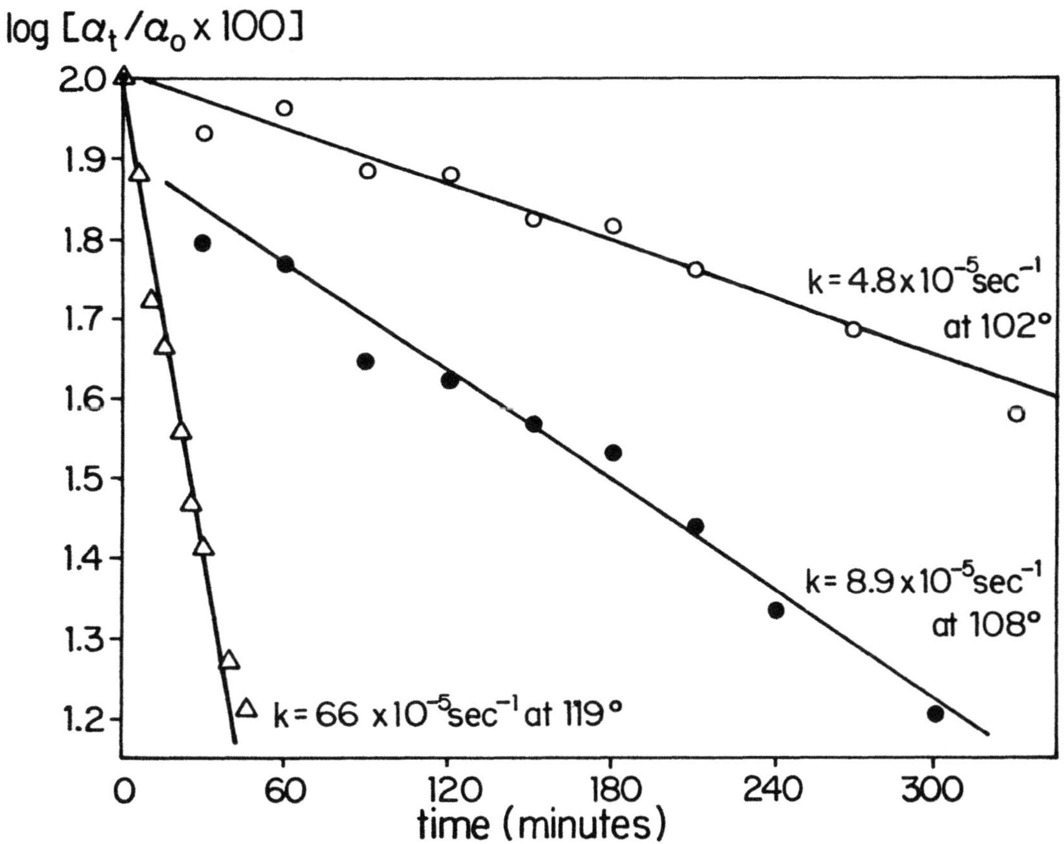

FIGURE I. First order kinetic plots for loss of optical rotation in polycrystalline samples of 4,4'-dimethyl-1,1'-binaphthyl.

One chiral agent was, however, successful in controlling the nucleation or growth of crystalline DMB to give a new crystalline and chiral form of this compound. Induction of chirality was also easily achieved and an excess of either chosen enantiomer readily produced. This agent is the eutectic form (i.e. a chiral form) of naphthidine, 3 X = NH_2, mp 204°. For example, a single crystal of the (+) form of naphthidine was powdered and distributed into several samples of DMB (ca. 0.3mg in 10mg). The samples were sealed in vials, heated to 150° to melt the DMB, then held unmolested at 98-113° while crystallization proceeded over 8 hrs. The samples were then dissolved in acetone and gave absolute rotations of +30, +27, +82, +100, +71, +99, +28, +67, +13, +69, +59, +25, +35, +25, +99, and +100°. Similar samples treated with a powdered (-)-naphthidine seed crystal gave $[\alpha]_D$ -5.5, -46, -75, -89, -19, -84, -42, -34, and -94°. (Control samples, i.e., without naphthidine, or with naphthidine completely dissolved in the melt of DMB, gave rotations of zero). Such results definately prove a high degree of stereospecific control by this added chiral agent in its solid form. Crystals of the eutectic form of DMB could be seen to grow around the naphthidine seeds; this form of DMB, obtainable only by this induced crystalization, was characterized by an I.R. spectrum different from that of racemate DMB.

The induced resolution gave high specific rotations at temperatures up to about 113°, but at 130° always gave the optically inactive racemate form of DMB. Also the optically active form of DMB gave a melting point identical to that of the racemate form, and indeed, was rapidly transformed to the racemate when heated above 125°. A kinetic study of the racemization was therefore carried out with polycrystalline samples of resolved DMB (purified by removal of naphthidine). Some of the results are shown in Figure I where loss of rotation in a set of individual samples from three different polycrystalline preparations of resolved DMB are given in first order kinetic plots. The rates of racemization varied with each batch and depended, for example, on the state of subdivision of the samples (e.g., $t_{1/2}$ changing from 150 min to 96 min at 108° when a sample of DMB was ground more finely). However, from the figure it is seen that there is at least an orderly loss of optical activity in samples having the same preparative background and that the rate increases greatly with temperature. Again, in contrast with results with binaphthyl where resolved samples are indefinately stable right up to melting at 159° (2), the optically active form of DMB is kinetically very unstable with respect to its racemate even some 25° below its melting point.

These kinetic studies also further point up the curious ability of naphthidine to induce the optical resolution of DMB under thermodynamically unfavorable conditions. The racemization of solid, polycrystalline DMB occurs at temperatures as low as 102° ($t_{1/2}$ ca. 4 hrs) and the racemate product is therefore the thermodynamically favored state. Nevertheless the induced production of optically active samples of DMB can be carried out at the same or even at higher temperatures (up to 113°) over long times (12 hrs or more). The kinetic control of the growth of eutectic form DMB produces well formed crystals that withstand conversion to the thermodynamically stable racemate form. As an illustration of this effect three well formed single crystals of eutectic DMB were heated at 108°. The half-lives for racemization were 33, 42 and 60 hrs. Similar crystals, when ground to a polycrystalline state gave kinetic samples with a half-life of 4 hrs at 108°.

4,4'-Dimethylbinaphthyl, relative to binaphthyl, thus has a greatly decreased ability to produce optically active samples. The major reason for this is a reversal in relative thermodynamic stability of polymorphic forms (optically active eutectic and racemic racemate forms) due to the presence of the methyl groups. However, a kinetic preference towards growth of optically active DMB can be created by introduction of solid naphthidine and this may be used to produce either enantiomer of the optically active, but thermodynamically unstable, form of dimethylbinaphthyl.

ACKNOWLEDGEMENT. We thank Mr. John Sanders for some initial experiments and The Natural Sciences and Engineering Research Council for support of this research.

REFERENCES

(1) Cooke, A.S. and Harris, M.M., J. Chem. Soc., 2365 (1963).
(2) Wilson, K.R. and Pincock, R.E., J. Am. Chem. Soc., 97, 1474 (1975).
(3) Pincock, R.E., Perkins, R.R., Ma, A.S. and Wilson, K.R. Science 174, 1018 (1971).
(4) Pincock, R.E. and Wilson, K.R., J. Chem. Educ. 50, 455 (1973).
(5) Pincock, R.E., Bradshaw, R.P. and Perkins, R.R. J. Mol. Evol. 4, 67 (1974).
(6) Fieser, L.F. and Seligman, A.M., J. Am. Chem. Soc., 61, 136 (1939).
(7) Clar, E., Sanigök, Ü. and Zander, M., Tetrahedron 24, 2817 (1968).
(8) Theilacker, W. and Hopp, R., Chem. Ber., 92, 2293 (1959).
(9) Lu, M.D. and Pincock, R.E., J. Org. Chem. 43, 601 (1978).
(10) Trotter, J. and Pauptit, R., unpublished results of an X-ray crystallographic study.

AMPLIFICATION OF OPTICAL ACTIVITY BY CRYSTALLIZATION IN THE PRESENCE OF TAILOR-MADE ADDITIVES. THE "INVERSION RULE"

L. Addadi, J. van Mil, E. Gati and M. Lahav

Department of Structural Chemistry, The Weizmann Institute of Science, Rehovot, Israel.

Experiments are described in which enantiomerically pure dimers, trimers and oligomers, generated from non chiral monomers by crystallization and topochemical reaction in suitably designed chiral crystals, induce preferential crystallization of the parent monomer in the enantiomorphous chiral phase of opposite absolute configuration. (We coin the term "inversion rule" for this effect). A general mechanism of amplification of optical activity by crystallization is proposed on the basis of these results, involving selective adsorption of resolved impurities on the surface of chiral crystals of similar stereochemistry, resulting in a decrease of the growth rate of the affected enantiomer and consequent preferential crystallization of the antipode. The implications of this mechanism to the generation and amplification of chirality in a closedsystem are discussed.

When considering a process in which an initially symmetrical, closed system is transformed spontaneously into a chiral one, two distinct stages must be accounted for. A generation step which introduces asymmetry via statistical fluctuations from the symmetrical state, followed by an efficient feedback mechanism through which the first asymmetric core is preserved and propagates asymmetry to the whole system. One possible model for such a process of generation and enhancement of chirality stems from the intrinsic property of certain non chiral molecules (A), to crystallize spontaneously in chiral crystals ($\{Ar\}_d$ or $\{As\}_\ell$ of scheme 1). Coupling of this process with a topochemical reaction which transforms the chirality of the crystal, in a suitable homochiral phase, into stable configu-

Y. Wolman (ed.), Origin of Life, 355–364.

rational chirality of the products (Pr or Ps) may lead, in a single experiment, to the generation of net molecular optical activity.[1-4]

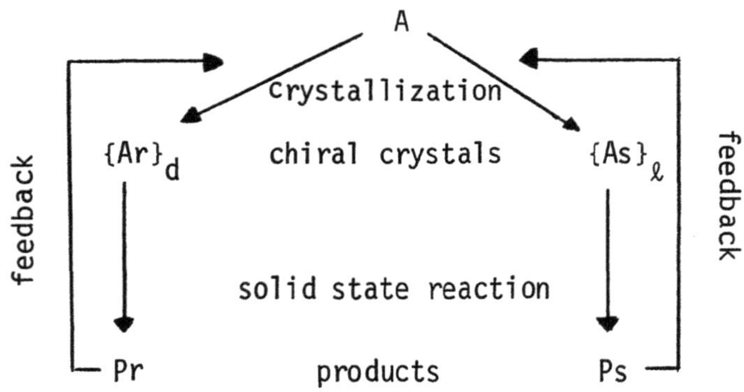

Scheme 1

However, if a large number of experiments is performed, inevitably a statistical distribution of induced chiralities of both signs will be obtained. In order to overcome this drawback, Green and Heller envisaged a cyclic process, through which the same products Pr or Ps influence a fresh crystallization of the starting material, driving it preferentially towards one of the two enantiomorphous phases (Scheme 1). The idea has been tested in the system composed of 4,4'-dimethyl chalcone and the chiral dibromide (see Table II) obtained by action of bromine vapor on chiral crystals of this chalcone.[1,5] Surprisingly, the product obtained from a "d" crystal of chalcone caused systematically preferential crystallization of the reagent in an "ℓ" crystal. The authors suggested however that "this situation occurs by chance" since it was a priori equally possible that the opposite would happen.

In this last case, as well as in previous examples of "absolute asymmetric synthesis" by the outlined route, and in some other attempts aimed at achieving induced asymmetric crystallization (see e.g. Pincock's experiments on binaphthyl[6] and Harada's on amino acid-copper complexes[7]), the approach has been empirical: systems have been found accidentally or chosen on the basis of preexistent knowledge of crystal structures, and this limited both the magnitude of the effect and the understanding of the mechanisms involved.

The aim of the work described here was to engineer from the beginning a suitable system, which undergoes the whole cyclic process of Scheme 1, with an enantiomeric yield as close as possible to quantitative.

The system must meet a number of requirements. First, there is needed an achiral substrate crystallizing in a chiral crystal structure. Further, the solid-state reaction must be entirely controlled by the crystal lattice, yielding chiral products of one single handedness. Finally, rigid products are needed, with a well defined stereochemical relationship to the parent phase, so that they will interact selectively with one of the two enantiomorphous crystals.

The various steps leading to the selection of a suitable substrate and to successful execution of the generation step have been extensively described elsewhere[8-10] and will be only briefly recalled. We shall rather concentrate here on the amplification process.

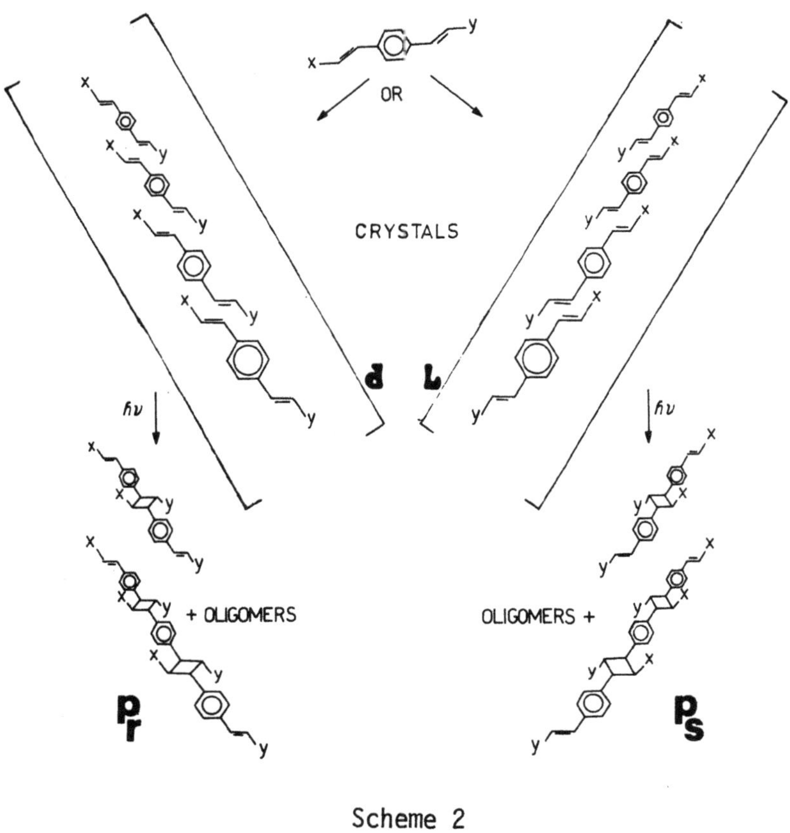

Scheme 2

Screening of the known topochemical reactions led us to the choice of $(2\pi + 2\pi)$ photocycloaddition as the best suited to the stated requirements; this kind of photochemical coupling reaction between substituted arylethylenes has been thoroughly investigated both in dimerization[11] and polymerization[12] and is known to be strictly dependent on the packing arrangement of the molecule in its crystal. We thus selected to study an asymmetrical photopolymerization of monomers containing two non equivalent double bonds, A. Following a systematic analysis of the possible crystalline motifs leading to asymmetric polymerization[13] we concentrated our efforts on the attainement of molecules crystallizing in the chiral arrangement shown in Scheme 2.

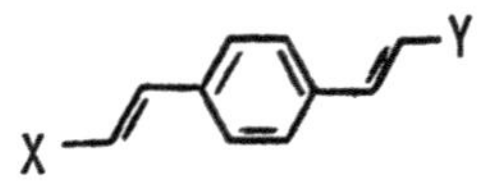

A

From this motif, under conditions of full topochemical control, only one enantiomeric chain can be produced in homochiral crystals.

The rigid products will furthermore resemble closely the crystalline matrix in which they were generated, in the sense that one dimer molecule can take up in the crystal, with only a slight deformation, the place of two monomer molecules; similarly the trimer will replace three monomers and the polymer n monomers. On the other hand, there will be no such clear relationship between the same products and the crystal of opposite chirality.

In order to build up the appropriate molecules, we applied then some empirical rules of "crystal engineering", coupled with symmetry considerations.

We got thus finally to seven systems in the family of compounds I (1-7 of Table I) which were found to pack in chiral structures characterized by the motif of scheme 2 and to give upon irradiation dimers, trimers, and oligomers of the correct stereochemistry (Table 1).

Not all of these systems are suited to the purpose of "absolute asymmetric synthesis" due to the difficulty arising from the metastability of their chiral phases(1, 4, 5). However, at least three of them 3, 6 and 7, have all the required characteristics. Due to the better quality of the crystals, we concentrated our efforts on the preparation of single crystals or homochiral phases from monomer 3 either by slow crystallization from the melt starting from a single nucleus, or from MeOH solution, exploiting autocatalytic seeding effects. High enantiomeric yields of dimerization, trimerization, and oligomerization were consistently obtained from such crystals, and were quantitative for the best samples.[4,14]

Table I

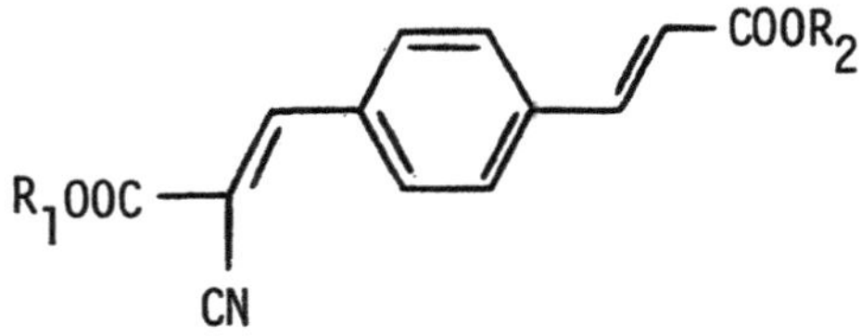

	1	2	3	4
R_1	Ethyl	n-Propyl	Methyl	Ethyl
R_2	(R,S)<u>sec</u>-butyl	(R,S)<u>sec</u>-butyl	3-Pentyl	3-Pentyl
	5	6	7	
R_1	Ethyl	Ethyl	n-Propyl	
R_2	ter. Butyl	1:1 Isopropyl: 3-Pentyl	3-Pentyl	

The generation step was thus successfully terminated with the achievement of an asymmetric polymerization with ~ 100% enantiomeric yield, starting from an achiral monomer and in the absence of any external chirality-inducing element.

Following the direction outlined in Scheme 1, we planned in the next step to test the induction power of the chiral products with respect to the asymmetric crystallization of the parent monomers (feedback step). Systems 1, 2, 3, 4, 6 and 7 were used as substrates; as inducing agents (additives) we used optically pure dimers, trimers and oligomers of 1, 3 or 4. The substrates were crystallized either from the melt or from solution in the presence of variable amounts of additive (1% - 15%), irradiated, and the enantiomeric purity of the products was used as a measure of the asymmetric induction during crystallization.[15]

As a typical example, we report in Fig. 1 some results obtained from monomer 3 in the presence of dimer of 1. All the other monomers 1, 2, 4, 6, 7 mentioned above behave in the same way. On the other hand, crystallization of many other systems

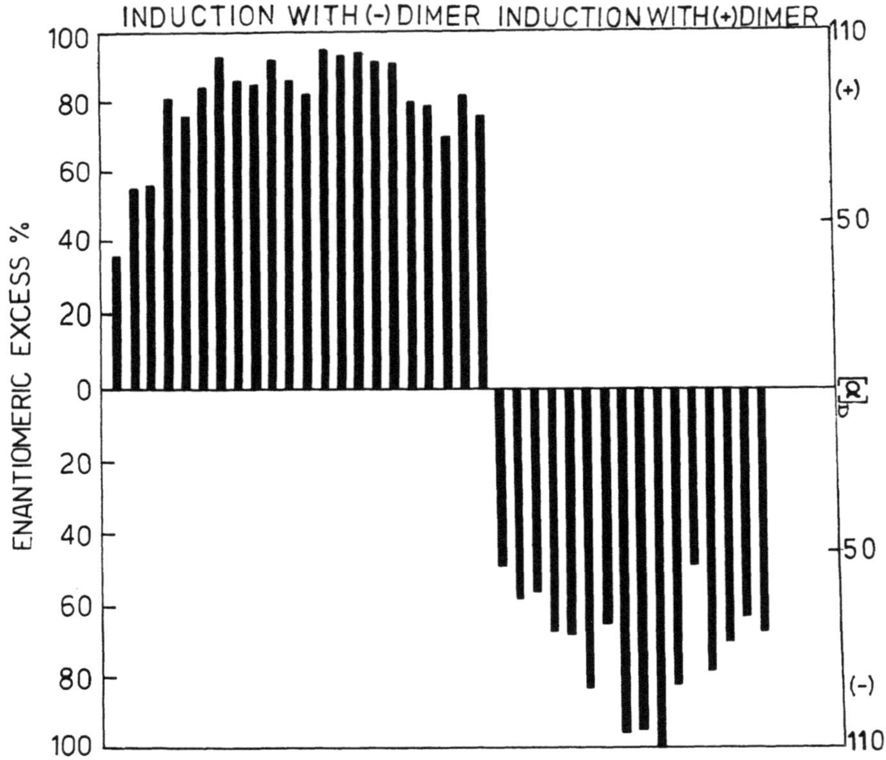

Fig. 1: Some results of induced asymmetric crystallization on monomer 3 by dimer of 1.

of the same family, which had different crystal structures, was not affected by the chiral additive. This confirmed the importance of a stereochemical relationship between the inducing molecule and the crystalline substrate.

This series of experiments unequivocally indicates that in all the affected monomers, independent of the experimental conditions and of the inducing agent used (1, 3 or 4) additive Pr favors the formation of Ps. (Fig. 1). (The inversion rule). This means that either it exercises an inhibitory action on the crystallization of $\{Ar\}_d$ or an accelerating one on the crystallization of $\{As\}_\ell$. Other parameters, like rate of crystallization, medium and temperature, influence only the absolute magnitude of the induction. Only small amounts of additive are required to optimize the effect, since approximately the same results were obtained with 1, 3, 8 and 15% of additive. This conclusion is supported by the fact that when more than 3% dimer is used in a crystallization from the melt, some of this additive is segregated in the form of amorphous oil, and therefore cannot play a part in the induction.

Some additional experiments were performed: resolved monomer R(-) **1** which by virtue of the chiral handle is forced to crystallize in the homochiral phase $\{Ar\}_d$[8], was crystallized from solution in the presence of small amounts of racemic dimer (Pr+Ps) (see Scheme 2).
The crystals were isolated and the additive cocrystallized with the monomer (∼ 1%) was separated, and was found to be 70% optically pure, with Pr always present in excess. The results demonstrate therefore that, as expected, Pr interacts preferentially with the phase $\{Ar\}_d$ having the same absolute configuration.

These results can be explained in the terms of a phenomenon well known in crystal growth, whereby selective adsorption of tiny amounts of impurities at the growing sites of a specific face of a crystal, decreases dramatically the relative rate of growth of this face with respect to the rates of growth of other faces of the same crystal. This results in a change in habit of the crystal, as well as in a decrease of the overall growth rate.[16,17]

The same principle can rationalize our results, if we take into account the preferential interaction of the dimer, trimer and oligomer with the enantiomorphous phase of the same absolute configuration; thus Pr adsorbed at the surface of crystals of $\{Ar\}_d$ prevents their growth more effectively than that of $\{As\}_\ell$.
Once this hypothesis is accepted, a quite general mechanism of kinetic resolution can be derived from it (Scheme 3), which can be applied to any racemic system crystallizing in the form of a conglomerate. The systems described here and by Green and Heller thus represent particular examples of this kind of resolution coupled with total asymmetric transformation.

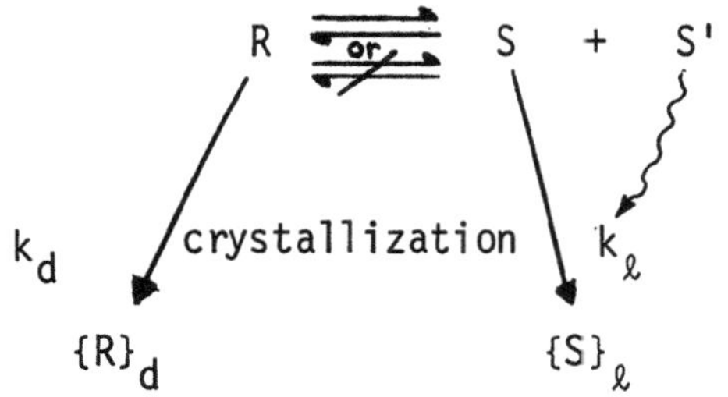

S'=impurity stereochemically similar to S

in absence of S' $k_d = k_\ell$

in presence of S' $k_d \gg k_\ell$

Scheme 3

TABLE 2
Examples taken from the literature of racemic mixtures resolved in the presence of impurities, fitting the"inversion rule".

Racemic mixture	Chiral additive	Preferentially cryst.enamtiomer	Ref.
D,L GLU $NH_3^+CH-COO^-$ $(CH_2)_2$ COOH	L ASP $NH_3^+CH-COO^-$ CH_2 COOH	D GLU	18
D,L GLU	L GLU Methyl-ester	D GLU	19
D,L $Cu(ASP)_2$	L GLU	D $Cu(ASP)_2$	21
D,L $Cu(ASP)_2$	L ALA $NH_3^+CH-COO^-$ CH_3	D $Cu(ASP)_2$	21
D,L $NaNH_4$tartrate $Na^+NH_4^+$ COO^- (CHOH) COO^-	D(+) malic acid $HOOC-CH(OH)-CH_2COOH$	D(-)$NaNH_4$tartrate	20
(±) narwedine 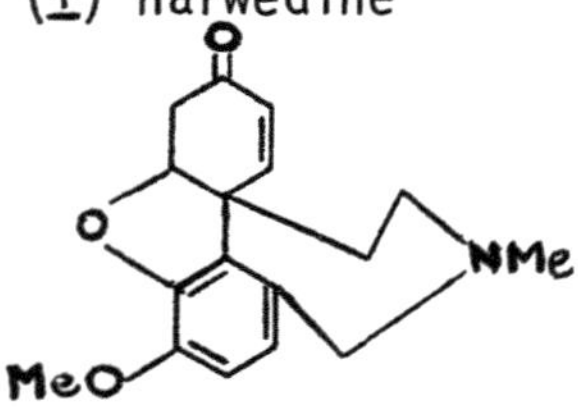 	(-)galamthamine 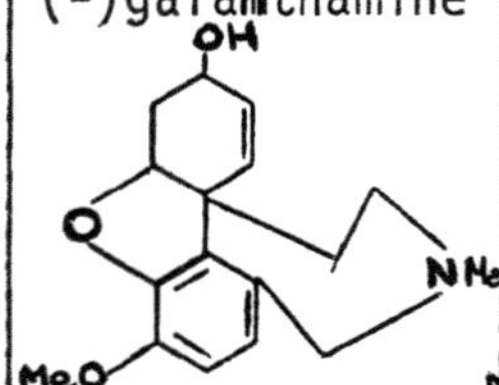 	(+) narwedine 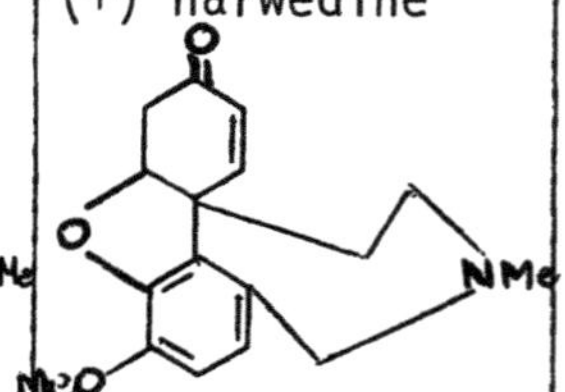 	22
p-p'dimethyl chalcone 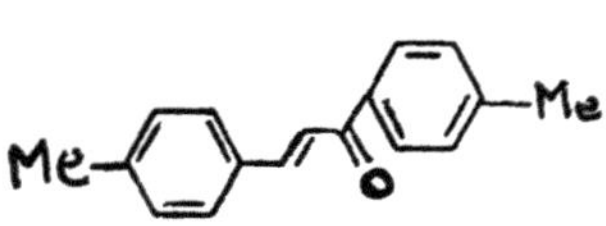 	dibromo p-p'diMe chalcone from"d" crystals 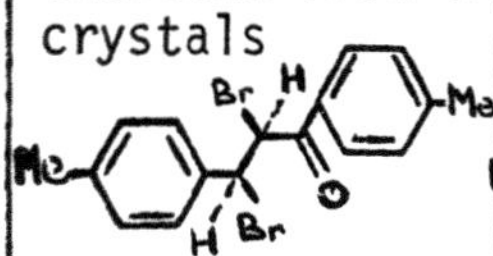 	p-p'diMechalcone "l"crystals 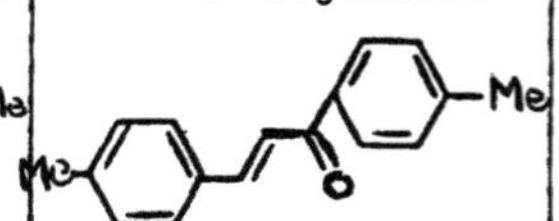 	

Following this idea, we searched the literature for kinetic resolutions of racemic conglomerates performed in the presence of chiral impurities, which can fit the "inversion rule" established here: a number of such examples were found, scattered in the literature of the last century. Thus, D-Glutamic Acid (GLU), was crystallized preferentially from a solution containing the racemic mixture and L-Aspartic Acid (ASP)[18] or L-γ-Methyl Glutamate,[19] $NaNH_4$-Tartrate was resolved in the presence of Malic Acid or Asparagine;[20] D-Cu$(ASP)_2$ separated from its racemic solution with L-GLU or L-Alanine,[7,21] and the alkaloid narwedine was resolved in the presence of traces of its resolved precursor, galanthamine[22] (Table II). Explanations, when given, for these results varied very much from case to case, including chiral solvent effects, ligand exchange interactions, adsorption and chiral seeding. In all these examples however, the relationship between the stereochemistry of the substrate and that of the chiral impurity is straightforward (Table 2). In some of them, as well as in the systems described in this paper and other new systems investigated for this specific purpose, we were able to demonstrate unequivocally, through coupled crystallization and dissolution experiments, preferential adsorption measurements and morphological studies, that indeed the mechanism we have proposed is responsible for resolution. These results will be the subject of separate communications.[23,24]

Going back to our initial aim of chiral amplification, we see clearly now that the sign of the obtained asymmetric induction is not a matter of chance, but the result of a well defined mechanism. Any amplification of this kind will always follow an alternating pattern. However, since less than 1% product is necessary to induce asymmetry with $\sim$ 100% enantiomeric excess, a repeated sequence of crystallization and reaction following an initial successfulstep, will lead all the same, in a closed system, to practically pure chiral products.

Nevertheless, we feel that, in the light of the above mechanism, it may still prove possible, by further planning, to design a system in which a direct amplification takes place. Studies in this direction are in progress.

ACKNOWLEDGEMENTS: We thank the Israel Commission for Basic Research and The Petroleum Research Fund, administered by the the American Chemical Society for support. We also thank our colleagues Profs. M.D.Cohen and L.Leiserowitz for most useful discussions.

REFERENCES

1. Penzien K. and Schmidt G.M.J., Angew.Chem.Int.Ed.8,608(1969).
2. Elgavi A., Green B.S. and Schmidt G.M.J., J.Am.Chem.Soc. 95, 2058 (1973).
3. Green B.S. and Lahav M., J.Mol.Evol.6,99 (1975).
4. Addadi L.and Lahav M., Stud. Phys. Theor. Chem. 1979, 7 (origins Opt. Act. Nat.) 179.
5. Green B.S. and Heller L., Science 185, 527 (1974).
6. Pincock R.E., Perkins R.R., Ma A.S. and Wilson K.R., Science 174, 1018 (1971).
7. Harada K., Naturwis. 57, 114 (1970).
8. Addadi L.and Lahav M., J.Am.Chem.Soc. 100, 2838 (1978).
9. Addadi L. and Lahav M., J.Am.Chem.Soc.101, 2152 (1979).
10. Addadi L. and Lahav M., Pure Appl. Chem. 51, 1269 (1979).
11. Schmidt G.M.J., Pure Appl.Chem. 27, 647 (1971).
12. Nakanishi H., Hasegawa M. and Sasada Y., J.Polym.Sci. Polym.Phys. Ed. 15, 173 (1977).
13. Green B.S., Lahav M. and Schmidt G.M.J., Liq.Cryst.Mol.Cryst. 29, 187 (1975)
14. Addadi L. and Lahav M., in preparation.
15. van Mil J., Gati E., Addadi L. and Lahav M., submitted for publication.
16. Cabrera N. and Vermilyea D.A. in "Growth and Perfection of Crystals", Eds. Doremus, R.H., Roberts R.W. and Turnbull. D. John Wiley & Sons Inc. London, 1958, p.393; Frank F.C., ibid, 411; Sears G.W. ibid, 441; Dunning W.J. and Albon N. ibid, 446; Dugua J. and Simon B; J. Cryst Growth 44, 265 (1978), Dugua J. and Simon B, ibid, 44,280 (1978).
17. Miles F.D., Proc. Roy. Soc. London A., 132, 266 (1931).
18. Purvis J.L., U.S. Pat. 2,790,001 (1957).
19. Fike H.L., U.S. Pat. 2,937,200 (1960).
20. Ostromisslensky I., Chem. Ber. 41, 3035 (1908)
21. Harada K. and Tso W., Bull.Chem.Soc.Japan. 45, 2859 (1972).
22. Barton D.H.R. and Kirby G.W., J.Chem.Soc. 806, (1962).
23. Addadi L., van Mil J. and Lahav M., in preparation.
24. Addadi L., Gati E. and Lahav M., in preparation.

GENERATION AND AMPLIFICATION OF OPTICALLY ACTIVE COMPOUNDS VIA INCLUSION IN CHIRAL TRI-O-THYMOTIDE CLATHRATES

R. Arad-Yellin[a], B. S. Green[a,b] and M. Knossow[c].

[a]Department of Structural Chemistry, Weizmann Institute of Science, Rehovot, Israel; [b]Israel Institute for Biological Research, Ness-Ziona, Israel; [c]Laboratoire de Physique, Centre Pharmaceutique, 92290 Chatenay-Malabry, France.

Tri-o-thymotide (TOT) forms channel-type and cage-type clathrate inclusion compounds with a wide variety of guest molecules. Since these clathrate crystals are chiral, any single crystal contains different amounts of enantiomeric guests. The clathrate type as well as the chiral discrimination (the preferred enclathration of one enantiomer over the other) depend on the guest used. Channel clathrates display an unexpected amplification on crystallization from partially enriched guest solutions: the final enantiomeric excess (ee) in the crystals is much larger than that anticipated on the basis of the ee of the starting guest in solution and the chiral discrimination found for crystals grown from racemic guest solutions.

Clathrate inclusion compounds comprise two different molecular species, guest and host, which associate on crystallization to form regular arrays. The guests are situated within channels or cages comprised of the host molecules with only weak van der Waals interactions between guest and host; no ionic or covalent bonds link the different species (1).

Tri-o-thymotide (TOT) has proved to be a versatile host molecule and affords clathrates of both the cage-type and channel-type. These clathrates are chiral, i.e. spontaneous resolution takes place on crystallization and each resulting single crystal is either right- or left-handed. All the TOT molecules in such a clathrate crystal are, barring disorder, homochiral and have either a P (right-handed) or M (left-handed) propeller-shaped, non-planar conformation. These conformations equilibrate

Y. Wolman (ed.), Origin of Life, 365–372.

in solution; on the basis of polarimetric studies with solutions prepared by dissolving single crystals (2) and from nmr studies as a function of temperature (3) the energy of activation for enantiomerization, M ⇄ P, was established as ca 21 Kcal/mole.

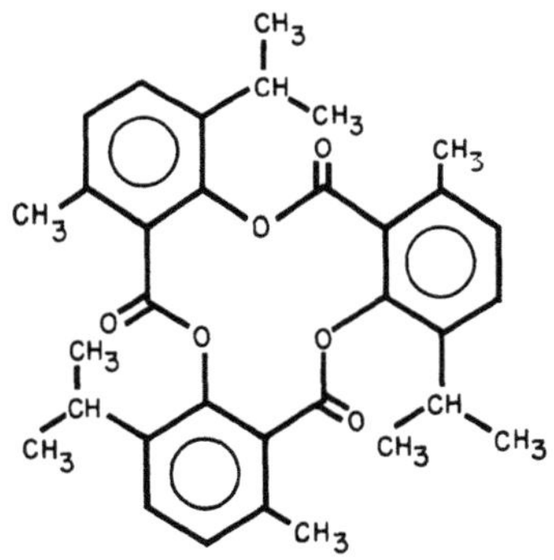

Tri-o-thymotide (TOT)

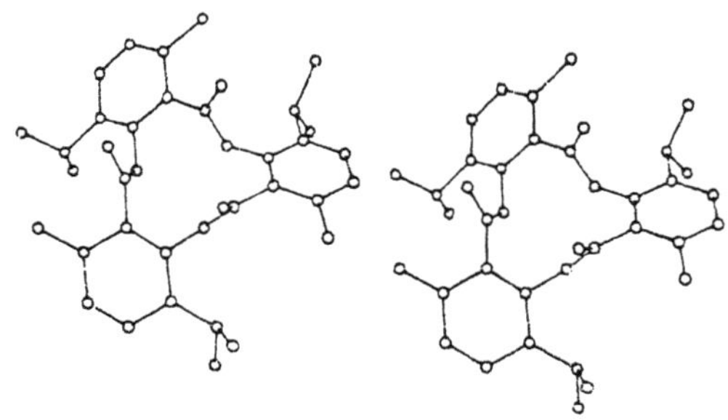

Stereoview of M-(-)-TOT

The structures and properties of the cage- and channel-type complexes of TOT were first studied by Powell (2, 4). More detailed descriptions of these structures are now available from recent crystal structure analyses (5, 6). (The structure of solvent-free TOT, which forms an achiral crystal, as well as additional, different clathrate structures of TOT (7) have also been reported). In addition, the absolute configuration of TOT has been firmly established (6) and corrects an earlier assignment (8).

It was recognized by Powell that the cavities occupied by the guest molecules in the TOT clathrate structures should display chiral discrimination for enantiomeric guest species (9). On the basis of some preliminary experiments Powell concluded that the chiral discrimination is nearly perfect; i.e., that single crystals of clathrate grown from racemic solutions of chiral guests will contain only one of the guest enantiomers (10). (More recent results indicate that perfect chiral recognition is rare (11, 14)).

The aims of our investigations are three-fold: to attempt the resolution of guest molecules which would otherwise be difficult to separate; to see if correlations of configuration between guest and host in single crystals might allow absolute configuration assignments for guests of unknown configuration; and, finally, to consider the chiral cavities as reaction sites for performing asymmetric synthesis. We have recently reported results which relate to the first two points; the third aspect is currently under investigation.

The results to date may be summarized as follows: All the chiral guests enclathrated in TOT crystals display enantiomeric preference, which varies from 2% to 47% in the cases studied. In these experiments, a solution of TOT in a large excess of racemic guest is allowed to crystallize. Both (+)-TOT and (-)-TOT clathrate crystals are formed so that, after measuring the guest ee in the crystals, there are two independent measurements for each system and the reproducibility of these values as well as the reliability of the correlation between host and guest configuration can be estimated. Because of the large excess of guest initially present, the growing clathrate crystal in such experiments is exposed to an essentially racemic solution of guest molecules throughout the crystallization process. A selection of results is presented in Table 1 (11).

TABLE 1

Enantiomeric Excess and Correlation of Configuration of Guest and TOT in (+)-TOT Clathrate Crystals.

Guest Molecule	Clathrate Type	TOT Guest Ratio	e.e. of Guest	Configuration of Guest
2-Bromobutane	Cage	2:1	34%	S-(+)
2-Chlorooctane	Channel	2.6:1	4%	S-(+)
2-Bromododecane	Channel	3.8:1	5%	S-
trans-2,3-Dimethyloxirane	Cage	2:1	47%	S,S-(-)
2-Methyltetrahydrofuran	Cage	2:1	2%	S-(+)
Methyl methanesulfinate	Cage	2:1	15%	R-(+)

The cage structures generally display appreciably higher ee than the channel structures. This is reasonable when one considers that, in the cage structure, each guest species is surrounded by eight homochiral TOT molecules; each guest experiences only guest...host interactions in this structure. In the channel structures, the guests are packed end-to-end and experience guest... guest interactions as well as guest...host interactions and therefore the influence of the chiral environment is less pronounced in the channel. In addition, there are greater possibilities for disorder in the channel structures: guest molecules can be aligned parallel or antiparallel to a channel axis; the guests may be located in different positions relative to the long axis of the unit cell; and for each of these relative displacements, the guest may adopt several rotational orientations in the channel. There is no possibility of coincidence of guest molecular symmetry and host cavity symmetry. For these reasons, crystal structure analyses of channel structures performed thus far do not allow one to see the guests. By contrast, in the cage structures the guests are necessarily restricted to one location relative to the TOT host and often adopt orientations which coincide with the two-fold symmetry of the cage cavity. Thus, trans-2,3-dimethyloxirane and 2-bromobutane lie on the two-fold axis of the cage; the latter achieves 2-fold symmetry through disorder involving two molecular orientations related to one another by 180° rotation.

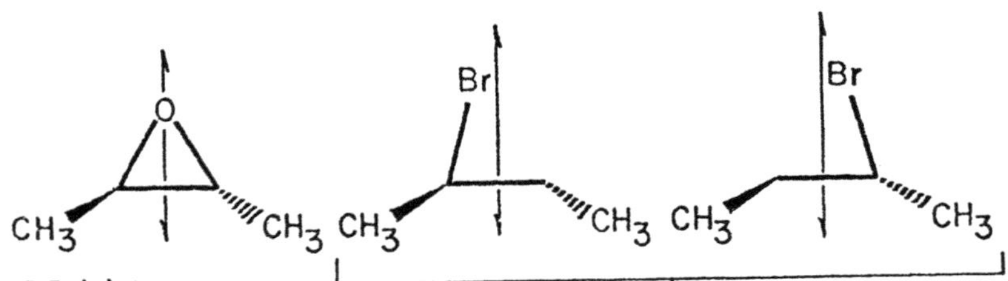

R,R-(+)-trans-2,3-dimethyloxirane

R-(-)-2-bromobutane

The chiral cavities might be expected to show systematic preferences for guests of the same configuration (9) and, if so, within a series of related materials one could deduce the configuration of unknown compounds from the configuration of known guests. The configuration of the TOT in a single crystal is readily assigned by dissolving a small portion of the crystal in cold solvent and noting the optical rotation; the sign of rotation and/or configuration of the guests of known configuration can be determined by methods such as polarimetry, gc on chiral phases, or nmr with chiral shift reagents. Within series of related compounds studied thus far (11, 12) an excellent correlation of guest and host configurations exists. Thus, as indicated above, the RR-(+)-trans-2,3-dimethyloxirane and R-(-)-2-bromobutane enantiomers both crystallize in the same, M-(-), TOT clathrate. Similarly R,R-(+)-trans-2,3-dimethylthiirane and R-(-)-2-chlorobutane crystallize preferentially with M-(-) TOT. Assuming that methyl methanesulfinate adopts the same orientation as related guests, e.g. 2-bromobutane, then the (-)-enantiomer which is enclathrated preferentially in a (-)-TOT clathrate has the S-configuration. Crystal structure analyses of some of these (13), and other (14), guests support the view that related guests indeed adopt closely similar orientations in the cage-type clathrates.

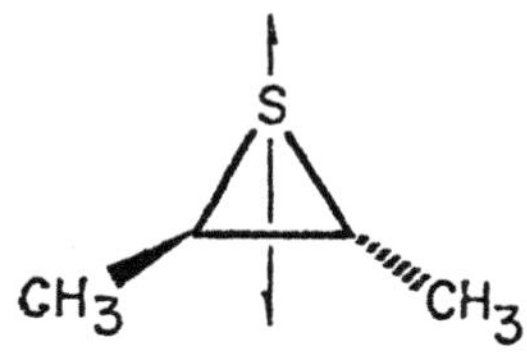

R,R-(+)-trans-2,3-dimethyloxirane

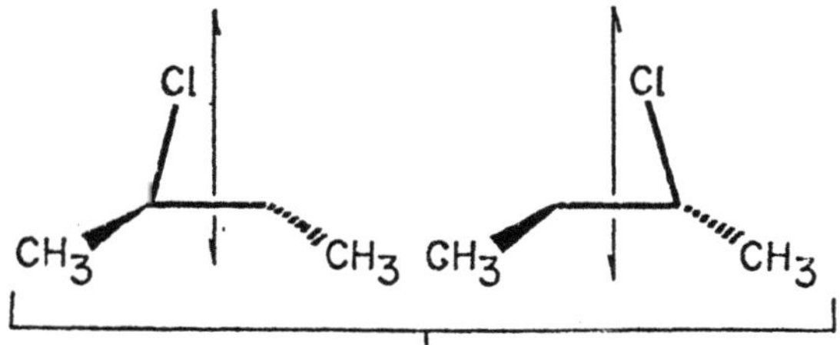

R-(-)-2-chlorobutane

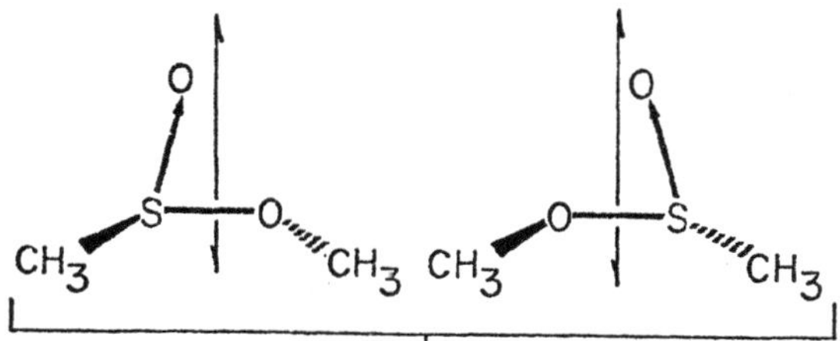

S-(-)-methyl methanesulfinate

In each crystallization of TOT from a racemic guest solution the initially achiral, optically inactive solution deposits chiral clathrate crystals sometimes comprising an excess of (+)- and sometimes an excess of (-)-TOT complex. When starting from non-racemic guest solutions, the resulting crystals invariably possessed an excess of the TOT having the chirality expected on the basis of the preferred guest-host configuration described above. During the course of these experiments it was noted that the enantiomeric composition of the guest molecules in crystals grown from initially non-racemic solution was appreciably greater than that anticipated from the results obtained for racemic guest solutions. The experiments were carried out using solutions of varying initial optical purity of 2-bromooctane. The enantiomeric ratio of the guest was determined by converting the 2-bromooctane to 2-aminooctane, acetylating with trifluoroacetic anhydride (16) and analysing the resulting N-trifluoroacetyl-2-aminooctane on a chiral gc phase (16).

Single clathrate crystals grown from racemic 2-bromooctane were analysed first; they were found to contain a 5% $\pm$ 1% ee (i.e. ratio of enantiomers between 52:48 and 53:47); (+)TOT preferentially incorporated the S(+)-enantiomer and (-)-TOT crystals contained an excess of the R-(-)-guest. This value may be considered the "intrinsic" chiral discrimination of a growing TOT clathrate crystal for this guest. Barring any additional factors, crystals of TOT formed from a non-racemic solution should contain guest with an optical purity which is a product of the initial guest enantiomeric ratio and the intrinsic chiral discrimination, ~52.5/47.5. In these experiments TOT (50 mg) was crystallized from solutions (0.3 ml) of 2-bromooctane, by slow cooling from 80° to 0°. The solutions had varying initial optical purity and were prepared by mixing optically active R-(-)-2-bromooctane (synthesized from R-(+)-2-octanol (17)) and racemic 2-bromooctane. The resulting polycrystalline clathrate samples were rinsed with acetone to remove any adhering liquid and the guest isolated by heating the crystals to 180° for 30 min at 0.1 mm Hg pressure and collecting the 2-bromooctane vapor in a dry-ice-cooled portion of the tube. Derivatization and gc analysis, as described above, gave the enantiomeric ratios, each value being the result of 1-2 crystallizations and 3-4 gc analyses per crystallization.

TABLE 2.

Amplification of Optical Purity Upon Crystallization of TOT From Non-Racemic 2-Bromooctane.

Initial 2-Bromooctane Optical Purity (Enantiomer Ratio) in Solution	Final 2-Bromooctane Optical Purity (Enantiomer Ratio) in Clathrates	
	Expected[a]	Found[b]
0% (50:50)		5% (52.5:47.5)[c]
12% (56:44)	17% (58.5:41.5)	30% (65:35)
33% (66.5:33.5)	37% (68.5:31.5)	66% (83:17)
50% (75:25)	54% (77:23)	80% (90:10)

a. Calculated from the 5% "intrinsic" chiral discrimination, 52.5/47.5 x the enantiomeric ratio in the starting solution, with the product normalized to a total numerator + denominator = 100; e.g. for 12% initial optical purity 56/44 x 52.5/47.5 = 58.5/41.5, a 17% enantiomer excess.

b. Maximum error limit $\pm$ 10% of value.

c. Single crystals.

The rationalization of these results is not simple because of the complexity of the system and the many stages where the chiral discrimination may be expressed during the process of solution → crystal. During the aggregation of guest and host molecules into precrystal nuclei and during the growth of these nuclei into clathrate crystals there may be disparate amounts of either guest or host. We do not know whether the TOT molecules deposit about the guests, whether guests are included into growing channels, or whether both processes occur during clathrate growth. Although the TOT host structure collapses if all of the guest is removed, it is possible to partially deplete the channel clathrate of guest and maintain the integrity of the TOT host structure; in several cases more than 50% of the guest molecules were removed from clathrate samples and the channel structures were not disturbed (12). For this reason, it may be reasonably assumed that host is deposited more rapidly that guest and that guest molecules are included into partially formed channels. If so, the molecules already present in the channel may interact with the approaching, about-to-be-included guest and express a chiral discrimination beyond that of the chiral host. We suggest that an RRR arrangement of the guests in the channel is more favorable than an RRS, RSR, or SRR arrangement. Thus, the chirality of the molecule which is included favors homochirality in the channel and, as the starting concentration of one enantiomer increases, the probability of a homochiral chain increases.

An alternative, or additional, "cooperativity" phenomenon may exist in the solution phase at the crystal surface: as the homochiral guest...guest concentration of one enantiomer increases, interactions may exclude the heterochiral enantiomer and favor the deposition of the predominant enantiomer far beyond the concentration factor. Some related results have been described in the literature (18).

Some of the experimental approaches now called for, include the following: seeding a solution of TOT in racemic guest with a clathrate seed containing optically pure 2-bromooctane (to see whether cooperativity will increase the chiral discrimination in newly deposited clathrate); to grow clathrate crystals, as in Table 2, but from solutions containing varying concentrations of achiral guest, e.g. 2-methyloctane (to see whether the decreased homochiral guest...guest interactions will decrease the amplification); and, finally, inquiring whether this amplification phenomenon occurs in cage complexes as well.

In conclusion, we have demonstrated a phenomenon of possible relevance to abiotic generation and amplification of optical activity. Although the present system does not contain biorelevant molecules, the principle has been demonstrated, that starting from an achiral, optically inactive sample one can generate, by the processes of crystallization, crystal separation, dissolution, and recrystallization, an increasing amount of optically active material. The system is self-reinforcing; the chirality produced favors the production of additional clathrate which, in turn, favors an even higher optical purity (19).

REFERENCES

(1) For a review see MacNicol, D.D., McKendrick, J.J. and Wilson, D.R., Chem. Soc. Revs. 7, 65 (1978).
(2) Newman, A.C.D. and Powell, H.M., J. Chem. Soc. 3747 (1952).
(3) Ollis, W.D., Stoddart, J.F. and Sutherland, I.O., Tetrahedron 30, 1903 (1974).
(4) Lawton, D. and Powell, H.M., J. Chem. Soc. 2339 (1958).
(5) Brunie, S. and Tsoucaris, G., Cryst. Struct. Comm. 3, 481 (1974); Williams, D.J. and Lawton, D., Tetrahedron Letters 111 (1975); Brunie, S., Navaza, A., Tsoucaris, G., Declercq, J.P. and Germain, G., Acta Cryst. B33, 2645 (1977).
(6) Gerdil, R. and Allemand, J., Tetrahedron Letters 3499, (1979); Arad-Yellin, R., Green, B.S., Knossow, M. and Tsoucaris, G., Tetrahedron Letters 21, 387 (1980).
(7) Arad-Yellin, R., Brunie, S., Green, B.S., Knossow, M. and Tsoucaris, G., J. Am. Chem. Soc. 101, 7529 (1979); see also footnote 45 in this paper.
(8) Downing, A.P., Ollis, W.D., Sutherland, I.O., Mason, J. and Mason, S.F., J.C.S. Chem. Comm. 329 (1968).
(9) Powell, H.M. Nature 170, 155 (1952).
(10) Powell, H.M. Endeavour 15, 20 (1956).
(11) Arad-Yellin, R., Green, B.S. and Knossow, M., J. Am. Chem. Soc., 102, 1157 (1980).
(12) Arad-Yellin, R., Green, B.S. and Knossow, M., Unpublished results.
(13) Arad-Yellin, R., Green, B.S., Knossow, M. and Tsoucaris, G., To be published.
(14) Gerdil, R. Unpublished results and private communication.
(15) Feibush, B. and Gil-Av, E., J. Chromatog. 5, 257 (1967).
(16) Weinstein, S., Feibush, B. and Gil-Av, E., J. Chromatog, 126, 97 (1976).
(17) Gerrard, W. and Hudson, H.R., J. Chem. Soc. 2310 (1964).
(18) Wynberg, H., Chimia 30, 445 (1976).
(19) Green, B.S. and Heller, L., Science 185, 525 (1974).

ASYMMETRICAL RADICAL FORMATION IN β-IRRADIATED D- AND L-ALANINES

Mitsuhiko Akaboshi, Masato Noda, Kenichi Kawai, Hirotoshi Maki and Keizo Kawamoto

Research Reactor Institute, Kyoto University, Kumatori, Osaka, 590-04, Japan

It was demonstrated that the relative radical concentration was distinguishably higher in D-alanine than in L-alanine when irradiation was carried out by using yttrium-90 β-ray source. On the other hand, when irradiation was carried out by using non-polarized β-rays derived from Van de Graaf generator, no such difference was observed in the radical formation of both alanines. The results suggested that some process being involved in the parity non-conservation in a β-decay must play an important role in the observed phenomenon.

In the preceding paper (1), We examined the conversion of D- or L-alanine to D- or L-cysteine by using recoiled sulphur-32 atoms obtained from the β-decay of phosphorus-32. The experimental results seemed to suggest an asymmetric conversion, but more definitive evidence is needed before a conclusion can be reached because of too much variation in the data obtained. One possible explanation for the fluctuating data may be that there are many factors which may mask the expected difference insofar as the chemical changes were observed by chemical means. To avoid these disturbing factors, We studied the radical formation in β-irradiated D- and L-alanines by ESR measurement.

D- and L-alanines were purchased from Wako Chem. Co., and were purified by dissolving in triply distilled water and reprecipitating with spectro grade of ethanol. The precipitations were repeated more than 3 times. DL-alanine was prepared in the same manner from the mixture (1:1) of the D-

Y. Wolman (ed.), Origin of Life, 373–378.

and L-alanines. They were dried and stored under vacuum until they were used. Before irradiation, the poly-crystal powders of alanines, thus obtained, were packed into the ESR tube and were sealed off at about 1 x 10^{-5} mm Hg pressure. ESR tube (Spectrosil) with a 3.5 mm inner diameter was used. At the beginning, the radical formation in cobalt-60 γ-irradiated ethanol (spectro grade) was examined at 77 K. Then 15 tubes out of 30, ranging within the mean value ± 1.8 % relative to radical concentration yielded, were selected for the experiments.

For β-ray irradiation, the two of β-ray sources, the one was yttrium-90 produced by reactor irradiation of yttrium-89 metal, and the other the Van de Graaf generator which was established in the Faculty of Engineering of Kyoto University were used. To produce the yttrium-90 source, about 10 g of a cylindrical Y-metal (12 mm in diameter and 30 mm in height with 5.2 mm diameter of a vertical hole for irradiation of the samples) was irradiated in a hydraulic conveyer of KUR (Kyoto University Reactor) with a thermal neutron flux of approximately 8.15 x10^{13} $n.cm^{-2}.sec^{-1}$ for 10 hr. Total radioactivity of the metal immediately after the irradiation was calculated to be about 8 Ci. The dose rate during the experiments was 5 x 10^4 – 10^5 $rad.hr^{-1}$. The radiation source was put into an Al-pipe which was placed in a Dewar flask with or without liquid N_2. To get non-polarized electron beam, the Van de Graaf generator was used. It was operated at 1.65 MeV. In this case 12 samples were irradiated at the same time.

To calibrate the differences in the size of ESR tubes, the different densities of alanines packed, and many other things (even different impurities, if any), irradiation with 50 – 100 fold (to neglect the effects due to pre-β-irradiation) of cobalt-60 γ-rays was carried out on the same samples immediately after the radical measurement. The ratio of the radical produced by the β-irradiation to that by the γ-irradiation (R_β/R_γ) was taken as the standard. The γ-irradiation was carried out using 10 000 Ci of cobalt-60 γ-ray source with a dose rate of about 1.5 x 10^6 $rad.hr^{-1}$.

ESR measurements were carried out with a Varian Model E-3 spectrometer operating in the X-band. The spectra were recorded as the first derivatives of the absorption curves, with a modulation frequency of 100 kc sec^{-1}. To avoid saturation effects, the incident microwave power was 0.5 mW. Under these conditions, no saturation effect was observed in the irradiated alanines within the dose ranges used.

Figure 1 shows the radicals dealt with in this study. It is known that both radicals were quite stable in the given conditions, and that, in contrast to valines and leucines, no differences could be detected in the resonances of irradiated

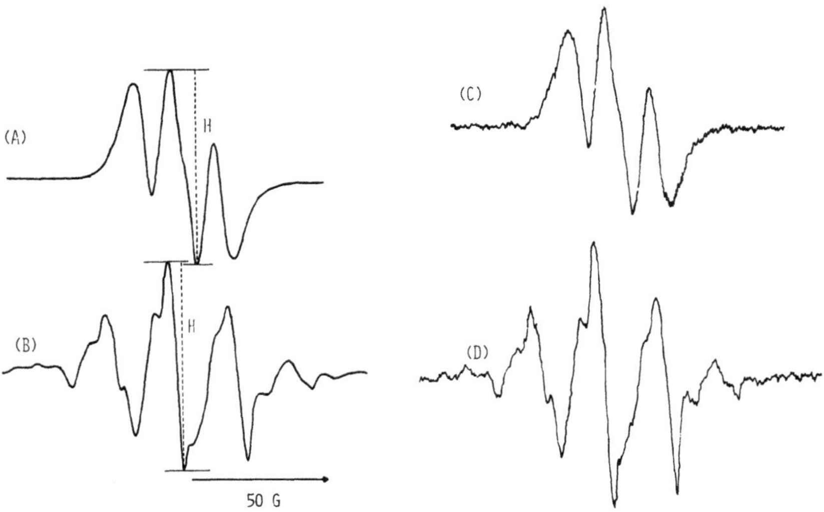

Fig. 1. ESR spectra of alanine irradiated at 77K and measured at 77K (A and C) and room temperature (B and D).

Sample: L-alanine
Irradiation: ^{60}Co-γ-rays, 1.5 x 10^6 rad for (A) and (B)
^{90}Y-β-rays, 2 x 10^4 rad for (C) and (D).

D-, L- and DL-alanines (2). This was also the case in the present study. These alanines gave exactly the same spectra with the same line width, peak position and ratio of each peak height when irradiated with a low dose of β-rays. In both radicals, the height of the central peaks was used as the measure of relative radical concentration.

TABLE 1 shows the relative radical concentrations produced by yttrium-90 β-ray and γ-ray irradiation (at 77 K), and the ratios R_β/R_γ. Radiation doses in this case were about 2 x 10^4 for β-rays and 1.5 x 10^6 rad for γ-rays. As seen in the TABLE, the radicals produced by β-irradiation were significantly higher in D-alanine. The ratios R_β/R_γ were 0.502 for D-alanine and 0.413 for L-alanine in the case of the radical measured at 77 K. The ratio was placed between both in the case of mixed alanine, while that of DL-alanine was considerably higher than those of D- and L-alanines. These tendencies applied also in the case of radicals observed warming up to R.T. (room temperature). After the first run, D-alanine in the ESR tube was replaced by L-alanine and vice versa. They were irradiated in the same conditions, and then ESR spectra were measured in the same geometrical condition. The results obtained were closely consistent with those given in TABLE 1, suggesting that observed difference originated by some difference in D- and L-alanines.

No such differences were observed when both irradiation and

TABLE 1. The Ratios (R_β/R_γ) of Radical Yielded by 77 k-Irradiation with Yttrium-90 β-Rays.

	Radical Observed at 77 K			Radical Observed at R. T.		
Samples	β-Ray	γ-Ray	(R_β/R_γ)	β-Ray	γ-Ray	(R_β/R_γ)
D-Alanine	4.82	9.64	0.500	9.99	10.32	0.968
	4.56	8.90	0.512	8.94	9.41	0.950
	4.52	10.13	0.495	9.72	11.10	0.942
			0.502 ± 0.009			0.953 ± 0.013
L-Alanine	3.92	9.70	0.404	7.96	9.49	0.839
	3.52	8.46	0.416	6.86	8.37	0.820
	3.87	9.21	0.420	9.35	10.78	0.867
			0.413 ± 0.008			0.837 ± 0.019
DL-Alanine	2.39	3.40	0.703	3.13	3.00	1.043
	2.14	3.04	0.704	2.87	2.70	1.062
			0.704 ± 0.001			1.053 ± 0.013
Mixed Alanine*	3.68	7.92	0.465	7.46	8.18	0.921
	4.39	9.66	0.454	10.04	10.94	0.918
			0.460 ± 0.008			0.920 ± 0.002

*Equal weight of D- and L-alanine powders were mixed.

ESR measurement were carried out at R.T..

In order to confirm the role of β-rays in the asymmetric radical formation, low dosage of pre-γ-irradiation, instead of the pre-β-irradiation, was carried out. Namely, the ratios of radical produced by a 1 min irradiation with cobalt-60 γ-rays to that by a 50 min irradiation were examined in each alanine. However, the ratios were closely consistent in the all alanines, suggesting that the difference in the radical formation between D- and L-alanines was brought about by β-irradiation.

For a physical interpretation of the phenomena, we proposed two possible mechanisms (3). One would be that there are some differences in radiation susceptibility of the two enantiomers of a molecule if irradiated with β-particles and/or their associated circularly polarized γ-rays. Another possibility is based on a difference in the quality of both radiations, though it sets the problem apart from the parity non-conservation. High LET radiation such as β-rays, may bring about different radical saturation and/or radical diffusion in the crystal lattices of the different enantiomers. To clarify which is the valid, irradiation using non-polarized β-rays which were derived from the Van de Graaf generator operated at 1.65 MeV. Practically this energy can be regarded as the same to that of yttrium-90 β-rays. TABLE 2 shows a result obtained from such experiments. The dose rate and irradiation time, in this case, were about 2 x 10^5 rad. hr^{-1} and 10 min, repectively for β-irradiation. It is obvious from the TABLE that no difference is observed in the radical formation in D- and L- alanines irrespective of observation conditions.

TABLE 2. The Ratios (R_β/R_γ) of Radical Yielded by 77 K-Irradiation with β-Rays Derived from Van de Graaf Generator.

	Radical Observed at 77 K			Radical Observed at R. T.		
Samples	β-Ray	γ-Ray	(R_β/R_γ)	β-Ray	γ-Ray	(R_β/R_γ)
D-Alanine	11.27	17.13	0.658	15.42	12.84	1.201
	10.00	15.75	0.635	13.11	9.78	1.341
	10.92	16.11	0.678	13.53	9.75	1.388
			0.657 ± 0.021			1.310 ± 0.097
L-Alanine	10.39	16.38	0.634	12.27	9.70	1.265
	8.52	12.78	0.667	12.75	9.15	1.393
	11.39	16.51	0.690	13.03	9.44	1.380
			0.664 ± 0.028			1.346 ± 0.070
DL-Alanine	2.90	2.85	1.018	3.45	1.64	2.104
	3.12	3.05	1.023	4.08	1.89	2.159
	3.92	3.91	1.057	4.26	1.93	2.207
			1.037 ± 0.021			2.157 ± 0.052
Mixed Alanine	7.50	11.35	0.661	10.52	7.92	1.328
	7.92	12.83	0.617	9.83	7.83	1.255
	9.63	14.51	0.664	10.95	8.55	1.281
			0.647 ± 0.026			1.288 ± 0.037

Thus it is clear from the present results that the difference in radical formation between D- and L-alanines was found only when β-irradiation was carried out by using polarized β-rays. This suggests that some process being involved in the parity non-conservation in a β-decay must play an important role in the observed phenomenon.

As the higher R_β/R_γ ratio in DL-alanine was found in both polarized and non-polarized β-irradiations, it may be attributed to the latter possibility described above because it is known that the crystal structure of DL-alanine is considerably different ($Pbn\ 2_1$) from that of D- and L-alanines ($P2_12_12_1$) (ref. 5).

ACKNOWLEDGEMENT

The authors wish to express their gratitude to Dr. H. Hase, Dr. T. Matsuyama and Mr. T. Warashina for their helpful advice and discussion. We also wish to thank Prof. H. Noda of Institute of Biological Chemistry, University of Tokyo for his great encouragement and help.

This work was supported in part by a Grant in Aid from the Ministry of Education, Science and Culture.

REFERENCES

(1). Akaboshi, M., Kawai, K., Maki, H. and Kawamoto, K.: Origin of Life, Prodeedings of 2nd ISSOL and 5th ICOL (ed. by Noda, H., Center of Academic Publication Japan), 1978, p343.
(2). Shields, H. and Gordy, W.: J. Phys. Chem., 62, 789, (1958).
(3). Akaboshi, M., Noda, M., Kawai, K., Maki, H. and Kawamoto, K. : Origins of Life, 9, 181, (1979).
(4). Akaboshi, M., Noda, M., Ito, Y., Kawai, K. and Maki, H.: Viva Origino, 8, 66, (1980).
(5). National Bureau of Standards: Crystal Data, 3rd Ed. Vol. 1, (ed. by Donnay, j. D. H. and Ondik, H. M.', Item No. 0.39-0.40, 1972.

A NEW EXPERIMENT TO INVESTIGATE THE ORIGIN OF OPTICAL ACTIVITY USING A LOW ENERGY POSITRON BEAM OF CONTROLLED HELICITY

D. W. Gidley, A. Rich and J. C. Van House

Department of Physics, University of Michigan
Ann Arbor, Michigan 48109
and
P. W. Zitzewitz
Department of Natural Sciences, University of Michigan
Dearborn, Michigan 48128

In recent years many experiments have been undertaken in search of a correlation between the origin of optical activity in biological molecules and the helicity of beta particles emitted in nuclear beta decay. These experiments have shown null or ambiguous results. We will discuss an experiment now in progress in which a low energy polarized positron beam of controlled helicity interacts with an optically active material to form positronium in vacuum. Initially we will be able to detect a helicity dependent asymmetry in triplet positronium formation of 1 part in 10^4. Improvements to better than 1 part in 10^5 should be possible. Our work may therefore have up to 10^4 times the sensitivity of prior experiments to the effect under consideration.

More than 20 years ago Vester (1) and Ulbricht (2) suggested that a causal connection might exist relating the origin of the asymmetry between optical isomers in biomolecules to the then recently observed (Lee and Yang, 1956; Wu et al., 1957)(3) violation of parity in nuclear beta decay. Since that time numerous experiments have searched for such connections, but with no reproducible positive results. Many of these experiments have exploited the fact that the beta particles emitted in nuclear beta decay have, as a direct consequence of parity violation, a net helicity i.e. their spin directions are correlated with the direction of their velocities on emission from the nucleus. Experiments of this type can be classified

Y. Wolman (ed.), Origin of Life, 379–384.

into two categories: radiolysis experiments that search for asymmetric degradation of one of the isomers of an initially racemic mixture by beta radiation; or electron capture experiments that search for preferential formation of positronium or muonium when positrons or muons with a net helicity interact with L vs. D isomers.

Reviews (4) of experiments through 1976 conclude that no positive effect had been observed up to that time with the exception of a radiolysis experiment by Bonner et al. (5). This work was however recently shown to be non-reproducible by Hodge et al. (6) so that the situation here remains ambiguous. Concerning positronium experiments, Garay et al. (7) reported a large asymmetry in the formation of triplet positronium in L vs. D isomers of various amino acids but this effect was also shown to be non-reproducible (8). In addition, it was shown (9) that the largest effect that could be expected in these experiments due to the residual helicity of the positron at positronium formation energies was far smaller than the result reported in reference 7. Finally, we note that experiments involving muons (10) have shown no effect.

This brief summary indicates that no conclusive demonstration of a connection between parity violating beta decay and biomolecular asymmetry has yet been presented in the literature. In the remainder of this paper we will describe an experiment currently underway in our laboratory which will search for a helicity dependent asymmetry in the formation of the triplet ground state of positronium when low energy polarized positrons interact with an optically active substance. This experiment has a potential sensitivity up to 10^4 times greater than previous positronium experiments.

The apparatus, shown schematically in Fig. 1, consists of a low energy polarized positron beam generator, a Wien filter spin rotator, the optically active positronium formation surface, and the annihilation γ-ray detectors. Polarized positrons emitted by a ^{58}Co source are incident over a 2π solid angle onto a cylindrically symmetric moderator consisting of a 0.2 mg/cm^2 gold foil coated with 4 mg/cm^2 of MgO. The source positrons thermalize in the MgO and approximately 1 in 10^4 are re-emitted as slow positrons. During the thermalization process the polarization $\vec{P} = \langle\hat{S}_i\rangle$ remains virtually unchanged. In this expression for $\vec{P}$ the brackets denote a quantum mechanical average over the particles of the beam and $\hat{S}_i$ is a unit vector in the spin direction of the ith particle. The helicity, h, is defined by $h = \langle\hat{S}_i \cdot \hat{v}_i\rangle = \langle\hat{S}_i\rangle \cdot \langle\hat{v}_i\rangle$, where $\hat{v}_i$ is the unit velocity vector of the ith particle. Since the positrons have essentially random velocities even prior to thermalization ($\langle\hat{v}_i\rangle \simeq 0$) the helicity approaches zero.

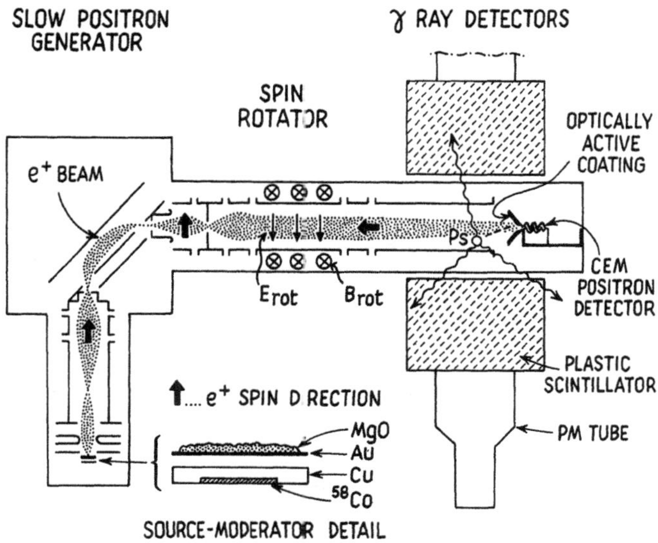

Fig. 1. The experimental arrangement.

Accelerating the slow positrons emitted from the moderator into an axially polarized beam restores the helicity of the beam. The rehelicitized beam is then transported to a cylindrical mirror energy analyzer where electric fields change the beam velocity by 90° but leave the spin direction unchanged. The then transversely polarized beam ($h=0$, $\vec{P}$ perpendicular to $\langle\hat{v}_i\rangle$) enters the Wien filter where $\vec{P}$ is rotated parallel to the beam velocity ($h>0$) or antiparallel to the beam velocity ($h<0$). Using this spin rotator we have measured (11) the polarization of the beam to be $P=0.22\pm0.02$. The positrons of a selected helicity and energy ($\simeq$ 100 eV) are focused onto the coated surface of a channel electron multiplier (CEM) and are detected through the emission of secondary electrons. We have succeeded in coating the CEM surface with the amino acids leucine, tryptophan, and alanine while maintaining high positron detection efficiency.

The incident positrons slow down to approximately 10 eV in the optically active coating and should sustain only minor loss of helicity while slowing down. Then, with 15-25% efficiency, they form positronium (Ps) in the triplet ground state. Ps so formed leaves the coated CEM surface and lives for approximately 140 nsec in the surrounding vacuum chamber before annihilating into three γ rays. Detection of one or more of the annihilation γ rays in large pilot B plastic scintillators allows the lifetime of each positron event to be directly measured and recorded with a conventional time-to-amplitude converter-multichannel analyzer system.

The positron lifetime spectrum so obtained consists of a prompt peak of freely annihilating positrons and singlet Ps whose decay occurs within one nsec of the CEM start pulse, a 140 nsec exponential component due to triplet Ps decays and a constant background due to random coincidences. Defining $N^+(N^-)$ to be the number of Ps events occurring in a background corrected time window from 10 ns to 300 ns for incident positron helicity positive ($h>0$, N^+) and then negative ($h<0$, N^-), the asymmetry in triplet positronium formation is defined to be

$$A = \frac{N^+ - N^-}{N^+ + N^-} .$$

With our anticipated slow positron rate of 3×10^4 sec^{-1} we should be able to set statistical limits of $\pm10^{-4}$ on A in one day of data acquisition, a factor of 100 improvement over previous experiments.

We assume that the only external factor responsible for an asymmetry in the positronium formation rate is the helicity of the positrons at the time Ps is formed. In addition, we assume that Ps formation occurs only at positron energies of order 10 eV. The measured asymmetry can then be related to the estimated residual helicity h' of the positrons undergoing electron capture by the relation $|A|=k|h'|$, where k is a constant which characterizes the interaction between the positron and the optically active isomer. The maximum value of k is unity.

Our initial experiment will measure $|A|$ at the level of 1 part in 10^4. Assuming $|h'|=0.2$, this will set an upper limit on k of 5×10^{-4}. By contrast, no asymmetry in Ps formation has been observed at better than the 1% level. Since we estimate h' for previous experiments to be less than 1%, the current experimental upper limit on k is about 1, the maximum possible value. Straightforward improvements in our beam intensity and longer data acquisition times will reduce the statistical error by a factor of 10. Assuming that systematic errors at this level can be overcome, we would set an upper limit of about 50 ppm on k. We thus hope to obtain the most stringent quantitative limit on the connection between nuclear parity violation and molecular asymmetry obtained by workers in this area to date.

In summary, the features of our experiment which represent major departures from, and improvements on, previous work are:

1) The use of a low energy positron beam. In previous

positronium asymmetry experiments the positron velocity randomizes in the slowing down process and thus the helicity is very nearly zero when positronium formation occurs. By contrast, we calculate that our 100 eV incident positron beam will have $h \approx 0.15$ when Ps is formed.

2) The helicity of the positrons is reversible. Therefore it is not necessary to change the optically active isomer from L to D in order to measure A. This eliminates one of the most serious systematic concerns in the earlier experiments associated with changes in the Ps formation fraction due to differences in the impurity levels of the two isomers. Of course, if an asymmetry is observed the opposite isomer will be tested to see if the sign of A changes as expected.

3) The measured lifetime of the Ps formed on the optically active coating is about 140 nsec instead of the pick-off shortened 1 nsec lifetime of positronium decaying inside solid amino acids. This long lifetime provides unambiguous separation of the triplet positronium decay component from the prompt annihilation peak in the lifetime spectrum. Consequently the sensitivity of our measured values of N^+ and N^- to systematic drifts in our timing system (such as t=0 drifts or shifts in system linearity) is reduced by at least a factor of 140 nsec/1 nsec $\simeq 10^2$.

4) The effect of timing system drifts can be reduced by a factor of 10 by means of an averaging process based on frequent helicity reversals.

ACKNOWLEDGEMENTS

We thank G. W. Ford, R. R. Lewis, and G. Karl for helpful discussions, A. Nudelfuden for laboratory assistance and R. H. Sands for major equipment loans.

This work was supported by the National Aeronautics and Space Administration and by the U.S. National Science Foundation.

REFERENCES

(1) Vester, F., Seminar, Yale University, February 7, 1957.
(2) Ulbricht, T. L. V., Quart. Rev. 13, 48 (1959).
(3) Lee, T. D. and Yang, C. N., Phys. Rev. 104, 254 (1956), and Wu, C. S., Ambler, E., Haywood, R. W., Hoppes, D. D., and Hudson, R. P., Phys. Rev. 105, 1413 (1957).
(4) Darge, W., Laczko, I., and Thieman, W., Nature 261, 523 (1976), and Bonner, W. A., Van Dort, M. A., Yearian, M. R., Zeman, H. D., and Li, G. C., Israeli J. of Chem. 15, 89 (1977).
(5) Bonner, W. A., Van Dort, M. A., and Yearian, M. R., Nature 258, 419 (1975).
(6) Hodge, L. A., Dunning, F. B., and Walters, G. W., Nature 280, 250 (1979).

(7) Garay, A. S., Kezthelyi, L., Demeter, I., and Hrasko, P., Nature 250, 332 (1974); Chem. Phys. Lett. 23, 549 (1973).
(8) Brandt, W. and Chiba, T., Phys. Lett. 57A, 395 (1976); Dezsi, I., Horvath, D., and Kajcsos, Z. S., Chem. Phys. Lett. 24, #4, 514 (1974); and Jean, Y. and Ache, H. J., J. Phys. Chem. 81, #12, 1157 (1977).
(9) Rich, A., Nature 264, 482 (1976).
(10) Lemmon, R. M., Crowe, K. M., Gygak, E. N., Johnson, R. F., and Patterson, B. D., Nature 252, 692 (1974).
(11) Zitzewitz, P. W., Van House, J. C., Rich, A., and Gidley, D. W., Phys. Rev. Lett. 43, 1281 (1979).

COMPETITION, COEXISTENCE AND IRREVERSIBILITY IN MODELS OF EARLY MOLECULAR EVOLUTION

C. Blomberg, G. von Heijne, O. Leimar.

Research group for Theoretical Biophysics
Royal Institute of Technology
S-10044 Stockholm, Sweden.

We have studied some mathematical models of prebiotic systems with monomer availability as a limiting factor in order to elucidate concepts of irreversibility and coexistence in early molecular evolution.

One has long searched for general principles in molecular evolution. These are sometimes formulated by more or less clear ideas but often, they are described in terms of mathematical models. This is of considerable importance since it gives a possibility of testing unifying principles. Our purpose here is to present and discuss some models that yield information, in particular, about the concepts of irreversibility and coexistence.

Irreversibility, closely connected with the concept of fitness as an arrow of evolution, lies in the very heart of evolution. In the lapse of time, species are replaced by more "fit" ones. There are many attempts to give a meaning of fitness in physico-chemical terms, such that there is an increase of fitness with each step of evolution. One would like to have a general principle, possibly of energetic character, with the same scope as the second law of thermodynamics, but rather describing the irreversible building up of an increased order. Attempts in this direction are made e.g. by Prigogine and his group (1).

One may also ask whether the irreversible evolution is deterministic in the same sense as the 2nd law of thermodynamics, i.e. if it proceeds towards a certain goal. In many mathematical models, this is indeed the case, but then this possibility is often implicit at the very onset by the formulation of the model. In a more general context, the picture is somewhat different as we will discuss.

Y. Wolman (ed.), Origin of Life, 385–392.

There is a mathematical way of expressing the irreversibility which, however, does not in itself give an insight into the "fitness"concept. As described in the formalisms of e.g. Prigogine and coworkers (1), a certain steady state situation can be changed for instance by the occurrence of a mutant, and the earlier steady state becomes unstable. The system will evolve towards a new state, which is stable as long as there is no change in external parameters or a more "fit" species appears that makes the situation unstable. This kind of description is not deterministic, as there, at each stage, may be many possible events that can yield instabilities, of which only one occurs. Unfortunately, it does not imply any general principle for the increase of fitness. It is our purpose here to use this picture as a basic approach, and look for further principles. As will be seen, for certain simple situations, there is a principle but we shall also show examples in which the order of events is crucial. We will conclude that the irreversibility in evolution is not a clear concept.

Another important concept is that of coexistence. In many models of molecular evolution, including those of Eigen (2) and Eigen and Schuster (3), there is in general only one dominating species at a time. The question is interesting from many aspects. In particular, the uniqueness of the basic molecular processes in all life forms suggest that there is a unique origin, which still bears strong evidence of evolution (e.g. the genetic code). One here encounters the question why alternatives have not arisen, and a theoretical description that shows that, at least up to some point, there should be only one competitive survivor would be very satisfying in this respect.Often, one discusses coexistence in terms of ecological niches and, as we shall show here, concepts of a similar nature may well appear in molecular evolution.

Our models will be similar in scope to those of Eigen. However, we will not use his constraint of "constant overall organisation", that seems to be somewhat artificial, but rather equations where the constraint (and non-linearity) comes from the monomer availability (which also may mean availability of some metabolic molecules).As noted by Eigen, this kind of approach yields qualitatively similar results as that of constant overall organization, but it allows to make other types of interpretations, and to consider energetic principles.

We will now discuss some explicit models. We assume a situation where there is one or several subsystems, called "species".These may involve several kinds of small polymers that are competing for the monomers. Each subsystem is represented by one variable, the concentration of the relevant self-replicating polymer. Basically, we have a scheme like this:

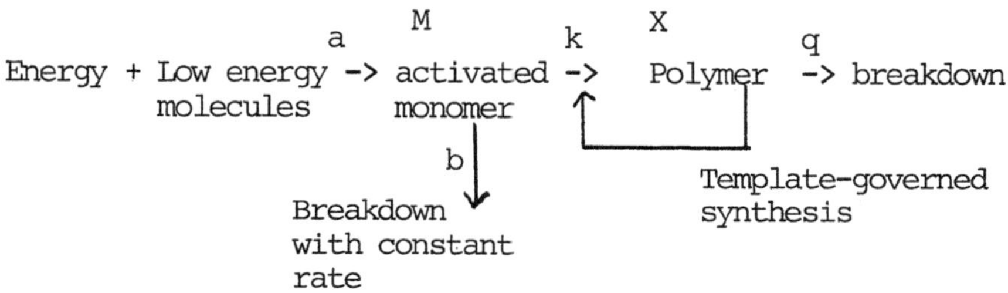

In the first models, we only consider one type of monomer, for which the species compete. The first three models are similar to those considered in refs. (2), (3) and (4) with the constant overall organization assumption. In their reformulated form here some new interpretations can be made. The last three models are new.

(i) linear growth. Let M be the concentration of monomer, and X that of the polymer. Then:

$$\dot{X} = k_1 MX - q_1 X$$
$$\dot{M} = a - bM - nk_1 MX \tag{1}$$

n is the number of monomers per polymer. There is a steady state if $a > bq_1/k_1$ for which $X \neq 0$. The monomer concentration is then q_1/k_1, which value is determined by the equation for $\dot{X}$. It is seen from (1) that the value of M determines whether X grows:

If $M > q_1/k_1$, then X grows, if $M < q_1/k_1$, X decreases.

A new species, Y may occur as a mutation. Its growth is described by

$$\dot{Y} = k_2 MY - q_2 Y \tag{2}$$

If X and Y in this way are competing for M, we also have;

$$\dot{M} = a - bM - n_1 k_1 MX - n_2 k_2 MY$$

Assume that we start from a steady state with only X where M is q_1/k_1. Then, if Y appears and $q_2/k_2 < q_1/k_1$, Y can grow, M will decrease, which causes X to decrease. Eventually, X vanishes and Y dominates. In this kind of model, only one species will survive, and the evolution leads to lower and lower steady state concentrations of the monomer M. Thus, there is in this case really an evolutionary principle with "fitness" given by the smallness of the quotient q/k (breakdown rate/synthesis rate).

(ii) Hypercycle growth. The conclusions of the linear growth case must be modified or changed if one allows for non-linear growth terms. The hypercycles of Eigen and Schuster (3) lead to equations:

$$\dot{X} = k_1MX + K_1MX^2 - q_1X$$
$$\dot{M} = a - bM - n_1k_1MX - n_1K_1MX^2 \quad (3)$$

The X^2-term arises since this in principle is a 2-component system where the components both contribute to their mutual synthesis.

Several cases can occur depending on the parameters, but we assume that there is a stable steady state with $X \neq 0$. The main features of (3) are the following. If X originally is at a low level, it can grow provided the value of M is greater than q_1/k_1, which we call the growth threshold. With increasing values of X, M decreases, and its steady state value is $q_1/(k_1+K_1)$, which can be considerably smaller than q_1/k_1. If a mutant Y appears at that stage, it can only grow if its growth threshold, q_2/k_2, is smaller than the steady state value of M. This requires a drastic reduction of the quotient. Mutants with rate constants not too much changed from those of X will not have a possibility to grow, although they may be able to decrease the M-concentration. This is a qualitatively similar result to that found by Eigen and Schuster (3) that a hypercycle will not allow competing species, and a further evolution must appear within the hypercycle itself. There is here no coexistence, and , when possible, species can be replaced only by ones that lower the concentrations of monomers.

(iii) Epstein's dimerization model. Epstein (4) treated some models in order to show the possibility of coexistence. One of these is in our terms equivalent to:

$$\dot{X} = k_1MX - K_1X^2 - q_1X$$
$$\dot{M} = a - bM - nk_1MX \quad (4)$$

The idea behind this model is that there may be a decrease of the active polymer X by a dimerization, yielding the term $-K_1X^2$. In this case, the concentration of M increases when X growths, and the steady state value, $(q_1+K_1X)/k_1$ is larger than the growth threshold. Then, a similar but new species with slightly different values of the rate constants can increase when X is at a steady state, even if its steady state value of M is higher than that of X. Y can completely destroy X only if the steady state concentration of M in the presence of Y is lower than the growth threshold for X, which means a drastic improvement of rate constants. Otherwise, X and Y will coexist although they are competing independently. Still, this coexistence state has lead to a lower steady state value of M.

There may be combinations of the behaviour of models (ii) and (iii). In general, any model of independently competing species as these always yield the result that the occurrence of a new stable situation leads to a decrease of the steady state concentration of the monomer, which means that it uses this energy source in a more efficient way.

(iv) The situation is more complicated if the competing species are not independent. We consider a model with, at the onset a species X, desciribed by (1), but where we for simplicity put n = 1. (This is by no means crucial). X is formed by the reaction M+X->2X. Then, there may occur a new species Y, which can use X as a precursor, although the synthesis is still template governed. There is a direct synthesis of Y:2M+Y->2Y, and one over X: M+X+Y->2Y.We choose particular rate constants and write:

$$\dot{X} = MX - 6MXY - Y$$
$$\dot{Y} = 2MY + 6MXY - 3Y \qquad (5)$$
$$\dot{M} = 2 - M - MX - 4MY - 6MXY$$

The features of the model are that Y is much more efficiently synthesized via X than directly. With only the direct pathway, Y would not be able to compete with X.Start with a steady state with only X: X=1, M=1. If now Y occurs, it can grow, primarily by the use of the high X-level. However, X will then vanish, and eventually, one gets a situation where Y=1/12, M=3/2, and X=0. (There cannot be any coexistence).In this new state, the value of M has increased! X cannot grow because of its immediate destruction by the Y-molecules. A new mutant Z can occur at this stage, which competes independently with Y, governed by the equation:

$$\dot{Z} = 3MZ - 4Z$$

which is of the same type as (1). By the arguments of case (i),Z will grow when M > 4/3 and it will replace Y. The system evolves to the state Z = 1/6, M = 4/3. Thus, the evolution over Y has lead from X to Z with an increase of the steady state value of M. On the other hand, if Z had occurred at the beginning, and then X appeared, X would have replaced Z by the arguments of model (i). This is a demonstration that, depending on the sequance of events there is no clear irreversibility principle; final and initial states may be inverted.

(v) We now investigate coexistence that is more close to a differentiation in ecological niches. The first possibility involves a spatial organization.Our model considers species of hypercycles (3), for which, as discussed for case (ii), coexistence and competition is particularly constrained.However,because of this, it is possible to get a niche organization.

Consider species X and Y distributed in a linear array of boxes. The notations are similar to that of model (ii), and the index k = 1,2,..N numbers the boxes. We have equations:

$$\dot{X}_k = D_1(X_{k+1} - 2X_k + X_{k-1}) + X_kM_k + X_k^2M_k - q_1X_k$$
$$\dot{Y} = D_2(Y_{k+1} - 2Y_k + Y_{k-1}) + Y_kM_k + Y_k^2M_k - q_2Y_k \qquad (6)$$
$$\dot{M}_k = D_3(M_{k+1} - 2M_k + M_{k-1}) + a_k - M_k - X_kM_k - X_k^2M_k - Y_kM_k - Y_k^2M_k$$

The first terms are diffusion terms between the boxes. For simplicity, we put rate constants b, k and K (cf. (3)) equal to 1 as we do for n. (This does not influence the qualitative results). We get no-flux boundary conditions by putting $X_0=X_1, X_N=X_{N+1}$, etc. We assume $q_2 < q_1$, which in the earlier description means that Y is a more fit species than X. A spatial structure is obtained if the monomer generation, a_k, varies from box to box. This can be the case if the activation of M depends on the availability of free energy in some form.

Consider a numerical example with 10 boxes and:

$q_1=1$, $q_2=0.8$, $D_1=D_2=0.01$, $D_3= 0.25$; $a_1=1.50$, $a_2=1.50$, $a_3=1.10$ $a_4=1.05$, $a_5=0.95$, $a_6= 0.90$, $a_7= 0.90$, $a_8=0.85$, $a_9=0.80$, $a_{10}=0.70$.

Start with a situation of only X. The steady state distribution is shown in figure 1. If now Y occurs, the situation is unstable since Y can grow in boxes 4 - 8, where the M concentration is above the growth threshold for Y. Still, in boxes 1 and 2, where X is already established, Y will not be able to compete. The new steady state, obtained by the integration of (6), is given in figure 2. X is concentrated to boxes 1 and 2, and Y dominates in boxes 3 - 8. This can be interpreted as species Y taking over the fringe of the ecological niche occupied by X.

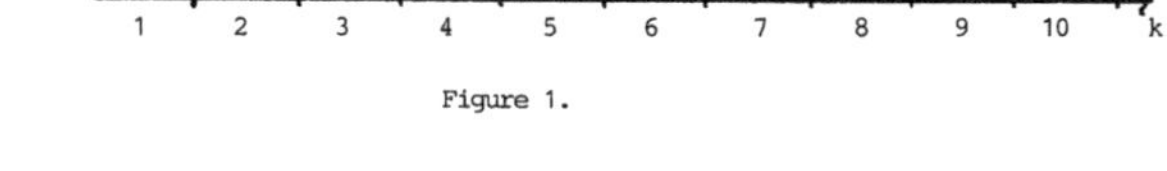

Figure 1.

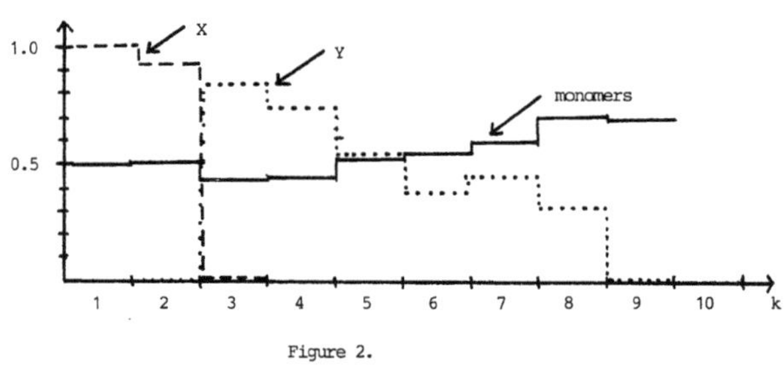

Figure 2.

A somewhat lower value for the rate of generation of M in boxes 1 and 2 will cause Y to take over the entire system, even if X is established at the onset. Thus, the effects of a spatially varying environment, represented by the variation of the a_k, can radically change the behaviour predicted from the study of one box systems. A conclusion from this analysis is that hypercycle systems may, because of their inherent features, easily organize into niches and coexist beside each other.

(vi) Finally, we consider a simple example of macromolecular coexistence based on what may be called "weak" competition. The case under consideration involves two self-replicating RNA-molecules with different gross compositions, i.e. two species only weakly competing for common resources (the four nucleotides). Our aim is to show that if, as has been asserted in e.g. ref (5), GC-rich sequences tend to replicate faster than AU-rich ones, it is not necessarily true that the latter will be eliminated.

We use the following notations:

$c_A = c_U$ concentration of A and U monomers. (assumed equal for simplicity).

$c_G = c_C$ concentration of G and C monomers.

$x_G^{(i)}$ frequency of G+C in species "i".

$x_A^{(i)}$ frequency of A+U in species "i"

$c^{(i)}$ concentration of species "i".

$N^{(i)}$ number of nucleotides in species "i".

The mean time needed to select and insert one nucleotide when replicating species "i" is given by $1/k_i'c_G$ or $1/k_i'c_A$ depending on the kind of base to be inserted. The total synthesis time for species "i" is

$$\frac{N_i}{k_i'}\left(\frac{x_G^{(i)}}{c_G} + \frac{x_A^{(i)}}{c_A}\right) \equiv \frac{1}{k_i} \tag{7}$$

With, as before, a linear rate of breakdown, the kinetic equations that describe the system are:

$$\begin{aligned} \dot{c}^{(i)} &= (k_i - q_i)c^{(i)} \\ \dot{c}_G &= k_G - k_G'c_G - \sum_i N^{(i)}x_G^{(i)}\, k_i c^{(i)} \\ \dot{c}_A &= k_A - k_A'c_A - \sum_i N^{(i)}x_A^{(i)}\, k_i c^{(i)} \end{aligned} \tag{8}$$

In calculations, we have, for simplicity, put $N_i = N$, and $k_i' = k'$ (all i), and $k_G = k_A = k_m$, $k_G' = k_A' = k_m'$. With only one species present, the steady state of (8) is easily obtained. A typical solution is shown in figure 3. $c^{(1)}$ is different from zero only if a threshold condition is fulfilled, viz.

$$q_1 < kk_m/Nk_m' \tag{9}$$

For two species, an explicit although more complicated solution for the steady state can be given. The most interesting feature of the system is that coexistence now becomes possible if the two species differ enough in their gross compositions. A representative situation is shown in figure 4.

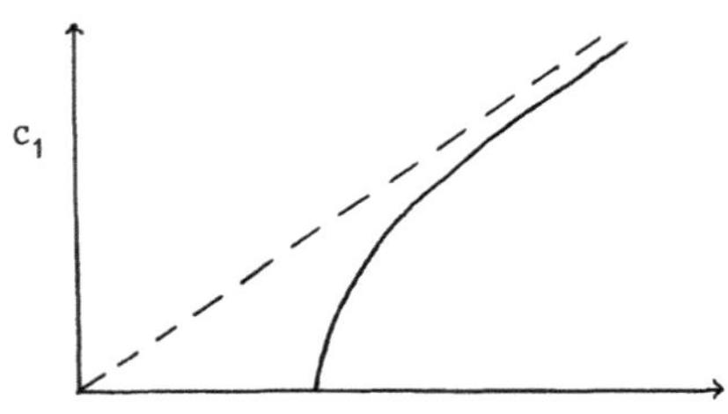

Figure 3

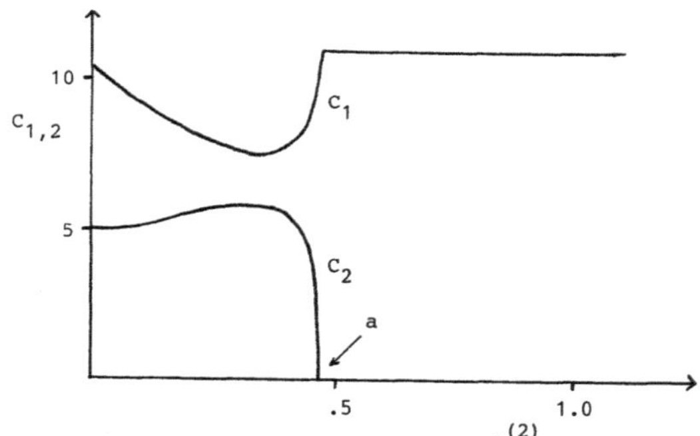

Figure 4

In figure 4, $x_G^{(1)} = 1$, $q_1 = 0.08$, and $q_2 = 0.16$. All other parameters =1. This means that species "1" is more "fit" in the same sense as in model (i). Nevertheless, if species "2" is sufficiently AU-rich, it will not be completely eliminated by species "1". (Species "1" in this case consists solely of GC-pairs). Smaller values of $x_G^{(1)}$ yield qualitatively similar results, although point "a" in figure 4 is moved to the left. If $x_G^{(1)}$ falls within a certain critical region around the value 0.5, coexistence is not possible.

Thus, it may well have been possible that GC- and AU-rich species coexisted under prebiotic conditions.

REFERENCES

(1) Prigogine,I.,Nicolis,G.,Babloyantz,A.,Physics Today,Nov 72,23.
(2) Eigen,M., Naturwissenschaften 58, 465, (1971).
(3) Eigen,M. and Schuster,P., Naturwissenschaften 64,541,(1977).
(4) Epstein,I.R., J.Theor. Biol. 78, 271, (1979).
(5) Eigen,M. and Schuster,P., Naturwissenschaften 65,7, (1978).

HISTIDYL-HISTIDINE CATALYSIS OF GLYCINE CONDENSATION IN FLUCTUATING CLAY ENVIRONMENTS

David H. White and Jeffrey C. Erickson

Chemistry Department, University of Santa Clara
Santa Clara, CA 95053, USA.

Histidyl-histidine is a remarkably effective catalyst of peptide bond formation in the reaction of glycine in a fluctuating (hot-dry, cold-wet) clay environment. It has shown turnover numbers (molecules of glycine incorporated in oligoglycines per molecule of catalyst) as high as 18 in a single cycle and as high as 52 overall. A number of other dipeptides were tested, as well as monomeric histidine, N-acetyl histidine, and imidazole, none of which showed turnover numbers greater than one. Histidyl-histidine is a model for a prebiotic protoenzyme, and implications for the development of a simple translation mechanism are discussed.

The reaction of glycine to form oligopeptides on dried clay surfaces has been shown (1) to be enhanced by fluctuations between hot, dry and cold,wet conditions, referred to as "cycling". The mechanism of peptide bond formation is not known for certain but can be postulated to involve an activated intermediate due to the fact that the lowest concentrations of glycine give the highest percent yield of product. Covalent acyl esters with clay surface hydroxyl groups are known (2); other evidence that supports this idea will be presented shortly.

Our major interest was in the effects of added peptides on the condensation of glycine, in order to provide a model for primitive protoenzyme catalysis. Some results with two different peptides are shown in Table 1. Histidyl-serine causes a slight decrease in yield, whereas histidyl-histidine causes a very significant increase. The yield seems to reach a plateau, in that increasing the amount of histidyl-histidine from 10-100 nmol gives only a small increase in yield. (Additional experiments

Y. Wolman (ed.), Origin of Life, 393–397.

TABLE 1
Effects Of Added Dipeptides on Glycine Oligomerization

Dipeptide[a]	Total Product[b]	Turnover Number[c]
none	41.6±10.3	
His-His (10 nmol)	82.9± 1.8	4.1±1.0
His-His (100 nmol)	98.2±23.6	0.6±0.3
His-Ser (10 nmol)	36.3± 2.0	
His-Ser (100 nmol)	38.9± 8.4	

[a]Reaction of 2000 nmol glycine on 10 mg kaolinite for six cycles at 94°C
[b]Total nmol glycine in oligoglycines ± standard deviation of the mean.
[c]Turnover Number = ((nmol product in presence of catalyst) - (nmol product in absence of catalyst)) ÷ (nmol catalyst)

show that 1000 nmol of catalyst actually gives less observed glycine oligomerization than 100 nmol.) The reasons for this plateau effect appear to be twofold. First, our analytical method does not normally detect products in which glycine is bound to the catalyst itself. In experiments where we have analysed for glycyl-histidyl mixed peptide products, we have found increasing amounts as the histidyl-histidine is increased, but not enough to fully explain the plateau effect. Thus, the second factor contributing the the plateau in yield appears to be the formation of only a limited amount of activated glycine which is capable of being trapped by the catalyst. Once enough catalyst is added to trap most of the activated glycine, additional catalyst does not produce more peptide.

In order to better understand the role of the catalyst, we have defined a turnover number as the moles of additional glycine incorporated into oligoglycine in the presence of catalyst divided by moles of catalyst. This gives a convenient, though approximate, measure of the number of times each catalyst molecule has carried out the reaction. Thus, with 100 nmol of histidyl-histidine, the turnover number is less than one, but with only 10 nmol it is 4.1. The fact that it is greater than one is a clear indication that histidyl-histidine is acting as a true catalyst, causing repeated reaction and by inference, being at least partially regenerated for further reaction. One would expect even better turnover numbers at lower amounts of histidyl-histidine and this is indeed what we observe. Typically, 2 nmol gives the best turnover number, with a value of 11.4 in a typical experiment.

Unfortunately, the uncertainty increases as the amount of catalyst decreases; in one case, 0.2 nmol of catalyst produced a turnover number of 44±45, which is not significant. So we observe increases in turnover number as the amount of catalyst decreases, with no maximum yet observed. Since we would expect the turnover numbers to reach a maximum as the rate of catalysis became the limiting step rather than the supply of activated intermediate, this means that we have not yet observed the maximum activity of the catalyst.

Other peptides which we have tested, not shown in the table, either gave a decreased yield of glycine oligomers (histidyl-tyrosine, aspastyl-serine and aspastyl-aspastate) or gave small increases in yield which were not statistically significant and which did not ever display turnover numbers greater than one (histidyl-lysine, glycyl-histidyl-lysine, poly-his and poly-lys). Monomeric histidine and two analogous structures (N-acetyl-histidine and imidazole) all gave small increases, which were also not statistically significant nor gave turnover numbers greater than one. Thus, histidyl-histidine is markedly more effective than any other catalyst yet tried in this system. It displays a cooperative effect between the two amino acid residues in the dipeptide much greater than that observed in those residues by themselves.

Some more recent results are shown in table 2. In this case larger amounts of glycine and clay with only 2 nmol of histidyl-histidine gave the largest turnover number we have yet observed, 52.3 after 5 cycles.

However, it is clear that the turnover numbers are increasing most rapidly during the first few cycles. The leveling off may be due to hydrolysis of some proportion of the catalyst during each successive reaction cycle. The result from 2 nmol of catalyst after a single cycle is especially interesting. The turnover number of 18.5, plus the fact that triglycine is observed after a single cycle with the catalyst but not in the control indicates that histidyl-histidine is remarkably dynamic, producing multiple reactions, even in a single cycle on the dried clay surface. The turnover rate (turnover number per unit time) for this result is 18.5 $cycle^{-1}$ or about 10 da^{-1}.

Histidyl-histidine demonstrates that even a dipeptide can have significant repeated catalysis. Thus, we suggest that it serves as a model of what primitive enzymes were like: extremely short peptides, rather than the long protein chains of contemporary biology. Histidyl-histidine may not be a prebiotically relevant compound, since histidine has not yet been observed in any monomer synthesis experiments. However, it illustrates the dynamic catalytic ability of which even dipeptides are capable.

TABLE 2
Change In Turnover Number As A Function Of Cycle Number

Histidyl-Histidine	1 cycle (1.8da)	3 cycles (5.5 da)	5 cycles (9.9 da)
	Total Product		
none	23.6±1.0	79.1±9.1	151.0±13.8
2 nmol	60.6±9.4	166.8±10.8	255.6±15.3
10 nmol	86.5±11.4	177.5±13.0	249.3±37.1
	Turnover Number		
2 nmol	18.5±4.7	43.9±7.1	52.3±10.3
10 nmol	6.3±1.2	9.8±1.6	9.8±4.0

All reactions were 20 nmol glycine on 50 mg kaolinite for the indicated number of cycles at 94°C.

It should be mentioned that the future direction of this research will be an attempt to demonstrate whether histidyl-histidine (or other short peptides) can play a role in a primitive template translation apparatus, rather than the non-templated reactions discussed above, which would presumably result in essentially random peptides from mixtures of amino acids. In the cycling clay environment, polyribonucleotides enhance the condensation of glycine as much as fourfold, whereas polydeoxyribonucleotides do not (4).

About a two-fold greater enhancement is observed with Poly G than with other bases. These results, in conjunction with other evidence, suggest but do not prove that the polyribonucleotide is serving as a direct template due to covalent bonding of amino acids to the 2-OH groups of the ribose residues. We will be investigating this system further to see whether it is capable of a crude singlet translation effect. Furthermore, the most interesting question of all is whether a simple peptide such as histidyl-histidine can play a catalytic role in such a system. As the next paper shows, neither the involvement of a catalytic peptide alone nor of a template alone is as significant for the development of a primitive genetic apparatus as their cooperative involvement. We have observed that histidyl-histidine gives further yield increases from those observed with polyribonucleotides in some cases but not in others, and we do not yet know whether this is a significant result.

ACKNOWLEDGEMENT

This work was supported by NASA-Ames University Consortium Interchange Nos. NCA2-OR685-703 and -806, and by Research Corporation.

REFERENCES

1. Lahav, N., White, D., and Chang, S., Science 201,67 (1978)
2. Theng, B.K.G., The Chemistry of Clay-Organic Reactions, Wiley, New York, 1974, p.13
3. White, D.H., and Erickson, J.C., J.Mol. Evol., in press
4. White, D.H., and Erickson, J.C., submitted for publication

A THEORY FOR THE ORIGIN OF A SELF-REPRODUCING CHEMICAL SYSTEM BY NATURAL SELECTION FROM SHORT, RANDOM OLIGOMERS

David H. White

Chemistry Department, University of Santa Clara, Santa Clara, CA 95053, USA.

ABSTRACT
A simple chemical system named an autogen, consisting of two short oligonucleotide sequences coding for two simple catalytic peptides, would be capable of genetic self-reproduction if certain boundary conditions are satisfied. Limited catalytic ability, short oligomer sequences, and relatively low accuracy are shown to be adequate for exponential increase in population due to autocatalysis in a hypercyclic organization. This theory may be capable of experimental verification, with spontaneous emergence of a self-reproducing system in the laboratory by natural selection.

The most critical event in the origin of life was the transition from chemical evolution to true Darwinian evolution due to genetic properties of informational reproduction, mutation, and selection for function.

This paper presents a theory which suggests that the first self-reproducing collections of molecules capable of both replication and translation were extremely simple and formed in a short period of time; that is, that genetic properties of life were among the most fundamental and earliest to appear. This theory is related in parts to many others, which are discussed in more detail in the full paper (1). It represents a new and extreme synthesis of the ideas that crude and inaccurate interactions with low rates among populations of very short peptides and oligonucleotides (at most decamers and possibly dimers or trimers)

Y. Wolman (ed.), Origin of Life, 399–404.

are adequate for a self-replicating system to emerge from the background of random, short oligomers.

The theory requires that an appropriate collection of monomers (amino acids and nucleotides) and usable energy be available to allow a low-yield background synthesis of essentially random short peptides and oligonucleotides. There must be a mechanism to at least partially localize collections of oligomers in the vicinity of each other. This theory was to some extent developed on the model of a fluctuating clay environment (2), and the computer calculations are based on this environment, but it is applicable to any environment in which oligomers have a reasonable probability of remaining nearby for a cycle of oligomer formation, such as micelles, liposomes, microspheres, and other mineral or dehydrated environments besides clay. It is the localization mechanism, rather than highly specific recognition processes such as Eigen proposes (3), that allow all of the molecules in the system to be reproduced cooperatively. The next requirement is for some kind of inaccurate translation effect such that peptides are statistically related to oligonucleotide templates. The previous paper suggests a possible singlet template mechanism which might conceivably be able to produce a crude translation effect in the clay environment, but any process which produces peptide sequences related to oligonucleotide sequences in greater than random abundance would be sufficient. The overall accuracy of translation need only be about 10%, and it may be that peptide catalysts can improve the accuracy of a much sloppier system (1% or 0.1%), which will be discussed below. Oligonucleotide replication must also occur and must be more accurate than translation, at least 90% overall, in order to prevent too rapid dilution of the useful sequences; a certain amount of inaccuracy can be tolerated due to the continual reemergence of the reproducing species from their non-functional neighbours. Once again, peptide catalysts might increase the accuracy of the process.

The most critical requirement for the theory is that two short peptide sequences have adequate (though very low) catalytic ability to increase the rate and possibly the accuracy of both replication and translation in comparison with the uncatalyzed rates. These peptides will be called protoenzymes. The length of the shortest peptides capable of accomplishing these tasks interacts with the accuracies of both replication and translation and with the rates of catalytic activities of the protoenzymes to produce boundary conditions for the system to be able to reproduce: the shorter the peptides, the less accuracy and rate of catalysis is required.

More than anything else, this theory will stand or fall on the question of whether short peptides acting as catalysts will provide (i) adequate rates, and (ii) adequate accuracy of translation and replication. The results with histidyl-histidine discussed in the previous paper, although that was not a translation system, suggest that short peptides <u>can</u> give adequate rates. The question of accuracy of replication and translation due to the influence of the catalyst is a more critical question. It may be that either process will be sufficiently accurate without the protoenzyme. The scanty experimental evidence that is presently available is not encouraging especially with respect to translation. It is easy to visualize the replication protoenzyme changing the selectivity of replication merely by constraining the already existing base-pairing specificities. The reasons for thinking that a similar process could occur with the translation protoenzyme are discussed in the full paper (1).

A number of catalysts are known to change reaction selectivities by factors of hundreds or more due to electronic, hydrophobic charge, or steric differences. It would be enough if already-existing weak interactions between amino acids and nucleotides can be increased by factors in the range of two to sixteen, depending on oligomer length; representing a change of only 0.5 - 1.5 kcal/mole in the transition state. The catch is that each nucleotide would have to prefer a different amino acid in order to have a translation effect. The answer to whether this is possible may be open to experimental determination in the near future.

The development of the self-reproducing system can be thought of in terms of stages of increasing complexity (not necessarily chronological). The first stage is essentially random background oligomer synthesis, as already mentioned, and the second is peptide catalysis in the absence of a template; the histidyl-histidine experiments done so far are models of this stage. The third stage is templating without catalytic peptides, representing the very crude processes of replication and translation discussed earlier. A typical oligonucleotide (N_o) is replicated through its complement ($-N_o$) and translated into a peptide (P_o) which is assumed to have no function. The lack of feedback from P_o to N_o means there is no genetic process

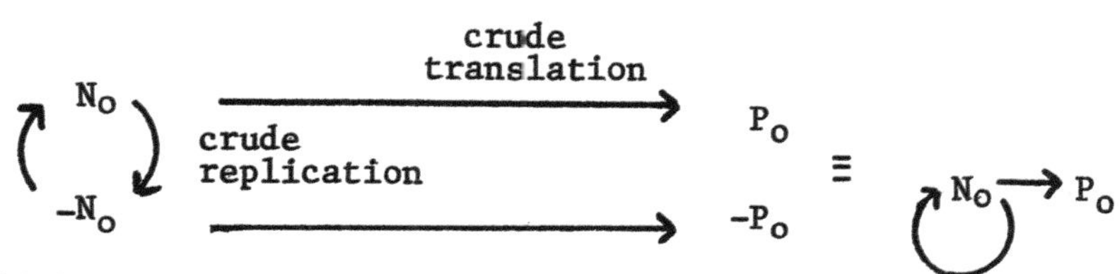

on which natural selection can operate.

The fourth stage arises from a combination of catalysis and templating, due to the existence of oligonucleotides that code by the crude translation process for functional peptides, the protoenzymes. When such an oligonucleotide, called a protogene (N_1) codes for a replication protoenzyme (P_1) and the protoenzymes that result are localized nearby, N_1 has a selective advantage.

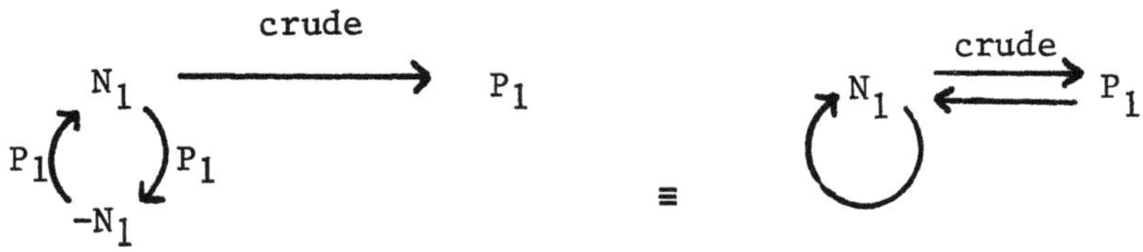

However, the combination of P_1 and N_1 is only mildly autocatalytic, since P_1 is made slowly by uncatalyzed translation. Similarly, N_2 codes for a translation protoenzyme, leading to autocatalytic growth of P_2 in the vicinity of N_2, but no selective advantage at all for N_2.

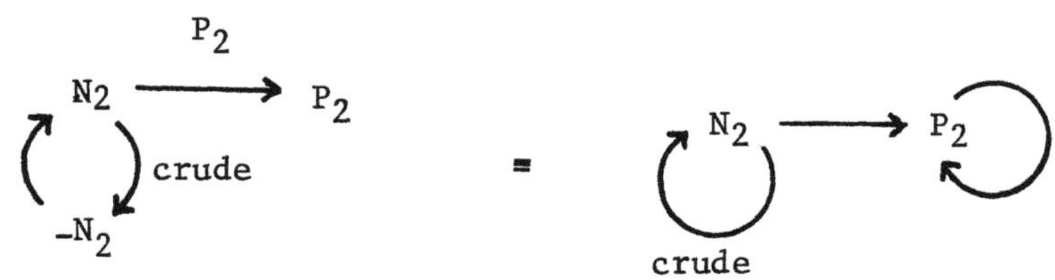

Computer modeling of stage four processes shows that N_1 and N_2 in different localized regions would have a very difficult time helping each other, with unrealistically high protoenzyme turnover rates required. A far more efficient system occurs when the two protogenes are localized in the same region. The protoenzymes act equally on both protogenes since there is no recognition; thus the replication protoenzyme catalyses the replication of both protogenes and the translation protoenzyme catalyzes the translation of both protoenzymes, including itself. Each component of the system is produced more rapidly than random oligomers, and the entire system is acted upon as a unit by natural selection. The resulting system is called an autogen (meaning

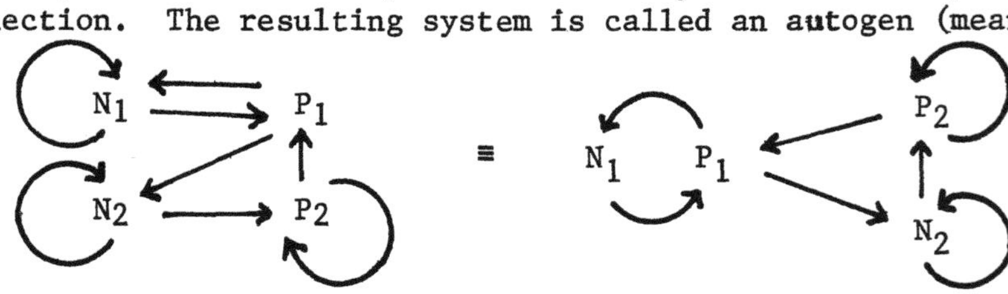

producing itself), and is to some extent related to Eigen's hypercycles (3) in being an autocatalytic cycle of autocatalytic cycles. This is not the only conceivable self-replicating system which could arise in this manner, but it seems to be the most likely candidate.

The components of the autogen can occur in the same localized region by a process of nucleation by chance from random background oligomers. Calculations based on the clay environment suggest that very short oligomers (2-5 units) could nucleate many times in a single cycle (day) in a single test tube. On the other hand, if oligomers greater than decamers are required for the necessary catalytic functions, then the nucleation is far too unlikely to see in a test tube and at slightly greater lengths become impossible to occur even on the primitive Earth over geological time periods, at least by the kind of chance combination assumed in this work.

Computer modeling shows that the autogen, once nucleated, can quickly grow to dominance over random oligomers if the boundary conditions are satisfied. Typical conditions are 90% overall replication accuracy, 10% overall translation accuracy, and turnover rates of 10-100 monomers incorporated per protoenzyme per reaction cycle, depending on the length of oligomers required. It is worth noting that histidyl-histidine shows a turnover number of 18 per cycle, which demonstrates that a dipeptide catalyst can produce the rates required for dimeric or trimeric autogen components. Although this is not yet a demonstration of translation protoenzyme activity, it does demonstrate that the required rates are reasonable. The question of whether sufficient accuracy can be achieved even for the very crude processes envisioned here cannot yet be answered.

The emergence of the autogen would make the transition from chemical to biological (Darwinian) evolution, since true genetic processes of reproduction and mutation would now be occurring. Once the autogen were nucleated and reproducing, the constraints on oligomer length would be removed, and natural selection between variants might lead to longer and more efficient oligomers. Further evolution and a possible transition to a triplet code (if the first code were singlet) are discussed elsewhere (4).

The computer calculations show that the emergence of the autogen essentially occurs either rapidly or not at all. With a turnover number of 10 for a dimer, growth to about 10% of the total oligomer mixture would be complete in 100 cycles, whereas a turnover number about half that great would fail completely. This implies that simple self-replicating systems may have emerged rapidly on the primitive Earth, with no need to postulate a long period of chemical evolution. It also implies that laboratory experiments may be possible to allow the emergence of self-replicating systems in reasonable time frames, such as several weeks. To do this would require that all of the postulated interactions were occurring simultaneously in the same environment, and no such environment is yet known. We are investigating specific aspects, particularly to model the critical behaviour of the translation protoenzyme, in the cycling clay environment; but there is no reason for thinking that this is necessarily the most appropriate environment to test this theory. A wide variety of environments will have to be tested before a final judgement can be made.

ACKNOWLEDGEMENTS

A special acknowledgement is due to Noam Lahav of the Hebrew University of Jerusalem, Rehovot, who shared the original inspiration that simple self-reproducing systems could form rapidly, and to Michael Raab for some of the computer modeling. Financial support or NASA-Ames University Consortium Interchange Nos. NCA2-OR685-708 and -806 and of Research Corporation is gratefully acknowledged.

REFERENCES

1. White, D.H., J. Mol. Evol., in press
2. Lahav, N., and Chang, S., J. Mol. Evol. 8, 357 (1976)
3. Eigen, M., Naturwiss. 58, 465 (1971)
4. White, D.H., submitted for publication

A MATHEMATICAL METHOD FOR THE ENUMERATION OF DOUBLET CODES

Colonel Georges CULLMANN

Institut Georges GAMOW
23,Rue Delarbre
63000 CLERMONT-FERRAND FRANCE

and 19,Boulevard de la Garonnette
31000 TOULOUSE FRANCE

ABSTRACT: The analysis of selection laws against nonsense mutations (1,2,3,4)is a fundamental idea in theoretical work of genetic decoding.A single nonsense mutation is a mutation of a codon outside its code,one base only changes under the mutation.

Any code is characterized by the number of single nonsense mutations which can be associated with a HAMMING 's distance in the coding theory.

The best code is the one which owns the minimum number of single nonsense mutations and,therefore, a definite HAMMING 's distance,that is the maximum number of codons which differ only in one base.It is possible by combinatory analysis to list all the codes for a given HAMMING 's distance.The present study concerns all the doublet codes with four bases.

INTRODUCTION: A fundamental idea for the understanding of the internal logic of the genetic code is represented by resistance to mutational effects(5).A static study of selection against the appearance of terminators made it possible to understand the laws of connection of the sense codons in the representative matrix of theoretical triplet codes using 20 sense codons and 44 terminators (1).This work can be developped in a more dynamical way (3,4).Several assumptions can be associated in order to obtain theoretical codes

Y. Wolman (ed.), Origin of Life, 405–413.

increasingly similar to the present genetic code (4).Thus using only two assumptions:selection against nonsense (1) and vocabulary extensions (6) the similarity becomes very stricking (4).On the other hand it is possible to propose a combinatorial approach for enumerating those theoretical codes.

Coding theory has been using since 1950 the notion of HAMMING 's distance (7) which can be applied to this problem.

Here we show the simplest example of theoretical doublet codes but this method can be extended to any length,in particular triplet codes.

ENUMERATION METHOD: A doublet code using 4 nucleotides:U,A,C and G is defined by its configuration, which is a Table including 4 boxes per line and 4 boxes per column.This Table is a matrix of order 4 (TABLE 1)

TABLE 1

X'	U 1	A 2	C 3	G 4
U 1	X	X	X	3
A 2	X	X	3	2
C 3	2	2	1	0
G 4	2	2	1	0

X

m14 = 3,m23 = 3

Each significant doublet:i.e.,assigned or sense theoretical codon is indicated by a cross in TABLE 1.A box without a cross is said to be empty (and corresponds to a nonsense doublet or terminator).The matrix of TABLE 1 is the configuration of the considered code:UU,UA,UC,AU,AA,of size 5(with five sense doublet codons).In each empty case we write the number m_{ij} of non significant single mutations(on one base):i.e.,the sum of the numbers m_i and m_j of sense doublets which are found on line i and on column j respectively of the empty boxe considered ij: $m_{ij} = m_i + m_j$

The total number $N = \sum_{i=1}^{4} \sum_{j=1}^{4} m_{ij} = \sum_{i=1}^{4} mi + \sum_{j=1}^{4} m_j$ in this example 18,which is a characteristic of the code.Two codes are equivalent when they have the same number N.As we know that the nature of a matrix,in particular N,does not change with a permutation of lines and columns,we have at least the same number of equivalent codes as possible permutations.If we are considering n columns (respectively n lines) among a are equal, b are equal but different from the preceding ones,1 is different from all the others,then the number of permutations of columns (or of lines) is:

$$R^{n}_{a,b,1} = \frac{n\,!}{a!b!1!} \quad \text{with } a + b + 1 = n$$

For the configuration 1 we have $R^{4}_{2,1,1} = \frac{4!}{2!1!1!} = 12$ permutations for the columns and the same number for the lines, that is: 12 $\times$ 12 = 144 doublet codes of size 5.

The assignement to nucleotidic bases U,A,C,G remains stable in the permutations of columns 1,2,3,4 and of lines 1,2,3,4.

In this demonstration we assume that the doublet UU must be used in each code.Then the number of codes will be less important but more difficult to determine.

In the case of configuration 1,we can only permute the last three columns and lines that is:

$$R^{3}_{1,1,1} = \frac{3!}{1!1!1!} = 6 \text{ permutations for columns}$$

and $$R^{3}_{2,1} = \frac{3!}{2!1!} = 3 \text{ for lines,that is}$$

6 $\times$ 3 = 18 codes.But as we have kept fixed the first column and line,we see on the Table 1 that there are still two acceptable configurations:the first one is obtained by permutation of the first and third columns and the second by permutation of the first and second line.We thus obtain 3 $\times$ 3 = 9 and 6 $\times$ 3 = 18 new codes,that is at all 18 + 9 + 18 = 45 codes.Finally if we note that we do not change the number N while turning the matrix around its principle diagonal XX',each column taking the place of its corresponding line and vice versa,the total number of codes of size 5 (each of them using UU) for N=18 is 45 $\times$ 2=90.

Let us note two important points.

1)If sense doublets are symetrical 2 by 2 in relation to the principle diagonal XX',the matrix is said to be symetrical. Then all the transposed matrices are included in the matrices obtained by permuting of the columns or the lines.We do not multiply by 2 the number of codes.

2)If all the doublets are on the principle diagonal XX', the matrix is said to be diagonal.For a matrix of order 4 and for a code of size p+1,the number of equivalent codes(having always the doublet UU) is:

$$\left(\frac{3\,!}{(3-p)!}\right)^{2} \times \frac{1}{p!}$$

For configuration 2(Table 2)we have :

TABLE 2

	U	A	C	G
U	X			
A		X		
C			X	
G				

$$\left(\frac{3!}{(3-2)!}\right)^2 \times \frac{1}{2!} = 18 \text{ codes}$$

A change in configuration being obtained by changing a certain number of crosses in the initial matrix,we will calculate the variation,positive or negative,of number N.Suppose we move a cross to an empty box.Let us examine the index of each significant doublet,that is the number $s_{ij} = s_i + s_j$ (sum of the numbers of the other sense doublets s_i on its line i and s_j on its column j).

The matrix 1 is written,underlining the s_{ij};Table 3

TABLE 3

	U	A	C	G	lin.
U	$\underline{3}$	$\underline{3}$	$\underline{2}$	3	1
A	$\underline{2}$	$\underline{2}$	3	2	2
C	2	2	1	0	3
G	2	2	1	0	4
Col.	1	2	3	4	

Let us calculate the variation A of N when the cross of box 12(UA) is going to box 33(CC).

$A = -m_{33} + (O_{33} = O_3 + O_3$ = number of the other empty boxes in line 3 and column 3 of box 33) + S_{12} (index of UA) - $(O_{12} = O_1 + O_2$ = number of empty boxes in line 1 and column 2 of box 12) that is:

$$A = -1 + (3 + 2) + 3 - (1 + 2) = 4$$

For the new configuration 1_{ter} (Table 4) the total number N' of non significant single mutations is:

$$N' = N + 4 = 18 + 4 = 22$$

TABLE 4

	U	A	C	G
U	X		X	
A	X	X		
C			X	
G				

The variation obtained by changing a doublet from box ij to an empty box kl is written: $A = -m_{ij} + O_{ij} + s_{kl} - O_{kl}$

When the shifting is made on the same line i or on the same column j,we respectively have:

$$A_i = -m_{ij} + O_{ij} + s_{il} - O_{il} + 2$$

and $$A_j = -m_{ij} + O_{ij} + s_{kj} - O_{kj} + 2$$

A configuration is optimal when its corresponding number N is minimal.

This calculation enables us to obtain equivalent configurations for A = 0,configurations of N' value higher or lower according to the positive or negative value of A.Let us simplifie this evaluation.Two u-plets(sequences of u letters or figures taken from a base,alphabet or given numeration system) are at the distance d if they differ on d positions (7).For u = 5, alphabet A,B,C and D,the two 5u-plets ABACC and ACBDC are at the distance 3.It is the same for 01022 and 02132 in the numeration system 0,1,2,3.It is the notion of HAMMING's distance. For a given d distance,we will call cumulated HAMMING's distance the number D(d) of couples of u-plets of a code(set of u-plets) which are at the distance d.For the triplet code of size 4;GAC,AGC,GCC and AAA,we have the Table 5 of distances.

TABLE 5

	GAC	AGC	GCC	AAA
GAC	0	2	1	2
AGC		0	2	2
GCC			0	3
AAA				0

D(1)=1 D(2)=4 D(3)=1

For a doublet code of given size and of base 4, for each cumulated distance relative to the distance 1, D(1), corresponds a particular number N.

We are going to show, for an optimal code with Nmin, that the cumulated HAMMING's distance relative to the distance 1 is maximal D max(1), which corresponds to a maximum of couples at the distance 1. Indeed let us calculate the D(1) variations when we are going from one configuration to another by moving one cross. For that let us consider successively each line i=1 to 4 and each column j=1 to 4 of the matrix relative to a given configuration.

If we have K_i crosses on line i and K_j crosses on column j, then we have:

$$h_i = \frac{Ki(Ki-1)}{2}$$ couples of doublets at the distance 1 on the line i,

and $$h_j = \frac{Kj(Kj-1)}{2}$$ couples of doublets at the distance 1 on the column j, that is:

$$D(1) = \sum_{i=1}^{4} h_i + \sum_{j=1}^{4} h_j$$

For matrix 1 we have: $h_1=3, h_2=1, h_3=h_4=0$, on the lines and $h_1=h_2=1, h_3=h_4=0$, on the columns, that is $D(1)=3+1+1+1 = 6$, which can be verified on Table 6.

TABLE 6

	UU	UA	UC	AU	AA
UU	0	1	1	1	2
UA		0	1	2	1
UC			0	2	2
AU				0	1
AA					0

Let us consider the value of the cumulated distance D(1) when we move a cross to an empty box.

If we add a doublet (a cross) to an empty box (i,j) already having K_i doublets on its line and K_j on its column, we initially had:

$$h_i = \frac{K_i(K_i-1)}{2} \quad \text{and} \quad h_j = \frac{K_j(K_j-1)}{2}$$ respectively on the line i and the column j.

After adding a doublet into the box ij we have:

$$h'_i = \frac{(K_i+1)K_i}{2} \quad \text{and } h'_j = \frac{(K_j+1)K_j}{2} \quad \text{,that is:}$$

$$\begin{cases} h'_i = h_i + K_i \\ h'_j = h_j + K_j \end{cases} \quad \text{since } \frac{K(K-1)}{2} + K = \frac{(K+1)K}{2}$$

If we remove a doublet(a cross) from the line K and from the column 1,already having K_k and K_1 doublets,we previously had:

$$h_k = \frac{K_k(K_k-1)}{2} \quad \text{and} \quad h_1 = \frac{K_1(K_1-1)}{2}$$

After cancelling the doublet we have:

$$h'_k = \frac{(K_k-1)(K_k-2)}{2} \quad \text{and} \quad h'_1 = \frac{(K_1-1)(K_1-2)}{2} \quad \text{,that is:}$$

$$\begin{cases} h'_k = h_k -(K_k-1) \\ h'_1 = h_1 -(K_1-1) \end{cases} \quad \text{since } \frac{K(K-1)}{2} -(K-1) = \frac{(K-1)(K-2)}{2}$$

So,if we move a cross from the box Kl to an empty box ij, we can calculate the increase variation of D(1):

$$A' = \left[(h'_i + h'_j) + (h'_k + h'_1)\right] - \left[(h_i + h_j) + (h_k + h_1)\right]$$

$$= + (K_i + K_j) - \left[(K_k-1) + (K_1 -1)\right]$$

and if we note that: $K_i + K_j = m_i + m_j = m_{ij}$ and that:

$(K_k-1) + (K_1-1) = s_k + s_1 = s_{kl}$,then we can write:

$$A' = m_{ij} - s_{kl}$$

If this movement is carried out along a line or along a column,we respectively have:

$$A'_{line} = m_j - s_1 \quad \text{and} \quad A'_{column} = m_i - s_k$$

We are going to show that this variation A' is equal to the half of those of N for the same movement,but with an inverted sign.

$$A' = -\frac{A}{2} \quad \text{thus} \quad N_{min} = D_{max} \quad (1)$$

In fact $A = -m_{ij} + O_{ij} + s_{kl} - O_{kl}$

Now for a given matrix of order n (here n=4),we have:

$$O_{ij} = O_i + O_j = n - (m_i+1) + n - (m_j+1)$$
$$= 2n - m_{ij} - 2$$
$$O_{kl} = O_k + O_l = n - (s_k+1) + n - (s_l+1)$$
$$= 2n - s_{kl} - 2$$

and then $O_{ij} - O_{kl} = -m_{ij} + s_{kl}$,i.e.:

$$A = -2m_{ij} + 2\,s_{kl} = -2\,A'$$

Now we can write a configuration with the help of(m_i,m_j) and (s_k,s_l) and make movements (hand made or by computer) which can obtain:

Either equivalent configurations A'=0,eventually using permutations relative to the first column and to the first line,which makes research easier,but also the cumulated HAMMING's distance.

Or configurations for which $A' > 0$ or $A' < 0$ trying each time to increase or decrease A' of 1 if it is possible,otherwise of 2,...etc.

Matrix 2 will be symbolized on Table 7.

TABLE 7

	U	A	C	G
U	2<u>1</u>	2<u>1</u>	2<u>0</u>	30
A	1<u>1</u>	1<u>1</u>	21	20
C	02	02	01	00
G	02	02	01	00

The movement of doublet UA in the empty box AC gives a configuration equivalent $A' = (2+1) - (2+1) = 0$.We obtain the same result by changing UC in AC: $A' = (2) - (2) = 0$.The movement of AA to UG gives an optimal configuration:

$$A' = (3+0) - (1+1) = +1 \qquad A = -2$$

It should be noted that we always obtain an optimal configuration of a code of given size by placing the doublets on the first line for sizes less than 5,then on the second line for size between 5 and 8,..etc.

To count all the codes of a given size according to the different values of N,we start from an optimal configuration N_{min} and we look for all the possible equivalent configurations, then we look at N_{max} + 2,i.e.; A'=-1 if possible,and we look for all the possible equivalent configurations and so on.

The enumerating of all those codes is then made searching for all the permutations.

CONCLUSION: This enumeration for triplet codes is possible.

The study of the structure of optimal codes(i.e.; those which best resist against the appearance of terminators under the influence of random mutations) shows the importance of the connections of the different sense codons in relation to the notion of cumulated HAMMING's distance which is another formulation of the concept of BUÇCION.

This study puts in evidence the particular shielding against noise of the present genetic code.

REFERENCES

1 SONNEBORN T.M. in:"Evolving Genes and Proteins",Eds.:BRYSON V. and VOGEL H.J. Academic Press:New York (1965),377
2 LABOUYGUES J.M.in:"Contribution critique et mathématique à l'étude du code génétique"Thèse Doctorat d'Etat en Médecine Université de CLERMONT-FERRAND - FRANCE (1973)
3 LABOUYGUES J.M. Agressologie 17 ,6,329 (1976)
4 LABOUYGUES J.M. and FIGUREAU A.The 6th International Conference on the Origins of Life,This volume.
5 LABOUYGUES J.M. 145 th Meeting of the American Association for the Advancement of Science,Houston,Abstracts 123(1979)
6 CRICK F.H.C. Nature (London)213,119 (1967)
7 HAMMING R.W. The Bell System Technical Journal 26,2,147(1950)

EVOLVING NUCLEOTIDE BINDING SURFACES

Thomas Kieber-Emmons and Robert Rein

Roswell Park Memorial Institute, Buffalo, New York
14263 U.S.A.

We present an analysis of the stability and nature of binding of a nucleotide to several known dehydrogenases. Our approach includes calculation of hydrophobic stabilization of the binding motif and its intermolecular interaction with the ligand. We follow the evolutionary changes of the binding motif by calculating the Euclidean deviation of the respective dehydrogenases. We then discuss the possible structural elements involved in the origin of nucleotide recognition by non-coded primordial polypeptides.

A structurally well-conserved nucleotide binding motif has been observed in dehydrogenases, kinases, flavodoxins and recently a synthetase (1,2). Such conservation suggests that this binding surface has evolved from a common ancestrial origin and is a representative result of gene fusion (1). Walker (3) has recently suggested a primordial sequence which may have formed a fundamental beta unit within the ancestrial binding motif. We address ourselves to the evolution of such a binding surface and the characterization of its fundamental structural elements and binding ability.

The analysis of relations between sequence and secondary structure shows that beta strands can be formed from a limited number of amino acid residues which include those suggested to be in abundance under primordial conditions (4). Sheet formation therefore only requires the connection of strands by intervening loops. It is of interest to note that topographical assessments of known protein structures (5) have indicated that intrastrand regions of 10 residues or less have a strong statis-

Y. Wolman (ed.), Origin of Life, 415–422.

tical preference to form antiparallel connections while intra-strand lengths of 15 to 25 residues are about three times more likely to make the parallel form. Recent studies on residue interactions which lead to the observance of β structures indicate that about one-third of all residue contacts in β parallel sheets are made by Val and Ile (6). This is approximately the proportion of these contacts found in the beta sheets in the AMP sites. In addition, Walker has suggested that these two residues were likely to be prime primordial precursors.

An evolving binding surface would however experience selective pressures such that amino acid substitutes would occur. To determine how varied the beta structures within the motif have become due to residue substitutions, a quantitative measure of their evolutionary distance as a function of their coding requirements is needed. Such a measure is provided by calculating the Euclidean deviation (7) of a set of residues.

We have calculated this deviation (δ(s)) (Eq. 1), for the beta sheet components of the AMP binding site of several binding proteins, (Table I).

Eq. (1) $$\delta(s) = \left[\sum_{i=1}^{n} \left(S_{(i)} - \frac{100}{61} D(i)\right)^2\right]^{\frac{1}{2}}$$

In Eq. 1 S(i) is defined as the percent frequency of the ith amino acid and D(i) its codon degeneracy.

It is seen from Table I that for the AMP site in the dehydrogenases the sheets have with the exception of Horse LADH, essentially stabilized very soon somewhere between subtilisin and the dehydrogenases (Horse LADH may be an exception due to localized radiation effects).

In order to further clarify how the overall integrity of the beta sheets have changed with evolutionary time we have estimated the interaction energy upon strand association (Table II). Hydrophobic (8,9) and non-covalent components (10,11) were calculated upon Van der Waals contact of the respective strands.

It is seen from Table II that the effects of substitutions as expected appear. However of significance is that even with substitutions the stabilization energies are fairly similar. This implies that for the most part residues that have been substituted are essentially equivalent in their stabilizing performance. Overall these results further indicate that the original beta structure constituted a sound molecular building block for utilization as a binding or catalytic surface.

TABLE I
Euclidean Deviations of AMP Binding Sheets

	δ(s)	time x 10^9 yrs ago(12)
Walker Peptide	32.45	3.2 - 4.5(3)
Subtilisin	30.66	3.2 - 4.5
Dogfish LDH	23.57	1.5 - 3.2
Pig GAPDH	23.31	.6
Bovine GLUDH	22.84	.3
Lobster GAPDH	22.10	.6
Yeast GAPDH	21.17	1.2
Horse LADH	17.56	1.5 - 3.2

TABLE II
AMP Binding Site Sheet Stabilities In Kcal/mole

	Hydrophobic	Non-Covalent		
		βA-βB	βB-βC	TOTAL
Lobster GAPDH	-23.5	-18.73	-22.45	-64.68
Dogfish LDH	-29.5	-29.06	-23.41	-81.97
Horse LADH	-25.2	-22.18	-17.56	-64.94

A survey of binding and active sites of various enzymes shows that when they occur in association with beta sheets, they tend to be located at the sheet edge rather than on their surfaces. Based upon this observation, Brack and Orgel (13) have suggested a possible model for a catalytically active primordial enzyme consisting of an anti-parallel sheet with a beta turn. This simple type of model, while sufficient for a catalytic proto-enzyme, is not entirely adequate for a proto-nucleotide binding surface.

From geometric considerations it is difficult to readily formulate a simple model for a binding motif whose basic structural component is an antiparallel sheet. Primarily, the difficulty arises in regard to the suitability of the conformation of the connecting loop or turn in ligand binding. This requires an intervening loop of somewhat longer length and implies that a functional binding motif may begin to emerge involving right-handed cross-over connections of 15 residue lengths to form a parallel sheet (Fig. 1). It is of further interest to note that recent studies involving an antiparallel nucleotide binding motif have in fact indicated that for the most part the antiparallel form exhibits weak binding but the motif does show relatively strong catalytic activity.

It is apparent from the previous discussion that the basic component of the binding surface would be a beta sheet comprised of parallel strands, with ligand binding occuring at an edge rather than at the sheet surface. Fig. 2 shows the spatial arrangement of the beta sheets contained in the AMP site for several dehydrogenases with the ligand in its binding location. It is evident that binding occurs at the sheet edge, with the strands running parallel. Fig. 2 shows the syn position of the adenine moiety in the "green" subunit of GAPDH(a) while the anticonformation is shown for LAD(b) and LADH(c).

It is readily seen that the Aspartic acid at the sheet edge in all cases is able to form a hydrogen bond with the sugar C2'-hydroxyl hydrogen (13). However, the magnitude of course depends upon various factors.

In Fig. 2 we include the residues which have been suggested to line the general Adenine binding pocket (1). It is important to note that of the suggested residues more than half in all cases comprise the sheet edge of strands A and B directly or residues immediately following this edge. In addition, many of the suggested equivalent residues which are considered not part of the generalized pocket but take part in binding directly (14) are as well located within several residues of the sheet edge comprised of strands A and B. It appears then that residues associated with strands A and B play a more important role in binding the ligand directly than strand C. However, from Fig. 1 it is as well apparent that strand C contributes to the formation of a cross-over connection from strand B. This analysis further suggests that a well-defined hydrophobic pocket may not have been necessary in a primordial binding surface and that the observed present day general pockets are a result of structure refinements with evolutionary time.

In Table III we show the component interaction energies between the ligand and the respective significant residues (14). Detailed inspection of this table indicates that with the exception of Aspartic acid of position 7, attractive London forces are dominant and not very specific while the electrostatic components seem to be alternating between attraction and repulsion. Because of the arbitariness of the assignment of the dielectric 2 to the electrostatic calculation, we in addition performed a calculation with an effective dielectric of 10. Such a scaling procedure may in fact be considered more representative so as to include the effects of bulk water at the sheet edge as well as counter ion condensation and screening of the respective charges. This also has the advantage that due to limited crystallographic accuracy, errors attributed to close contacts are reduced.

TABLE III

SELECTED EQUIVALENT RESIDUE-LIGAND COMPONENT ENERGIES

	GAPDH			LDH			LADH		
	βA ASP 6(6)-E			VAL 27			PHE 198		
	ELEC	NON-BOND	TOTAL	ELEC	NON-BOND	TOTAL	ELEC	NON-BOND	TOTAL
P	9.53	-.06	9.47	.09	-.02	.07	-.02	-.01	-.03
S	-1.88	-.21	-2.09	.34	-1.36	-1.02	.06	-.17	-.11
A	1.69	-1.32	.37	-.52	-2.84	-3.36	-.12	-.19	-.31
			7.75			-4.31			-.45
	βA GLY 7(7)-E			GLY 28			GLY 199		
P	-.45	-.43	-.88	-.15	-.09	-.24	-.28	-.03	-.31
S	.13	-2.11	-1.98	.12	-2.18	-2.06	.06	-.79	-.73
A	-.22	-2.81	-3.03	-.23	-1.34	-1.58	-.09	-.09	-.18
			-5.89			-3.88			-1.22
	βA PHE 8(8)-L			CYS 29			LEU 200		
P	1.22	-.41	.81	.68	-.38	.30	-.02	-.02	-.04
S	-.36	-3.17	-3.53	-.28	-1.99	-2.27	-.08	-.30	-.38
A	.37	-.50	-.13	.05	-.17	-.12	-.07	-.05	-.12
			-2.85			-2.09			-.54
	βA ARG 10(10)-L			ALA 31			GLY 201		
P	-8.77	-.22	-8.99	-1.67	-.79	-2.46	-.84	-.04	-.88
S	-.45	-.15	-.60	.13	-.16	-.03	.08	-.05	.03
A	-.68	-.09	-.77	-.02	-.01	-.03	-.004	-.005	-.009
			-10.76			-2.52			-.86
	βA ILE 11(11)-L			VAL 32			VAL 203		
P	-.94	-.37	-1.31	-.93	-.35	-1.28	-.54	-.03	-.57
S	.10	-.05	.05	.14	-.19	-.05	.07	-.03	-.10
A	-.06	-.02	-.08	-.10	-.03	-.13	-.04	-.06	-.10
			-1.34			-1.46			-.77
	βB ASN 31(6)-E			VAL 52			VAL 222		
P	-.27	-.01	-.28	.14	-.004	.136	.24	-.005	.235
S	.06	-.20	-.14	-.14	-.13	-.27	-.12	-.16	-.28
A	-.10	-2.42	-2.52	.21	-.57	-.36	.16	-.19	-.03
			-2.94			-.50			-.08
	βB ASP 32(7)-E			ASP 53			ASP 223		
P	11.53	-.06	11.47	9.41	-.02	4.39	10.45	-.03	10.42
S	-5.62	-3.95	-9.57	-2.87	-2.79	-5.66	-3.38	-3.31	-6.69
A	3.33	-3.18	.15	3.73	-2.69	1.04	2.79	-.92	1.87
H-BOND		-3.63	-1.58	H-BOND	-.35	4.42	H-BOND	-.29	5.31
	PHE 34(9)-L			MET 55			ASN 225		
P	.46	-.04	.42	.40	-.04	.36	-.50	-.03	.47
S	-.46	-1.12	-1.58	-.49	-1.62	-2.11	-.65	-1.49	-2.14
A	.17	-.30	-.13	.14	-.37	-.23	.19	-.21	-.02

Table IV represents the total energies obtained by this procedure and the non-bonded values from Table III. These values further demonstrate that the less specific dispersion forces are dominant and sufficient for binding in the rudimentary model.

The remainder of Table IV further depicts the evolutionary contribution to enhance binding by the formation of the general hydrophobic pockets. This component is realized upon pocket formation (upon folding) and ligand insertion.

TABLE IV
SUMMARY OF LIGAND BINDING ENERGY

	Total Elec	Total Non-Bonded	Hydrophobic
Lobster GAPDH	1.21	-23.02	-45.69
Dogfish LDH	.818	-20.13	-40.94
Horse LADH	1.17	-8.10	-31.95

Our results then indicate that a beta sheet of a minimum of 2 parallel strands connected by a 15 residue loop would be sufficient to bind a nucleotide. In so much that it has been argued by others that primordial residues can form beta sheets it seems plausible that nucleotide binding to these sheets could have originated without a code.

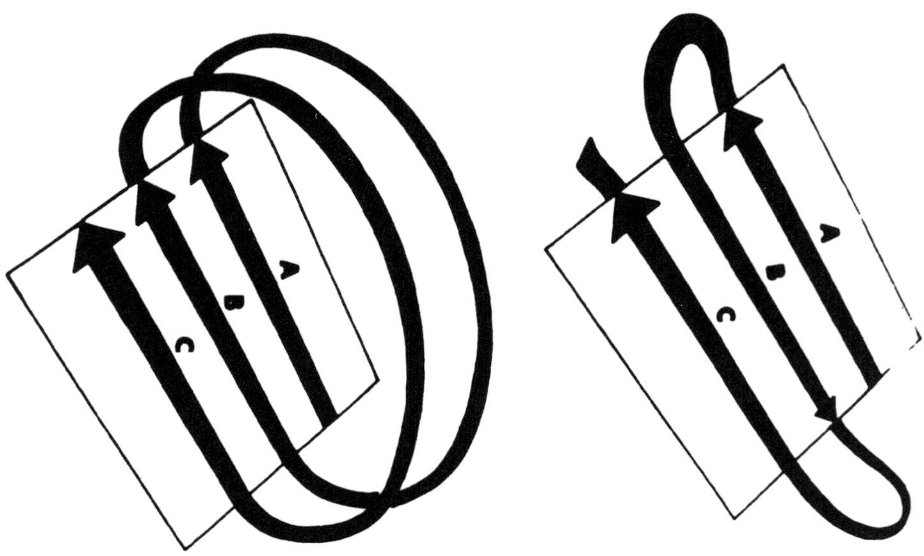

Figure 1

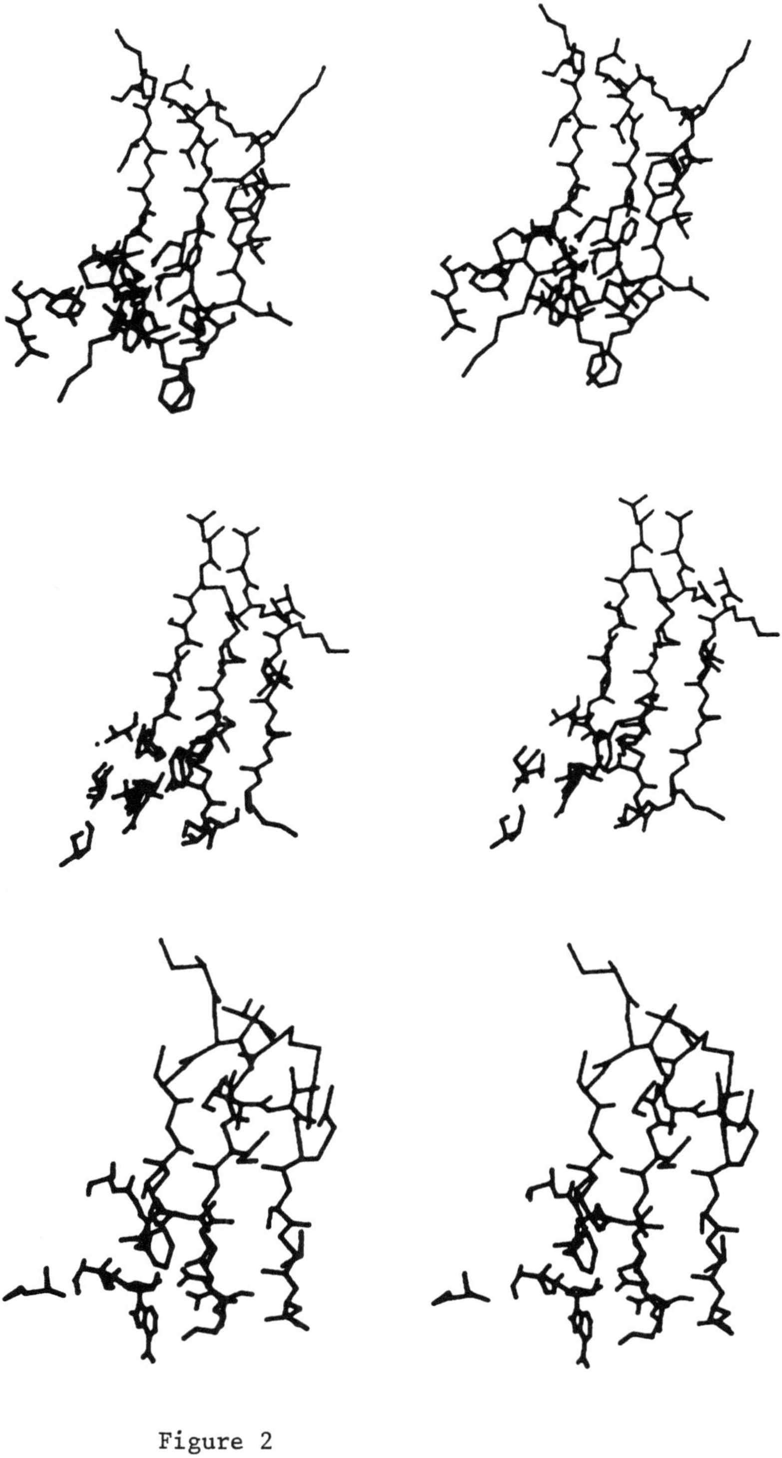

Figure 2

ACKNOWLEDGEMENT - This work was supported by NASA Grant NSG-7305. We thank R.Parthasarathy for Crystallographic co-ordinates obtained from the Protein Data Bank. We thank Cheryl McGuire for her technical assistance and Deborah Ross for her fine typing of this manuscript.

REFERENCES

1. Rossmann, M.G., Liljas, A., Branden, C.I., Banaszak, L.J. "The Enzymes" Boyer, D.D. ed. Vol. XI, Academic Press, N.Y., 1975, 61.
2. Irwin, M.J., Nyborg, J., Reid, B.R., Blow, D.M. J. Mol. Biol. 105:577(1976).
3. Walker, G.W.R. in Origin of Life, Noda, H. ed. Center for Academic Publications Japan/Japan Scientific Soc. Press, Tokyo, 1978, 479.
4. Robson, B., Suzuki, E. J. Mol. Biol. 107:337(1976).
5. Sternberg, M.J.E., Thornton, J.M. J. Mol. Biol. 110:285 (1977).
6. Lifson, S. and Sander, C. Nature 282:109(1979).
7. Hiller, J. J. Theor. Biol. 74:599(1978).
8. Lee, B., Richards, F.M. J. Mol. Biol. 55:379(1971).
9. Chothia, C. Nature 248:338(1974).
10. Momany, F.A., McGuire, R.F., Burgess, A.W., & Scheraga, H.A. J. Phys. Chem. 79:2361(1975).
11. Hopfinger, A.S. in Conformational Properties of Macromolecules, Academic Press, N.Y., 1973, 38.
12. Rossmann, M.G., Moras, D., Olsen, K.W. Nature 250:194(1974).
13. Brack, A. & Orgel, L.E. Nature 256:383(1975).
14. Olsen, K.W., Moras, D., Rossman, M.G., & Harris, J.I. J. Biol. Chem. 250:9313(1975).

ORIGIN AND EVOLUTION OF THE GENETIC CODE

Mikio Shimizu

Institute of Space and Aeronautical Science
Univeristy of Tokyo
Komaba 4-6-1, Meguro-ku, Tokyo, 153, Japan

ABSTRACT A hole on a complex of anticodon and discriminator base on a tRNA has a key hole - key relation to the side chain of the corresponding amino acid. The protein synthesis system might have evolved so that it could have continuously utilized the above stereochemical discrimination scheme. 'Half tRNA' might have been the form of the primitive tRNA and the primitive ribosome might have had a hairpin structure with two CAAGR on both sides of it.

The dictionary of 'nucleic acid language', the genetic code, has experimentally established in early 1960's. However, a central question on the genetic code, why is there a correspondence between codons and amino acids, is not yet answered, although it has been known that the information transfer mechanism from DNA to protein contains mRNA, tRNA, ribosome, aminoacyl tRNA synthetase (will be referred to synthetase later on), etc..

Indeed many questions occur naturally at the first glance of the genetic code.

(1) Why are the codons composed of three nucleic acids and not of two or five ones?
(2) How had the twenty kinds of protein amino acids been selected?
(3) Why is the pentose in tRNA is of d type and the amino acids of l type in optical activity?
(4) Why do the third codons frequently degenerate?
(5) Why is serine coded by completely different sets of bases such as UC and AG?
(6) Lysine has four $-CH_2-$ between the amino radical and α

Y. Wolman (ed.), Origin of Life, 423–430.

carbon, but no protein amino acid with one to three $-CH_2-$ exists. Why?

(7) Why are the amino acids, whose second codons are U or C, all non polar?

etc. etc.

By using the CPK molecular model, we have suggested a possibility (1) that a hole on a complex of four nucleotides (C4N) composed of anticodons and the corresponding discriminator base discussed by Crothers et al. (2) is in a key hole- key relation to the side chain of the corresponding amino acid (Fig. 1 and Fig. 2). The nucleic acid bases in the anticodon are assumed to be in anti form and to use their three hydrogen bonds (one bond from each base) to combine with three sites on the discriminator base. Therefore the genetic code should be written in terms of three bases. Protein amino acids were selected from more than 100 amino acids to fit the above key-key hole relation. This also restricts the d-l relation between pentoses and amino acids. Sometimes, the first anticodons slightly contribute to the form of the hole. Consequently the third codons are degenerated in some cases.

There is a possibility that the hole may be occupied by another false amino acid which does not exactly correspond to the C4N but the structure of whose side chain is chemically similar to the true one. In reality, it is known that the living system has frequently made mistakes in discriminating amino acids. For instance, a confusion among valine, isoleucine, and methyonine etc. occurs (3).

It was found recently that the mitochondria has a slightly different genetic code from that of E.Coli (4). For instance, UCA in the genetic code of the mitochondria specifies tryptophane, instead of terminator in the E.Coli case. In the latter case, the discriminator of terminator may be G as for tryptophane. By changing it to U as in the case of cysteine, however, the anticodon AGU (corresponding to UCA codon) can discriminate tryptophane. Therefore, the existence of 'ancient' and 'modern' nucleic acid language is explained in terms of discriminator evolution (eventually related to a change in the enviroment).

An experimental evidence (5) for the above C4N model is already discussed in my previous paper (1). Another supporting evidence is a good correlation between the hydrophilicity of the protein amino acid and that of the corresponding anticodon (or dinucleotides) (6) (7). Crudely speaking, this means that, the larger the hole, the larger the side chain of the corresponding amino acid, in accord with our C4N theory.

We are studying the physical interaction between the C4N and the amino acid by taking into account the Van der Waals force, electrostatic force, hydrogen bond, etc., in a computor experiment.

FIGURE 1. C4N of UUA≡G with asparagine on it. The discrimination base is shown only in the form of nucleic acid base. The CCA bond which starts from the nucleotide bond of the discriminator base is so flexible that it can easily reach the peptide bond of the amino acid at any configuration on the C4N. The acceptor stem of tRNA is located in the lower right corner of the figure without disturbing the aboce configuration.

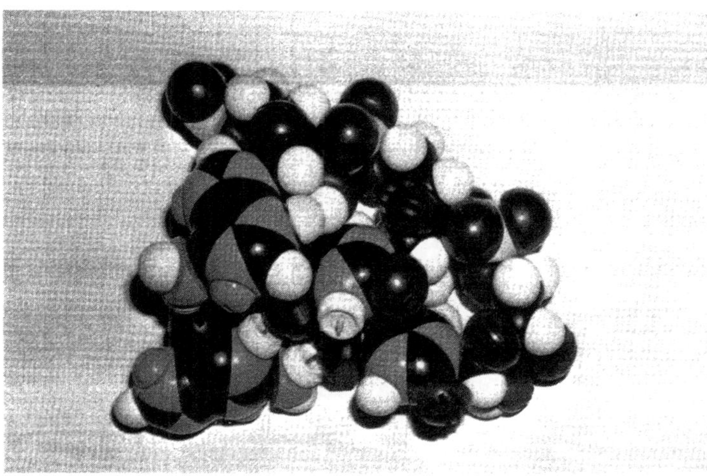

FIGURE 2. C4N of UUA≡G without asparagine.

In the present structure of the tRNA, the anticodon is separated far away from the discriminator base. However, there is a possibility that a 'half tRNA' is an ancestral form of tRNA. Hopfield (8) has already pointed out such a possibility, proposing a cage formed by the anticodon and the CCA tail to enclose the corresponding amino acid. His discussion may not directly be accepted, since the CCA tail is not written in the gene and the presence of discriminator base has been neglected. It is likely, however, that there is a symmetry between 3' and 5' end sides of tRNA (9). We shall adopt the idea of 'half tRNA' in the following discussions, since it can easily form a C4N on its end.

The primitive ribosome might also have simpler form than the present one. It is argued by using the statistical analysis (Ohnishi, 1980, a private communication) that proteins in the ribosome might be derived from a common ancester. Spirin (10) argued that, eventually, no protein were contained in the primitive ribosome. 5SrRNA is known to have two conformation causing a switching mechanism to move tRNA from one side to another on the ribosome (11). On the other hand, the structure of 16S rRNA is composed of many 'hairpin' base sequence. A careful study of the structure reveals that these hairpins might have a common ancester of a hairpin form, which has CAAGR (corresponding to YCψTG common to all T loops of tRNA) on both sides of it. It is likely that this primitive hairpin rRNA had attracted the two T loops on the half tRNAs on mRNA as to combine the amino acids on the tails of the half tRNAs to form peptide bond. See Fig. 3.

The synthetase is known to be a dimer composed of two same monomers. The monomer appears to have been the primitive synthetase at the stage of 'half tRNA'. It would have worked to raise the efficiency to form C4N on the half tRNA. Later on, the primitive synthetase might have combined in a anti-parallel way to become dimer (the present tRNA). This is to form two C4Ns on the dimer, by catching two L letter type tRNA on it, as explained in our previous paper (1). Ref.(15) supports the idea.

Before the stage of half tRNA- primitive synthetase-primitive ribosome (stage 2), there might exist a more primitive tRNA, which might have composed of an anticodon loop plus a discriminator base. In this period, no T loop existed on tRNA and therefore the presence of any form of synthetase and ribosome was unnecessary. This stage may be called as the stage 1.

It is usual to assume a simple form similar to that proposed here for the primitive tRNA (for instance, in (12), but note that we take here the stereochemical theory). In this case, the discrimination of amino acids is done by nucleotides. If the discrimination at present is done by the synthetase alone, there should have been a revolutional change in the discrimination mechanism in past, from nucleotides to protein. This is difficult, although not impossible. Consequently, the usage of

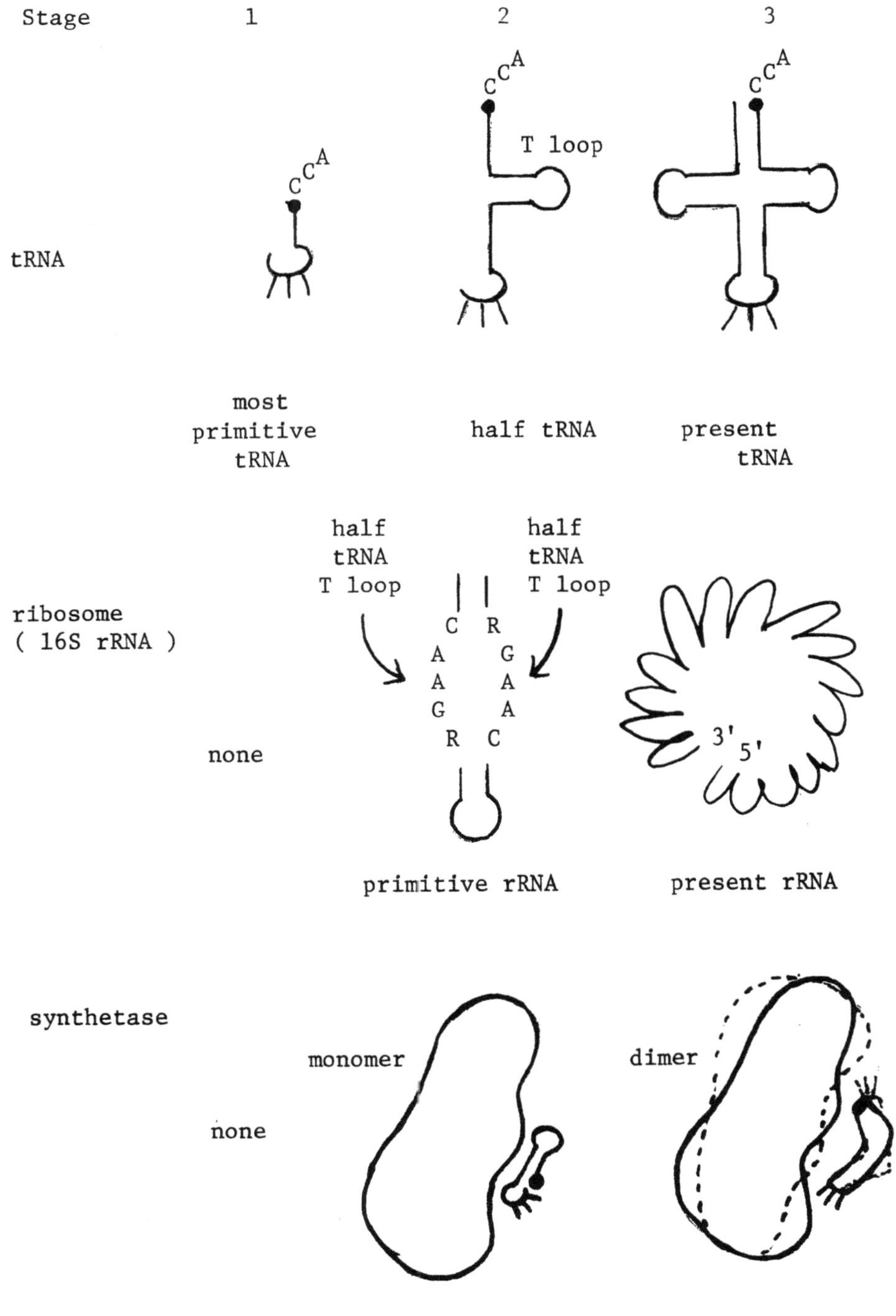

FIGURE 3. Evolution of Protein Synthesis System in Three Stage

the C4N throughout the evolution of protein synthesis system appears to be more natural.

It may be claimed that the above scheme is too simple. For instance, it was found that the precursor of tyrosine tRNA of yeast contained a long leader at 5' end and a complex form of anticodon loop (13). However, it was found later that the splicing of these base sequence occurred in the cell nucleus alone (14). It appears that these base sequence would have been deviced to avoid the attack by poisonous oxygen gradually accumulated in the enviroment.

Our proposed scheme for the evolution of protein synthesis system is not unique even within the C4N scheme. The scheme could be improved in future, to deciphur the genetic code and to study the origin of life. The evolutional route of the genetic code will be discussed in another paper in this conference.

Addendum   C4N on the synthetase

The structure of the tRNA-enzyme-amino acid complex is still a matter of debate. By using a laser scattering data (16) and an NMR data (17), the presence of an U type conformation of the tRNA has been proposed. In this case, the 3' end and anticodon become close to each other (even without the tRNA synthetase). However, the data and the interpretation has not yet been confirmed. The crystal structure of tyrosyl-tRNA synthetase (15) has almost been established by the X ray analysis. On the other hand, the form of tRNA-enzyme complex has been under an active discussion. The number of tRNAs on the enzyme (dimer) could be one to four according to X ray and neutron small angle scattering experiments (18, 19). Kim (20) proposed a complex form by coinciding the pseudo twofold axis of tRNA with the exactly twofold axis of enzyme. However, the model is not in accord with the low angle X ray scattering work by Osterberg et al. (18) and neutron one by Giegé et al. (19). We emphasize that the enzyme has a dimer structure, the monomer being combined in an antiparallel way. Why so? It is most easily interpreted in terms of the C4N theory: The usage of crossed two L type tRNAs on the synthetase is necessary to form C4N's by the present living system.

We have already confirmed that all the CO and NH radicals in the side chain of amino acids, if they exist (as asparagine, arginine, glutamic acid etc.), as well as the NH_2 radical of the peptide bond can always find its recombination sites on the C4Ns to form H bonds (just in the exact position, no excess and no deficiency), stabilizing the amino acid-C4N complex. If the side chain is composed of C and H alone, the corresponding hole on the C4N is so deep that the Van del Waals forces can work well. It is found that a tryptophanyl tRNA, whose anticodon has changed from ACC to AUC, thus becomes to be a sort of terminator tRNA although no tRNA for terminator exists, can discriminate (not only tryptophane but also) glutamine (21). This is easily be explained in terms of the C4N theory, since the hole on AUC≡G has the same conformation as that on GUC≡G for glutamine. The inactivation and the new discrimination ability of glutamic acid of tyrosyl tRNA by the replacement of its discrimination base from A to G (22) is also interpretable in terms of the C4N theory. These phenomena are the evidences to confirm the C4N model. It is also possible to interpret the problem of 2' OH and 3' OH specificity in tRNA aminoacylation (23) in terms of geometry of adenine at 3' end to the carboxyl radical of the amino acid. We predict that the role of the synthetase is to form the C4Ns on it, rather than to recognize the amino acid and tRNA independently by its two recognition sites and to bring them together. The author thanks to Dr. S. Nishimura of National Cancer Center for giving some newest informations on this research field.

Origin of the optical activity

The optical activity of nucleotides cannot be racemic as a carrier of heredity information. Then, pentose in the nucleotides, and thus saccharides in general (hexoses converts to pentoses by oxidation), should also be of either l or d type. The combination of optical activities of protein- nucleic acid is restricted to be either l-d or d-l in the C4N model. We emphasize that the tRNA is the most typical interaction area of the amino acid with nucleotide. Finally the choice of an l-biosystem (l-protein and d-nucleic acid and saccharides) to a d-biosystem is a matter of chance with a high probability of 1/2 (not 1/4 by virture of the C4N model).

It is also to be noted that the C4N model choose the ribose for the nucleotides, not the arabinose, xylose, nor hexoses, due to the stereochemical reason.

REFERENCES

(1) Shimizu, M., Proc. Japan Acad. Ser. B., 55, 387 (1979)
(2) Crothers, D.M., Seno, T., and Soll, D.G., Proc. Natl. Acad. Sci., USA, 69, 3063 (1972)
(3) Wetzel, R., Origin Life, 9, 39 (1978)
(4) Hall, B.D., Nature, 282, 131 (1979)
(5) Hayashi, H., and Miura, K., Nature, 278, 137 (1966)
(6) Weber, A.L., and Lacey, J.C., J. Mol. Evol., 11, 199 (1978)
(7) Jungck, J.R., J. Mol. Evol., 11, 211 (1978)
(8) Hopfield, J.J., Proc. Natl. Acad. Sci. USA, 75, 4374 (1978)
(9) Eigen, M., and Schuster, P., Nature, 7, 341 (1978)
(10) Spirin, A.S., Origin Life, 7, 109 (1976)
(11) Jagadeeswaran, P., and Cherayil, J.D., J. Theor. Biol., 83, 369 (1980)
(12) Crick, F.H.C., Brenner, S., Klug, A., and Pieczenik, G., Origin Life, 7, 389 (1976)
(13) DeRobertis, E.M., and Olson, M.U., Nature, 278, 137 (1979)

(14) Melton, D.A., De Robertis, E.M., and Cortese, R., Nature, 284, 143 (1980)

(15) Irwin,M.J.,Nyborg,J.,Reid,B.R.,and Blow,D.M.,J. Mol. Biol. 105, 577 (1976)

(16) Olson, T., Fournier, M.J., Langley, K.H., and Ford, N.C., J. Mol. Biol. 102, 193 (1976)

(17) Reid, B.R., In Nucleic Acid-Protein Recognition, H.J. Vogel ed. (New York, Academic Press) p. 375 (1977)

(18) Osterberg, R., Sboberg, B., Rymo, L., and Lagerkvist, U., J. Mol. Biol. 99, 383 (1975)

(19) Giege, R., Jacrot, B., Moras, D., Thierry, J.C., and Zuccai, G., Nucl. Acid. Res., 4, 2441 (1977)

(20) Kim, S.H., Nature, 256, 679 (1975)

(21) Yaniv, F., Folk, W.R., Berg, P. and Soll, L., J. Mol. Biol. 86, 245 (1974)

(22) Shimura, Y., Aono, H., Ozeki, H., Sarabhai, A., Lamfrom, H., and Abelson, J., FEBS Letters, 22, 144 (1972)

(23) Hecht, S.M., Tetrahedron, 33, 1671 (1977)

THE ORIGIN AND EVOLUTION OF THE GENETIC CODE

Alain FIGUREAU[+] and Jean-Michel LABOUYGUES [++]

[+] Institut de Physique Nucléaire de LYON
Université Claude BERNARD
69622 VILLEURBANNE Cedex FRANCE

[++] Institut Georges GAMOW
23,Rue Delarbre
63000 CLERMONT-FERRAND FRANCE

ABSTRACT: We are studying a quantitative model which suggests that the present genetic code appeared under the influence of mutations and evolved in optimizing its resistance to their effects.The number of sense codons grew progressively whereas the number of terminators decreased.The most important constraint is due to the selection against the appearance of nonsense codons.The competition between ancestor codes favorizes those which best resist mutational effects leading to nonsense. The structure of the selected systems converges towards that of the present genetic code.

INTRODUCTION

The idea that the genetic code has an internal logic is very attractive and has inspired many efforts in various works (1,2,3,4,5,6).However no theoretical model has been able to obtain the general characteristics of the present genetic code. Some fundamental ideas have nevertheless been expressed,which we group and exploited in a coherent scheme allowing us to reach this goal.In particular from the notions of selection against mutations leading to nonsense or BUCCION (2,4,5),of vocabulary extension (7) and of synonymy (8),we deduced a relatively small number of theoretical codes which have a structure ressembling that of the present genetic code.

Y. Wolman (ed.), Origin of Life, 431–437.

Some years ago, the study of theoretical codes using 20 sense codons and 44 terminators or nonsense triplets made it possible to show that the repartition of these two categories among the 64 possible triplets strongly influenced the resistance against nonsense mutations of these systems, the most important variable being the relative connection between the 20 chosen codons (2,9). This study was criticized because it was not very selective and because a very large number of codes having optimum resistance were possible (10). Furthermore, the notion of an evolution for the code, such as was later considered (3,5,9), remained unanswered. Our model precisely incorporates these two basic postulates (connection of codons in relation with mutations and evolution with time) in a quantitative manner, where mutations play a dual role: they are at the origin of vocabulary extensions and they cause the operation of the selection against non-optimized systems. We will here describe the simplest version of this model that can be imagined, other variations will be the subject of a forthcoming publication.

We are postulating that the genetic code has evolved from an initial state in which few aminoacids were translated. The number of sense codons increases under the pressure of mutations, the resultant vocabulary extension being selected by optimization against the occurence of terminators.

The association of these postulates leads to a relatively small number of possible codes. These have a priori no relation with those that would be selected in a static model, i.e. in which the 20 codons are directly chosen. Our step by step optimization does not automatically lead to the same results as the global, direct optimization.

Our scheme is nevertheless very general and many calculations are feasible, which depend on the detailed postulates chosen for the evolution of our code. We present below the simplest one, which allows us to deduce the general structure of the present genetic code. Consequently the nearly absolute universality of the mechanisms for protein translation seems to result from the optimization of resistance against the effects of mutations.

MODEL FOR THE EVOLUTION OF THE GENETIC CODE:

We will now formulate the detailled hypothesis which allows us to make a calculation: they have been chosen for their simplicity and could easily be modified if experimental facts or theoretical developments made them obsolete.

1) The vocabulary extensions are made codon by codon, from a theoretical code consisting of a unique sense primitive codon.

2) Single base mutations play a dominant role: we will neglect the effects of more complex errors in the translation mechanisms.

3)These point mutations on nucleotides are equiprobable.

4)At each step of the evolution,the various sense codons are used with the same frequency.

We will then define BUCCION as the probability that a given set of (point) mutations of any sense codon gives another sense codon belonging to the same code (5).

In this first model,we will not consider the relation between aminoacids and codons,and will leave aside all possible hypothesis for the mechanisms of synonymy.This study is to be compared to the general concept of HAMMING's distance (11).At the same Meeting Colonel Georges CULLMANN will present the matrix calculation for the simplest doublet theoretical codes (12),but this other formulation is applicable to any length of n-uplets,in particular for triplet codes.Each step of our model consists then in selecting from many possible codes (which consist of an initial code previously selected enlarged by any codon not yet translated)that or those which have the largest BUCCION.These optimized codes are then considered as the initial codes for the next step.let us note that we are not considering the possibility for more than one code to coexist and that we are looking for all the alternative solutions,the choice between them being left for more detailed hypotheses.We determine the structure of all these codes and measure their resistance against the occurence of terminators.As this study is relatively complex,the calculation was made by a computer.

As an illustration of our scheme,we first present in TABLE 1 the simple example of one of the choices which are made in the first step of our calculation:having arbitrarily taken the codon UUU as the only codon in the code.We show 2 (among the possible 63) codes with 2 sense codons and 62 nonsense triplets. Their 18 single mutants are also shown.The comparison between the two figures indicates the role of the relative position of codons in the matrix.In our model,the code on the right will be selected.Eight more in the 63 will have the same BUCCION. For all of them the enlargement consists of the addition of one of the 9 "neighbours" of UUU (Note that this neigbourhood directly depends on the choice of dominant mutations).

TABLE 1

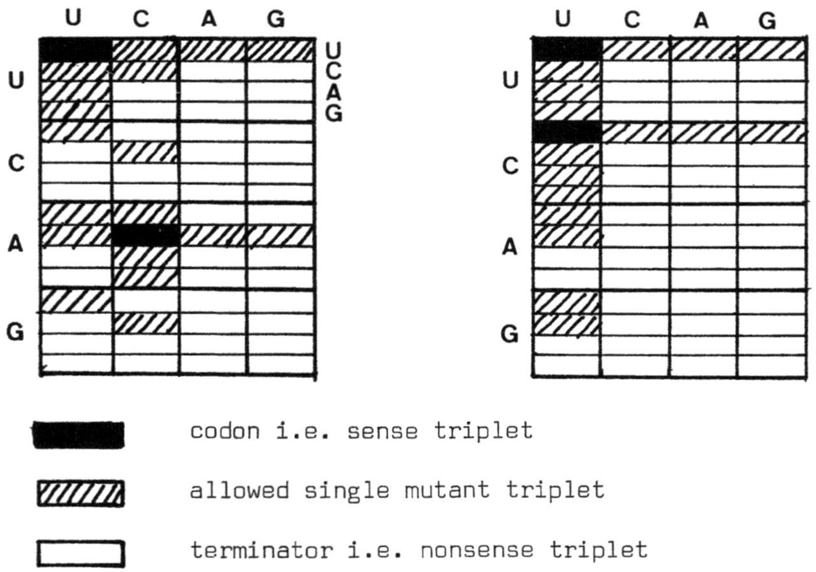

In TABLE 2,we give now the number of optimized codes in each step of the extension process,when UUU has been chosen as the initial state,and the value of the corresponding BUCCION. We have arbitrarily written only the 20 first numbers,but the calculation can continue untill 64 codons have been used for the code.We notice the relatively low values for the number of codes.The largest,936,is to be compared to $C_{64}^{20} \simeq 10^{34}$, which represents the total number of possible codes with 20 codons,built without any constraint.The reduction is considerable.

The numbers in TABLE 2 do not depend on the initial codon;in fact,many codes found from UUU are also selected if one starts with another codon.For instance,if we consider the 64 different calculations,we find in the end only 12 optimized codes with 16 codons,which all have the same structure:in each of them,one of the three positions is occupied by the same base for all the 16 codons,the two other positions giving all the necessary latitude.It is then immediatly obvious that only 4 of these 12 solutions,i.e.:those for which the third base is fixed,may,after the addition of four more codons,be compared favorably with the present genetic code.The hypothesis that the third base has a particular role allows then to divide the number of our solutions by 3.

TABLE 2

Number of codons	Number of codes	Detailled BUCCION
2	9	2
3	9	6
4	3	12
5	72	14
6	180	18
7	93	24
8	207	32(18);28(189)
9	522	36(144);34(378)
10	306	42
11	684	50(144);42(540)
12	639	60(18);56(621)
13	402	66(78);64(324)
14	126	74
15	45	84
16	3	96
17	144	98
18	576	102
19	336	108
20	936	116(72);112(864)

In TABLE 3 we present as an example one of the 312 solutions with 20 codons, elaborated from the UUU initially chosen (in fact it can be constructed from any initial state with an uracil in the third position).

TABLE 3

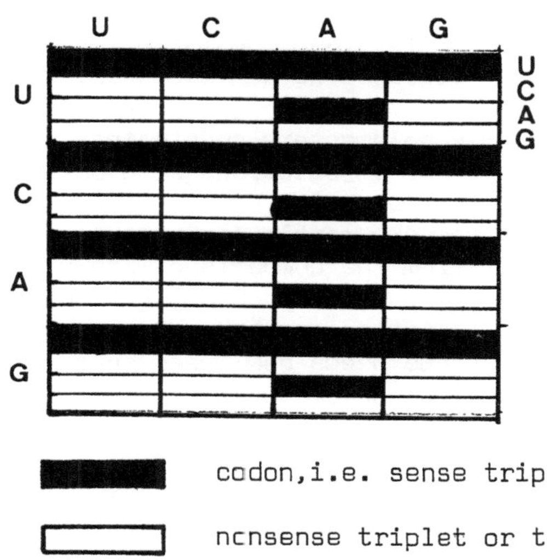

This example is one of the two solutions which show the best similarity to the present genetic code.18 different aminoacids are represented by at least 1 of their codons.The 310 others have slighly less overlap with the actual code, though their structure is identical (when BUCCION is 116),or very similar (when BUCCION is 112) to that represented in TABLE 3.This difference is related to the asymmetry of the genetic code which contrasts with the high symmetry of our assumptions.The relatively irregular structure that is observed will be obtained only if our hypotheses on the code evolution are modified accordingly:we will show in a future study some changes we made and their effect on our present results.Already, we have seen above that the third base had apparently a particular role,if included in our calculation,this constatation allows us to get rid of codes in which the 1st or the 2nd base play the same role.These precisely give the worst approximation by far to the genetic code.

Finally,for each optimized code,we can study how each sense codon resists mutational effects,i.e.,what proportions of sense codons and of terminators are found among its mutants (by single or point mutations,in this model).For the theoretical code presented in TABLE 3,one finds that 4 codons resist best (those ending with AU,which give by mutation 7 sense codons and 2 nonsense triplets)and 4 others worse (those ending with AA,which give 4 sense codons and 5 terminators),the remaining 12 giving 6 sense codons and 3 terminators.One can then infer that some codons might play a particular role,possibly not tractable with our simplified assumptions.

CONCLUSION: We have seen that our scheme,based on two fundamental postulates:the progressive extension of the translated vocabulary,and the maximization of resistance to nonsense mutations,allowed us to find theoretical codes with a structure very similar to that of the present genetic code. The consideration of supplementary constraints should lead to a reduction in the number of solutions with optimized resistance. Modification of the initial state or of the evolution laws, creation of palindromes are particularly envisaged.On the other hand,we have assumed from the comparison with the present genetic code that each new anticodon may give access to a new aminoacid.Other possibilities are however open,and the complete understanding of the code needs new ideas on the synonymies. For instance,the association of our scheme and of the PreWobble hypothesis (13) should point to ambiguities in the occurence of some families of codons.

Finally the existence of more sophisticated mechanisms can also be considered,such as dynamic recognition,detection, repairing and correction against mutational effects,in which the supernumerary codons of Arginine and Serine would play a role (5,9).

These complements to our model presented in this work are necessary,and we began the systematic extension of our calculations to more complex assumptions.However it is remarkable that using such simplistic hypotheses as presented here allow us to obtain results with general characteristics very similar to those that were found looking at the "Cell-free system" results.

ACKNOWLEDGEMENTS: We thank Mrs M.GOURGAND and Colonel G.CULLMANN for useful discussion.

This work is dedicated to the memory of Mrs Louis BÉRARD nee Magdou GARANÇON and to George GAMOW who introduced the general idea of logical "codology"in Genetics.

REFERENCES

1 GAMOW G.,RICH A. and YČAS M. Advan.Biol.Med.Phys.4,23(1956)
2 SONNEBORN T.M. in:"Evolving Genes and Proteins" Eds.:BRYSON V.and VOGEL H.J. Academic Press:New York(1965)377
3 CRICK F.H.C. J.Mol.Biol.38,367(1968)
4 LABOUYGUES J.M. in:"Contribution critique et mathématique à l'étude du Code Génétique"Thèse Doctorat d'Etat en Médecine Université de CLERMONT-FERRAND FRANCE (1973)
5 LABOUYGUES J.M. Agressologie 17,6,329(1976)
6 BERTMAN M.O. and JUNGCK J.R. J.of Heredity 70,379(1979)
7 CRICK F.H.C. Nature(London)213,119(1967)
8 CRICK F.H.C. J.Mol.Biol.19,548(1966)
9 LABOUYGUES J.M. 145 th Meeting of AAAS,HOUSTON Abst.123(1979)
10 WOESE C.R. in:"The Genetic Code"Eds.:HALVORSON H.O.,ROMAN H.L. and BELL E.Harper and Row,Publishers:New York,Evanston and London(1967)
11 HAMMING R.W.The Bell System Tech.J.26,2,147(1950)
12 CULLMANN in the 3 rd Int.Meeting of ISSOL,This volume
13 LABOUYGUES J.M. Agressologie 18,4,173(1977)

We are glad to publish in JERUSALEM the work initiated by Tracy SONNEBORN.We know that the SONNEBORNS were a help to ISRAEL during its dramatic creation(Golda MEIR's memories).

EVOLUTIONARY PROCESSES OF THE GENETIC CODE

Tamaji Noguchi

Tokyo College of Economics, Kokubunji, Tokyo, Japan

Evolution of the genetic code has been analysed by dividing it into seven stages according to the perceptivity of codon-anticodon interaction. One of our basic rules, which is supported by evidences, may have answered the question that what cause is responsible for the difference in the coding behavior between pyrimidines and purines. We also have obtained the evolutionary trees of tRNAs which would tell us how the tRNAs are related with one another.

The evolution from the primitive tRNAs up to the sophisticated protein synthesizing apparatus involving ribosome, aminoacyl-tRNA synthetase and so on, may be evaluated on the basis of two fundamental abilities: i.e., "codon-reading ability," and "amino-acid-recognizing ability." Concerning the codon-reading ability, we will set forth three basic rules, which may enable us to classify the evolution of this ability into seven stages according to the degree of perceptivity of codon-anticodon interaction. Contrarily, the experimental knowledge about the amino-acid-recognizing ability is scarce, preventing us from considering such a classification. Hence, we are compelled to deal with this aspect more or less intuitively.

A molecular "quasi-species" defined by Eigen and Schuster (1), i.e., a population of replicating molecules whose sequences are distributed around one or several most fitted sequences, would arise if the preciseness of replication reached the critical value. I will adopt this concept as one of the bases of our discussion. I also suppose that the background of the earliest stage of the evolution was as follows: (1) the common ancestral RNA had already

Y. Wolman (ed.), Origin of Life, 439–446.

differentiated into primitive mRNAs and primitive tRNAs; (2) the mRNAs were long chain molecules and provided the region where amino acids were assembled by the help of tRNAs; (3) the tRNAs, the most important molecular species in the evolution of the genetic code, had to constitute a quasi-species, so the selfreplication of these molecules was the crucial process in the earliest stage; (4) since no specific enzymes could exist in this period, it is natural to suppose that random polypeptides would have played the part of "replicase" of these molecules; (5) we do not specify whether the scene was on the surface of clay or inside the protocell. Along these lines we will adopt the presumption that in this stage the selection pressure directed the choice of amino acids toward those having capability to accelerate polymerization.

Let us investigate into the evolution of the codon-reading ability. At first, we set forth three basic rules as follows which will be used for classifying the codon-reading ability: (1) the number of hydrogen bonds in the first two codon positions shall be considered the primary criterion; (2) the base pair composed of pyrimidine in the codon and purine in the anticodon is more stable than the converse case in the last two codon positions (see the following paragraph); (3) the third base of the codon shall be read only in later stages.

Rule 2 needs some remarks. As a matter of fact, I introduced this rule speculatively, intending to explain the difference in coding pattern between pyrimidines and purines. But, later, I have found some evidences which support this rule. First, the recent success in identifying the sequences of viral genomes as well as eukaryotic mRNAs made it possible to unveil which bases are actually used in the degenerated codons. I examined the third bases of five amino acids which have 4-fold degeneracy and also 4-fold degenerated part of three 6-fold degenerated amino acids.

The genome of ϕX174, a single-stranded DNA bacteriophage, was sequenced by Sanger et al.(2). As the mRNAs of this phage are shown to be the same strand as the genome, base T of the genome sequence shall be replaced by U in the mRNA sequences. We can now examine which bases are used as the third base of their 4-fold degenerated codons. Using the data listed in Table 4 in reference 2, we have the conclusion that U, C, A, and G are used 340, 136, 77, and 112 times, respectively. Hence, prymidines are used 476 times, while purines only 189 times. In case of MS2, a RNA bacteriophage, the pyrimidines and purines we are concerned with are found in the proportion of 326 : 238 (3). Similarly, in SV40, a simian virus, the corresponding proportion is 342 : 236 (4).

The same phenomenon is observed in some vertebrate mRNAs as well. In case of chicken ovalbumin the proportion has been found to be 92 : 72 (5). In case of rabbit and human β globins, it has been shown to be 44 : 32 in either case (6). To summarize, all the six cases so far examined have shown the preference of pyrimidines over purines in the third position of 4-fold degenerated codons. This outcome is congruent with the statement of rule 2.

Second, I will refer to the experiments on template synthesis, in particular, those conducted in Orgel's laboratory (7). It has been shown there that poly U and poly C forms helical structures with complementary monomeric or short oligomeric purine nucleosides or their derivatives. On the contrary, the converse synthesis, i.e., with a poly A or poly G template, have not been successful. It is concluded from these experiments that pyrimidines are preferable to purines as template bases. So, if we recognize that both templates and mRNAs are playing the same role of "information donor," this conclusion provides another evidence for the validity of rule 2. The reason for the choice of pyrimidine as the directing part and purine as the directed part may be ascribed to the former's more flexibility in conformation under the constraint of information carrying molecules.

Based on the three basic rules described above, we now classify the codon-reading ability of tRNAs into seven stages. In the first stage, we assume that codons could be recognized by tRNAs when either G or C occupied the first two codon positions, meaning that the number of hydrogen bonds formed in the first two positions was equal to six. The third base could not be read at all. So, 16 triplets, i.e., GGN, GCN, CGN, and CCN, where N denotes any one of the four bases, fulfill the requirements. However, because of inability of the tRNAs to read the third base, they recognized these 16 codons as though they were four distinct species. So, amino acids which could be encoded by these tRNAs was four at the most. We will call "effective codons" the codons which can be

TABLE 1
Seven Stages in the Evolution of Codon-Reading Ability

Stage	(1)	(2)	(3)	(4)	(5)	(6)	(7)
Hydrogen bonds at first two positions	6	5	5	5	4	4	/
First position	G,C	N ( )	N ( )	N ( )	U,A	U,A	/
Second position	G,C	Y	R	R	U,A	U,A	/
Third position	/	/	$\bar{Y}$	$\bar{R}$	$\bar{Y}$	$\bar{R}$	R

$\bar{Y}$: pyrimidines not discriminated between U and C, $\bar{R}$: purines not discriminated between A and G, ( ): constrained by the number of hydrogen bonds.

TABLE 2
"Effective Codons" and Encoded Amino Acids

(1)	(2)	(3)	(4)	(5)	(6)	(7)
gly (GGN)	leu^{1} (CUN)	asp (GAY)	glu (GAR)	phe (UUY)	leu^{2} (UUR)	ileu^{2} (AUA)
ala (GCN)	val (GUN)	his (CAY)	gln (CAR)	tyr (UAY)	term^{2} (UAR)	met (AUG)
arg^{1} (CGN)	ser^{1} (UCN)	ser^{2} (AGY)	term^{1}* (UGR)	ileu^{1} (AUY)	ileu^{2}* (AUR)	term^{1} (UGA)
pro (CCN)	thr (ACN)	cys (UGY)	arg^{2} (AGR)	asn (AAY)	lys (AAR)	trp (UGG)

Superscript numbers indicate whether the encoding was the first or second one in certain multiply encoding cases; star marks mean tentative encoding which would afterwards be reduced to half by "invasion."

discriminated by tRNA. Then, we may say that there existed four effective codons in the first stage. The question why specific amino acids were chosen so as to be bound with these effective codons is another question which will be discussed later on.

In the second stage, adopting rule 2, we assume that only pyrimidines were allowed for as the second base. We will denote Y pyrimidines. The first base could be any one of the four bases so far as the number of hydrogen bonds formed in the first two positions was five. We denote such a constraint as ( ). The third base could not be read again. Hence, in the second stage four more codons, i.e., CUN, GUN, UCN, and ACN became effective. At the third stage, those codons which had a purine in the second and pyrimidine in the third position became effective, although the third pyrimidine could be read only as a pyrimidine in general, i.e., not being discriminated between U and C. We denote $\bar{Y}$ pyrimidines when recognized as a pyrimidine in general. The newly added codons were GAY, CAY, AGY, and UGY.

It is important to note that these codons as well as those which would be introduced in the next stage (see Table 1&2) have a purine in the second position and twofold degeneracy in the third position. In contrast, the effective codons in the second stage have a pyrimidine in the second position and fourfold degeneracy in the third position. These considerations may answer the question why most of the codons having pyrimidines in the second position have fourfold degeneracy and codons having purines in the same position have twofold degeneracy. In other words, fourfold and twofold degeneracies originated from different stages in the evolution of codon-reading ability. We also view that the sixfold degeneracy originated from twice encodings which are accompanied with fourfold and twofold degeneracies, respectively. The superscript numbers in Table 2 indicate whether the encoding concerned was the first or the second one.

Following the rules indicated in Table 1, we obtain effective codons four by four in every stage from the third through the sixth (Table 2). The number of effective codons at the end of the sixth stage was 24 (4×6), so if the encoding was done to the best advantage, 23 amino acids and one terminator would have been realized.

Concerning the seventh stage, some special remark is needed. After a brief discussion along the continued line, we will take up certain strange phenomena occurred specifically in the seventh stage. The codon-reading ability was further improved in this stage and it became possible to discriminate between the two purines, i.e., A and G, in the third position. The effective codon AUR which had coded for ileu at the sixth stage differentiated into AUG and AUA, coding respectively for met and ileu. Likewise, UGR which had coded for term differentiated into UGG and UGA, coding respectively for trp and term. However, it should be noted that here the encoding occurred in a way of "invasion" by a stranger amino acid. Furthermore, note that the invaded codons have a guanine in their third position. So, we will call this phenomenon "guanine invasion." No alike events had occurred in the previous stages.

The second strange phenomenon was a sudden "freezing" in the evolution of the genetic code. The following case may be in accord with this view. That is, it is known that GUG, a valine codon, codes for the initiator in certain cases. A reasonable interpretation of this fact would be that the freezing occurred when the process of "guanine invasion" was just going on. The freezing occurred so abruptly that we cannot help believing that it was forced by some outside accident. It might be understandable if the evolution had terminated either at the end of the sixth or the seventh stage. In the latter case "guanine invasions" might have occurred 16 times in all.

Now we will take up another important question, i.e., why specific amino acids were chosen to bind with the effective codons which were available at that time. Why at the first stage gly, ala, pro, and arg were chosen among a wide range of species then existed in the primitive soup. The molecular background of that time was described earlier, in which we pointed out the important role of random polypeptides as a "replicase" of primitive tRNAs.

This idea led to the conclusion that the selection pressure must have directed the choice of amino acids toward those which have a capacity to accelerate polymerization. Then, suppose that the four amino acids, gly, ala, pro, and arg, were accumulated in a narrow region near the mRNA, where oligomers such as gly-pro-arg may have been formed perhaps by the help of certain inorganic catalysts. It is known that the sequence gly-pro-arg is the active site of fibrinogen α chain, which will be exposed after fibrinopeptide A is cut off by the attack of thrombin, leading to the formation of a peptide bond between two side-chains of α and β chains, i.e., glutamine and lysine, respectively. So, it may be suspected that this basic oligomer has a capacity to accelerate polymerization. Furthermore, in cases of fibrous proteins such as collagen, it is shown that they comprise an abundance of these four amino acids. These proteins are mostly large molecules with stable α helical structures. So, these findings together with the above one make it reasonable to suspect that this set of amino acids were selected for the capacity to accelerate polymerization. Here we have a degenerated quasi-species of tRNAs having four most fitted sequences.

The encoding process of other amino acids may be as follows: (1) a mutation occurs somewhere in the sequences which are related to the function of amino acid recognition, leading to the binding with a moderately different amino acid; (2) codon-reading ability may be potentially improved either by the evolution of tRNA itself or by introducing new devices, but its anticodon still remains unchanged; (3) a single mutation in the anticodon changes it into one which can read one of the effective codons belonging to the next stage; (4) the coexistent state differentiates into two distinct species of tRNAs, one coding for exclusively the predecessor amino acid and the other for the recent one. These processes constitute one fundamental step of the evolution of tRNA, in other words, that of the genetic code.

The evolutionary trees of tRNAs are shown in Figure 1. Each tRNA is represented by the amino acid and the codon which are recognized by it. Time is shown only in its order indicated by the numbers of the stage. Each dotted arrow shows one fundamental step of the evolution of tRNA as described above. Those tRNAs which are connected by the arrow may recognize moderately different amino acid, i.e., limited changes due to one mutation, yet detectable by the ability of tRNAs of that time. Their anticodons, hence their codons, are related with each other by a single base change. Those tRNAs which are enclosed by dotted rectangles belong to the same stage of codon-reading ability. Therefore, they must have some structual similarity with each other so that they can recognize the codon-anticodon interaction with the same subtleness. This may be achieved, not phylogenetically, but competitively in such a way as "convergence."

FIGURE 1: Evolutionary Trees of tRNAs

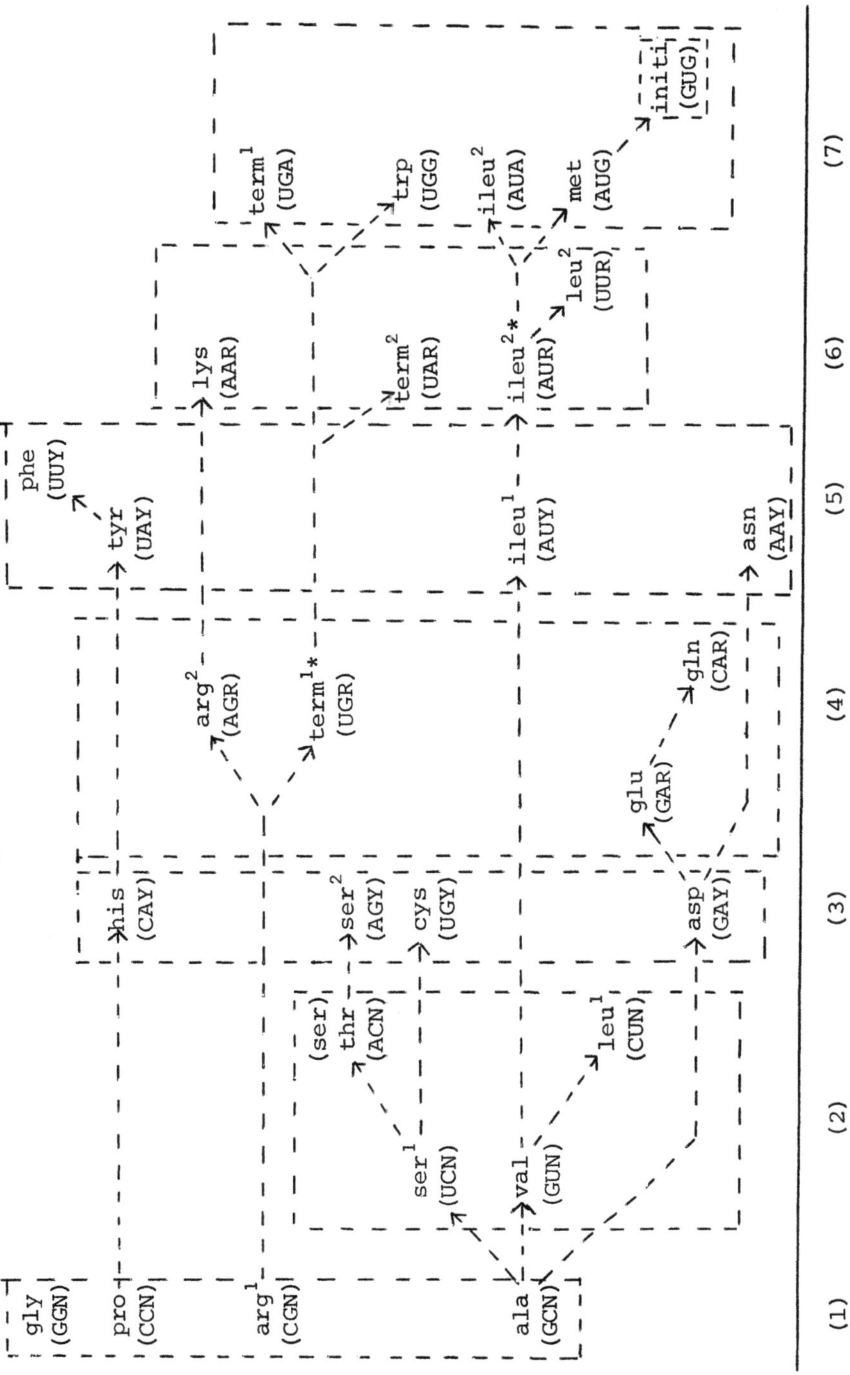

The figure provides a number of suggestions which may be teste by laboratory experiments. It may also be useful to interprete some features observed in the contemporary genetic code. Let us conside the origin of 6-fold degeneracy. First, let us discuss the encoding of argenine. Arg^1 with 4-fold degeneracy was discussed earlier. Arg^2 is related to arg^1 with one fundamental step, however this ste was an isolated case where mutation failed in catching a different amino acid. The failure may perhaps be due to the scarceness of ot species of similar size in the primitive soup. In case of serine, the tRNA recognizing ser^1 was derived from that of alanine and enco ed with 4-fold degenerated codons. After another fundamental step threonine became recognizable. However, in those earlier stages the only available apparatus was the primitive tRNA which may not have discriminated serine, a similar but a little bit smaller molecule, from threonine. Such a mixture is denoted by parenthesis in the fi ure. From this situation another step involving a back mutation at the amino-acid-recognizing site led to the encoding of ser^2. This reasoning also uses the assumption that serine existed abundantly. Concerning leucine, its encoding may have occurred twice independen ly as shown in the figure. The assumption of abundance is also nee

Now we conclude our discussion with a short word. Although we discussed only the 6-fold degeneracy, the figure provides a numb of interpretations and predictions on the relatedness of tRNAs. Mu experimental work is required to provide evidence for or against th proposed relatedness. Some of them, rather theoretical ones, have already been obtained by the author, however I would like to discus them fully in a subsequent paper.

REFERENCES

1. Eigen,M. and Schuster,P., Naturwissenshaften, 65, 341 (1978)
2. Sanger,F. et al. Nature, 265, 687 (1977)
3. Fiers,W. et al. Nature, 260, 500 (1976)
4. Reddy,V.B. et al. Science, 200, 494 (1978)
5. Gannon,F. et al. Nature, 278, 428 (1979)
6. Kafatos,F.C.,Efstratiadis,A.,Forget,B.G.,Weisman,S.M., Proc. Nat Acad. Sci.(USA), 74, 5618 (1977)
7. Miller,S.L. and Orgel,L.E., The Origins of Life on the Earth, Prentice-Hall, Englewood Cliffs (1974), pp.157-158.

SUMMARY OF EVIDENCE FOR AN ANTICODONIC BASIS FOR THE ORIGIN OF THE GENETIC CODE

James C. Lacey, Jr. and Dail W. Mullins, Jr.

Laboratory of Molecular Biology
University of Alabama in Birmingham
University Station
Birmingham, Alabama 35294 USA

This article summarizes the data available from our own laboratory and others supporting the hypothesis that the genetic code origin was based on relationships (probably affinities) between amino acids and their anticodon nucleotides. Selective activation seems to follow from selective affinity and consequently, incorporation of amino acids into peptides can also be selective. We believe that these selectivities in affinity and activation, coupled with the base pairing specificities, allowed the origin of the code and the process of translation.

Certainly the origin of living matter on earth required an information storage system, and that need was uniquely supplied by the polynucleotides. Because of their base pairing specificities we can readily understand how the transcriptional and replicative aspects of the genetic apparatus were made possible even if we do not yet know exactly how, in a primitive system, those operations might have taken place. While the character of the nucleotides suits them, ideally, for information storage and readout, the resulting polynucleotides, because of their limited conformations, limited sequence variability and monotonous ribose phosphate backbones, can do little else but store and transmit information. Consequently, a second type of molecule was required, one that could have many possible sequences with various characteristics, many conformations, and consequently many functions. The proteins uniquely meet the requirements. Evolution then had the problem of coupling the two, i.e. how is potential function in the nucleic acids converted into the actual functional molecules, the proteins? The importance of this translational process can be made clear by realizing that,

Y. Wolman (ed.), Origin of Life, 447–456.

although a system might have information in its nucleic acid component, that information could not be selected per se until it was expressed as protein. Put more straightforwardly, natural selection of DNA for the information it contained required the existence of the protein equivalent of the DNA.

The fundamental requirements for the process of protein synthesis are three, not necessarily in chronological order: (1) activation, the amino acids must be activated so they have sufficient energy for formation of the peptide bond; (2) organization, they must be organized into specific sequences as dictated by a template; and (3) polymerization, they must be polymerized while so activated and organized.

Some information is now available on all three of these aspects, and it appears that, although it is convenient to attempt to work with one isolated aspect, e.g. organization, those principles which govern one aspect also influence the others. For example, as will be subsequently shown, the rate of activation of hydrophobic amino acids by ATP seems in part a direct function of the affinity of the amino acids for ATP.

Other workers suggested before us (1-3) that the genetic code might be based on relationships between the amino acid and anticodonic nucleotides. One of the appealing arguments for this is that, in contemporary protein synthesis, the amino acid is selected and attached to tRNA (containing the anticodon) before the mRNA (containing the codon) is present, i.e. recognition processes developed around the molecule containing the anticodon, and the anticodon is still directly involved in some of the contemporary recognition processes (4). Since the exact mechanism of recognition in the contemporary is not known, one cannot say, for example, whether the amino acid is "seen" directly by the anticodon. We must presently rely on other information.

There are basically two schools of thought relating to the origin of the code. One, including ours, believes that the code origin was based on selective molecular interactions and reactions between amino acids and nucleotides, and that these can be discerned by experiment. The other school includes a broad spectrum of workers ranging from those who believe the code is based on molecular specificities whose nature is more subtle than simple direct interaction, to those who believe the code evolved through stochastic processes based on nuances we shall never be able to elucidate through experiment.

As Crick pointed out years ago (5), what is needed are binding constants of the amino acids for the various nucleotides. Fortunately, that data is now beginning to appear with increasing frequency, utilizing a variety of techniques. Road signs toward an anticodon basis, however, began to appear some years ago. Woese et al. (6) were the first to report on the physical properties of the amino acids and their relationship to code assignments. Through chromatographic experiments they were able to make polarity assignments to the amino acids and emphasized the importance of hydrophobic-hydrophilic properties. They did not have corresponding data for the nucleotides. Otherwise they might have made then the anticodonic correlations that only appeared much later. Certainly the importance of Woese's contribution was great, not only in delineating an important aspect of the problem, but in his penetrating study (7) of the code in general and, in particular, his in-depth analysis of what the parameters of the code evolution were (8). Certainly Woese's work influenced our own and probably the studies of Fendler et al. (9) and Armstrong et al. (10), who showed differential uptake of amino acids and nucleotides into micelles. Their data allowed one to arrange the nucleotides in order of decreasing uptake into some micelles, generally A > G > C > U, but a corresponding ordering of the amino acids did not seem to fit a discernible code-related pattern.

Our own incursions into this area of amino acid-nucleotide relationships began with studies of mononucleotides interacting with polybasic amino acids (11). Unfortunately, our studies, and those of other workers (12, a review), did not reveal code-related specificities but rather, at sufficiently high concentrations, interactions based on (1) electrostatic attraction of the negatively charged nucleotides and the polycations, followed by (2) a cooperative stacking of the nucleotides with each other. This latter feature resulted in the heirarchy of G > A > C > U (the same as the order of nucleotide self-stacking ability) for the nucleotide-polybasic amino acid complex stability, regardless of which polybasic amino acid was used.

In the development of our thinking on the code problem, the next bit of a suggestion came from work with S.W. Fox at the University of Miami. Preparation of lysine-rich thermal proteinoids showed that these materials, at lysine contents > 15 mole%, generally will form complexes with the homopolyribonucleotides and, although Mg^{++} will displace the proteinoids from the complexes, the proteinoid was displaced less readily from a complex with polyuridylic acid. Furthermore, the lysine-rich proteinoid under most conditions seemed to prefer to coprecipitate with polyuridylic acid over the other polyribonucleotides. Further work in collaboration with Fox (13) confirmed the lysine-U preference and suggested that phe-rich proteinoids preferred polyadenylic acid. These two preferences are anticodonic.

It was, in fact, these results with thermal proteinoids

which led to a thorough search of the literature for data that might bear on the question of amino acid-anticodon relationships. We found that Garel et al. (14) had determined the partition coefficients between an oil and water phase (and thus an estimate of hydrophobicity) for all amino acids and mononucleotides. We further found, using their data, that a direct correlation existed between these partition coefficients for the four homocodonic amino acids (phe, pro, gly and lys) and those of their anticodon nucleotides (Fig. 1) (15). Utilizing their mononucleotide data and a three dimensional plot (Fig. 2), we then showed the correlation for all amino acids and the estimated properties of their anticodonic dinucleotides. Not fitting the correlation were three sets of anticodons, for arg, leu and ser, each of which has a second set of anticodons which do fit the correlation, and trp, which was later found not to fit any correlations. We suggested that these particular anticodon assignments were not based on physical properties of the two types of molecules, but were late evolutionary assignments.

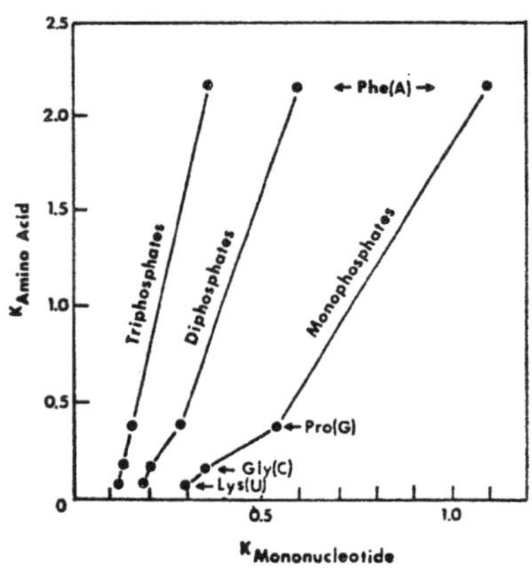

Fig. 1. Plot of relative hydrophobicity of the homocodonic amino acids versus that of their anticodon nucleotides (mono-, di- and tri-phosphates). Data from (14) and graph from (16).

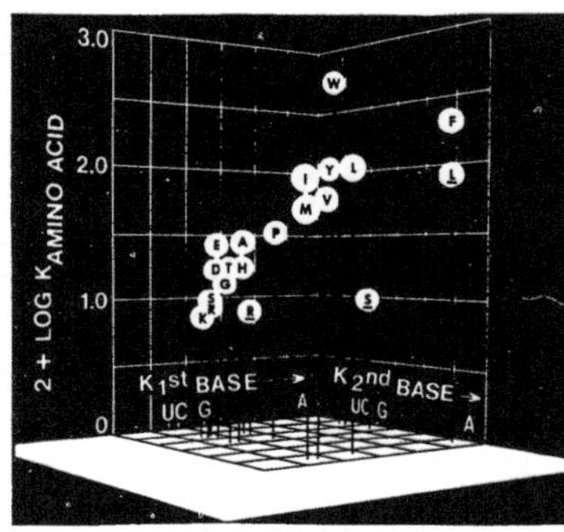

Fig. 2. Photograph of three dimensional plot of Garel et al (14) partition coefficient data. The data are from mononucleotides and amino acids. The horizontal axis on the left represents the nucleotide at the 3′ end of the anticodon, the horizontal axis on the right is the data for the nucleotide at the 5′ end of the dinucleotide anticodon. The vertical axis is the partition coefficient data for the corresponding amino acids. The letters on the circles are the one letter symbols for the amino acids. The letters which are underlined represent amino acids that have a second set of anticodons which fit the correlation better (15).

Since the solvent (butoxyethanol) used by Garel et al. (14) was not geologically relevant, we decided to study the relative hydrophilicities of the amino acids and nucleotides, using paper chromatography with high salt concentrations. Thus, the more hydrophilic compounds could compete better for water molecules and move up the paper faster. R_f was then an estimate of hydrophilicity. Again, with the homocodonic amino acids, there was a direct correlation of the properties of these four amino acids and their anticodonic nucleotides (Fig. 3)(16). The heterocodonic amino acids and the dinucleoside monophosphates also showed an anticodonic correlation (Fig. 4), but seemed to be divided into two families, one related to charged amino acids and all having at least one uridylic acid in their anticodons (lower curve), and the other showing a more straightforward relationship of hydrophilicity. Trp, tyr and one ser anticodon (AG, 3' → 5') did not fit the correlation at all. Trp and this same ser assignment failed to correlate in the earlier plot, thus strengthening the suspicion that their assignments were perhaps made on a different basis than the bulk of the other amino acids.

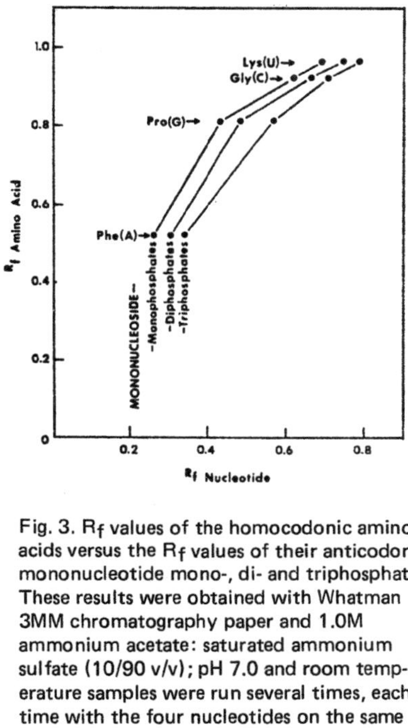

Fig. 3. Rf values of the homocodonic amino acids versus the Rf values of their anticodon mononucleotide mono-, di- and triphosphates. These results were obtained with Whatman 3MM chromatography paper and 1.0M ammonium acetate: saturated ammonium sulfate (10/90 v/v); pH 7.0 and room temperature samples were run several times, each time with the four nucleotides on the same chromatogram. Although there was day to day variation in the absolute Rf values, the Rf ordering was invariably U > C > G > A (16).

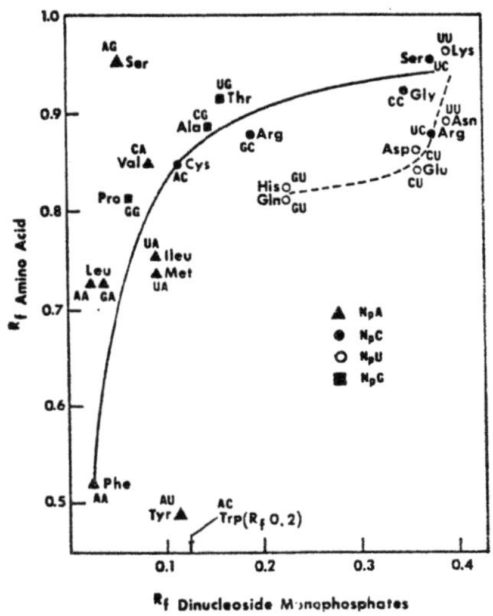

Fig. 4. Rf values (same procedure as in Fig. 3) of amino acids plotted versus the Rf values of the dinucleoside monophosphates representing the first two letters (3' → 5' direction) of their anticodons (16).

These correlations were analyzed by Jungck (17) and found to be statistically significant. Furthermore, Jungck extended these correlations by gathering other information from the literature and showed that, in addition to hydrophobicity and hydrophilicity, the properties of polarity and bulkiness also correlated anticodonically.

We suggested that the genetic code is fundamentally the genetic anticode (18), and a logical arrangement of it, based on variations in the properties of the anticodons, is shown in Table 1.

TABLE 1

The genetic anticode

Decreasing hydrophobicity →
middle letter

Decreasing hydrophobicity ↓ (3' end)

3' end	A	G	C	U	5'end
A	phe	ser	cys	tyr	A
	phe	ser	cys	tyr	G
	leu	ser	try	term	C
	leu	ser	term	term	U
G	leu	pro	arg	his	A
	leu	pro	arg	his	G
	leu	pro	arg	gln	C
	leu	pro	arg	gln	U
C	val	ala	gly	asp	A
	val	ala	gly	asp	G
	val	ala	gly	glu	C
	val	ala	gly	glu	U
U	ile	thr	ser	asn	A
	ile	thr	ser	asn	G
	met	thr	arg	lys	C
	ile	thr	arg	lys	U

The genetic anticode is presented 3' → 5' so that the anticodons can be more easily imagined as base pairs of their codon equivalents i.e. the codon-anticodon strands pair in an antiparallel fashion.
This tabulation is actually a simplification of the real anticodons which appear in the anticodonic loop of tRNAs. The difference resides principally in the nucleotide of the 5' end which is most often not the one predicted from Watson-Crick base pairing with the codons, but usually a modified base, frequently inosine. However, since the first two letters in the anticode are the most important this simplification does not alter the major thesis here.

There are three sets of anticodon assignments. The bulk of them fit the generalized curve shown in Fig. 3. The vertical box in the Table includes charged amino acids (except one set of anticodons for arg) and derivatives of charged amino acids which fit the lower curve in Fig. 3. The third set all have an A residue at the 3' end. We have shown (16) that the presence of the A residue at the 3' end drastically influences the properties of AC and AU (3' → 5') to overlap those of the more hydrophobic dinucleoside monophosphates. We suggest that AC and AU were not initially usable as anticodons. They are, in fact, the anticodonic equivalents of the terminator codons. The trp (AC), tyr (AU), cys (AC) and perhaps ser (AG) anticodons were perhaps late evolutionary assignments made possible by the fact that they were not used previously. Perhaps the continuing study of the evolution of the tRNA's will reveal whether this is so.

These above correlations are quite suggestive of the fact that the code is based on, at the very least, relationships between amino acids and anticodon nucleotides. Somehow, in considering the origin of the code and the process of protein synthesis, we must have a basis for the organizational component. It would seem a complex between amino acids and nucleotides would uniquely fill the requirements. The preliminary data with thermal proteinoids interacting with homopolyribonucleotides suggested an amino acid-anticodon affinity. Pursuing that idea further, we produced radioactive (for easier quantitation) thermal proteinoids rich in each of the four homocodonic amino acids and studied their coprecipitation with the four homopolyribonucleic acids. In the case of lys-, gly- and pro-rich proteinoids, an anticodonic preference was seen (19). Data with the phe-rich proteinoids was inconsistent, but showed a preference for poly U, if anything. We believe the phe-rich polymer did not display its true character because the phe residues would tend to interact with each other in the interior, leaving the hydrophilic residues toward the solvent. Hydrophilic residues should prefer poly U.

These data with thermal proteinoids, in the main, support the belief in an amino acid-anticodon affinity. Other data, in fact, show that hydrophobic residues have an affinity for the most hydrophobic nucleotide, A. For example, Razka and Mandel (20), using NMR, showed an ordering of decreasing affinity of three amino acids trp > phe > his for poly A. This ordering is the same as the ordering based on hydrophobicity. In adenosine solubility experiments we found that phe increased the solubility of adenosine, whereas pro, gly and lysine did not (18). The inference is that only phe, of these homocodonic amino acids, has appreciable affinity for adenosine. Reuben and Polk (21), using an interesting NMR technique, showed an ordering of binding constants of phe (5.1) > lys (2.8) > pro (2.7) > gly (1.7) (all in reciprocal molarity) between the methyl esters of these amino acids and AMP. Thomas and Podder (22), using an adenosine solubility experiment, found phe (3.1) > lys (0.55) > gly (.039) (all in reciprocal molarity). These measurements leave little question but that, of the homocodonic amino acids, phe has the greatest affinity for A, its principal anticodonic nucleotide. Toulme (26) also showed an heirarchy of interactions (trp > tyr > phe > ala > his) for ATP in a ternary complex with Zn^{++}, and suggests the importance of metals in these interactions.

In addition to these solution studies, the only cocrystals of small peptides and nucleosides which have been satisfactorily analyzed by x-ray diffraction are those from glycylglycine and cytidine (23). Glycine and cytidine are anticodonically related.

We have recently begun a more systematic study of those hydrophobic amino acids having A as their middle, and most important anticodonic letter. These we shall call the A amino acids, since they are in the A column in the anticode (Table 1). The ordering of amino acids in this column is based on the hydrophobicity of their anticodons: AA (phe) > GA(leu) > val(CA) > ile(met) UA. This ordering is not the same as the hydrophobic ordering of the amino acids (phe > leu > ile = met > val). We would expect differences in the interaction of these amino acids with A, phe showing the greatest, leu next, etc. Data from Reuben and Polk (21) allows the association constants to be calculated phe (5.1) > leu (3.1) > ile (2.9) > val (2.5, all in reciprocal molarity) for AMP without a divalent cation. We have used Toulme's UV technique and found an ordering phe (51) > ile (44) > val (40) > leu (38) > met (36) for ATP with Zn^{++} (all in reciprocal molarity). In both of these studies, phe, the most hydrophobic amino acid in the set (and having the most hydrophobic anticodon (AA)), interacts with A(TP) to the greatest extent. Using three other measures of the affinity of this set of amino acids for A derivatives (described and summarized in Table 2), we find in every case that phe has considerably greater affinity for the A derivatives. The ordering of the other A amino acids varies with the type of experiment.

TABLE 2

SUMMARY OF THE RELATIVE ACTIVATION RATES OF SOME HYDROPHOBIC AMINO ACIDS BY ATP AND VARIOUS ESTIMATES OF THEIR AFFINITIES FOR ADENINE DERIVATIVES

Amino Acid	Activation Studies							Estimates of Interaction				
	(a)	(b)	(c)	(d)	(e)	(f)	(g)	(h)	(i)	(j)	(k)	(l)
Phe	1.00	1.00	1.00	1.00	1.00	1.00	3.115	50.8	63.2	5.10	0.863	0.035
Leu	0.89	0.56	0.78	0.51	0.48	0.66	---	37.8	20.0	3.13	0.387	0.023
Val	0.47	0.49	0.59	0.50	0.51	0.43	0.213	39.7	39.9	2.48	0.425	0.022
Ile	---	0.48	0.53	0.56	0.46	0.43	---	44.4	38.3	2.87	0.393	0.022
Met	0.18	0	0	0	0	0.16	---	36.0	41.8	3.08	0.547	0.024

The activation studies were carried out at 50° C, pH 5, using 0.15M amino acid, 0.1M ATP, 0.4M hydroxylamine and a divalent metal cation as indicated below. The values given relate to the relative amounts of activated amino acid formed with time, as determined by the hydroxamate assay of Lipmann and Tuttle (24): (a) Be^{++}/ATP = 1; (b) Mg^{++}/ATP = 1; (c) Zn^{++}/ATP=2; (d) 2 x seasalt (Mg^{++}/ATP = 1); (e) Mg^{++}/ATP = 2; (f) 4 x seasalt (Mg^{++}/ATP = 2), Mullins and Lacey (26);
(g) association constants (K_a, M^{-1}) for amino acid-adenosine complexes at 20° C, as determined from the adenosine solubility studies of Thomas and Podder (22);
(h) association constants (K_a, M^{-1}) determined by us for amino acid-ATP-Zn^{++} complexes using the UV difference method of Toulme (25) with 0.5 mm quartz cuvettes in 0.5M $NaClO_4$ and 0.03M Tris buffer at pH 7.6, ATP conc. 10^{-3}M, Zn^{++} 2 x 10^{-3}M with variable amino acid concentrations many times greater than ATP. The study with phe was made directly, determining the hypochromicity induced @ 255 nm by increasing concentrations of the amino acid. For the other amino acids, hypochromicity was induced with phe at 10^{-2}M and the ability of the second amino acid to eliminate that hypochromicity was studied. The equations suggested by Reuben (26) were used in determining the association constants and assuming that all the ATP exists as the Zn^{++} salt;
(i) % retardation at the origin by ATP of amino acids following paper chromatography on Whatman 3mm in 5% $MgCl_2$.
$6H_2O$ in absolute ethanol. The applied sample contained on ATP/amino acid ratio of 1, and all values are quench corrected for ATP, Mullins and Lacey (26);
(j) association constants (K_a, M^{-1}) for amino acid methyl esters and AMP as calculated from the average dissociation constants (K_d, M) determined by Reuben and Polk (21);
(k) apparent partition coefficients of hydrophobic amino acids in a Mg^{++} ATP phase separation system. Solutions containing 0.25M ATP, 0.5M $MgCl_2$ and 0.05M [^{14}C]-amino acid (~ 7 x 10^{-3} μCi μmole $^{-1}$), pH 5, were allowed to undergo phase separation overnight at 3° C. The resultant phases were separated, and the amino acid concentration in each determined by liquid scintillation counting. The apparent partition coefficient is the ratio of the concentration of the amino acid in the bottom (ATP-rich) phase to that in the top (aqueous) phase, Mullins and Lacey (26);
(l) adenosine solubility (M) in 1 x seasalt - 0.05M Tris, pH 6.5, in the presence of 0.15M amino acid at room temperature. Saturated solutions of adenosine in the above solvent were incubated (with shaking) for 72 hours at room temperature in the presence of varying concentrations of each of the hydrophobic amino acids. After filtration through Whatman 1 filter discs, the concentration of adenosine in the filtrates was determined spectrophotometrically, Mullins and Lacey (26).

Now if we presume that the affinity of amino acids for mononucleotides can be responsible for their organization on a template, we must then explain how such organization would assist in incorporating the amino acids into peptides.

We had earlier proposed (15) a mechanism for a primitive translation system in which the four mononucleotide triphosphates served as both activators and adaptors for different classes of amino acids. For example, ATP was postulated to have activated all hydrophobic amino acids, UTP all charged amino acids, etc. In such a system, recognition (organization) and activation are interrelated. We then began a study of the non-enzymatic activation process, using ATP. The main features of this study are described separately (26), and show that pH, carboxyl pK_a, divalent metal cations and salt concentration all have effects. Most important to the present discussion, however, is the fact that in the series of hydrophobic(A) amino acids (phe, leu, val, ile, met), and under a wide variety of conditions, phe was invariably activated the fastest. Interestingly, the results in Table 2 (also see ref. 26) show that the ordering of activation is often the same as the ordering in the A column of the genetic anticode, i.e. phe, leu, val, ile(met).

As a result of the data summarized above, we would like to propose the following, albeit tentative, hypothesis - namely, that the extent of the non-enzymatic activation of an amino acid by a nucleoside triphosphate is largely a function of the affinity which these two molecules have for one another. If this is the case, then we can point to at least two factors which should operate in concert to favor the incorporation of specific amino acids into peptides: (1) affinity, which will help organize amino acids into nucleotide complexes, and (2) selective activation, which will enhance the formation of peptide bonds between amino acids. We believe that it was these phenomena of selective affinity and activation, together with the well characterized base pairing properties of the nucleic acids themselves, that allowed for the origin and evolutionary development of a protein synthesizing system.

ACKNOWLEDGMENT We gratefully acknowledge the continued support of the National Aeronautics and Space Administration (NGR 01-010-001).

REFERENCES

1. Dunnill, P., Nature 210: 268 (1966).
2. Ralph, R.K., Biochem. Biophys. Res. Commun. 33: 213 (1968).
3. Nagyvary, J. and Fendler, J., Orig. of Life 5: 357 (1974).
4. Singhal, R.P., Fallis, P.A.M., Prog. Nuc. Acid Res. 23: 227 (1979).

5. Crick, F.H.C., J. Mol. Biol. 38: 367 (1968).
6. Woese, C.R., Dugre, P.H., Saxinger, W.C. and Dugre, S.A., Proc. Natl. Acad. Sci. U.S.A. 55: 966 (1966).
7. Woese, C.R., The Genetic Code, Harper and Row, N.Y. (1967).
8. Woese, C.R., Naturwiss. 60: 447 (1973).
9. Fendler, J.H., Nome, F. and Nagyvary, J., J. Mol. Evol. 6: 215 (1975).
10. Armstrong, D.W., Seguin, R. and Fendler, J.H., J. Mol. Evol. 10: 241 (1977).
11. Lacey, J.C., Jr. and Pruitt, K.M., Nature 223: 799 (1969).
12. Lacey, J.C., Jr. and Pruitt, K.M., J. Pharm. Sci. 64: 473 (1975).
13. Fox, S.W., Lacey, J.C., Jr. and Nakashima, T., Nucleic Acid Protein Interactions, (Ribbons, D.W., Woessner, J.F. and Schultz, J., eds) North Holland Pub. Co., Amsterdam,(1971) p. 113.
14. Garel, J.P., Filliol, D. and Mandel, P., J. Chromat. 78: 381 (1973).
15. Lacey, J.C., Jr. and Weber, A.L., Protein Structure and Evolution (Fox, J., Deyl, Z. and Blazej, A., eds) Marcel Dekker, Basel (1976) p. 213.
16. Weber, A.L. and Lacey, J.C., Jr., J. Mol. Evol. 11: 199 (1978).
17. Jungck, J.R., J. Mol. Evol. 11: 211 (1978).
18. Lacey, J.C., Jr. and Weber, A.L., Precamb. Res. 5: 1 (1977).
19. Lacey, J.C., Jr., Stephens, D.P. and Fox, S.W., BioSyst. 11: 9 (1979).
20. Razka, M. and Mandel, M., Proc. Natl. Acad. Sci. U.S.A. 68: 1190 (1971).
21. Reuben, J. and Polk, F.E., J. Mol. Evol. (in press)
22. Thomas, P.O. and Podder, S.K., FEBS Lett. 96: 90 (1978).
23. Szalda, D.J., Marzilli, L.G. and Kistenmacher, T.J., Biochem. Biophys. Res. Commun. 63: 601 (1975).
24. Lipmann, F. and Tuttle, L.C., J. Biol. Chem. 159: 21 (1945).
25. Toulme, J., Bioinorg. Chem. 8: 319 (1978).
26. Mullins, D.W. and Lacey, J.C., Jr. (this volume, page).
27. Reuben, J., FEBS Lett. 94: 20 (1978).

PRIMITIVE TRANSFER RNA AND ORIGIN OF DARWINIAN SYSTEM

Masahiro Ishigami and Masataka Kinjo*

Laboratory of Biology, Jichi Medical School,
Minamikawachi-machi, Tochigi-ken, 329-04
*Laboratory of Agricultural chemistry,
Utsunomiya University, Utsunomiya, 320 Japan

ABSTRACT

Stepwise progress of the relation between amino acids and nucleotides in the origin of protein synthesis mechanism is speculated. As an experimental support for the speculation, the effect of polynucleotide and polypeptide on the polypeptide formation is tested. Basic protein, histone, promotes the reaction of polypeptide formation from aminoacyl adenylate. The relation between origin and development of genetic code and incerasing of invariable site in protein in the course of molecular evolution is discussed. Learning system is thought to have some analogy with Darwinian evolution.

1. LIFE BEGAN WHEN A SYSTEM DEVELOPING THROUGH BIOLOGI CAL EVOLUTION WAS APPEARED.

The landmark of origin of life might be corresponded to appearance of a system which developed through Darwinian evolution mechanism in primordial broth. Appearance of a system evolving in Darwinian way means that of selfreproducing system. Small fluctuations were introduced into reproduction of the system by mutation. Natural selection occured among these fluctuated systems. These evolutionary mechanisms produced more adapted and self-organized systems. Selfreproducing systems containing polynucleotides appeared only when the systems were supplied with both raw materials and free energy.

Y. Wolman (ed.), Origin of Life, 457–464.

2. EVOLUTION OF PROTOCELLS AND FORMATION OF PROTEIN SYNTHESIS MECHANISM

Polypeptides and polynucleotides must be the basic materials of the first evolving systems just as modern organisms. Organic droplets was formed with polypeptides as suggested by Oparin(1938) and Fox et al. (1959). These droplets concentrated amino acids and other organic substances in primordial broth(Evreinova, 1966) and formation of polypeptide was accerelated by concentration. Probably the formation of polypeptides in chemical evolution should be the most primitive feedback process(Fig.1 I).

Riboses and bases of nucleotides along with amino acids were incorporated into the droplets. Various kinds of nucleotides and their derivatives was formed under the influence of condensing agent which had also been incorporated into or formed in the deoplets(Fig.1 II). Paecht-Horowitz and Katchalsky(1967) succeeded in the formation of polypeptides by polycondensation of aminoacyladenylate anhydride in aqueous solution. It must be the first step of the mutual contact between amino acids and nucleotides(Fig.1 III). The evidences on the synthesis of polypeptides and polynucleotides and on their mutual dependence under prebiotic conditions were given by Orgel and his coworkers(Lohrmann and Orgel, 1973; Lohrmann et al. 1975; Sawai and Orgel, 1975; Weber and Orgel, 1978). Polynucleotide as well as nucleotide monomer activated amino acids and accerelated the formation of polypeptides. On the other hand Sulston et al.(1968 a,b,) succeeded in the template directed formation of polynucleotides. This formation represents the reproduction of polynucleotides in the droplets of primordial broth(Fig.1 IV).

Protein synthesis mechanism must be established through the stepwise evolution of polynucleotides which not only activate amino acids but also evolve to produce "better" polypeptides for probionts(Ishigami et al., 1977). The outline of mechanism postulated is as follows. In the first step, polynucleotides evolved to promote the reaction of polycondensation of amino acids. In the second step, they evolved to produce polypeptides which contained more variegated amino acids than were expected from the mere proportionality to the environmental medium. In the third step, polynucleotides evolved to produce polypeptides with "better" amino acid sequence(Fig.1 IV).

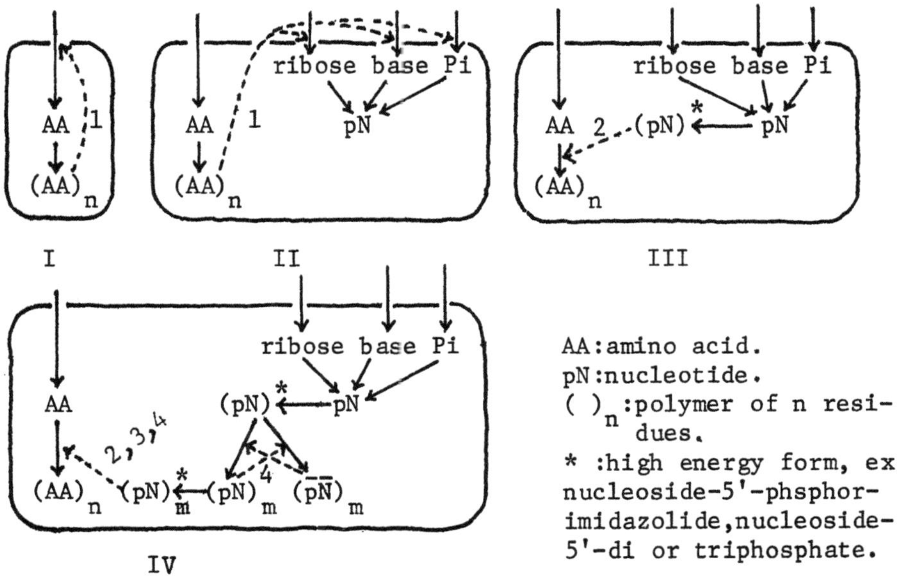

Figure 1. Evolution of Protocells.
⟶: flow of matter, - - - -→: regulation
1: increasing of the rate of incorporation. 2: accerelation.
3: regulation of composition. 4: regulation of sequence.

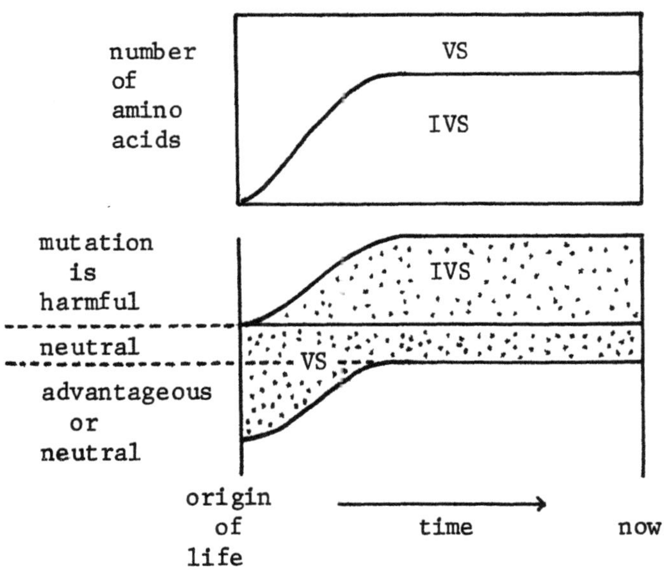

Figure 2. Diagramatic explanation of increasing of invariable site and harmfulness of mutation.
VS : variable site, IVS : invariable site.

3. IMAGINAL PRIMITIVE TRANSFER RNA(tRNA)

Among various kinds of replicable polynucleotides in probionts, some had a conformation of single hairpin loop with two specific sites. One of the sites had a high affinity for amino acid and the other for polynucleotides by base pairing. These polynucleotides bound and activated amino acids, and when two of them came together, they enhanced polycondensation of amino acids by bringing them close each other. The probionts containing these kinds of polynucleotides grew more quickly. It may represent a kind of primitive tRNA.

4. SELECTION OF PRIMITIVE tRNA AND FIRST DARWINIAN SYSTEM

Sueoka(1961) has found that small amino acids, i.e. glycine or alanine, which are thought to be abundant in primordial broth, contained much guanine(G) and/or cytosine(C) in its cognate codons and anticodons but hydrophobic or hydrophylic large amino acids contain much adenine(A) and/or uracyl(U). This inclination was explained as the result of natural selection of primitive tRNAs(Ishigami and Nagano, 1975. Ishigami et al., 1977). A set of tRNAs which made "better" polypeptides in their ratio of various amino acids was selected.
The relationship that the physicochemically akin amino acids are coded by similar codons was explained by Sonneborn(1965) to be the code system evolved to minimize mutability since mutation is generally harmful for modern organism. Noguchi(1978) found a tendency of incerase in invariable sites(IVS) of cytochrome c with the molecular evolution on the basis of neutral mutation theory of molecular evolution. At the time of origin of the protein synthesis mechanism, there might be no IVS in any proteins(see Fig.2). Some variable site(VS) changed to IVS through Darwinian evolution. Mutations in IVS were almost always harmful for a organism but those in VS were neutral or even advantageous. A mutation is generally harmful for modern organisms but it might be neutral or advantageous for a probiont in which any protein did not evolve yet.

5. FREEZING OF GENETIC CODE AND ITS UNIVERSALITY

Eigen(1972) interpreted the universality of genetic code as the result of the selection among hypercycles.

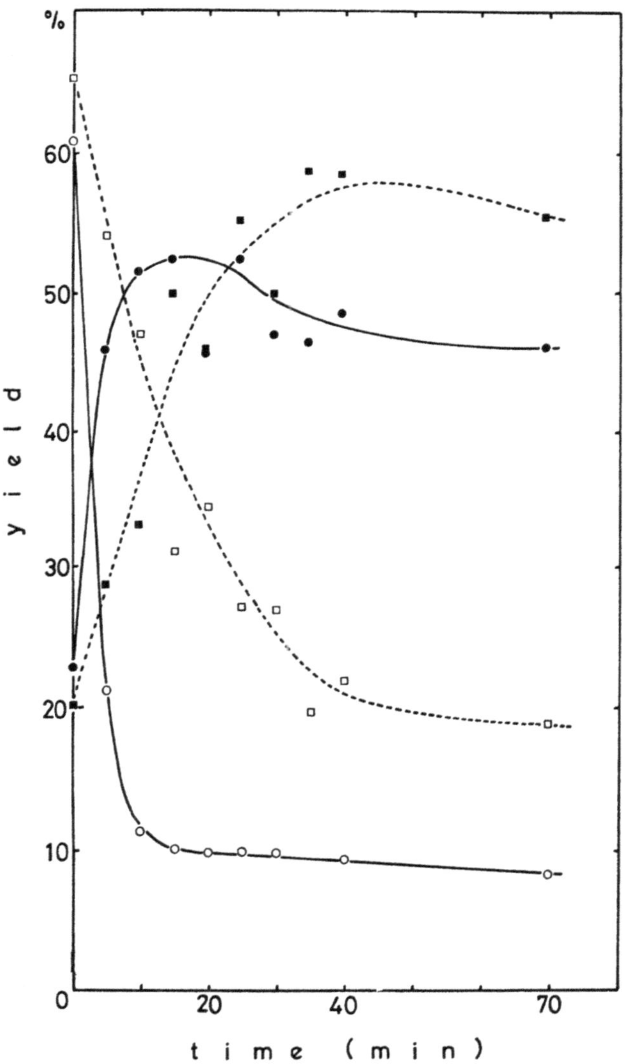

Figure 3. Time Course of the Relative Amount of Reaction Products from Phenylalanyl adenylate.

------ □ ------ : Phe-pA + Phe, without histone.
------ ■ ------ : Phe_2 + Phe_3, without histone.
—— ○ —— : Phe-pA + Phe, with histone.
—— ● —— : Phe_2 + Phe_3, with histone.

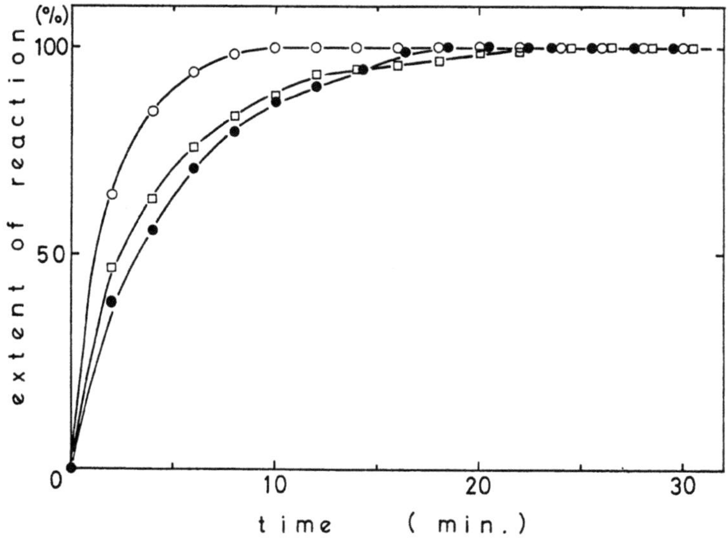

Figure 4. Effect of Histone on the Polycondensation of Phenylalanine Activated as Phenylalanyl adenylate.
○ : with histone, ● : with bovine serum albumin,
□ : no addition.
Reaction rates were obtained by the observation of released proton.

TABLE 1. Analogy of Learning with Darwinian Evolution.

	(Darwinian evolution)	(Learning)
Trial	Mutation.	Random trial.
Feedback system	Function of protein. Selectively advantageous.	Desire is satisfied. Obtain good result.
Memory	Juxtaposition of base in polynucleotide.	Mutual relationship of nerve cells by conduction of specified synapsis.
Keeping	Duplication and division.	Education.
Result	Adaptation. Apparent purposefulness	Recognition of necessity. Reflection of environment. Purposeful behavior.

Another interpretation is that the modern organisms evolved from single ancestor.
Freezing of genetic code into a given combination would have resulted from appearance and gradual increase of IVS in proteins, since a mutation of amino acid sequence in IVS would be generally harmful and change in the translation principle, i.e. the genetic code, would bring about a great confusion on every protein sequence. In the nascent phase, several lines of organisms with slightly different coding systems might compete each other before final survival of the line with the "modern" genetic code.

6. EFFECT OF POLYPEPTIDE AND POLYNUCLEOTIDE ON THE POLYCONDENSATION REACTION OF POLYPEPTIDE FROM AMINOACYL ADENYLATE ANHYDRIDE

The effect of polynucleotides and polypeptides on the reaction rate of polymerization of amino acid activated as phenylalanyl adenylate anhydride(Phe-pA) was tested. Polyuridylate(poly(U)) which made base pairing with adenine showed no effect. Histone accelerated the rate of polycondensation while neutral protein, bovine serum albumin, produced no effect. This result suggests a possibility of self-catalyzing or positive feedback loop in the primordial state, where formation rate of polypeptides could be accerelated by the formation of at least some kind of polypeptides themselves(Fig.3 and 4, Ishigami et al. 1980).

7. DARWINIAN EVOLUTION AND LEARNING

Among various kinds of polynucleotides present at first, specific ones were selected out in the Darwinian way through function of proteins which had been formed under the direction of these polynucleotide. We can find no system which have a common Darwinian feature outside the biological world. But we can find other examples among living world, the learning system of brain. Organisms evolve and adapt to their environment and reflect environment by Darwinian selection. By this way organisms become to have apparent purposefulness. On the other hand, animals can learn and recognize their environment and reflect it. Behavior of learned animal has become to have apparent purposefulness as the result of random trials and errors and selection of behavior pattern. Random fluctuations by mutation in Darwinian evolution system correspond to the random trials in the animal learning. The suitability of mutations in a polynucleotide is determined through the function of the protein which was formed under the direction of the polynucleotide itself. Thus the result of polynucleotide mutation is fed back to itself via the product(polypeptide) according to the degree of it being more or less advantageous. By contrast, a behavioral pattern of an animal is determined by random trials which would or would not result in satisfaction of a desire. The mechanism of memory is possibly changes in network connections of synapses.

Thus, juxtaposition of bases in polynucleotide may be considered to be analogous to the arrangement of nerve cells in the nerve network. The information accumulated in the course of biological evolution is transmitted by duplication and division of genes; the information obtained and accumulated by learning is transmitted through education. The relationship is summarized in Table 1.

REFFERENCES

Eigen, M.(1971) Quart.Rev. Biophys. 4, 149

Everinova,T.N.(1966) Condensation of substances and action of enzyme in coacervates, Izd., Nauka, Moscow.

Fox,S.W., Harada,K., Kendrick,J.(1959) Science 129, 1221

Ishigami,M., Nagano,K.(1975) Origins of Life, 6, 551

Ishigami,M., Nagano,K., Tonotsuka,N.(1977) Biosystems 9, 229

Ishigami,M., Kinjo,M., Tonotsuka-Ohta,N., Nagano,K.(1980) Origins of lige 10, in press.

Lohrmann,R., Orgel,L.E.(1973) Nature 244, 418

Lohrmann,R., Ranganathan,R., Sawai,H., Orgel.L.E.(1975) J.Mol.Evol. 5, 57

Noguchi,T.(1978) Sixth International Biophysis Congress, Kyoto.

Oparin,A.I.(1938) Origin of Life, Macmillan, New York

Paecht-Horowitz,M., Katchalsky,A.(1967) Biochim.Biophys. Acta, 140, 14

Sawai,H., Orgel,L.E.(1975) J.Am.Chem.Soc., 97, 3532

Sueoka,N.(1961) Proc.Natl.Acad.Sci.USA. 47, 1141

Sulston,J., Lohrmann,R., Orgel.L.E., MillesH.T.(1968 a) Proc.Natl.Acad.Sci.USA. 59, 726

Sulston,J., Lohrmann,R., Orgel,L.E., Milles,H.T. (1968 b) Proc.Natl.Acad.Sci.USA. 60, 409

Weber.A.L., Orgel,L.E.(1978) J.Mol.Evol. 11, 9

EVOLUTION OF THE TERRESTRIAL ATMOSPHERE AND ITS FOSSILS IN BIOSYSTEMS

Mikio Shimizu

Institute of Space and Aeronautical Science
University of Tokyo
Komaba 4-6-1, Meguro-ku, Tokyo, 153, Japan

ABSTRACT A comparison of the terrestrial atmosphere with those of Mars and Venus suggests the presence of CO in the primitive atmosphere on the Earth, resulting the formation of C_3O_2 and HCHO polymers in the primitive oceans. Various fossil vestiges of these polymers can be found in the living system. The most important one is the evolutional route of the genetic code, which is derived by combining the discussion of Eigen and Schuster for the formation of mRNA, that of Egami for the classification of amino acids and nucleic acid bases in relation to the genetic code, and our C4N model described in another paper of this conference.

The terrestrial atmosphere would have formed by sudden degassing from its molten mantle. The atmospheric composition would have been determined by equilibrium with molten basalt at 1500 K to be H_2O, CO_2, H_2, CO, and N_2. Water would have formed oceans, carbon dioxide would have been dissolved in the oceans to form calcites, and H_2 would have quickly dissipated to space. The residual CO and N_2 would have formed the primitive atmosphere (1). It has been argued that the present atmospheric composition of Venus and Mars are in accord with this discussion (2). Recent finding of the excess of non radiogenic rare gases such as ^{36}A and ^{86}Kr on Venus by the Venera 11 and 12 and the Venus Pioneer 1 has given another support for the secondary origin of the terrestrial atmosphere. These rare gases may be the remnant of the primitive primary atmosphere, mixed with the atmosphere of secondary origin such as CO_2, N_2, etc.(3). On the other hand, there is no excess of rare gases of this type on the Earth. It has been also argued that these rare gases might have been degass-

Y. Wolman (ed.), Origin of Life, 465–471.

ed from the interior of Venus (4). However, in this case, the solar nebular density at the orbit of Venus should have been denser than that at the orbit of the Earth by a factor of 50 and the nebular temperature at both Venus and Earth should have been very cool, similarly to the liquid nitrogen temperature. Such condition has never been derived from any theory on the evolution of the solar system (5).

By taking into account of the balance of the released oxygen by phytolysis with the possible sinks for it, Hart (6) suggested a possibility that a gas of the amount of 10^{22}g (corresponding to the surface partial pressure of 1 atm.) might have been contained to reduce the oxygen gas in the primordial atmosphere. CH_4 and H_2 of this amount would have dissipated by photolysis very quickly in a geological time (2,7) and cannot be the candidate of this gas. CO is sometimes argued to have been unstable due to the dissolution to water. However, the ancient oceans were not composed of pure water, but it was acidic due to the dissolved CO_2 and Cl^- etc., and contained many kinds of metallic ions. The formation of formates in the oceans from CO might have been inhibited under this condition. Our previous estimate of the primordial CO amount is about 1 atm., in accord with his estimate. However, this may merely be coincidence. The real estimate of the amount of reducing material on the surface of the primitive earth is not a easy task.

In the early history of the Earth, oxygen, and thus ozone, was absent. Then, the cold trap of water vapor in the stratosphere would have been quite strong and no water vapor was present in the upper atmosphere. By the photochemical reactions caused by the solar ultraviolet radiation, CO could have converted to C_3O_2 polymer in the upper atmosphere. The carbon suboxide precipitated on the primitive oceans would have formed a polymer soup of high concentration, perhaps 0.1 % by weight (2). A part of the polymer might have been injurred by the irradiation of the solar ultraviolet radiation to form C, C_2O radicals. After precipitating into oceans, these radicals would have reacted with water to form aldehydes.

Egami (8) pointed out a close correlation of bioelements with minor elements in the primeval oceans, whose composition would have been essentially similar to the present ones. Hatanaka and Egami (9) and Ochiai et al. (10) investigated the catalytic effect of transition elements on the formation of abiotic molecules from simpler molecules such as formaldehyde and hydroxylamine, by using the 'modified sea medium' containing various kinds of metallic ions (MnO_4^{--}, Zn^{++}, Co^{++} etc.) and detected almost all natural amino acids in the medium. Yanagawa and Egami (11) found some particles of protocellular structure in the modified sea medium, when glycine and formaldehyde or glycine and other amino acids mixture were chosen as starting materials. There particles, 'mariglanules' could constitute a third model of the protocell, following the 'coacelvate liquid droplet' model of

Oparin (12) and the 'protenoid microsphere' model of Fox and Harada (13).

When carbon suboxide polymer reacted with hydroxylamine in the modified sea medium, some amino acids such as glycine, alanine, and lysine were formed (14). In addition to the formation of amino acids, Yanagawa and Egami found that C_3O_2 polymers reacted with urea to afford unidentified compounds in the medium with maximum absorption spectrum at 258 nm, which was similar to those of barbituric acid or uracil, Consequently it is likely that carbon suboxide polymer had yielded amino acids and nucleic acid bases in the primitive oceans.

Various kinds of biomolecules have been found in the living system and its enviroment. Many of them may retain the fossil vestiges of C_3O_2 in their molecular structure and their functions. At first, we shall point out that the vestiges of C_3O_2 can be seen in the structure of many small biomolecules as shown in the figure 1. Of course, a mere enumeration of simple examples of this type could be criticized as a matter of sophistic taste. At least, however, some explanations would be required to understand the frequent presence of C_3 compounds in the central structure of biomolecules which appear in fundamental biological phenomena such as energy requiring metabolism and structure of food and cellular membrane.

More important, the crucial roles of C_3O_2 and HCHO in the biosystem should be seeked in a central part of the biological function such as the genetic code, the symbol of the central dogma of the molecular biology. In my separate paper in this conference, I have tried to decipher the genetic code by using the C4N model. In this model, the correspondence between anticodons and amino acids is unique (except some degeneracies), since the hole on the C4N has a key hole-key relation to the side chain of the corresponding amino acids. However, in spite of this uniqueness, the adoption of twenty protein amino acids by the living system might have not been simultaneous due to various reasons. In this meaning, the genetic code might have experienced an evolutional history. We shall discuss some correlated factors with this point one after another.

(I) Enviroment and amino acids

Even if a tRNA was already formed, it could not have worked without the formation of the corresponding amino acids in the enviroment. From the previous discussion in this paper, however, we may assume that this situation has not occurred.

In this concern, it should be referred to the elegant working hypothesis proposed by Egami (15) on the birth of nucleic acid bases, protein amino acids, and the primitive genetic code. He pointed out that protein amino acids could be classfied to purine group and pyrimidine group, being each group again devided to guanine (Gly + nC_1) and adenine (Gly + nC_2) group and uracil

FIGURE 1. Some Examples of Hidden C_3O_2 in Biomolecules

1a

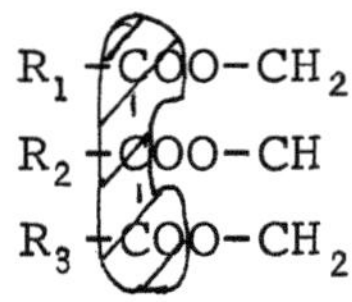

1b

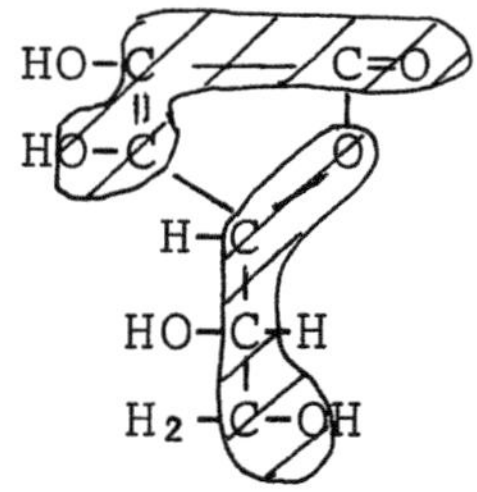

1c

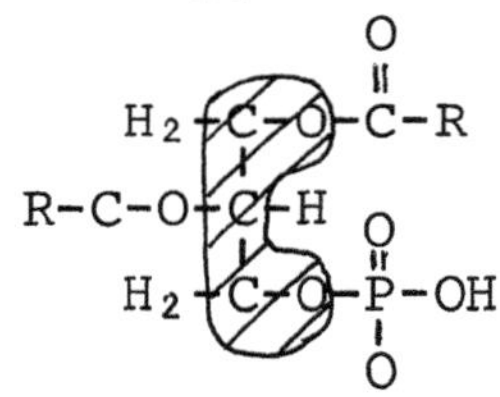

1d

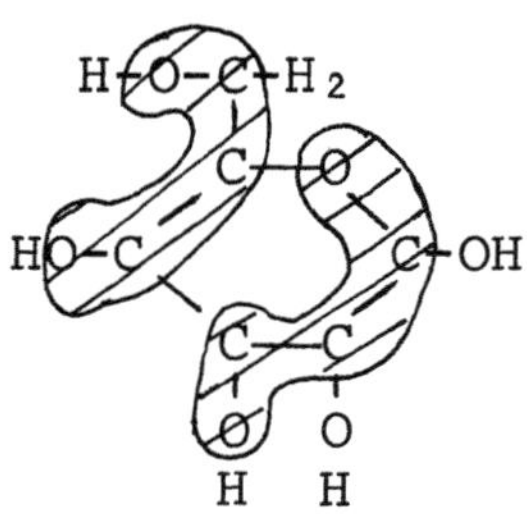

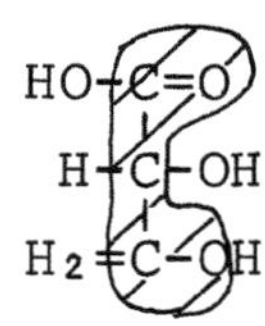

FERMENTATION

C_6 COMPOUNDS ⟷ C_3 COMPOUNDS

PHOTOSYNTHESIS

1a, fatty acid 1b, ascorbic acid

1c, lipid

1d, hexose and its derivertives in sugar metabolism

($C_3C_6C_9$) and cytosine (C_5 chain) group, respectively. The codons of each nucleotide group usually contain the nucleotides of the group. The purine and pyrimidine groups of amino acids correlate at the same time with purine and pyrimidine nucleotides, respectively, since glycine (easily formed from formaldehyde) plus (C_1N_1) compounds make purine and C_3 compounds (C_3O_2) plus urea make pyrimidine, respectively. We shall use this ingenious idea indirectly to trace the evolution of the genetic code later.

(II) Amino acid-tRNA interaction

The strength of interactions between amino acids and the C4N varies case by case. In some cases, the interaction contains the hydrogen bonds. Then, the adoption of the amino acid to the biological system might have occurred easily than the non hydrogen bond case.

(III) Degeneracy in the combination of anticodon and discriminator base

Sometimes a similar hole will be formed by the different combination of anticodons and discriminator bases. For instance, both AGY≡G and UCN≡G specifies serine. On the other hand, it is also possible that a change of discriminator base will deform the size of the C4N hole to fit another side chain of amino acid.

(IV) tRNA-mRNA interaction

Jukes (16) found that modification of the nucleic acid in the neighbor of the third anticodon would have increased the fidelity of translation of codons, avoiding ambiguity due to the wobble phenomena. Consequently, even if the correspondence between anticodon and amino acid is unique, that between codon and amino acid was not necessarily unique.

(V) mRNA evolution

Crick et al. (17) assumed that the primitive mRNA sequence started from a repeating family of sequence RRYRRY--, while Eigen and Schuster (18) argued that the repeating family of sequence RNYRNY-- was a better choice, pointing out some difficulty of the RRY sequence. We shall adopt the RNY sequence. In this case, G, C, A, U would have been used in this order by biosystem, since G≡C bond is stronger than A=U bond.

Generally speaking, the factors (I) and (V) suggest that small amino acids were adopted earlier by the living system, while the factor (II) favors large amino acids for it.

Another important factor to discuss the evolution of the genetic code is the chemical continuity rule, namely, it is known that two amino acids whose codons differ at one nucleotide alone are chemically similar to each other. In the C4N theory, this rule is instantly explained in terms of the stereochemical property. On the other hand, in the frozen accident theory, this tendency is a result of optimal selection to minimize the frequency of deleterious mutation. In such theory, however, the preferrence of a protein amino acid to non protein amino acid, say, arginine to ornithine, is not obvious.

FIGURE 2. Evolutional route of the genetic code (anticodons are used)

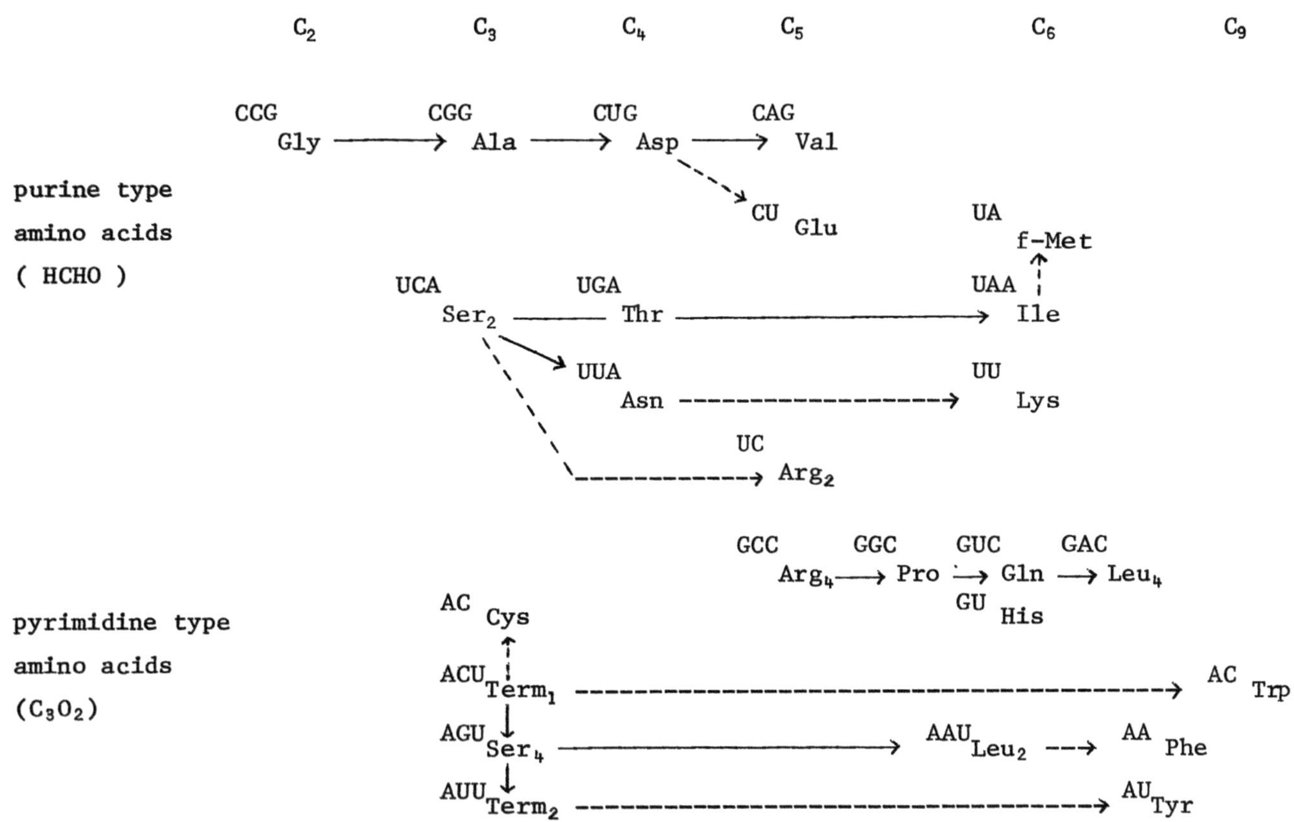

By taking into account the chemical continuity rule and the above five factors, we have constructed a possible route for the evolution of the genetic code as shown in the figure 2. In this figure, the codes are specified by the series of 3rd-2nd-1st anticodons. The C4N holes for the terminators ($term_1$ and $term_2$), whose discriminator base are assumed to be the same as that of tryptophane (G) and as that of tyrocine (A), respectively, are both small. Consequently, their location in the figure 2 are placed at the left side of the group. It can also be seen that the fossil vestiges of C_3O_2 appears in the clearest form for the sequence, serine-leucine-phenylalanine, while that of HCHO in the sequence of glycine-alanine-asparatic acid etc.. Consequently, the above well classified evolutional route of the genetic code appears to be the best evidence (fossils) for the primitive (rather oxidative) enviroment, especially that for the primitive atmosphere consisted of CO and N_2.

REFERENCES

(1) Shimizu, M., 'Origin of Life', ed. by H. Noda, Center Acad. Publ. Japan, Tokyo, p.35 (1978).
(2) Shimizu, M., Precamb. Res., 9, 311 (1979).
(3) Shimizu, M., Moon Plan., 20, 317 (1979).
(4) Pollack, J.B. and Black, D.C., Science, 205, 56 (1979).
(5) Shimizu, M., Interdisc. Sympo. IUGG, Canberra, Dec. p.29 (1979).
(6) Hart, M.H., Origin Life, 9, 261 (1979).
(7) Shimizu, M., Precamb. Res., 3, 463 (1976).
(8) Egami, F., J. Mol. Evol., 4, 113 (1974).
(9) Hatanaka, H. and Egami, F., Bull. Chem. Soc. Japan, 50, 1147 (1977).
(10) Ochiai, T., Hatanaka, H., Ventilla, M., and Egami, F., 'Origin of Life', ed. by H. Noda, Center Acad. Publ. Japan, Tokyo, p.135 (1978).
(11) Yanagawa, H. and Egami, F., ibid, p.385 (1978).
(12) Oparin, A.I., 'The origin and initial development of life', Meditsina, Publ. House, Moscow (1966).
(13) Fox, S.W., and Dose, K., 'Molecular evolution and the origin of life', W.H. Freeman and Corp., San Francisco (1977).
(14) Yanagawa, H., and Egami, F., Proc. 12th Lunar Plan. Sympo., Tokyo, July (1979).
(15) Egami, F., Proc. Internat. Sympo. New Horiz. Biol. Chem., Nagoya, Nov. (1979).
(16) Jukes, T.H., Nature, 246, 22 (1973).
(17) Crick, F.H.C, Brenner, S., Klug, A. and Pieczenik, G., Origin Life, 7, 389 (1976).
(18) Eigen, M. and Schuster, P., Naturw, 65, 341 (1978).

ORGANIC GEOCHEMISTRY OF THE ISUA SUPRACRUSTALS

Clifford Walters, Akira Shimoyama,[1] Cyril Ponnamperuma

Laboratory of Chemical Evolution, Dept. of Chemistry
University of Maryland, College Park, MD. 20742

The Isua supracrustals of western Greenland are the oldest terrestrial rocks known, dated at > 3,750 m.y. Metamorphosed to lower amphibolite facies, they still can provide indications of their original environment of deposition. Graphite is widely dispersed throughout the meta-sediments, particularly within the ironstones. Most of the graphite is very well ordered. However, several samples show a marked disorder of crystallinity. These samples, when pyrolyzed at high temperatures, liberate organic fragments of low molecular weight (m/e < 200). The fragments suggest that at least some of the Isua graphite is derived from condensation of kerogen. Carbon isotope data has been interpreted as indication of photosynthetic fractionation. Whether the original organics, which are now seen as graphite, were biologically produced has not yet been unequivocally established. The pyrolyzed organics detected within the meta-sediments may well be the oldest molecular fossils yet found on the Earth.

The Isua supracrustal formation is located ∿150 km north-east of Godthab (Nuk), West Greenland and borders the inland ice. (Figure 1) A semicircular arc of meta-volcanics and meta-sediments surrounded by quartzo-feldspathic Amitsoq gneiss, the formation is recognized as the oldest known terrestrial rocks. Radiometric dating of the Isua meta-sediments (1,2) and the

[1]present address: Mining College, Akita University
Akita 010, Japan

Y. Wolman (ed.), Origin of Life, 473–479.

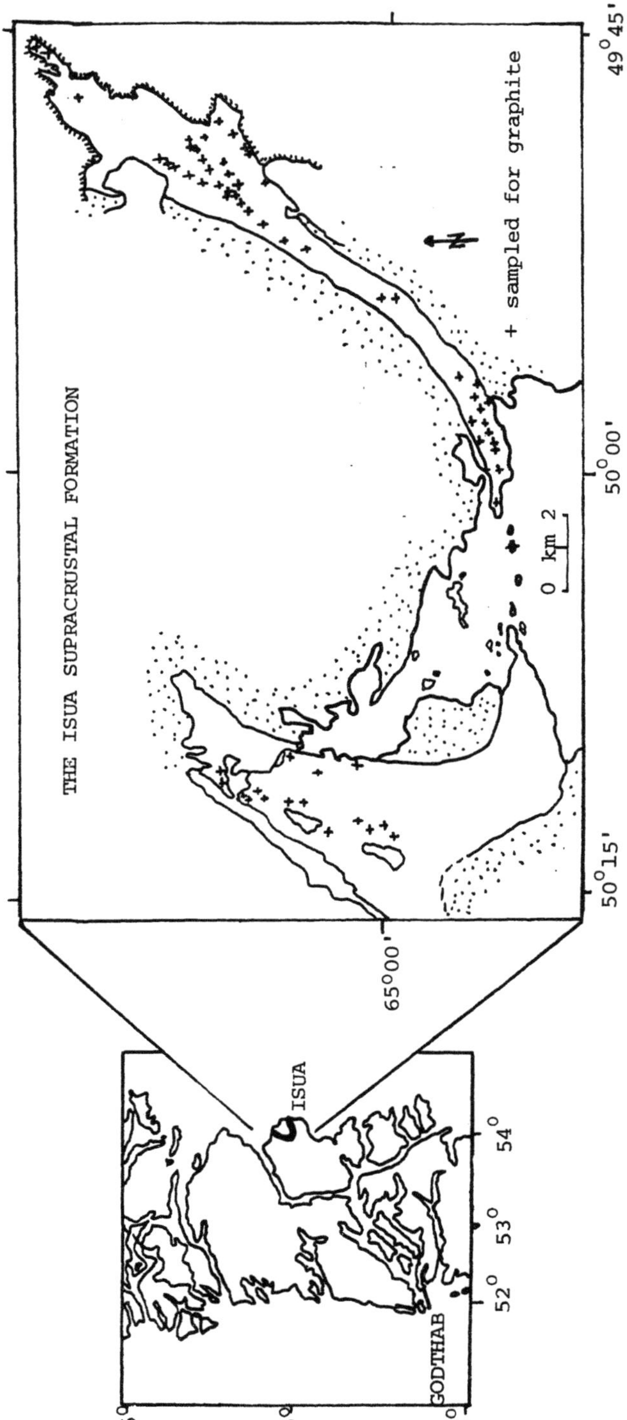
THE ISUA SUPRACRUSTAL FORMATION
+ sampled for graphite
N
0 km 2
49°45'
50°00'
50°15'
65°00'
ISUA
GODTHAB
54°
53°
52°
66°
65°
64°

Figure 1

Amitsoq gneisses yield (3,4) essentially identical ages, the earliest age determination being 3.824 $^{+.012}_{-.009}$ b.y. by U-Pb analysis of single zircons from acid boulders in a conglomerate unit (5).

The geology and major geochemical features of the Isua supracrustals have been described elsewhere (6,7). The manner in which the formation was developed is not known. Windley (8) has suggested that batholiths of active continental margins has similar lithographies. Others, however, suggest the various lithological units at Isua are unrelated and the present assemblage is a result of tectonic activity (9).

An immediate problem to any use of the Isua meta-sediments for studies concerning prebiotic or early biological evolution is the degree of metamorphism. Up to four periods of post-depositional recrystallization have effected the supracrustal rocks (10). This amphibolite facies regional metamorphism is well established, and earlier descriptions of greenschist facies mineralogy is a result of retrogressive metamorphism. It is therefore highly unlikely, if not impossible, that any morphological features of abiotic structures or biological organisms could survive. Reports of microfossils termed "Isuasphaera" (11,12) have been shown to be mineral stained fluid inclusions (13).

Since micropaleontological studies, which have provided the best evidence for living organisms in younger Archean - relatively unmetamorphosed sedimentary rocks- cannot be applied, only organic geochemical investigations remain. While the degree of metamorphism severely limits the amount of information which can be discerned, it is hoped that such geochemical studies can prove the existance of syngenetic organic matter within the Isua metasediments and whether this organic matter is abiotic or biological in origin.

Present studies are now centered on the graphite found within the Isua metasediments. A survey of over fifty ironstone outcrops has shown graphite to be most pervasive. Abundances range from <.01% for samples collected from the major magnetite-quartzite horizon in the far northeast to 2.73% for a particularly rich siderite-magnetite-quartz ironstone outcrop from the amphibolitic unit on the western branch of the formation. The average abundance for the iron formations is approximately 0.4% though the value is certainly high as the samples are not a random collection but centered on those which are from major horizons or those which were obviously graphite rich. Graphite is also found in appreciable amounts in other Isua meta-sediments (6) and some pelites contain up to 10% graphite (14).

Marked variation in the graphitic abundance can occur over short distances within a single outcrop. In a single ironstone unit, abundances varied more than 1% over a distance less than 10m. Furthermore, the distribution pattern is suggestive of some correlation between graphite abundance and fine-scale mineralogical layering.

The degree to which the graphite is ordered also varies greatly throughout the Isua formation. The degree of crystallinity can be compared by examining the peak intensity and shift of the (002) x-ray powder diffraction (15,16). The graphite from Isua meta-sediments varies from a totally ordered crystalline structure to a relatively disordered state (D-1a) (16). Again, the inhomogeneous nature of single Isua ironstone outcrop is demonstrated as such variations in crystallinity can occur between samples collected less than 10m apart.

If the graphite represents metamorphosed organic material which was deposited along with the sediments, those samples which possess the least ordered graphite may still have organic fragments which have survived metamorphism, though they would be in a highly altered state. The disordered crystalline structure may be due to organic fragments bound to the condensed aromatic centers which form the graphitic crystal structure. In our investigation, we have subjected extracted graphite from Isua meta-sediments to high temperature pyrolysis (up to 1000°C) and have evidence of some organic fragments liberated from the crystalline matrix.

In order to be certain that recent and laboratory contamination is removed from the samples to be studied, a detailed cleansing procedure is followed. After mechanical cleaning of the exterior, the sample is subjected to a series of solvent extractions and crushings until a powdered sample is obtained in which no organics are detected in further solvent extracts by gas chromatography. The sample is then treated in HCl and HF solutions to dissolve the rock matrix. Usually, these solutions are organic free; however, one ironstone showed a surprising amount of organic compounds which were liberated from the rock matrix. The origin of these organic molecules is uncertain and though it is perhaps unlikely that many of the compounds detected date to the time of metamorphism, some compounds, such as several biphenyls, are hard to explain in terms of recent contamination.

After treatment with HCl and HF solutions, the acid insoluble residue is collected, and further solvent extractions are made until no organics are evident. The residue consists primarily of graphite but may contain some sulfides or acid

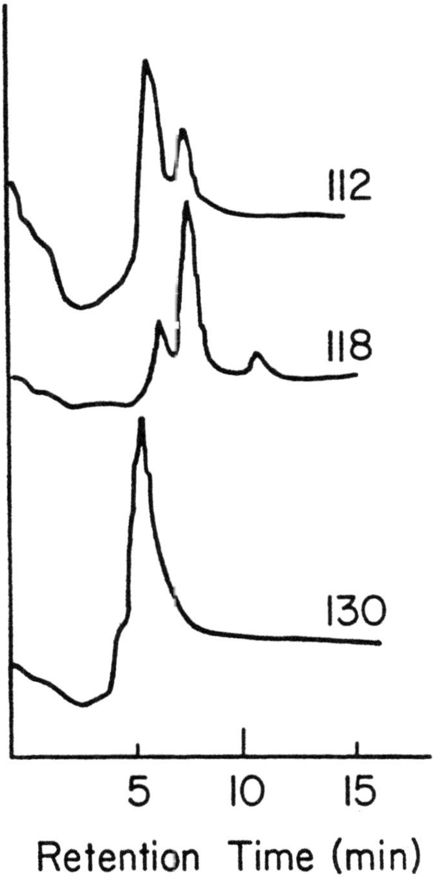

Figure 2

resistant minerals. The graphite is then subjected to stepwise pyrolysis-gas chromatography-mass spectrometry with the pyrolysis temperatures varying between 250-1000°C and usually lasting only for several minutes.

Previous studies of pyrolyzed Isua graphite have found only small fragments $\leq C_2$ (17,18). Our studies have confirmed these observations, however, some higher molecular fragments (up to m/e = 175) have been observed in subnanogram/gram levels. Unfortunately, detection of these fragments require that the mass spectrometer be operated on a single-ion monitoring mode which prevents further immediate identification. A benzene fragment has been identified.

The stepwise pyrolysis shows that the fragments are not merely absorbed on the surface of the graphite, but a truly chemically bound. At 250°C, no fragments can be detected above the background. At higher temperatures, fragments are broken free of the graphite and detected. Although of extremely low quantities, at least two possible three compounds, or groups of compounds, of relatively high m/e have been separated by a 50' OV-1 capillary column and detected by the mass spectrometer (HP-5992) (Figure 1). Positive identification is pending on current research.

While such studies seem to prove that at least some of the Isua graphite condensed from organic material during metamorphism, it is impossible to prove that these fragments are of biological or abiotic origin. The best evidence for life at 3.8 billion years ago are the carbon isotope values for Isua graphite (19,20). The values range from δC^{13}_{PDB} - 26.9 to -5.9°/oo and while the metamorphism has made interpretation very difficult, the values suggest that photosynthetic fractionation may have taken place, a model in agreement with a biological origin for the banded iron formations (4). Present research is centering on whether any correlations can be made between the presence of the banded iron, organic fragments and carbon isotope values. Even though the complex metamorphic history of the Isua formation has robbed Precambrian researchers of the usual techniques and has made interpretation of any data difficult, the secrets locked away for 3.8 b.y. are slowly being revealed.

ACKNOWLEDGMENTS: This research was supported by NSF grants DPP 7706993 and DPP 7900991. The authors also wish to thank Ms. J. Kemper for editorial assistance.

REFERENCES:

(1) Moorbath, S., O'Nions, R.K. and Pankhurst, R.J. Nature, 245:138 (1973)
(2) Hamilton, P.J., O'Nions, R.K., Evensen, N.M., Bridgwater, D. and Allaart, J.H. Nature, 272:41 (1978)
(3) Moorbath, S., O'Nions, R.K., Pankhurst, R.J., Gale, N.H. and McGregor, V.R. Nature Phys. Sci. 240:78 (1972)
(4) Moorbath, S., Allaart, J.H., Bridgwater, D. and McGregor, V.R. Nature, 270:43 (1977)
(5) Michard-Vitrac, A., Lancelot, J., Allégre, C.J. and Moorbath, S. Earth Planet Sci Lett, 34:449 (1977)
(6) Allaart, J.H. in The Early History of the Earth Windley, B.F. (ed.) John Wiley and Sons, London (1976), p.177

(7) Bridgwater, D., Keto, L., McGregor, V.R. and Myers, J.S. in Geology of Greenland, Escher, A. and Watt, W.S. (eds.) Geological Survey of Greenland (1976), p. 19
(8) Windley, B.F. The Evolving Continents, John Wiley and Sons, London (1977), p.59
(9) Bridgwater, D., Allaart, J.H., Baadsgaard, H., Collerson, K.D., Ermanovics, I., Gorman, B.E., Griffin, W., Hanson, G., McGregor, V.R., Moorbath, S., Nutman, A.P., Taylor, P., Tveten, E. and Watson, J. Rapp. Grnolands geol. Unders, 95:66 (1979)
(10) Bridgwater, D., Collerson, K.D. and Myers, J.S. in Evolution of the Earth's Crust Tarling, D.H. (ed.) Academic Press, London (1978), p.19
(11) Pflug, H.D. Naturwissenschaften, 65:611 (1978)
(12) Pflug, H.D. Nature, 280:483 (1979)
(13) Barghoorn, E.S. pers. comm.
(14) Bridgwater, D. pers. comm.
(15) French, B.M. Science, 146:917 (1964)
(16) Landis, C.A. Contr Mineral Petrol, 30:34 (1971)
(17) Nagy, B., Zumberge, J.E. and Nagy, L.A. Proc Nat Acad Sci, 72:1206 (1975)
(18) Leventhal, J., Suess, S.E. and Cloud, P. Proc Nat Acad Sci, 72:4706 (1975)
(19) Oehler, D.Z. and Smith, J.W. Precambrian Res, 5:221 (1977)
(20) Schidlowski, M., Appel, P.W.V., Eichmann, R. and Junge, C.E. Geochim Cosmochim Acta, 43:189 (1979)
(21) Cloud, P.E. Paleobiol, 2:351 (1976)

SECONDARY STRUCTURES OF POLYPEPTIDES AS EVOLUTIONARY "GROWING POINTS"

H. Baltscheffsky

Department of Biochemistry, Arrhenius Laboratory, University of Stockholm, S-106 91 Stockholm, Sweden.

Our theoretical predictions on secondary structure content in Clostridia-type ferredoxin apoproteins and our subsequent experimental results indicate that these polypeptides show much higher tendency to form β-pleated sheet structures than has been evident from X-ray data on the iron and labile sulfur containing ferredoxins. Such β-structures may be useful in orienting appropriate cysteine sulfur residues to bind two Fe atoms in each of the two 4(Fe + S) cubes which together with the apoprotein form the active ferredoxin molecule. In this process the β-structures would essentially disappear. The concerted hydrogen bond changes would constitute a functionally useful secondary structural change. Such useful concerted hydrogen bond changes may occur much more frequently during protein (and nucleic acid) function than is currently recongnized. The 3-D structures formed by concerted hydrogen bond formation (α- and β-structures in polypeptides), in particular those which are capable of reversible, functionally useful changes, may constitute important evolutionary "growing points", which are conserved and elaborated upon during evolution to more and more complex 3-D patterns, found in various present-day proteins.

Recently, we have predicted (1) that ferredoxin apoproteins of the Clostridia-type tend to form β--pleated sheet structures, and shown (2) that such structures indeed appear to be induced, when ethanol

Y. Wolman (ed.), Origin of Life, 481–486.

in suitable concentrations is added to a weakly buffered aqueous medium of ferredoxin apoprotein from Clostridium pasteurianum. The alcohol was added to decrease the polarity of the medium and to thus allow the predicted secondary structurization of this low molecular weight polypeptide. At high alcohol concentrations, α-helix formation became dominant.

As has been pointed out earlier, the expected β-structures of ferredoxin apoprotein would seem to a) serve a functional purpose (1) and b) bridge a previously apparent gap in the expected continuity at the 3-D structural level between prebiological and biological polypeptides (3). The predicted usefulness of the ferredoxin apoprotein antiparallel β-structures lie in what seems to be an ideal spacing of the sulfurs of two cysteine residues for binding two iron atoms in a 4(Fe + S) cube. This advantageous spacing of two cysteine residues would occur at two different locations in the ferredoxin apoprotein, as is shown in Fig. 1, and facilitate the initiation of the binding of both the 4(Fe + S) cubes which the active ferredoxin molecule contains. Thus a more or less shortlived temporary β-structure would serve to elicit the formation of

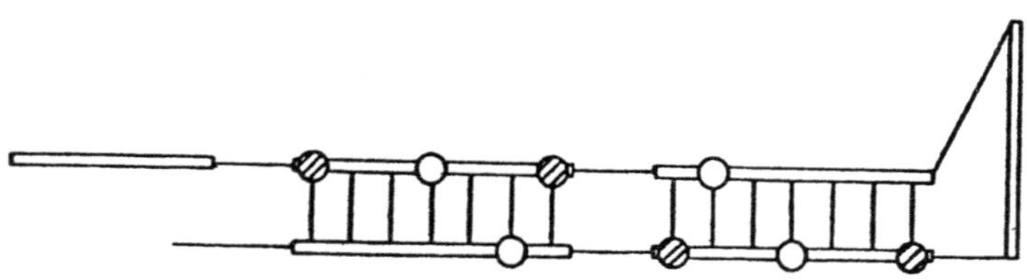

Figure 1. Schematic drawing of a predicted Clostridial ferredoxin apoprotein structure. The six heavy bars along the polypeptide chain represent stretches of extended structure, the first bar beginning at the free amino end. The filled and unfilled circles represent cysteine residues with side-groups projecting out from different sides of the two antiparallel β-pleated sheets, which are indicated by the thin connecting lines representing hydrogen bonds. The cysteine residues represented by unfilled circles appear to be ideally spaced to bind Fe-atoms in 4(Fe + S) clusters and thus to initiate their binding to the apoprotein.

the ferredoxin molecule, and essentially disappear in the process. This is in agreement with the published 3-dimensional structure of Peptococcus aerogenes ferredoxin (4), which shows very little secondary structurization.

Table 1 shows some data obtained from our circular dichroism measurements on the ferredoxin apoprotein from Clostridium pasteurianum. As had been expected, a weakly buffered water solution appeared to be a medium of too high polarity to allow any appreciable secondary structurization of this low molecular weight polypeptide (55 amino acids) to occur. Decreasing the polarity with increasing concentrations of ethanol gave more β-pleated sheet than α-helix structure at intermediate concentrations of alcohol, whereas the former structure decreased and the latter increased drastically when the ethanol concentration became very high. A strong tendency for α-helix formation in lipid media was no surprise, but the novel and essential observation to us was that β-pleated sheet structures

TABLE 1

Secondary Structures Of Clostridium pasteurianum Ferredoxin Apoprotein At Various Concentrations Of Ethanol

% Ethanol	% β-Pleated Sheet	% α-Helix
0	-	-
25	10	10
50	30	20
75	30	30
90	20	50

The data were obtained by running samples of apoprotein in 10 mM Tris-buffer pH 7.8, at 2°C, in a Jasco CD Instrument, as described in Ref. 2. Ethanol was added to decrease the polarity of the medium and thus allow secondary structurization of the small polypeptide. The experiment was performed in collaboration with the other authors given in Ref. 2, at University of London King's College (in March 1979).

could be elicited under appropriate conditions. This result would seem to support our prediction that the apoprotein is capable of β-pleated sheet formation and to further substantiate our hypothesis that a functionally useful secondary structural change may occur when the apoprotein binds iron-sulfur clusters to form ferredoxin. We have earlier (5) discussed the possibility that concerted hydrogen bond changes involving α- and β-structures of polypeptides may play a much more general and important role in protein function than has hitherto been assumed.

What is the evolutionary relevance of the ferredoxin apoprotein and its predicted and observed β--structurization? Well known amino acid contents and sequences of several Clostridia-type ferredoxins indicate their unique "primitiveness":
a) their content of those amino acids found in meteorites and in Miller-type prebiotic amino acid formation experiments is very high;
b) a similar parallelism holds for the amino acids in the ferredoxins and those first made by a primitive, evolving genetic code, according to Eigen and Schuster (6);
c) an early gene duplication event can be deduced from the current sequences.
But the function and also the evolution of proteins are most closely related also to their 3-D structures. Thus it is necessary to discuss in some detail, against the background given above, the evolutionary aspects of secondary structure in ferredoxins and, more generally, in proteins - so important for the origin and evolution of life. The treatment of these aspects will be focussed on the following questions:
1. Does the β-structurization of ferredoxin apoprotein indeed seem to bridge an apparent gap in evolutionary continuity in 3-D structure between prebiological and biological polypeptides?
2. Is secondary structural change capacity a generally useful property in protein function and thus a fundamental parameter in polypeptide evolution?
3. Can secondary structures and their capacities for functional changes be considered to be "growing points" in the molecular evolution of polypeptides (and polynucleotides)?

Let us retain the above numbers in our discussion:

1. In the early seventies Orgel (7) and Brack and Orgel (8) gave theoretical and experimental support to the concept (7) that prebiotic oligo- and polypeptides should have shown a high tendency to form β-structures. It is encouraging that the "most primitive" biological polypeptides seem to form anti-parall-

el β-structures, but one should not over-emphasize their evolutionary significance, as different types of secondary structure may frequently appear and disappear in the course of biological evolution.

2. The capacity of a polypeptide to reversibly change its secondary structural pattern in a functionally useful manner, so that the concerted hydrogen bond changes thus elicited is an essential part of the function, appears to exist in ferredoxin apoprotein as well as in other proteins, as has been discussed earlier (9). Both soluble and membrane-bound proteins may have this capacity. Small temporary changes in lipophilicity at an active site of a membrane-bound protein could have a profund influence on a functional secondary structure, for example in a case such as was recently discussed for membrane-bound energy conversion mechanisms (5). It has been pointed out (5) that a polypeptide 3-D structure with capacity for functionally useful, concerted hydrogen bond changes may well possess extremely high "survival" property during biological evolution. In addition, it may be expected to be elaborated upon, stepwise, in the process of biological evolution, to give, in various evolutionary directions, new variations on the secondary structure and secondary structural change themes, often leading to additional structural complexity at the protein molecular level. More about this in the next paragraph.

3. Although molecular evolution primarily may be seen to work on sequence in the structures of polynucleotides and polypeptides, the structure of a polypeptide at the 3-D level has long been recongnized to be the functionally important form on which protein evolution, in the final analysis, operates. In line with this is the clear tendency of secondary structure in polypeptides to be more strongly conserved during evolution than primary structure. Together with stronger covalent bonds and configurations formed by the actions of "hydrophobic forces", the hydrogen-bonded α- and β-structures of polypeptides are important elements in the 3-D structures of proteins. They may well be significant "growing points" in the molecular evolution of polypeptides, which form and evolve, for various structural and functional reasons, to give the abundance of variations seen in present-day proteins. It would appear that tightly hydrogen-bonded 3-D structural elements, whether short- or long-lived, in polynucleotides and polypeptides may have important and even related functional properties, based on the speci-

ficity and the versatility of the hydrogen bond.

ACKNOWLEDGEMENT This work has been supported by The Swedish Natural Science Research Council (Grant Nr. 2292-100).

REFERENCES

1 von Heijne, G., Blomberg, C., and Baltscheffsky, H. Origins of Life 9, 27 (1978).
2 Baltscheffsky, H., Rao, K.K., Ryan, M., Hall, D.O., and Drake, A.F. (in preparation).
3 Baltscheffsky, H., Living Systems as Energy Converters, R. Buvet et al. eds., Elsevier/North Holland, Amsterdam, 1977, 81.
4 Adman, E.T., Sieker, L.C., and Jensen, L.H. J. Biol. Chem. 248, 3987 (1973).
5 Baltscheffsky, H., Energy Conservation in Biological Membranes, G. Schäfer and M. Klingenberg eds., Springer, Berlin, 1978, 3.
6 Eigen, M., and Schuster, P. Naturwiss. 65, 341 (1978).
7 Orgel, L.E. Isr. J. Chem. 10, 287 (1972).
8 Brack, A., and Orgel, L.E. Nature 256, 383 (1975).
9 Baltscheffsky, H., From Cyclotrons to Cytochromes, N.O. Kaplan ed. (in press).

SEARCH FOR PRIMITIVE REPLICATIVE PROPERTIES ON EARLY POLYPEPTIDES.

André BRACK and Gérard SPACH

Centre de Biophysique Moléculaire, CNRS, 1A, avenue de la Recherche Scientifique, 45045 Orléans, France.

The sequences of polypeptide chains which are built up with alternating hydrophilic-hydrophobic residues concentrate themselves in the β-pleated sheet conformation when dissolved in salted water. The sheets are more stable than the α-helix and are shown to constitute a pathway to the enrichment in one enantiomer. They are considered as a single system with potential template activity.

One of the most crucial step during chemical evolution was undoubtly the emergence of the first self reproducing molecules. The fact that only one type of enantiomer (L-amino acids and D-sugars) is used by the present day living systems suggests that an asymmetric template system appearing at once but self-sustaining by selection of one enantiomer and able through its catalytic activity to orient the synthesis of polypeptides or polynucleotides in a given direction could have plaid this key role. If such an easily expanding process took place, one must be able to reproduce part of it in the laboratory.

With regards to the contemporary biological replicating systems, it is tempting to look at nucleic acids as candidates for such a process. However, it is well known that nucleic acids are produced less easily in mimicking laboratory experiments than are amino acids, for example. They are not very stable and their accumulation on the primitive Earth could have been difficult. A way to overcome these difficulties is to invoque alternatives to nucleic acids, which were displaced by the biopolymers later in chemical evolution. Amino nucleotides were checked by Lohrmann and Orgel (1). More recently, Nelsestuen(2) obtained

Y. Wolman (ed.), Origin of Life, 487–493.

polymers that contain purines, pyrimidines and amino acids, acting as macromolecular catalyst and capable of base pairing and stacking with ribonucleic acids. He started with α, β unsaturated compounds and various nucleophiles. Even more far away from the biopolymers is the genetic crystal proposed by Cairn-Smith (3). Another approach consists to look for a template activity among the polypeptide material, the spontaneous formation and accumulation of which seem possible under prebiotic conditions.

β-sheet structures of water soluble polypeptides with D- and L-residues

A special attention has been devoted to the conformation of simple polypeptides and more especially to the formation of β-structures since they are flat charged surfaces. In this structure, peptide chains are almost completely extended and hydrogen bonds bind neighbouring chains leading to the formation of a pleated sheet. The side-chains point alternately up and down while the C_α proton is roughly in the plane of the sheet. On models, it appears difficult to replace the C_α proton by a bulkier side-chain, situation which would happen if one tries to incorporate D-residues in a pleated sheet made of L-residues or if one tries to incorporate non-natural amino acids such as α-amino isobutyric acid.

β-structures accomodate nineteen of the twenty natural amino acids (except proline). In addition, we have previously shown that when hydrophilic and hydrophobic residues alternate along the chain, the chains are forced to adopt the β-structure in salted water (4)(5)(6). Due to alternative repetition of the side-chains, one gets a polypeptide bilayer with a hydrophobic interior and a charged hydrophilic exterior schematically represented on figure 1.

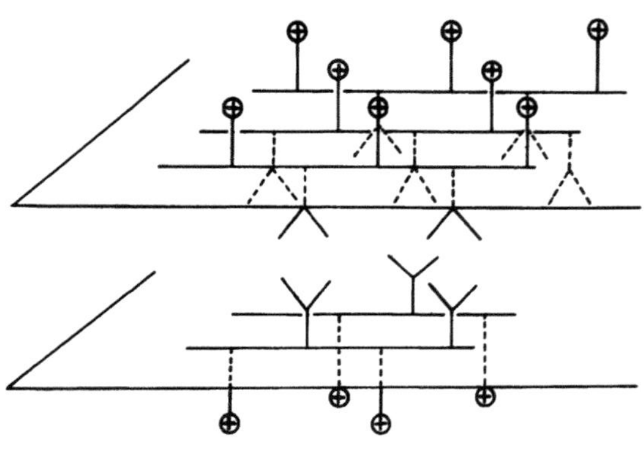

Fig. 1 - Schematic drawing of a bilayer of alternating polypeptide with charged hydrophilic (⊕) and hydrophobic (Y) residues.

Fig. 2 - Scheme of a poly DL peptide chain showing β-sheet nuclei (circled) surrounded by random coil segments.

We have also shown that the β-structures cannot be formed with disubstituted unnatural amino acids which are formed under prebiotic conditions like α-amino isobutyric acid (Aib) : poly (Aib-Gly-Gly) does not exhibit any β-structure.

To check if the β-structure is an asymmetric surface discriminating the D-residues when made of L-ones, we have prepared samples of alternating poly(Leu-Lys) with increasing amounts of D-residues randomly distributed along the chain but evenly shared out amongt leucyl and lysyl residues (7). In salted water, the samples exhibit a mixture of random coil and β-sheet conformation, the random coil contribution becoming dominant when the composition is closed to the racemic mixture. The good agreement between the experimental data and the results of a statistical analysis based on a model were only those segments having seven adjacent residues of identical chirality can associate to form β-sheets suggests the image of figure 2. Nuclei of β-sheets are formed surrounded by random coil segments containing both L- and D-residues.

On the other hand, the pool of β-sheet nuclei is rapidly enriched in L-residues when the L,D composition departs from the racemic mixture (figure 3). If this image is correct, random coil segments will be more sensitive to soft hydrolysis than β-nuclei. This has been checked experimentally (8). Kinetic measurements show two pseudo first order rate constants, in agreement with the existence of two conformational species. The results are examplified on figure 4. The change in the slope happens very rapidly for samples rich in L-residues and therefore rich in β-structures whereas it occurs lately when the random coil contribution is important. The remaining polymers were analyzed at these transition points : all of them are enriched in L-residues as predicted from the model.

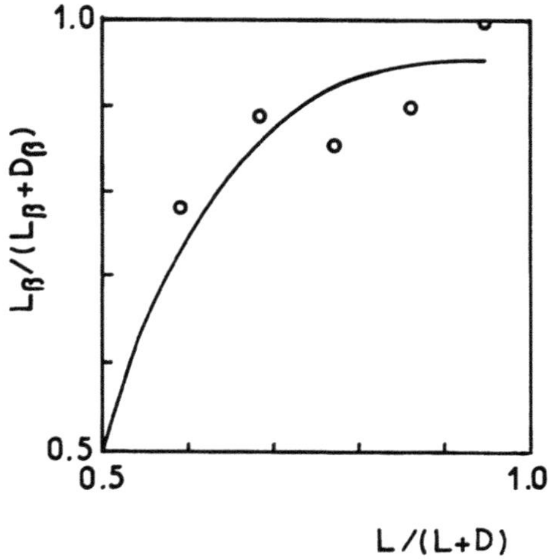

Fig. 3 - Fraction of L-residues in the pool of β-sheet nuclei as a function of the fraction of L-residues in the chains.

This model suggests a mode of amplification of the dominant enantiomer if the β-nuclei fraction could be isolated, then eventually depolymerized and polymerized again (9).

In addition to their enantiomeric discriminating property, β-sheet structures were found to be thermally stable whereas the α-helices are broken. This difference in behaviour appears when alternating poly(Leu-Lys) and random poly(Leu^{50}, Lys^{50}), which exhibits an α-helix at room temperature, are treated under the

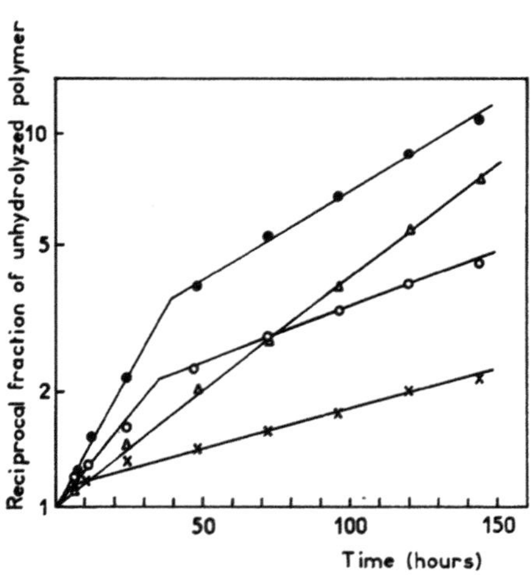

Fig. 4 - Hydrolysis kinetic of different polymers : alternating poly(Leu-Lys) with 95 % L-residues (x), with 86 % L-residues (O) and with 77 % L-residues (●) ; random poly(Leu,Lys) (Δ). HCl 0.1 M, 110°C.

same soft hydrolytic conditions. Random poly(Leu,Lys) is degraded at the same rate as the random coil fraction of poly(Lys-Leu) (figure 4), much faster than the β-fraction. Thus, at least at high temperature, sequences able to adopt the sheet conformation will be protected against hydrolysis.

Multiconformational water soluble polypeptides

Since random poly(Leu,Lys) transconforms to the α-helix in presence of salt, it was thought that polypeptides containing simultaneously random and alternating sequences of leucyle and lysyle residues may undergo transconformations modelling the self-organization of "primitive" proteins. Seven samples with varying amounts of alternating sequences were prepared and their conformations were studied in aqueous solution of varying ionic strength and temperature. In $NaClO_4$ solution (figure 5), the samples poor in alternating sequences take mainly the α-helical conformation. When the fraction of these sequences reaches 70 %, α- and β-structures coexist in comparable amounts. Above 70 %, the β-fraction dominates. Heating the sample containing 70 % of alternating sequences at 60°C induces a higher amount of β-structure with a loss of α-helix. Using NaCl instead of $NaClO_4$ promotes the appearance of an additional unordered conformation. Thus, it is possible to built up β-sheets surrounded by α-helices and random coil segments with two amino acids only just by playing with the sequence, at least when the amino acids have the same chirality.

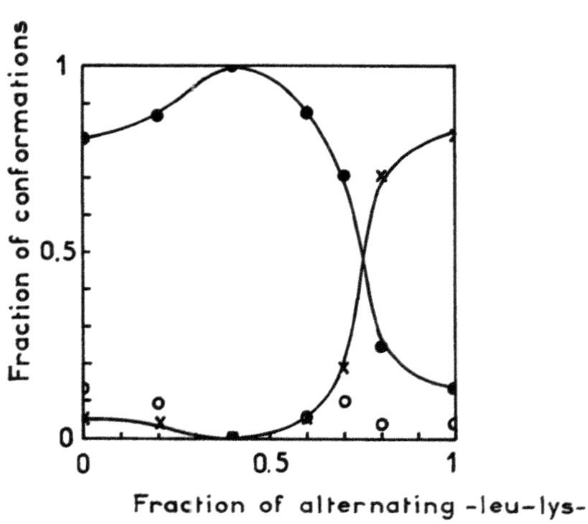

Fig. 5 - Fraction of different conformations, random coil (o), α-helix (●) and β-sheet (x) as a function of the fraction of alternating sequences in different samples of poly(Leu,Lys).

Prebiotic significance

The following scenario can be imagined : small alternating peptidic segments concentrated in highly discriminating, thermally stable β-nuclei, even if diluted in random sequences. These β-surfaces are good candidates for the search of a template activity as discussed in the introduction. Therefore we tried to polymerize activated nucleotide derivatives or activated amino acids on β-surfaces. At first, we tried to induce a coil → β transition of poly(Leu-Lys) induced by the negative charges of adenosine-5'-phosphorimidazolide (ImpA), then to polymerize the activated nucleotides kept close to the polymer. In fact, addition of ImpA precipitates the polypeptide under a β-form. However, no polymerization could be obtained at different pHs and temperatures. ImpA seems to be covalently bound to the polypeptide through a ε-lysyl phosphoramidate which did not lead to polynucleotides in our hands. To prevent this side reaction, we have prepared samples of poly(Arg-Leu), the synthesis of which is rendered particularly tedious by the presence of free arginine. Polymerization experiments with this polypeptide are under the way.

Concerning the polymerization of activated amino acids, a preliminary work has consisted in the search of a type of activation efficient in water. We tried water-soluble carbodiimide, azides, N-hydroxysuccinimid and p-nitrophenyl esters. The best activation was found to be p-nitrophenyl esters at pH 8.5. Experiments with glutamic acid p-nitrophenyl ester and poly(Arg-Leu) are under the way.

These systems could perhaps be more efficient if associated with compounds such as clays or phospholipids. Preliminary experiments showed that when mixing alternating polypeptides with sodium-montmorillonite, the clay becomes hydrophobic, 90 % of the sodium salt is released and the polypeptide exhibits the β-conformation. On the other hand, alternating polypeptides interact strongly with lysolecithin to form a mixed complex in which the peptide has the sheet conformation. Polymerization reactions with such more complicated systems will be undertaken.

Conclusion

The work described in this paper is based on the search for template activity on polypeptides. During the course of chemical evolution such a template would have had two advantages : firstly to maintain sequences of peptides emerging at once and necessary for further evolutionary steps ; secondly, to promote the synthesis of nucleic acids, giving rise to a relationship between peptides and nucleic acids. A template activity based on peptide is not relevant to contemporary processes but could have been displaced by better ones when a replicating system using polypeptides and polynucleic acids was formed. This scenario is supported by the fact that it is hard to imagine how nucleic acids and polypeptides could have evolved separately toward sophisticated systems able to interact and to give rise to the first controlled synthesis of peptides. Although we have no definitive answer concerning the catalytic properties of β-sheet surfaces, we bring evidences for their plausible existence under probiotic conditions and for their potentiality to concentrate elements having the same sequence and the same chirality.

REFERENCES

(1) Lohrmann, R., Orgel, L.E., Nature 261, 342 (1976)
(2) Nelsestuen, G.L., J. Molec. Evol. 15, 59 (1980)
(3) Cairns-Smith, A.G., The life puzzle. Edinburgh : Oliver and Boyd (1971)
(4) Brack, A., Orgel, L.E., Nature 256, 383 (1975)
(5) Brack, A., BioSystems 9, 99 (1977)
(6) Brack, A., Caille, A., Int. J. Peptide Protein Res. 11, 128 (1978)
(7) Brack, A., Spach, G., J. Molec. Evol. 13, 35 (1979)
(8) Brack, A., Spach, G., J. Molec. Evol., in press (1980)
(9) Spach, G., Brack, A., J. Molec. Evol. 13, 47 (1979)

EMERGENCE OF FLAVIN CATALYSIS. AN APPROACH BASED ON THE CONCEPT OF BIOORGANIC EVOLUTION

C.M. Visser

Department of Organic Chemistry, University of Groningen, Nijenborgh 16, 9747 AG Groningen, The Netherlands

Bioorganic chemical analysis of present-day coenzyme mechanisms (via chemical models) can reveal something about the stage of evolution of life in which those cofactors were introduced (and vice versa). Some cofactors date back to the time before the origin of the genetic code (RNA-like metabolism) whereas others are relatively recent. Flavin prosthetic groups are anomalous however, in that they have pre-genetic code structures but a great number of post-genetic code functions (oxygen handling) and reaction mechanisms. The recent isolation of a deazaflavin cofactor from one of the Archaebacteria suggests a possible way out of this dilemma. The significance of this for present-day flavoenzyme mechanisms is discussed.

Enzymes catalyze reactions through binding of substrates, intermediates and the transition states connecting them by a multitude of ways (hydrogen bonding, hydrophobic bonding, salt bridges, charge transfer and dipole-dipole interactions, general acid-base catalysis, covalent catalysis) to such an extent that often the binding of substrates or the release of products becomes rate determining (1). All these contributions to catalysis can be encompassed in the requirement that the active site of an enzyme be closely complementary (2) to the transition state of the reaction that has to be catalyzed (hydrogen bond donor located against an acceptor, cation against anion, etc.). The difference between bonding of the transition state and of substrates and products explains the catalysis. One can assume that free energy and structure of the substrate in the transition state are comparable to those of the transition state of the non-enzymatic reaction. The free energy of the substrate in the

Y. Wolman (ed.), Origin of Life, 495–504.

enzymatic reaction will not be expected to be much higher, since enzymes evolved for lowering overall activation energy, nor can it be lower, since any reaction (and thus also the non-enzymatic one) proceeds by definition via the transition state with the lowest free energy (3). This resemblance between enzymatic and non-enzymatic transition states is the basis for bioorganic chemistry in general. The demonstrated success of the bioorganic approach can be seen as an indication that evolution fundamentally is an additive process on this submolecular level too (4).

In metabolism, catalysis by enzymes is often additionally facilitated by the use of a substrate that is a relatively "high energy" compound in itself and has a high potential for group transfer, like ATP, NADH, pyridoxalphosphate etc. Because such substrates are recycled in a second enzymatic reaction, they can be considered as part of a catalyzing system (coenzymes). In later stages of evolution such cofactors can become attached to an enzyme or complex of enzymes, which then catalyzes both half-reactions. The coenzyme (which is a substrate) has then become part of the catalyst (prosthetic group), and its role has become analogous to that of other functional groups (amino acid side chains) on the active site. A great deal of bioorganic chemistry is concerned with the chemistry of these cofactors. By research on chemical models (thus the enzymatic reaction minus the polypeptide enzyme) it is determined what part of catalysis can be described to the cofactor itself. Since, however, evolution on this level too is an additive process, one is in fact reconstructing primitive metabolic reaction mechanisms that were used in the time that those cofactors were introduced in metabolism for the first time. In this sense bioorganic chemistry is research into the origin of metabolism on the submolecular level of reaction mechanisms. This process could be called bioorganic evolution and research into this level of complexity can be expected to contribute considerably in the future to the understanding of how chemical evolution became biochemical evolution. Central items in this bioorganic evolution are origin of the genetic code, of bioenergetics and of biocatalysis.

Judging from their chemical structure, their role in metabolism and their bioorganic chemistry one can conceptionally divide existing cofactors into one of the four following groups(5): I. Coenzymes and prosthetic groups with a partial or complete ribonucleotide-like structure (nitrogenous heterocycle, sugar or alcohol, phosphate). To this group belong most of the familiar group transferring cofactors. Their chemistry is relatively easily accessible via non-enzymatic models. Both characteristics (structure and chemistry) indicate that this group of cofactors is the eldest (6) in evolution and could be taken as metabolic "fossils" from the time before the origin of the genetic code, when RNA-like molecules still combined the functions of replication, of energy processing and of catalysis (4,7,8). To this group one could count for instance ATP and the other nucleotides, coenzyme A, thiamine diphosphate, pyridoxal phosphate, NADH, NADPH, tetrahydrofolic acid, vitamin B_{12}-coenzymes, adenosylmethionine, t-RNAS and the UDP- and GDP- sugars. It is striking that this group contains no light harvesting molecules. Seemingly, daylight in this pre-genetic code era was too harmful to be of any use. One can speculate therefore that evolution started with RNA replication in areas protected by an absorbing film of (abiotically synthesized) fatty molecules. Probably most cofactors from the next group are derived from these molecules.

II. Cofactors absorbing light energy and useful from the standpoint of protection against damaging radiation or harvesting of light energy. Examples: carotenoids, retinol (vitamin A), vitamin D, xanthophylls, fucoxanthin, chlorophylls, possibly some porphyrins, phycocyanobilins etc. Many of these cofactors have a hydrophobic polyprenyl part (9). It is conceivable that this group entered metabolism simultaneously with or immediately after the origin of the genetic code.

III. This group comprises a number of cofactors presumably introduced as oxygen detoxificators at a still somewhat later time, but definitely before the atmosphere contained significant amounts of oxygen. To this group belong undoubtedly many of the hemes, other iron and copper containing proteins (superoxide dismutases, for instance) tetrahydrobiopterin-iron complexes and possibly vitamin C and the vitamins E and K.

IV. Prosthetic groups that gained their functions at a presumably still later time and have no independent catalytic power but must be associated with the polypeptide part of the enzyme. To this group belong for instance α-ketoacid prosthetic groups, derived from an amino acid of the enzyme (pyruvyl and α-ketobutyryl), and biotine, a typical "prosthetic substrate" (11).

The above tentative subdivision of cofactors shows that new catalytic groups were introduced throughout the whole process of the origin of life, but perhaps predominantly at crisis points (group I: origin of biocatalysis; group II: origin of photosynthesis; group III: origin of oxygen metabolism). A comparative bioorganic analysis of the entrance of these cofactors seems thus very promising for the understanding of evolution. On the other hand, placing cofactors with unknown functions or unknown reaction mechanisms in this evolutionary framework can give hints for new hypotheses for their functions, relatively independent of experimental biochemical research (12). Two of the most important prosthetic groups have not yet been mentioned, [FeS] clusters and flavins. The former are formed spontaneously, are found in all life forms and have considerable catalytic properties of their own; they are probably therefore as old as the cofactors of group I. But on trying to fit the flavin prosthetic groups FAD and FMN in this scheme, serious problems arise.

The numerous metabolic half-reactions (13) in which a flavin prosthetic group is reduced or oxidized again can for convenience be divided in four types: a, one-electron transfers; b, a (formal) hydride transfer from NADH; c, oxygen reactions; and d, two-electron transfers other than from and to NAD(P)H. a. This half-reaction is part of numerous dehydrogenases (electron transfer to hemes, [FeS] clusters, to other flavin prosthetic groups). Only the flavodoxins (14,15) are mentioned here. These are small flavin containing proteins with no other prosthetic groups or coenzymes. In this they are comparable with ferredoxins and a number of c and b cytochromes. They shuttle between the fully reduced state and neutral radical (scheme I). They have consid-

Scheme I

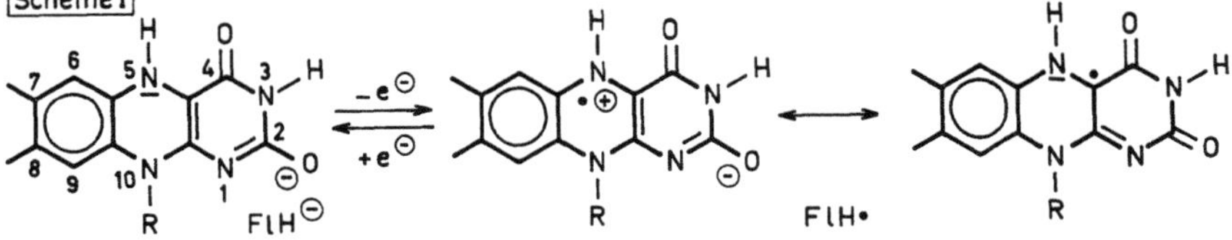

erably lower redox potentials than the free flavins and can therefore replace ferredoxins in a number of functions (photosynthetic reduction of NAD⊕). They function in anaerobic oxidation of pyruvate (phosphoroclastic oxidation) and transfer electrons to, for instance, hydrogenases and nitrogenases. These are probably very old functions (preceding even photosynthesis). Since flavins are formed abiotically (16) and since the reduced form under conditions of the group I cofactors (no oxygen, no excessive light) is stable, the one-electron transfer function can originate from pre-genetic code type metabolisms, provided that flavin is surrounded by an isolating polypeptide (or polymeric sugar) which prevents self-pairing and stacking and stabilizes the semiquinone. b. Reduction by NADH can take place non-enzymatically (17). Enzymatic reductions with NADH are generally considered to involve hydride transfers, but the non-enzymatic mechanisms are uncertain (18). Moreover, flavins are known to be bad hydride acceptors in dark reactions with obligate hydride donors (19). Still, the earliest origin of, for instance, the present-day NADH dehydrogenase reaction (electron transfer from NADH to a [FeS] cluster) is a combination of flavin half-reactions that in principle can date back to the pre-genetic code era. In this sequence flavin has already the function of "splitting" the electron pair in the hydride of NADH. Since however, the redox mechanisms of models for both cofactors is still uncertain, it is difficult to estimate the probability of such an evolutionary scheme. c. Reactions of reduced flavins and oxygen have also been subjected to a detailed analysis (20,21,22,13). A relatively simple scheme arises, applicable both to enzymatic as well as non-enzymatic reactions of appropriate model systems (scheme II). Oxidation starts with a one-electron transfer as

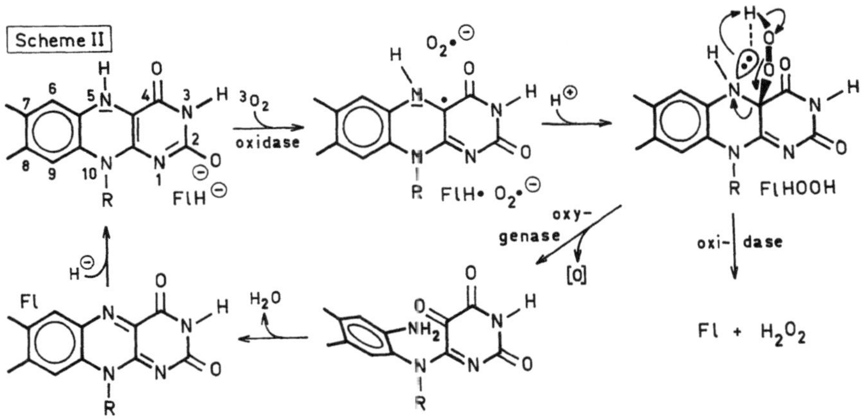

in a; a direct reaction leading to FLHOOH is spin-forbidden. Flavin containing superoxide generating oxidases are known (23). Moreover, many dehydrogenases produce in vitro superoxide when an "unnatural" oxygen reaction is allowed (24). In a next step a flavin hydroperoxide is formed that can split off H_2O_2 (oxidase) or can give hydroxylation reactions via an oxenoid mechanism (monooxygenases), similar to those of peracids and α-aminohydroperoxides. The ring-opened flavin is a good candidate for the light emitter in bacterial luciferase, an aldehyde monooxygenase (25), and will quickly reclose to flavin. The broad analogy between enzymatic and non-enzymatic reactions with oxygen marks the reduced flavins as excellent group III oxygen detoxificators. Thus far there are no great problems in imagining the bioorganic evolution of flavin cofactors. d. The remaining great number of half-reactions are for the most part reductions of flavins by substrates like succinate, fatty acylcoenzyme A, α-hydroxyacids and α-aminoacids. A great deal of the discussions on flavin mechanisms centers around the mechanisms of these substrates (20,21,26). For the enzymatic oxidation (27,28,29,30) of D- and L-α-aminoacids, α-hydroxyacids and amines by flavoproteins the generally held opinion favors formation of substrate α-carbanions, followed either by formation of a covalent adduct to flavin, or by one-electron transfer (scheme IIIA). The alter-

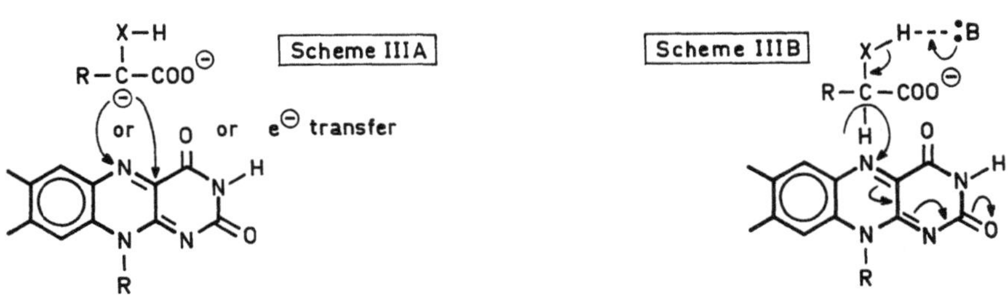

native has always been a base catalyzed hydride transfer, analogous to NAD$^{\oplus}$-dependent lactate dehydrogenase. In favor of scheme IIIA are a number of enzymatic experiments with β-halosubstrates (26) (dehydrohalogenation). However, this interpretation has been challenged (31) because non-enzymatic dehydrohalogenation of β-haloamines and -alcohols proceeds by means of X-H rather than C-H deprotonation affording aziridines and epoxides. In this new interpretation the experiments with β-halosubstrate analogs would plead for scheme IIIB. Apart from direct experimental evidence, the general opinion has drifted away from hydridoid mechanisms also on the ground of the bad hydride donating and accepting ability of non-prosthetic flavins (19). Moreover, a scheme IIIB mechanism cannot apply to succinate dehydrogenase and fatty acylcoenzyme A dehydrogenases.

A most important manner in which protein active sites will have evolved is by binding a potential substrate in an apolar surface cleft by hydrogen bonding. If such binding has a catalytic effect on a useful reaction, a rapid evolution can be expected to occur leading to a more perfected adjustment of the hydrogen bonds for general acid-base catalysis (32). However, since C-H bonds do not take part in hydrogen bonding (nor do carbanions) such a gradual evolution of carbon acid deprotonation is not possible. Evolving active sites will tend to avoid such "sudden" evolutionary pathways. All this means that for both reaction mechanisms of scheme IIIA and B the model chemistry that would be needed for envisioning a possible bioorganic evolution, is lacking. In spite of the overall RNA-like structure it has to be accepted that for these reactions flavins can only function on sophisticated enzymes with refined active sites and have always done so. A possible way out of this dilemma - a pre-genetic code structure, but post-genetic code functions - is offered by the recent recognition of a cofactor (33) in the enzymology of one of the methane forming Archaebacteria, a group of organisms that differ greatly from prokaryotes and eukaryotes in their RNA, DNA, cell membranes, cell walls, metabolism and cofactors (34,35). They must have split off from other bacteria very early and can probably be considered as bacteria in which part of the anaerobic metabolism from the time before the origin of photosynthesis has "fossilized". The cofactor (called F420) has a structure reminiscent of flavin, with the notable exception that N-5 is replaced by a carbon atom. The model chemistry of 5-deazaflavin is already fairly well known, because similar compounds have amply been used as flavin analogs. Their chemistry can be characterized as NADH-like, but in flavin shape (36). So now it is conceivable that first (NADH-like) active sites evolved with deazaflavins as coenzymes, and thereafter deazaflavins were exchanged by flavins (for instance when oxygen came into metabolism), which automatically turned into prosthetic groups (rapid back oxidation by oxygen). Such an exchange could

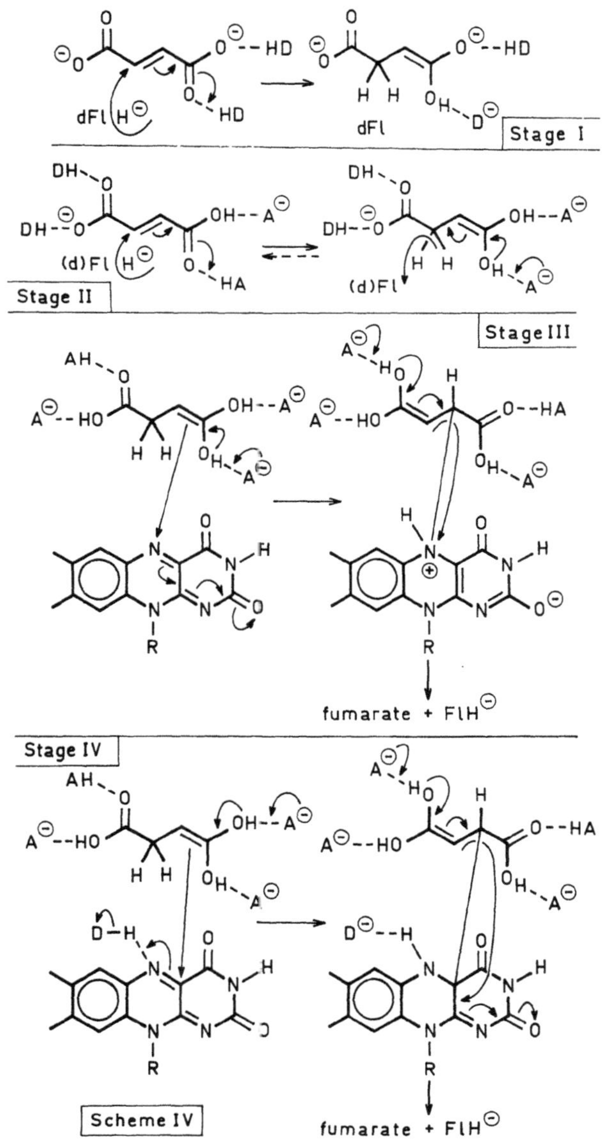
Stage I
Stage II
Stage III
fumarate + FlH⊖
Stage IV
fumarate + FlH⊖

Scheme IV

have marked the branching-point between archaebacteria and eubacteria, all of which contain flavins. The exchange is evolutionarily allowed because there is added a whole new type of chemistry (oxygen sensitivity, one-electron chemistry) while the "old" deazaflavin two-electron hydridoid chemistry is not cut off completely. The difference between the two systems lies predominantly in the polarization of the 4a-5 double bond, which is acted against in the case of the flavins by the normal C=N polarization, which means that there is now a second place for nucleophilic attack (C-4a). Additionally, for the flavins the middle ring becomes formally antiaromatic, which gives the system its one-electron chemistry and oxygen sensitity on reduction. This relatively late entrance of flavins in metabolism (except perhaps the flavodoxins) can explain why there up to now has not been found one clear cut case of a functional light absorbing or a functional flavin-metal complex in metabolism although the flavin system is well suited for such functions.

Since the use of fumarate as an electron sink is widespread among anaerobic bacteria (37) it can be assumed that fumarate reductase originated before archaebacteria and eubacteria separated. A probable bioorganic evolution of flavin function within succinate dehydrogenase can now be imagined as in scheme IV. In stage I a rapid evolution can be expected towards an active site that is complementary to the enolform of succinate. Thus, hydrogen donors HD turn into acids HA, by exchange for more acidic side chains or evolution of a proton-relay system. After exchange of deazaflavin (dFl) for flavin (Fl) (stage II), the reaction reverses (transition to oxygen metabolism). Hydride transfer becomes now less convenient than a covalent intermediate, first to N-5, because the active site was developed for electron input at that site (stage III). The other hydrogen donors for the substrate turn now into acids too and there develops a need for a general acid catalyst at N-5. In stage IV the general acid at N-5 is present and the covalent intermediate is now binded to C-4a, which gives a better-leaving reduced flavin. The active site has become perfect.

A similar smooth transition towards covalent intermediates is, however, not imaginable at this moment for the fatty acyl-CoA dehydrogenases (lack of the second carboxylic function) and for α-aminoacids and α-hydroxyacids (imino- and ketogroups are much better electron sinks than carboxylate groups). This means that it is entirely possible that these enzymes still function by the archaic hydride transfer mechanism.

REFERENCES

(1). Jencks, W.P., Catalysis in Chemistry and Enzymology, McGraw-Hill, New York, 1969.
(2). Pauling, L., Am. Sci. 36, 50 (1948).

(3). Lienhard, G.E., Science, 180, 149 (1973).
(4). Visser, C.M. and Kelogg, R.M., J. Mol. Evol. 11, 163 (1978)
(5). Although most important from the very beginning of chemical evolution, metal ions and (active site) water molecules are only discussed via their ligands, since there is not much to evolve in the catalytic power of an isolated inorganic ion or molecule.
(6). In this, "eldest" has to be measured from the moment of their first entrance into the "metabolism" of the replicating system, from which life ultimately descended, and not from their first abiotic synthesis.
(7). Orgel, L.E. and Sulston, J.E., Prebiotic and Biochemical Evolution, Kimball, A.P. and Oró, J., eds., North-Holland Publ. Cy, Amsterdam 1971, p. 89.
(8). White III, H.B., J. Mol. Evol. 7, 101 (1976).
(9). Interestingly, the Archaebacteria, recognized only recently as being "very old" (see later) have membranes made of polyprenyl ethers. One species has a light harvesting, proton pumping rhodopsin (10).
(10). Danon, A. and Stoeckenius, W., Proc. Natl. Acad. Sci. USA, 71, 1234 (1974).
(11). Visser, C.M. and Kellogg, R.M., J. Mol. Evol. 11, 171 (1978)
(12). Such has been recently tried for the vitamins C, E and K. (Visser, C.M., Bioorg. Chem., in the press (1980)).
(13). A more complete and detailed discussion of bioorganic chemistry and reaction mechanisms of the flavin catalyzed reactions from an evolutionary point of view will be published elsewhere.
(14). Mayhew, S.G. and Ludwig, M.L., The Enzymes, Boyer, P.D., ed., 3rd ed. Acad. Press, New York, 1975, vol XII, 57.
(15). Adman, E.T., Biochim. Biophys. Acta, 549, 107 (1979).
(16). Heinz, B., Ried, W. and Dose, K. Angew. Chem. Int. Ed. Engl. 18, 478 (1979).
(17). Singer, T.P. and Kearney, E.B., J. Biol. Chem. 183, 409 (1950).
(18). H. Sund, ed., Pyridine Nucleotide-Dependent Dehydrogenases, de Gruyter, Berlin 1977.
(19). Müller, F., Massey, V., Heizmann, C., Hemmerich, P. and Lhoste, J.M., Eur. J. Biochem 9, 392 (1969).
(20). Bruice, T.C., Progress in Bioorganic Chemistry, Kaiser, E. T. and Kézdy, F.J., eds. Wiley, New York 1977, vol. 4, 1.
(21). Hemmerich, P., Progress in the Chemistry of Organic Natural Products, Hertz, W., Grisebach, H., Kirby, G.W., eds. Springer, New York 1976, vol 33, 451.
(22). Massey, V., Palmer, G. and Ballou, D., Oxidases and Related Redox Systems, King, T.E., Mason, H.S., Morrison, M., eds. Univ. Park Press, Baltimore 1973, p. 25.
(23). Gabig, T.G. and Babior, B.M., J. Biol. Chem. 254, 9070 (1979).
(24). Massey, V., Strickland, S., Mayhew, S.G., Howell, L.G.,

Engel, P.C., Matthews, R.G., Schuman, M. and Sullivan, P.A. Biochem. Biophys. Res. Commun. 36, 891 (1969).
(25). Chemical energy for excitation could then come from the one-step dioxirane to acid rearrangement, analogous to the well-known dioxetane light producing reactions.
(26). Walsh, C., Enzymatic Reaction Mechanism, Freeman, San Francisco 1979, p. 358.
(27). Flashner, M.S. and Massey, V., Molecular Mechanism of Oxygen Activation, Hayaishi, O., ed., Acad.Press, New York 1974, p. 245.
(28). Massey, V. and Hemmerich, P., The Enzymes, Boyer, P.D., ed., 3rd ed., Acad. Press, New York 1975, vol 12, p. 190.
(29). Bright, H.J. and Porter, D.J.T., The Enzymes, Boyer, P.D., ed. 3rd ed., Acad. Press, New York 1975, vol 12, p. 421.
(30). Hatefi, Y. and Stigall, D.L., The Enzymes, Boyer, P.D., ed. 3rd ed., Acad. Press, New York 1976, vol 13, p. 175.
(31). Visser, C.M., Naturwissenschaften, 67, in the press (1980).
(32). An analogous evolution can be expected for the other mechanisms of catalysis. All will tend towards perfect complementarity to the transition state.
(33). Eirich, L.D., Vogels, G.D. and Wolfe, R.S., Biochemistry, 17, 4583 (1978).
(34). Balch, W.E., Magrum, L.J. Fox, G.E., Wolfe, R.S., and Woese, C.R., J. Mol. Evol. 9, 305 (1977).
(35). Thauer, R.K., and Fuchs, G., Naturwissenschaften, 66, 89 (1979).
(36). Hemmerich, P., Massey, V. and Fenner, H., FEBS Letters, 84, 5 (1977).
(37). Kröger, A., Biochim. Biophys. Acta 505, 129 (1978).

SELECTIVE ACYLABILITY OF AMINO ACIDS
BY NON-ENZYMIC MODEL REACTIONS
OF BIOCHEMICAL TRANSACYLATIONS.

BRICE D., LE PORT L. and BUVET R.
Laboratoire d'Energétique Electrochimique
et Biochimique
Université PARIS VAL DE MARNE
Avenue du Général de Gaulle
F94010 CRETEIL Cédex (France)

ABSTRACT

It is commonly admitted that the genetic information involved in the definition of sequences of peptidic biopolymers operates from a situation such as all aminoacyl group have roughly the same chances at equal concentration for being incorporated in the chain by transacylation.

Following previous studies of non-enzymic model reactions of biochemical transacylations, we have studied comparatively the acylability of many aminoacids by several acyl donors (acetylthioesters, acetylphosphate). The obtained results clearly show that the acylability of the amino group of different aminoacids is strongly dependent on the nature of the aminoacid. Any increase of the hydrophobicity of the side chain strongly decreases this acylability, in all cases. More specific effects depends on the nature of the acyl donor.

These results imply that effects due to the selective reactivities of the substrates themselves should play at least a far from negligible role in the determination of the sequences of peptidic chain and that genetic information does not operate from equal chances of incorporation of the different aminoacids.

It is currently admitted that the genetic information which determine the sequences of peptidic biopolymers operates from a situation where all aminoacids have approximately the same chan-

Y. Wolman (ed.), Origin of Life, 505–511.

ces at equal concentrations for being incorporated in peptidic chains.

This view is roughly supported by results of artificial peptidic syntheses in non aqueous media and in the presence of strong dehydrating agents which permit to form peptidic chains with comparable yields from different aminoacids. However, these reactions involving relatively high energy balances cannot be considered as a good basis for comparing aminoacids reactivities regarding the formation of peptidic bonds in biological circumstances which occur through transaminoacylations from aminoacylesters in aqueous media :

$$R_1NH\text{-}CHR\text{-}COX + H_2N\text{-}CHR'\text{-}COX_1 \longrightarrow R_1NH\text{-}CHR\text{-}CO\text{-}NH\text{-}CHR'\text{-}COX_1 + XH$$

In these reactions, aminoacylated residues, (or N-substituted derivatives), carboxyl group of which has been previously engaged in an ester function are transfered on the nitrogen atom of another aminoacid (or N-substituted derivatives).

Starting from previous studies of non enzymic model reactions of acyl transfer in aqueous solutions (1,2,3,4,5,6,7,8,9,-10,11,12,13,14,15) we have compared the acylability of amino-group of different aminoacids and derivatives in the simpler case of their non-enzymic acetylation by different moderate energy acetyl donors at ambient temperature in aqueous solutions.

All studied reactions have been carried out in aqueous solutions in the absence of enzyme, but in presence of hydroxyl ions in small excess (0.1 M) with regard to the quantity which should be used for hydrolysing and neutralizing the acetyl donor if no transfer occurred. In such conditions, all reactions are fast and practically instantaneous, but their yield is limited by the simple hydrolysis of the acetyl donor.

The acetylation of aminoacids were performed by adding different acetyl donors :

- acetylthioglycolate $CH_3\text{-}CO\text{-}S\text{-}CH_2\text{-}CO_2^-$ ($AcSCHCO_2^-$)
- acetylthiocholine $CH_3\text{-}CO\text{-}S\text{-}(CH_2)_2N(CH_3)^+$ ($AcSCh^+$)
- acetylphosphate $CH_3\text{-}CO\text{-}OPO_3H^-$ (AcOP)

at various concentrations $|D|$ to alkaline solutions of the aminated acceptor at concentration $|A|$ typically taken equal to 10^{-1} M.

The quantities of N-acetylated aminoacids produced were determined by precise analysis of the shape of the titration curve of the obtained solutions taking into account all the possible bases formed. (16,17,18)

Results are given in terms of yields Rc of production of the acetylated derivatives with regard to the acceptor quantity used. Results presented here concern first reactions between free amino-acids and acetyldonors mentioned above.

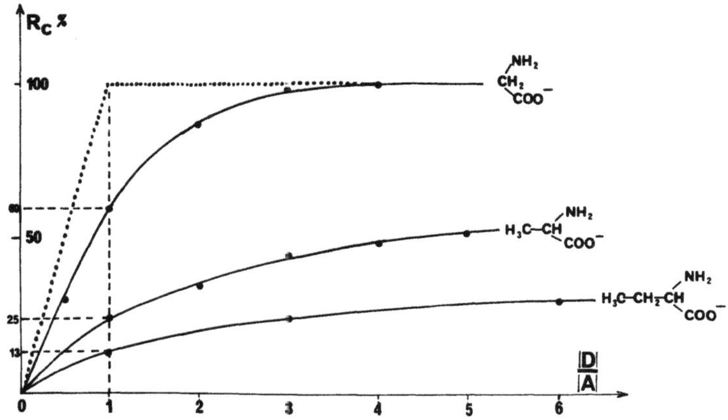

Figure 1 - Acetyl transfer from acetylthioglycolate to hydrocarbonated side chain aminoacids.
Acetylation yields Rc versus ratio $\frac{|D|}{|A|}$ between donor and amino-acid concentrations.

Experimental conditions

$$|H_2N\text{-}CH(R)\text{-}CO_2^-| = 0.1\ M$$
$$|AcSCH_2CO_2^-| = C(M)$$
$$|OH^-| = (2C + 0.1)\ M$$

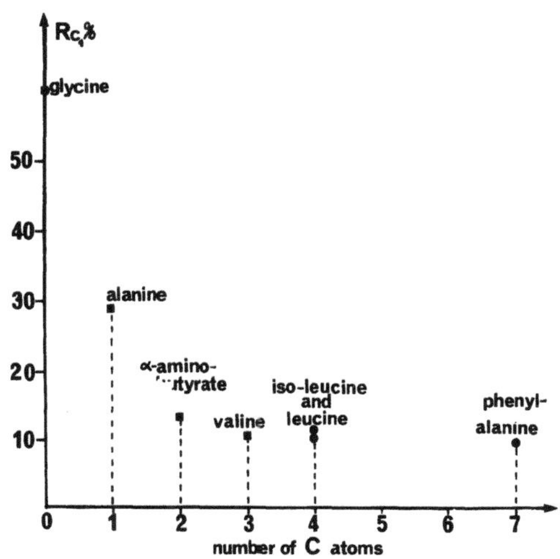

Figure 2 - Aminoacid acetylation yields Rc_1 versus number of carbon atoms contained in the hydrocarbonated side chain :

Experimental conditions :

$$|H_2N\text{-}CH(R)\text{-}CO_2^-| = 0.1\ M$$
$$|AcS\text{-}CH_2\text{-}CO_2^-| = 0.1\ M$$
$$|OH^-| = 0.3\ M$$

Figure 1 shows the results concerning acetyl transfer from acetylthioglycolate to glycine, to alanine and to αaminobutyric acid respectively. The shapes of these three curves which give the condensation yield Rc versus $\frac{|D|}{|A|}$ for the same hydroxyl ions excess are qualitatively similar.

On the other hand Rc values strongly depend on the aminoacid. In fact, these acetylation reactivity of these aminoacids strongly decreases here with the length of the linear hydrocarbonated chain. Glycine has the shortest lateral chain and presents the strongest reactivity.

The yield of the formed acetylated aminoacid depends upon the ratio $\frac{|D|}{|A|}$ between the acetylthioglycolate donor and aminated acceptor concentrations. It reaches a limit Rc for $\frac{|D|}{|A|}$ values betweeen 3 and 5 depending upon the acceptor and the donor involved.

For further comparison between different aminoacids, we have currently considered Rc_1 values for $\frac{|D|}{|A|} = 1$ and Rc_∞.

Figure 2 presents results concerning values of Rc_1 versus the number of carbon atoms in the side chain of some hydrocarbonated aminoacids. The donor used here is still acetylthioglycolate but similar results obtained with other acetyl donors, especially acetylthiocholine and acetylphosphate, permit to state that in the series of aminoacids with a simply hydrocarbonated side chain, Rc_1 regularly decreases until about 10 % for all acetyl donors studied, when the side chain mass increases.

Table I summarizes the main quantitative results concerning the acetylation yields Rc_1 and Rc_∞ obtained for various aminoacids from acetylthioglycolate, acetylthiocholine and acetylphosphate.

It shows that :

- Rc of hydrocarbonated side chain aminoacids, such as glycine, alanine, αaminobutyric acid and valine, are lower from acetylcholine and acetylphosphate than from acetylthioglycolate. But, for these three acetyl donors, the same classification is found for the different aminoacids. Thus, the NH_2 group is less reactive with respect to its acetylation when the aminoacid side chain is longer, more hydrophobic and ramified.
- for aminoacids which have a thiol function (S^-) in their side chain, an inversion of the acetylation yields between the two thioesters $AcSCH_2CO_2^-$ and $AcSCh^+$ is observed. NH_2 attack is here easier from $AcSCh^+$ than from $AcSCH_2CO_2^-$.

On the contrary, for aminoacids having an acidic side chain, this inversion is not found since $AcSCh^+$ Rc values are lower than those obtained with $AcSCH_2CO_2^-$.

- When the reactions occur from hydroxylated side chain aminoacids, such as serine and threonine, Rc values are lower with $AcSCh^+$ than with $AcSCH_2CO_2^-$ and decreases when methylen hydrophobic groups are added in the side chain.

- Finally, specifically high acetylation yields are observed in the case of cysteine and homocysteine acetylation by acetylthioesters, probably in connection with an internal catalytic effect of the thiol function.

Table I - Acetylation yields Rc_1 and Rc_∞ of various aminoacids from different acetyl donors.

Experimental conditions :

$$|H_2N\text{-}\underset{}{\overset{R}{CH}}\text{-}CO_2^-| = 0.1\ M \qquad\qquad |H_2N\text{-}\overset{R}{CH}\text{-}CO_2^-| = 0.1\ M$$

$$|AcS\text{-}CH_2\text{-}CO_2^-| \text{ or } |AcSCh^+| = C(M) \quad \text{and} \quad |AcOP| = C(M)$$

$$|OH^-| = (2\,C + 0.1)\ M \qquad\qquad |OH^-| = (3C+0.1)M$$

Aminoacids (DONORS)	side chains	$AcSCH_2CO_2^-$		$AcSCh^+$		AcOP	
		Rc_1	Rc_∞	Rc_1	Rc_∞	Rc_1	Rc_∞
Glycine	H -	60	100	30	75	30	80
Alanine	H_3C -	25	48	10	18	15	20
αAminobutyrate	$H_3C\text{-}CH_2$-	13	30	7	14	12	
Valine	$(CH_3)_2$-CH-	11	15	3	5	0	
Cysteine	$^-S\text{-}CH_2$-	60	94	80	100	14	
Homocysteine	$^-S\text{-}CH_2CH_2$-	80	100	85	100		
Penicillamine	$^-S\text{-}C(CH_3)_2$-	10	25	30	50		
S-methylcysteine	$CH_3\text{-}S\text{-}CH_2$-	10	14	0			
Serine	$HO\text{-}CH_2$-	28	70	15			
Threonine	$H_3C\text{-}CH(OH)$-	5	30	2			
Aspartate	$^-O_2C\text{-}CH_2$-	5	10	2.5			
Glutamate	$^-O_2C\text{-}(CH_2)_2$-	6	20	1			
Asparagine	$H_2N\text{-}CO\text{-}CH_2$-	3	7				

With acetylthioglycolate we have also compared the ability of some small peptides such as :

- diglycinate $H_2N\text{-}CH_2\text{-}CO\text{-}NH\text{-}CH_2\text{-}CO_2^-$

- triglycinate $H_2N\text{-}CH_2\text{-}CO\text{-}NH\text{-}CH_2\text{-}CO\text{-}NH\text{-}CH_2\text{-}CO_2^-$

- tetraglycinate $H_2N\text{-}CH_2\text{-}CO(NH\text{-}CH_2\text{-}CO)_2\text{-}NH\text{-}CH_2\text{-}CO_2^-$

to be acetylated in the same experimental conditions.

Figure 3 presents the acetylation yields RC_1 of these compounds versus the number of N-terminal aminoacid residues of glycine oligopeptides.

It clearly shows that the acetylation yields of the N-terminal aminoacid residue decreases in series of oligopeptides when the polymerization degree increases. Beyond four residues, Rc_1 remains roughly constant.

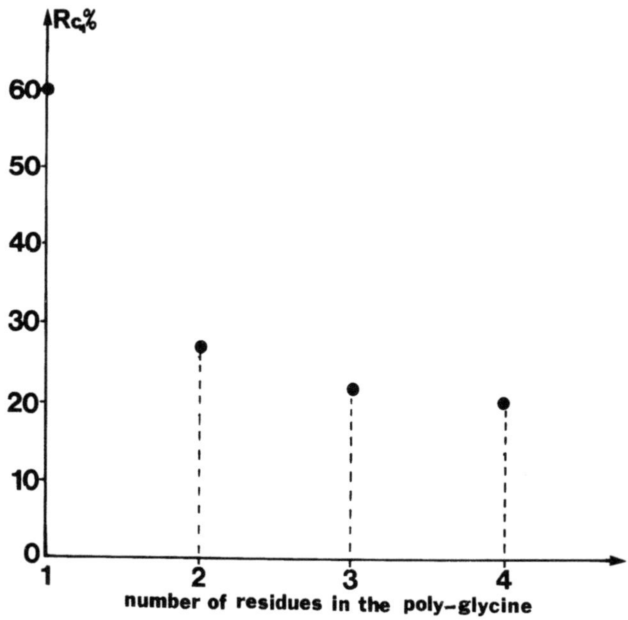

Figure 3 – Acetylation yields Rc_1 versus number of residues contained in the polyglycine.

Experimental conditions : $|\text{polyglycinate}| = 0.1$ M

$|AcS\text{-}CH_2CO_2^-| = 0.1$ M

$|OH^-| = 0.3$ M

These results show that in the non-enzymic conditions involved in primitive syntheses of polypeptides, the different aminoacids had very different reactivities with regard to the acylation of their amino group and consequently with regard to their possible incorporation in peptidic chains. As a general rule, the hydrophobic character of the aminoacid side chains very strongly influences the NH_2 reactivities of these compounds with respect to their acetylation with acetyl donors. Comparative data concerning the reactivities of condensed derivatives of the carboxyl group should be also obtained to compare aminoacids more completely on this respect.

REFERENCES

(1) LIPMANN F. and TUTTLE T.C. J, Biol. CHem. 159,21 (1945)
(2) LYNEN F., REICHERT E. and RUEFF L. Ann. 574, 1, (1952)
(3) NODA L.H., KUBY S.A. and LARDY H.A.
J.Amer. Chem. Soc. 75, 913 (1953)
(4) CONNORS K.A. and BENDER M.L. J.Org.Chem. 26, 2498 (1961)
(5) JENCKS W.P. and CARRIUOLO J.
J.Amer.Chem.Soc.82, 1778 (1960)
(6) JENCKS W.P. and CARRIUOLO J.
J.Amer.Chem.Soc.82, 675 (1960)
(7) BLACKBURN G.M. and JENCKS W.P.
J.Amer.Chem.Soc.90,2638 (1968)
(8) KENFULL D.G. and JENCKS W.P.
J.Amer.Chem.Soc. 93, 178 (1971)
(9) BRUICE T.C. and WILLIS R.G.
J.Amer.Chem.Soc. 87, 531 (1965)
(10) GREGORY M.G. and BRUICE T.C.
J.Amer.Chem.Soc.89, 2121 (1967)
(11) SHONBAUN G.R. and BENDER M.L.
J.Amer.Chem.Soc.82, 1900 (1960)
(12) OGILVIE J.M., TILDON J.T. and STRAUCH B.S.
Biochemistry 3, 754 (1964)
(13) LIENHARD G.E. and JENCKS W.P.
J.Amer.Chem.Soc. 87, 3863 (1965)
(14) FERSHT A.R.
J.Amer.Chem.Soc. 93, 3504, (1971)
(15) LE PORT L. Thèse de Doctorat es Sciences, Paris VI (1974)
(16) BRICE D. Thèse de doctorat 3e cycle, Paris VI (1978)
(17) LE PORT L. and BUVET R.
Compt.Rend.Acad.Sci. PARIS 270, 1753 (1970)
(18) LE PORT L., ETAIX E., GODIN F., LEDUC P. and BUVET R.
Chemical Evolution and the Origin of Life, eds BUVET R.
and PONNAMPERUMA C. (1971)
North-Holland Publishing Company.

EXPERIMENTS ON TRANSFER OF ORGANIC MOLECULAR INFORMATION INTO CRYSTAL LATTICE SUPERSTRUCTURES

Peter Decker and Horst Schweer

Chemical Institute, Veterinary School, Hannover, FRG.

Evolution in "bioids", open systems which can exist in several steady states (terrestrial life being an example of such) involves accumulation of information relevant to its selection value, i.e. in the simplest case the choice, by "mutation" of the most stable steady state ("species"). Therefore it appears problematic how "alien" information like nonstatistical structures arising in mineral crystallization or in interplanetary carbon compounds found in meteorites could become incorporated into ("understood by") the terrestrial biosphere. In contrast, an interaction of evolving biosystems with inorganic matter, involving temporal storage of bioinformation in crystal structures appears better compatible with the bioid concept. A model how 3 independent messages can be stored in a 3-dimensional lattice is presented. As a first experimental approach attempts to induce information from linear biomolecules into a nonstatistical distribution of components of mixed crystals are reported.

The Golem hypothesis.

The idea of a participation of clay and similar minerals in origins of life has two facets:

i. Specific catalytic effects producing ordered polymerization of organic monomers along crystal surfaces. Many interesting results have been obtained along this alley opened by the pioneer work of Katchalsky (1-3). Moreover, stereoinduction of optically pure polymers has been obtained using polymerisation of monomers inside a chiral crystal lattice (4).

Y. Wolman (ed.), Origin of Life, 513–520.

ii. Specific information transfer by isomorphic replacement of ions e.g. in silicate sheet crystals. This requires, that specific ion patterns, and, in the case of ions of different charge (say through replacement of Si^{4+} by Al^{3-}) also specific charge patterns, are copied onto a growing parallel crystal sheet. This idea implies the tempting speculation, that information in such matrices might be transfered to or imprinted by specific information contained in the base sequences of nucleic acids or in the amino acid sequence of proteins. Such hypotheses have been discussed by Smith-Cairns (5) at Glasgow and by A. Weiss at Munich (6). Weiss attempted to multiply specific kinds of mica ion patterns by crystallization. Related speculations consider a modern version of the classical panspermia idea, the origin of bioinformation from interstellar or meteoritic carbon compounds.

In this paper we discuss two different possibilities for a role of such a hypothetical transfer effect from non-biological matrices into life-like systems and vice versa: i. The origin of life, or of some information significant for its origin, from alien matrices originating in clay minerals or in interstellar space, or ii. of a participation, by information transfer, of inorganic matrices like clay minerals in the evolution of life-like systems.

On a theoretical model we demonstrate that in principle up to three independent messages can be imprinted onto a rather simple crystal-like lattice. Finally we report some preliminary attempts to imprint information of biomolecules into a nonstatistical distribution of molecules in mixed crystals.

Information from alien environments?

The acquisition of chemical information from inorganic minerals or from organic matter formed in the interplanetary or in the interstellar space i.e. of "alien" information as a possible precursor or part of biological information accumulated in the biosphere, the "biochemical Golem hypothesis" and the cosmic panspermia idea, appears hardly compatible with the most essential feature of life and life-like processes: Such processes thermodynamically represent dissipative structures in chemically reacting open systems(7), "bioids" (8,1o,11) which acquire, conserve and accumulate information through a generalized Darwinian mechanism. The terrestrial system produced a coding mechanism, where coded information is used in the nucleoprotic system for the production of enzymes as standardized catalysts.

We have shown that "information",as acquired through such mutation and selection processes in bioids,represents a mapping (by the selection process) of classes of relevant properties of the respective environments onto the states (thermodynamical steady states or discrete dissipative structures or, biologically, "species" or "mutants") available to the system (9,10,12,13).

Therefore, in all such systems, evolutionary relevant "information" primarily is equivalent with "function", the interface between the system and its environment as shaped by the Darwinian mechanism.

Extraneous, "alien", information, i.e. nonstatistical features which originate without connection with the function of the system in its environment, which is optimized through "Darwinian" acquisition of information from environments, is of no evolutionary use for the system. Mineralogic informational features of clay minerals and the very complex interstellar organic compounds, which recently have been claimed by Hoyle and Wickramsinghe (14) as a possible source of flue virus epidemies and even of life in the sense of an interstellar panspermia, contain only such alien information.

Therefore we do not believe that such information will be accepted ("understood") by the terrestrial life-environment system, i.e. that it cannot improve the selection value of the system. In this conclusion we rely on the bioid concept - a physical generalisation of the Darwinian principle. Besides the generalization of he concept of information, making it independent of the possibly spcifically terrestrial "nucleoprotic" mechanism, this concept gave us new insights into the origin of "biological" features like homeostasis, adaptation, self-organization (13), excitability (15), rhythmic behaviour (9,15), the origin of molecular asymmetry (16,17), standardization, the coupling to solar energy (18,19,20) and the evolution of reducing planetary atmospheres (21,22).

Information in minerals?

A possibility, which appears more compatible with the concept of evolution as a process of accumulation of information in open systems, is the participation by exchange of information between the mineral world and organic macromolecules. It extends the conventional scope of biological evolution, implying the possibility of storage of information in crystals over unlimited time, where, during that time, the information becomes independent of the energy-linked dynamical memory-refreshing mechanism(e.g. by self-replication of DNA),characteristic for a "living" system. Such information in crystals would represent a mineral "seed", a molecular energy-independent material information channel capable to transfer information over unlimited time like a written document. However, since this idea also implies a complementary mechanism of information retrieval back into a biosystem, it supposes in fact the conjunction of two hypotetical effects at once, making it quite speculative. Therefore, pragmatically we have to begin with attempts to find at least some experimental support for one of its parts - mineral-bio or bio-mineral information transfer.

First, let us discuss a theoretical model, which demonstrates that in principle up to three independent information messages can be imprinted onto a crystal-like structure(13).

Three messages in one crystal.

Let us suppose molecule-like elements, which in three dimensions exert three different kinds of conjugate interactions, say + and -; a and b; and r and s, each pair producing a specific bond between molecules, and each molecule exerting the conjugate forces towards opposite directions. Suppose further that (+-)-bonds always are directed along the x-axis, (ab)-bonds and (rs)-bonds along the y and z axis respectively.

Let us build a "crystal" from such elements. Beginning with an one-dimensional array, say along the x-axis, in the first element we have to decide whether we put its (+)-side or its (-)-side to the left. Having once made this choice, say (-) to the left and (+) showing to the x direction (Fig. 1,a) all further elements must be added in the same position respective to the x-axis, i.e. "(-)-(+)". However, in each element there remains a degree of freedom, allowing us to turn the element 180° around the x-axis, inverting the directions of a,b and r,s (Fig. 1,b). Using such inversions we can put an arbitrary message into such an one-dimensional array of elements.

Now, beginning a second row along the first one, we have to align the first element such that it fits with its r,s-side to the first element of the first row. Thus, we have no freedom to turn it around the x-axis. However, we have yet the choice to turn it arbitrarily around the z-axis, chosing again, as in the first element of the first row, whether its (+) or (-) side will look to the left. But having done this choice with the first element, we have no further freedom of choice with the remaining elements of the row, which, now, will now represent a redundant copy of the message we have put in x-direction into the first row. Therefore, whith each new row added in z-direction we can add one new bit of information, resulting in an independent message in z-direction.

Completing our crystal in the third direction, towards y, again we have one degree of freedom in the choice of the first element, which we may turn 180° around the y-axis. However, this time the choice of the first element decides an unique position for all subsequent elements of the whole two-dimensional sheet added in y-direction, and every new sheet can hold and make redundant only one additional bit of information, resulting in a third independent message in the y-dimension.

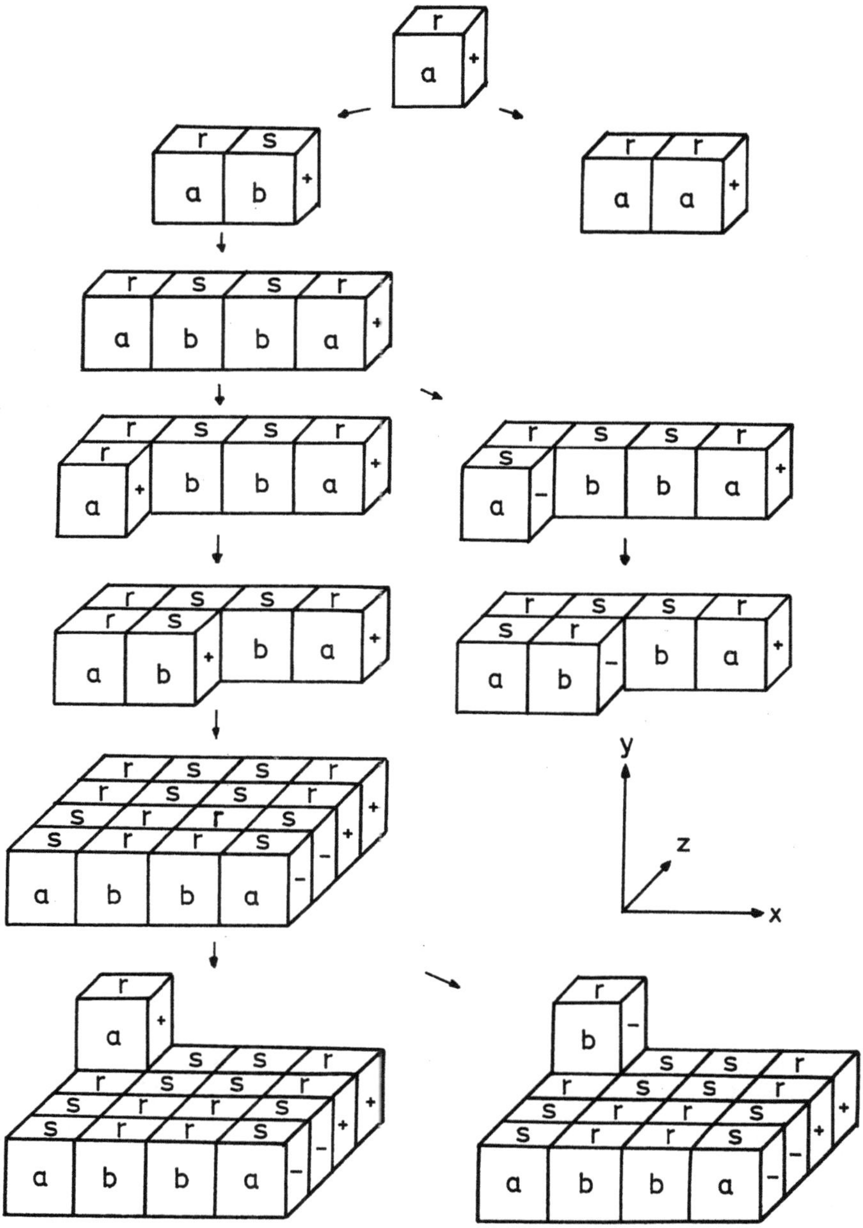

FIGURE 1
Three independent messages in direction x, y, and z through different orientations of elements in a three-dimensional crystal-like lattice.

Potential candidates for such coding by inversion of molecules in a lattice, at least in two dimensions, may be found among benzene derivatives interacting by different polar groups. Further crystallographic studies should show in which kinds of compounds such effects can be expected. However, at present we have not enough information how to find compounds with promising properties and how to introduce several different messages into a crystal lattice. Therefore we undertook a preliminary study on possibilities to transfer at least onedimensional information into a crystal lattice.

Bio-mineral information transfer.

How can we transfer information from biopolymers to crystals? If, during crystallization, such molecules are present, they may become adsorbed forming a regular twodimensional lattice on a growing crystal surface. If the adsorbed molecules would form a matrix for further growth of the adsorbed species, we had the well known phenomen on of epitaxial growth. In our case, however, the adsorbed molecules are supposed to act only as a matrix imprinting the spatial distribution of its own polar and other specific sites onto the distribution of guest molecules in the growing mixed crystal.

Experiments.

In order to test this hypothesis, we chose as original molecular information the information contained in the discrete amino acid sequence of purified proteins.

As a first crystal system in which we attempted to demonstrate the existence of such effects we report a search for superstructures in the regular pattern of mixed crystals of the type $Me^{2+}NH_4XO_4 \cdot 6H_2O$ grown in the presence of different proteins, where Me^{2+} and $XO_4{}^{3-}$ represents a mixture of Mg^{2+} and/or Mn^{2+}, Co^{2+}, Ni^{2+}, Zn^{2+}, and $PO_4{}^{3-}$ and/or $AsO_4{}^{3-}$.

$NH_4MgPO_4 \cdot 6H_2O$: Solutions of 0.2 g $(NH_4)_2HPO_4$, 0.12 g H_3PO_4 (85%) and 0.1 g $(NH_4)_2SO_4$ in 0.8 ml H_2O (solution 1) and 0.16 g $MgSO_4 \cdot 7H_2O$ in 1.2 ml H_2O (solution 2) were combined. Crystals of $NH_4MgPO_4 \cdot 6H_2O$ were obtained in some days.

$NH_4MgAsO_4 \cdot 6H_2O$: A solution of 0.2 g $Na_2HAsO_4 \cdot 7H_2O$, 0.2 g $(NH_4)_2SO_4$ and 0.06 g H_3AsO_4 (Density: 1.35 g/ml) in 0.8 ml H_2O (solution 3) was given to 0.16 g $MgSO_4 \cdot 7H_2O$ in 1.2 ml H_2O. $NH_4MgAsO_4 \cdot 6H_2O$ crystallized in about one day.

$NH_4Mg_{1-x}Me_xPO_4 \cdot H_2O$: Solution 2 was partly replaced by one of the following solutions: 0.145 g $MnSO_4 \cdot 4H_2O$, 0.182 g $CoSO_4 \cdot 7H_2O$, 0.182 g $NiSO_4 \cdot 7H_2O$ or 0.187 g $ZnSO_4 \cdot 7H_2O$ in 1.2 ml H_2O.

$NH_4MgP_{1-y}As_yO_4 \cdot 6H_2O$: Solution 1 resp. solution 3 were replaced by a mixture of these.

Preparation in the presence of proteins: The preparation was made in the same way, but 0.002 g Edestin, Gliadin, ß-Lactoglobulin, Albumin or Hemoglobin were added to solution 2.

Results and discussion.

If Mg^2 was partly replaced by Mn^{2+}, Co^{2+}, Ni^{2+} or Zn^{2+}, in most cases besides of the species $NH_4Mg_{1-x}Me_xPO_4 \cdot 6H_2O$ one or two additional crystal species where obtained, which were separated manually into the discrete kinds. (Table 1)

Table 1:

Me^{2+}	Mg^{2+} - Me^{2+} ratio (in solution)	Products
Mn	9:1	$NH_4Mg_{1-x}Mn_xPO_4 \cdot 6H_2O$ $NH_4Mg_{1-x}Mn_xPO_4 \cdot H_2O$ $Mg_{1-x}Mn_xHPO_4 \cdot 3H_2O$
Co	9:1	$NH_4Mg_{1-x}Co_xPO_4 \cdot 6H_2O$ $Mg_{1-x}Co_xHPO_4 \cdot 3H_2O$
Ni	9:1	$NH_4Mg_{1-x}Ni_xPO_4 \cdot 6H_2O$
Zn	9:1	$NH_4Mg_{1-x}Zn_xPO_4$

Using X-ray powder diagrams cell dimensions and cell volumes were found to lie between the values of the pure components. No reflections which could be interpreted as resulting from a superstructure, i.e. of some regular distribution of the guest component were found in comparing crystals obtained in the presence of the different proteins with control samples grown in the absence of such.

In the evaluation of those negative results one should consider, that so far only one kind of crystal lattice has been investigated. The adsorption effect may be dependent of the similarity of molecular distances in the crystal and in the inductor molecule. Moreover, the diffraction method used, powder diagrams, is not so sensitive as e.g. single crystal diffraction methods, which technically were not available to us. Therefore, those results should not be understood as a decision pro or contra the existence of a hypothetical possibility of a transfer of bioinformation into crystal lattices. We report it as an first experimental approach suggesting many similar experiments along this line of speculation.

REFERENCES

(1) Paecht-Horowitz, M., Berger, J. and Katschalsky, A., Nature 228, 636 (197o).
(2) Paecht-Horowitz, M., Origins of Life 5, 173 (1974), 7, 369 (1976).
(3) Lahav, N., J.Mol.Evol. 5, 243 (1975).
(4) a) Pentzien, K. and Schmidt, G.M.J., Angew. Chem., Int.Ed. 8. 6o8 (1969); b) Green, B.S., Lahav, M. and Rabinovich, D., Acc. Chem. Res., in press.
(5) Cairns-Smith, A.G., J.Theor.Biol. 10, 53 (1966);"The Life Puzzle", Oliver & Boyd, Edinburgh 1971.
(6) Weiss, A. in I.Th. Rosenquist ed., Proc. III. Europ. Clay Conf., Oslo, 2-5 June 1977, p. 229.
(7) Glansdorf, P. and Prigogine, I., Thermodynamics of Structure, Stability and Fluctuations, Wiley-Interscience, New York, 1971.
(8) Decker, P. and Speidel, A, Z.Naturforsch. 27b, 257 (1972), (Bioids 1).
(9) Decker, P., Ann. New York Acad.Sci. 316, 236-250 (1979) (Bioids VII).
(10) Decker, P. and Heidmann, W., in: H.Noda, Ed., "Origin of Life", Proc. 2. ISSOL Mtg, Jap.Sci.Soc.Press, Tokyo, 1978.
(11) Decker, P., in: DFG-Kolloquium "Evolution der Planetenatmosphären", Schliersee, FRG, 2.-4.Okt. 1979, in press.
(12) Decker, P., Speidel, A. and Nicolai, W., in:Proc Ist European Biophysics Congr., Baden/Vienna,E.Broda et al., Eds., Verl. Wiener Med.Akad.Vienna, Vol. 4, pp. 515-521, 1971.
(13) Decker, P., Evolution in offenen Systemen: prize essay submitted to the Bavarian Academy of Sciences; reprints (170pp) available from Documentation Center on the Origin of Life, Universitätsbibliothek, Ulm, FRG.; cf. Decker, P., Nachr.Chem.Tech. 23, 167 (1975).
(14) a) Arrhenius, S., Worlds in the Making, Evolution of the Universe, Harper Brothers, London 1908; b) Crick, F.H.C. and Orgel, L.E., Icarus 19, 341 (1973); c) Hoyle, F. and Wickramsinghe, N.C., Nature 266, 241 (1977); d) Brooks, J. and Shaw, G., in Noda, H. Ed., "Origin of Life", Center Academic Publications, Tokyo 1978.
(15) Decker, P. and Saygin, Ö., Z.Naturforsch. 34c, 649-51 (1979).
(16) Decker, P., in: D.C. Walker Ed., Origins of Optical Activity in Nature, Elsevier , Amsterdam, pp. 1o9-124, 1979.
(17) Decker, P., J. Mol. Evol. 4, 49 (1974), (Bioids VI).
(18) Saygin, Ö., Submitted to Zschr. Naturforschung (Bioids IX) in press.
(19) Saygin, Ö. and Decker, P., this Congress.
(20) Decker, P. and Saygin, Ö., this Congress.
(21) Decker, P., in: H.Noda Ed., "Origin of Life", Proc. 2. ISSOL Mtg., Jap.Sci.Soc.Press, Tokyo, pp. 631-637, 1978.
(22) Decker, P., Pure & Appl. Geoph. 112, 865 (1974).

PHYLOGENETIC SEQUENCE OF METABOLIC PATHWAYS IN PRECAMBRIAN CELLULAR LIFE

J. Barnabas,* R.M. Schwartz, and M.O. Dayhoff

*Ahmednagar College, Ahmednagar (M.S.) India,
National Biomedical Research Foundation, and
Georgetown University Medical School
Washington, D.C. 20007 USA

We present here a sequence of major metabolic events as they may have appeared during prokaryote evolution. This is based on (i) the phylogenetic schema derived from sequences of bacterial ferredoxin, 2Fe-2S ferredoxin, 5S ribosomal RNA, and c-type cytochromes; (ii) metabolic settings in which these macromolecules are found; and (iii) metabolic capabilities of the prokaryotes that carry these molecules.

By combining information from the phylogenetic schema derived from relevant protein and nucleic acid sequences (1,2), and from the metabolic capabilities of prokaryotes in which these molecules occur, we infer a phylogenetic sequence of major metabolic events appearing in Precambrian cellular life (Fig. 1). Our schema is based on the assumptions that, during prokaryote evolution, gene transfer followed by permanent acceptance is insignificant for genes of basic metabolic importance and that gene products whose structures are conserved have remained relatively unchanged in their basic functions.

The lower portion of Fig. 1, which is based largely on ferredoxin sequences, places strictly anaerobic fermentative bacteria such as clostridia, Megasphaera elsdenii, and Peptococcus aerogenes next to the root. The position of the root was deduced by taking into account a gene doubling present in all ferredoxins (5,6). These results are consistent with what is generally believed—that the primitive atmosphere was reducing (7) and that fermentative obligate anaerobes were the first organisms to appear on earth (8), some 4 to 3.5 Gyr (10^9 years) ago (9,10).

Y. Wolman (ed.), Origin of Life, 521–528.

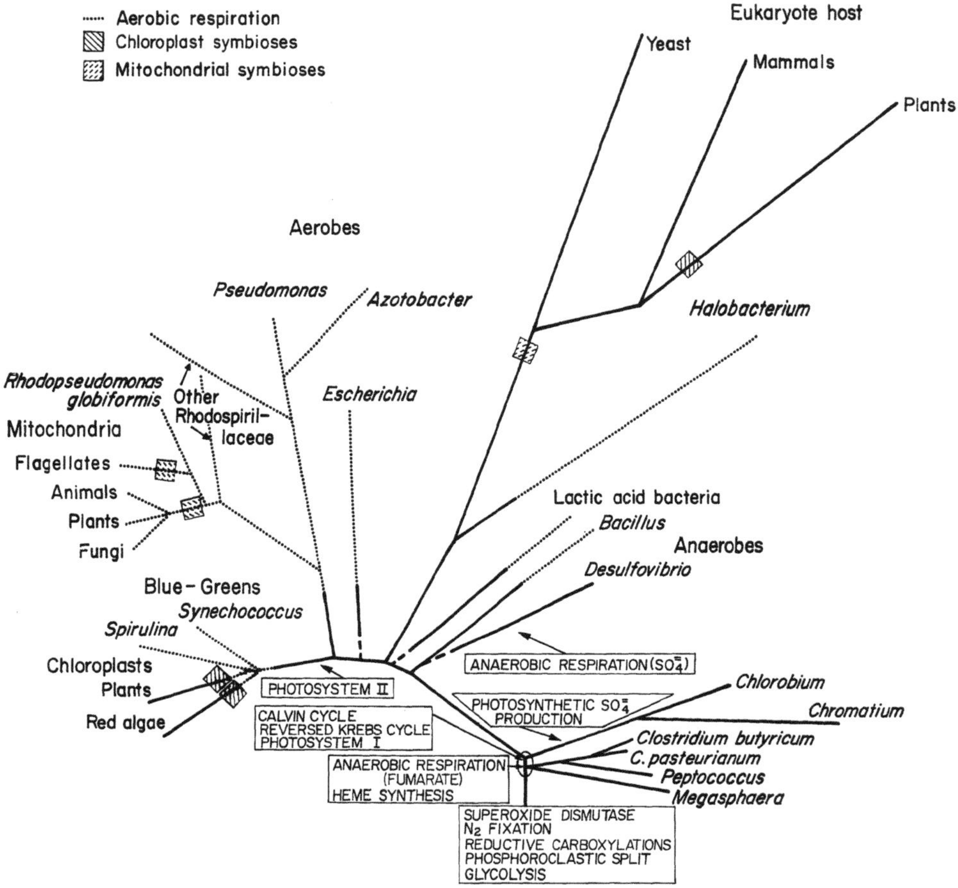

Fig. 1. The phylogenetic tree of prokaryotes and eukaryotes. The sequence of metabolic innovations over evolutionary time is shown. The tree is based on bacterial ferredoxin, 2Fe-2S ferredoxin, 5S rRNA, and c-type cytochrome sequences. Individual trees for these four sets of sequences were constructed using the least-squares matrix method (3,4). The trees derived for each set were scaled so that overlapping positions were comparable and the composite tree was constructed as described earlier (1,2).

Ferredoxins, iron-sulfur proteins that have an extremely electronegative redox potential, around -400mV (11), mediate a number of oxidation-reduction reactions (12). Common metabolic pathways in fermentative anaerobes in which ferredoxins take part are phosphoroclastic splitting of pyruvate leading to acetyl phosphate (13) and reductive carboxylation of an acyl-CoA derivative to an α-keto acid (14). It is possible that, during prokaryote evolution, glycolysis followed by phosphoroclastic split met the demands for ATP and reducing power needed for biosynthetic reactions. It is also possible that the ability to synthesize heme and respire anaerobically using fumarate as terminal electron acceptor were early acquisitions of the anaerobic stem. This is suggested by the finding that two species of Clostridium, C. thermoaceticum and C. formicoaceticum, like bacteria higher on the tree, contain cytochrome b and menaquinone. The role of these electron carriers at least in C. formicoaceticum is in the reduction of fumarate (15).

Two nitrogen-fixing clostridial species, C. pasteurianum and C. butyricum, emanate from a branch near the base of the tree in Fig. 1. Among the obligate anaerobes, numerous clostridia, a few species of Chlorobium, and a number of species of Desulfovibrio have the ability to fix N_2 (16,17). Nitrogen fixation is a well-understood process and is costly in terms of its energy requirements. The observation that one nitrogenase component from a bacterial species reconstitutes an enzymatically active hybrid with the other component from a different bacterial species (18) would suggest that the structures of the nitrogenase components from different organisms have been conserved. Furthermore, interspecies homology of nitrogenase genes has been established through hybridization of Klebsiella pneumoniae nitrogen fixation (nif) gene fragments to DNA's from a variety of widely divergent nitrogen-fixing bacterial species (19). From these results, we infer that these interacting types of iron-sulfur proteins appeared early and thus that nitrogen fixation is a primitive characteristic on this tree.

Figure 1 shows Chlorobiaceae (green sulfur bacteria) and Chromatiaceae (purple sulfur bacteria) separating from the anaerobic stem early in prokaryote evolution. The presence of stromatolites in the 3.4-3.5 Gyr old Pilbara Block of Western Australia and the possibility that these stromatolites came from anoxygenic photosynthetic flexibacteria would support an early evolution of anaerobic photosynthesizers (20). Chlorobiaceae and Chromatiaceae are able to carry out an anoxygenic photosynthesis using photosystem I with cyclic and noncyclic electron-delivery systems (21,22); reduced sulfur compounds in general act as electron donors. Chromatium, in addition, has the ability to carry out a thermodynamically unfavorable electron flow from a donor such as succinate to NAD^+ (23,24). For the assimilation of CO_2, these photosynthetic bacteria may use either the reductive carboxylic acid cycle (14) or the reductive pentose phosphate (Calvin) cycle (24). Clearly, these results imply that photosystem I, including cyclic, noncyclic, and reverse electron-delivery systems, as well as the pathways for CO_2 assimilation were early innovations.

Figure 1 (middle portion) shows Desulfovibrio gigas separating from the anaerobic stem after the divergence of other obligate anaerobes. This group is unique in being able to obtain energy by the reduction of sulfate. This dissimilatory sulfate reduction is a complex process involving the intermediary formation of adenosine 5'-phosphosulfate and is mediated by a specific electron carrier protein called cytochrome c_3 (25). Clearly, in agreement with the suggestion by Peck (25), Fig. 1 indicates that sulfate-producing photosynthesizers arose prior to sulfate-reducing *Desulfovibrio.* Recently, Schidlowski (26) has obtained isotopic evidence in the upper Archean of the Aldan Shield, Siberia (3 Gyr old) and in the Michipicoten and Woman River banded iron formations of Canada (2.8 Gyr) to show that sulfate reducers arose between 2.8 and 3 Gyr ago.

Figure 1 (middle portion) shows three groups of facultative anaerobes, *Bacillus*, lactic acid bacteria, and *Escherichia coli*, diverging from the trunk prior to blue-greens (cyanobacteria) and the aerobic radiation. In recent years, the notion that photodissociation of water could have generated small but significant amounts of oxygen in early Precambrian times (27,28) has gained some acceptance. The high degree of similarity in the N-terminal sequences of superoxide dismutases (SODs) between Fe- and Mn-enzymes of *E. coli*, mitochondrial Mn-enzyme, and *Bacillus stearothermophilus* Mn-enzyme (29) makes it likely that the ancestor at the base of the *Bacillus* branch possessed a functional SOD (29,30). The widespread occurrence of Fe-SODs in obligate anaerobes suggests that this enzyme may have evolved even earlier in the anaerobic stem (31-33).

It is generally believed that enough free oxygen to support respiration began to appear only because of the development of oxygen-releasing photosynthesis (34). This assumption implies that the common ancestor at the base of the blue-green branch was an anaerobe whose descendants eventually acquired aerobic respiration when the atmosphere became oxidizing. This in turn means that ancestors of *Bacillus*, lactic acid bacteria, and *Escherichia*, as well as the blue-greens, were anaerobes that later in evolution independently adapted to aerobic life. The diversity of terminal oxidases suggests an independent adaptation to oxygen. Sequence data from terminal oxidases could establish whether or not there was a conserved structural basis for this function.

The adaptation to oxygen involves a cytochrome-dependent electron transport chain in *Bacillus* (35) and a flavin-terminated respiratory system in lactic acid bacteria (36). In Fig. 1 (middle portion), *Lactobacillus* and *Streptococcus* lie on the lactic acid bacterial branch. This is consistent with the observation that there is a close immunological relatedness individually among aldolases (37), lactate dehydrogenases (38), and malic enzymes (39) in some species belonging to these two genera.

Although the exact position of the Escherichia branch is unclear, it is nevertheless probable that this branch acquired oxygen-terminated electron transport pathways independent of those of either Bacillus or cyanobacteria. In fact, there are two such pathways in E. coli: one operating at conditions of high aeration and terminating in an oxidase, cytochrome o, and the other operating at lower oxygen tensions and terminating in an oxidase, cytochrome d (40). Clearly, consistent with the suggestions of Broda (41) and Dickerson (42), these results suggest that at least the diverse terminal components in aerobic respiration are polyphyletic in origin.

It seems likely that blue-greens appeared in the fossil record between 2 and 3 Gyr ago (9). Geological evidence indicates that free oxygen began to accumulate in the atmosphere around 2 Gyr ago (34). From Figure 1 (upper portion), it is seen that the blue-greens and Rhodospirillaceae (purple nonsulfur bacteria) arose from the anaerobic stem as a branch distinct from that of anaerobic photosynthetic bacteria. However, it is known that a number of blue-greens (43) as well as Rhodospirillaceae are capable of anoxygenic photosynthesis. Clearly, this property is an attribute of the anaerobic stem that both of these branches have retained. Bacillus, lactic acid bacteria, coliform bacteria, and pseudomonads have lost this property.

Eventually, the ancestral stock of blue-greens began to use water as an indispensable electron donor for the photoassimilation of CO_2. This oxygen-releasing photosynthesis in blue-greens involves two photosystems (44) with chlorophyll a and phycobiliproteins acting as light-collecting pigments (45). The chemical identities and order of participation of redox carriers in these photosystems are fairly well understood (45). Whereas photosystem I was an early acquisition of the anaerobic stem, photosystem II, which is unique to this group, utilizes a b-type cytochrome, plastoquinone, a c-type cytochrome, and plastocyanin as redox carriers. In blue-greens, the reductant of CO_2 appears to be NADPH rather than NADH as in other photosynthetic bacteria. Whereas blue-greens do utilize oxygen as the terminal electron acceptor in respiration (46), many of them do not possess a complete Krebs cycle (47).

Figure 1 (upper portion) encompasses two bacterial groups: Rhodospirillaceae and Pseudomonadaceae. The Rhodospirillaceae possess a photosystem similar to that of the anaerobic purple sulfur bacteria. Both of these groups have energy-linked photosystem I involving, at the least, ubiquinone, a b-type cytochrome, and a c-type cytochrome; both have cyclic, noncyclic, and reverse electron-supply systems (21). The major difference is that in purple sulfur bacteria cyclic and noncyclic electron-delivery systems converge at the photooxidizable reaction center, and in purple nonsulfur bacteria the point of convergence appears to be at cytochrome c_2 (21).

Because purple nonsulfur bacteria and pseudomonads are aerobes, it can be assumed that the common stem of these organisms, which already possessed a photosynthetic electron transport system, acquired an oxygen-terminated electron transport chain as well. In fact, both of these processes have been shown to be similar in Rhodopseudomonas capsulata (48,49) in that they share a common electron-carrying arm consisting of ubiquinone, b- and c-type cytochromes, iron-sulfur proteins, and cytochrome c_2. However, in aerobic respiration, reduced flavoprotein generated by the entry of NADH + H^+ starts the electron cascade, and a cytochrome oxidase complex, either cytochrome o or aa_3, terminates it.

Once aerobic respiration and the Krebs cycle had become fully established in the prokaryotic stem, the potential was provided for deriving many times more ATP than from fermentation. This innovation opened many new niches to the aerobic stock. Furthermore, much larger and more complex cell types developed, giving rise to eukaryotes probably around 1.4 Gyr ago (50). There have been two major theories concerning the origin of eukaryotes. One is that eukaryotes arose as a result of serial endosymbiosis among prokaryotes (51), and the other is that they arose by compartmentalization of the DNA within the cytoplasm of a single line of prokaryotes (52,53).

The sequence data clearly support the endosymbiotic theory. A eukaryotic branch emanating from the middle portion in Fig. 1 represents the host line for the endosymbioses. The 5S rRNAs used in constructing this portion of the tree were from cytoplasmic ribosomes, one of the three ribosomal systems found in eukaryotes. Figure 1 (upper portion) places purple nonsulfur bacteria near eukaryote mitochondria. In particular, Rhodopseudomonas globiformis diverges from the flagellates Euglena and Crithidia close to their point of divergence from one another. Clearly, the ancestors of R. globiformis qualify as free-living prokaryotes that gave rise to eukaryote mitochondria, probably in more than one symbiotic event (54). Similarly, phylogenetic histories based on 2Fe-2S ferredoxin and cytochrome c_6 sequences identify blue-greens as the free-living organisms that gave rise to chloroplasts and indicate that more than one symbiotic event has occurred (55).

ACKNOWLEDGMENTS: We thank Dr. L.T. Hunt for directing our attention to a number of relevant articles, and K. Lawson for drafting the figure. This work was supported by NIH grant GM08710 and NASA contract NASW 3317. J.B. is a senior Fulbright-Hayes grantee.

REFERENCES

1. Schwartz, R.M. & Dayhoff, M.O. Science 199, 395 (1978).
2. Dayhoff, M.O. & Schwartz, R.M. in Endosymbiosis and Cell Research, Schwemmler, W. & Schenk, H., eds., Walter de Gruyer, Berlin, 1980, in press.
3. Dayhoff, M.O. Atlas of Protein Sequence and Structure, Vol. 5 and suppls. 1-3, National Biomedical Research Foundation, Washington, D.C., 1972, 1973, 1976, 1978.
4. Dayhoff, M.O. Fed. Proc. 35, 2132 (1976).
5. Eck, R.V. & Dayhoff, M.O. Science 152, 363 (1966).
6. Tanaka, M., Nakashima, T., Benson, A.M., Mower, H.F. & Yasunobu, K.T. Biochemistry 5, 1666 (1966).
7. Miller, S.L. & Urey, H.C. Science 130, 245 (1959).
8. Oparin, A.J. The Origin of Life on the Earth, Academic, New York, 1957.
9. Schopf, J.W. Biol. Rev. 45, 319 (1970).
10. Ponnamperuma, C. The Origins of Life, Thames & Hudson, London, 1972.
11. Tagawa, K. & Arnon, A.J. Nature 195, 537 (1962).
12. Yasunobu, T. & Tanaka, M. in Iron Sulfur Proteins, Vol. 2, Lovenberg, W., ed., Academic, New York, 1973, pp. 27-130.
13. Mortensen, L.E. Proc. Nat. Acad. Sci. USA 52, 272 (1964).
14. Buchanan, B. in The Enzymes, Vol. 6 Boyer, P.D., ed., Academic, New York, 1972, pp. 193-216.
15. Gottwald, M., Andreesen, J.R., LeGall, J. & Ljungdahl, L. J. Bacteriol. 122, 325 (1975).
16. Postgate, J.R. in Evolution in the Microbial World, Carlile, M.J. & Skehel, J.J., eds., Cambridge University Press, London, 1974, pp. 263-292.
17. Mortensen, L.E. & Thorneley, R.N.F. Annu. Rev. Biochem. 48, 387 (1979).
18. Emerich, D.W. & Burris, R.H. Proc. Nat. Acad. Sci. USA 73, 4369 (1976).
19. Ruvkun, G.B. & Ausubel, F.M. Proc. Nat. Acad. Sci. USA 77, 191 (1980).
20. Walter, M.R., Buick, R. & Dunlop, J.S.R. Nature 284, 443 (1980).
21. Dutton, P.L. & Prince, R.C. in The Photosynthetic Bacteria, Clayton, R.K. & Sistrom W.R., eds., Plenum, New York, 1978, pp. 525-576.
22. Knaff, D.B. in The Photosynthetic Bacteria, Clayton, R.K. & Sistrom, W.R., eds., Plenum, New York, 1978, pp. 629-654.
23. Olson, J.M. Evol. Biol. 11, 1 (1978).
24. Fuller, R.C. in The Photosynthetic Bacteria, Clayton, R.K. & Sistrom, W.R., eds., Plenum, New York, 1978, pp. 691-705.
25. Peck, H.D. Jr. in Evolution in the Microbial World, Carlile, M.J. & Skehel, J.J., eds., Cambridge University Press, London, 1974, pp. 241-262.
26. Schidlowski, M. Origins Life 9, 299 (1979).
27. Berkner, H.V. & Marshall, L.C. J. Atmos. Sci. 22, 225 (1965).

28. Towe, K.M. Nature 274, 657 (1978).
29. Harris, J.I. & Steinman, H.M. in Superoxide and Superoxide Dismutases, Michelson, A.M., McCord, J.M. & Fridovich, I., eds., Academic, New York, 1977, pp. 225-230.
30. Brock, C.J. & Harris, J.I. in The Evolution of Metalloenzymes, Metalloproteins & Related Materials, Leigh, G.J., ed., Symposium, London, 1977, pp. 85-98.
31. Hewitt, J. & Morris, J.G. FEBS Lett. 50, 315 (1978).
32. Gregory, E.M., Moore, W.E.C. & Holdeman, L.V. Appl. Environ. Microbiol. 35, 988 (1978).
33. Lumsden, L. & Hall, D.O. Nature 257, 670 (1975).
34. Cloud, P.E. Science 160, 729 (1968).
35. Miki, K. & Okunuki, K. J. Biochem. (Tokyo) 66, 845 (1969).
36. Dolin, M.I. in The Bacteria, Vol. 2, Gunsalus I.C. & Stanier, R.Y., eds., Academic, New York, 1961, pp. 425-460.
37. London, J. & Kline, K. Bacteriol. Rev. 37, 453 (1973).
38. Gasser, F. & Gasser, C. J. Bacteriol. 106, 113 (1971).
39. London, J., Meyer, E.Y. & Kulczyk, S. J. Bacteriol. 106, 126 (1971).
40. Haddock, B.A. & Jones, C.W. Bacteriol. Rev. 41, 47-99 (1977).
41. Broda, E. in Chemical Evolution and the Origin of Life, Buvet, R. & Ponnamperuma, C., eds., North-Holland, Amsterdam, 1971, pp. 446-452.
42. Dickerson, R.E. Nature 283, 210 (1980).
43. Garlick, S., Oren, A. & Padan, E. J. Bacteriol. 129, 623 (1977).
44. Hill, D. & Bendall, F. Nature 186, 136 (1960).
45. Krogmann, D.W. in The Biology of Blue-green Algae, Carr, N.G. & Whitton, B.A., eds., University of California Press, 1973, pp. 80-98.
46. Fogg, G.E., Stewart, W.D.P., Fay, P. & Walsby, A.E. The Blue-green Algae, Academic, New York, 1973, pp. 161-179.
47. Stanier, R.Y. in The Biology of Blue-green Algae, Carr, N.G. & Whitton, B.A., eds., University of California Press, 1973, pp. 501-518.
48. Mars, B. & Gest, H. J. Bacteriol. 114, 1045 (1973).
49. Mars, B. & Gest, H. J. Bacteriol. 114, 1052 (1973).
50. Schopf, J.W. & Oehler, D.Z. Science 193, 47 (1976).
51. Margulis, L. Origin of Eukaryotic Cells, Yale University Press, New Haven, 1970.
52. Raff, R.A. & Mahler, H.R. Science 177, 575 (1972).
53. Uzzel, T. & Spolsky, C. Amer. Sci. 62, 334 (1974).
54. Dayhoff, M.O. & Schwartz, R.M. Ann. N.Y. Acad. Sci., in press (1980).
55. Schwartz, R.M. & Dayhoff, M.O. Ann. N.Y. Acad. Sci., in press (1980).

COUPLING TO SOLAR ENERGY: SENSITIZED PHOTOREACTIONS - THE PRIMARY SOURCE OF SELF-ORGANIZATION.

Peter Decker

Chemical Institute, Veterinary School, Hannover, FRG.

Using a screening test for phosphate binding reactions, a sensitized photooxidation of 2-methylimidazole was found, which transforms up to 40% of inorganic phosphate into a crystallizable alkali-labile phosphate ester. Our rationale for this search was that life and all life-like systems thermodynamically are dissipative structures in chemically reacting systems far from equilibrium, and tha such systems should arise at or near the site of maximal energy dissipation, i.e. the degradation of the 6000° "hot" visible light quanta, providing the driving force for a generalized Darwinian evolution - a cumulative self-organization and dynamic conservation of information resp. structure in the open system. This raises the question of what primary photoreactions, sensitizers, first metabolic cycles, feedback structures, and energy storing or group transfer reactions could occur in a primeval environment. One such scenario involves a chemical feedback structure based on an "inverse assimilation" - the anaerobic photochemical H_2-evolution from methane and autocatalytic condensation of the resulting formaldehyde into sugars and other compounds, including sensitizers of the photoreaction.

Rationale

Since life thermodynamically is an open system (1,2), in all theories on its origin, the energy source gains first importance. Indeed, other things being equal, "evolution" representing a cumulative self-organization into chemically reacting dissipative structures (3), i.e. acquisition and accumulation of information (4,5), primarily should be proportional to the sum of energy turnover per mass unit. Beyond terrestrial life (the sole known

Y. Wolman (ed.), Origin of Life, 529–536.

element of a supposed general class, logically undefined, since it would require "common properties" for definition (6)) this applies to all bioids (a "bioid" being defined as "open system which can exist in different steady states" (6)), i.e. to systems capable of at last one step of a generalized Darwinian evolution - a transition ("mutation") from one state ("species") into a more stable one, representing the aequisition of one bit of information, which becomes conserved in the new dynamical structure.

This concept leads to conclusions quite different from the guidelines of many experiments done on origins of life during the last two decades (7,8).

i. That sensitized photochemical reactions involving visible light served as driving force of the very first steps of self-organization into primeval "metabolic" processes, because the dissipation of the 6000° "hot" visible light quanta at a planetary surface represents the site of maximal energy dissipation and therefore the most probable site of spontaneous self-organization.

ii. That the first such bioids involved material readily available in the primeval environment, and reaction cycles stabilized by nonenzymatic catalytic feedback into steady (or even defined oscillating) states - an "informal" storage (4,7) on noncoded information; in order to resist the dispersion by diffusion such reactions must have been fast and therefore experimentally well observable and reproducible. Moreover, because reactions and not cells are the competing units and the number of chemically feasible feedback structures certainly is rather limited, the result of "selection" of the fastest pathways should be experimentally observable and reproducible up to the stage, where the nucleoprotic system (or something like) would provide so many(9) alternative pathways that the resulting "Monte Carlo" regime will become practically irreproducible (4,8).

iii. That enzymes represent a standardization of catalysts involving self-replication of information coded in nucleic acids. Of course this might be chosen as defining property of terrestrial life. However, since only something existing can become standardized, this step must evidently have been preceeded by an already running system using nonenzymatic catalysis. i.e. metabolism was first (10) and enzymes may be considered as improved versions of prenucleoprotic catalysts - metal ions and their complexes, coenzymes and even H^+ and OH^- ions (11).

We consider the nucleoprotic system as one possibility of a set of chemically feasible evolution stages in a given environment. Many entirely different possibilities may become implemented in other, e.g. extraterrestrial individual courses of bioid evolution.

iv. That independently of such specific use,macromolecules can enhance selection value at a very early stage, improving and modifiying reactivity through concentration of catalytic metal ions, slowing down their own diffusion and, by the formation of a gel, the diffusion of other reactants, thus easing the way toward individuation into separate reacting areas (dissipative structures (4)), and phase separation (12,13) into micelles and more complex coacervates like scrolls, tubules and vesicles involving compartmentalization and allowing vectorial processes(14).

v. That self-organization into open systems far from equilibrium should favor "chemical dissipative" structures involving stepwise entropy production and anabolic reactions tending to retain negentropy in "energy-rich" compounds at each step, coupling group transfer to energy yielding reactions, and finally to the primary photoreactions.

Implications for experimental approaches

What are the implications concerning further experimentation resulting from this thermodynamical concept? First, that reactions are more interesting than compounds. Indeed, reactions, and especially catalytic reactions are elements of autocatalytic feedback, the basic prerequisite of self-organization. Therefore, instead of screening for compounds formed in abiogenic syntheses, we try to develop screening tests for reactions and specific catalysts thereof (15,16.17), as part of our "Hannover program",the search for the first self-organizing open systems.

The first kind of reaction to look for is the harvesting of light through photochemical reactions; it involves two facets: The search for photosensitizers including appropriate compounds which could occur in a prebiological scenario, and the study of appropriate sensitized primary reactions.

The life bias, a fallacy resulting from neglecting the uniqueness of terrestrial system (7,18) until recently imposed the belief that in every life-like system chlorphyll derivatives should be the sensitizer. The discovery of bacteriorhodopsin destroyed this belief: But now it should tell us that all rather those two sensitizers are now eligible for chanelling light energy into bioids. In this context we should note that the energy of light quanta is independent on distance, and therfore, the same primary photoreactions as on earth can occur in the jovian clouds and on its satellites, whereas the different thermal and chemical environment of course would impose quite different "metabolic" processes.

Besides of sensitizers we have to ask for sensitized reactions, which always basically represent a charge separation producing an oxidant and a reductant. A "useful" photoreaction results if an immediate reversal of this process is avoided and the light energy is channeled into secondary products driving some kind of metabolism involving components of the respective chemical scenario. Since such processes should be viable over geological times, the balance of matter requires us to ask for the possibility of an organization into cycles for every element, resulting, from the very beginning, in a steady state scenario with only secular changes.

One kind of primary photoreactions involves the hydrogen abstraction from a substrate, where, in an anaerobic environment, the hydrogen may be evolved as H_2 in the presence of appropriate catalysts (19). It is part of a prebiological scenario which, besides of a photochemical origin of life, also affords a novel aspect of origin of photosynthesis and its impact on the oxidation state of the Earth's crust. I proposed, that the present carbon cycle between biomass and CO_2 has been preceeded by a carbon cycle between biomass and methane, and that the present photoreduction of CO_2 and O_2-evolution were preceeded by photooxidation of CH_4 and H_2- evolution, "inverse assimilation" (Fig. 1), where the H_2 escaped from the atmosphere, which explains better than previous theories the present oxidative state of the Earth's crust (20,21,22).

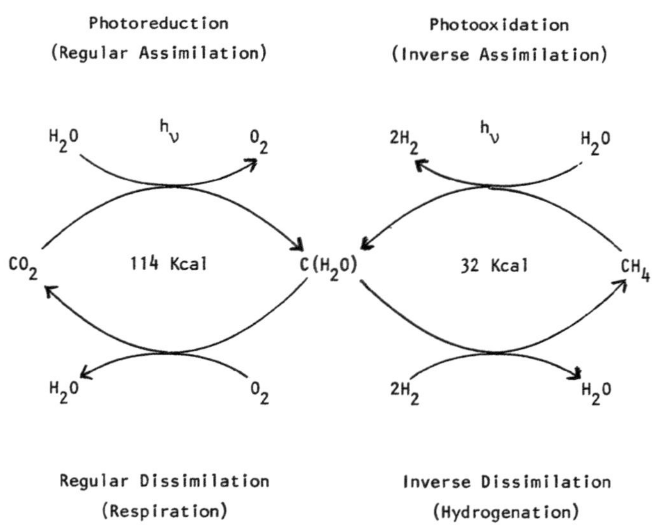

FIGURE 1

Regular assimilation (photoreduction) of CO_2 and "inverse" assimilation (photooxidation) of CH_4.

Oxidation of methane produces formaldehyde and the autocatalytic formation of sugars from formaldehyde possibly was the first "bioid" and the ancestor of all selforganizing systems in reducing planetary methane atmospheres. Reaction of ammonia or amino acids with the sugars (known as nonenzymatic browning or as Maillard reaction) affords a great diversity of metabolites, which may include sensitizers which could create a new feedback by enhancing the photooxidation of methane entailing the production of more formaldehyde and energetic intermediates, thus coupling the system to solar energy.

Pteridine derivatives, strongly fluorescent and probably photosensitizing compounds found by Dose (23) among the products of thermal condensation of amino acids, are also candidates for such functions.

A model of such inverse assimilation is presently often discussed as part of photochemical water-splitting systems: methylviologen if illuminated in the presence of EDTA and a sensitizer as e.g. proflavine, gives a strongly reducing blue radical with a redox potential of +0.45 V (24), which produces H_2 by coupling with a hydrogenase of collidal platinum (19). Our problem is now to look for catalysts which may allow to replace the EDTA (which becomes oxidized in this process by H-abstraction) by methane.

Interestingly, derivatives of deazaflavine, which are among the most effective sensitizers of the methylviologen reaction(19) also have been identified as constituent of the highly fluorescent coenzyme F_{420}, uniquely occuring in methanogenic bacteria (25). This is one of the oldest bacterial pyla; however, presently its species lack photosynthetic properties and perform the reverse reaction - fermentations involving "inverse dissimilation" of biomass producing methane.

A nonenzymatic photophosphorylation (26).

Following No.v. of our rationale, screening tests for the formation of energy rich intermediates as acyl (15) and phosphoryl (16,17) esters in the course of catalytic and oxidative reactions have been successfully used in earlier work. The working hypothesis in the previous use of Fenton's reagent was the conjecture that the great reactivity of radical intermediates might play a role in the ubiquitous biological oxidative and photo-phosphorylations.

Therfore, the same substrates as in our experiments with Fenton's reagent were illuminated by two 150 watt lamps with mirror reflectors in a water-jacketed 100 ml flask at 12°C in the presence of inorganic phosphate and different sensitizers

TABLE I

Substrate,	M	10^4(sensitizer),	M	PO_4^{3-} M	yield(%)
1-methylimidazole	0.14	methylene blue	1.0	0.26	0.7
2-methylimidazole	"	" "	"	"	7.2
pyrimidine	"	" "	"	"	0.7
histamine	"	" "	"	"	3.1
histidine	"	" "	"	"	1.9
"	"	hematoporphyrine	2.0	"	2.7
2-methylimidazole	0.8	"	0.5	0.5	40.0[a)]
"	0.32	rose bengal	1.0	0.18	11.4
"	"	safranin T	3.0	"	2.7
"	"	proflavin	1.0	"	2.1
"	0.25	-	-	0.25	1.4[b)]

Yields of bound phosphate in % of inorganic phosphate after 45min of illumination of 5 ml while gassing with O_2 at 12°C; initial pH 6.5; a) 20h illumination of 100ml, pH 6.5 (pH stat), 0°C, continuous replacement of bleached sensitizer (1.5 10^{-4}moles) and substrate (0.14 moles) end volume 150ml. b) 5h illumination with a Hg immersion lamp at 0°C.

Bound phosphate was determined as previously (16,17). Binding of phosphate was found to require O_2 and could be suppressed by singlet oxygen quenchers like Ni(II)-salts and sodium acide. Among sensitizers only those known to produce $^1O_2$ were effective.

In a preparative experiment, methylimidazole gave yields of bound phosphate of up to 40% of the added 0.5 m inorganic phosphate during overnight illumination, while gassing with O_2, continuously replacing bleached sensitizer (hematoporphyrine), and holding pH at 6.5 using a pH-stat.

By ion exchange chromatography the same alkali-labile crystalline compound Fp. 130° (decomp.)) as in our experiments with Fenton's reagent was obtained. According to elementary analysis and H-, P- and ^{13}C-NMR-spectra it represents the O-monophosphate of 4,5-dihydroxy-2-methyl-imidazoline (1,2). One mole of H_2O_2 per bound phosphate accumulates during the reaction.

Thus it turns out that the reaction involves a photooxygenation rather than an anaerobic photooxidation of the methylviologen type, which we were searching for as a model of photophosphorylation in a reducing atmosphere. Nevertheless, the reaction appears remarkable in its capability to scavenge up to 40% of inorganic phosphate from dilute aqueus solution producing a 0.2 m solution of bound phosphate. These results should encourage further experiments since the synthesis of energyrich intermediates involving oxidative and photochemical reactions may occur not so rarely as supposed at first glance.

The beginning of life as anaerobic fermentation has been proposed by Oparine (12) and Haldane (27), probably because in the twenties fermentation just had been explored as the first biochemical pathway. Modern bioenergetics started only 1929 with Lohmann's discovery of ATP (28) and the understanding of life as a process of thermodynamics of open systems was far away. In the light of modern knowledge,in our experimental approaches we should give more credence to the few authors as Dauvillier (29), Granick (30) or Gaffron (31) which early payed regard to the thermodynamic aspects of life and stressed the role of light as a primary driving force,opposing the ingrained paradigm of fermentation of a primeval soup in the warm little pond which, by a lucky chance, produced all at once, the first self-reproducing nucleic acid-protein system, and just the right enzymes needed for that fermentation.

ACKNOWLEDGEMENT

The experiments reported in the last section have been performed by Dr. Oemer Saygin and Dr. W. Schmidt. I thank Dr.Füldner Inst. f. biophysikalische Chemie, Göttingen, for this cooperation in NMR-spectroscopy, Mrs. R. Pohlmann and Mrs. R. Fallisch for technical assistance, and Mrs. H. Büsing for preparing this manuscript.

REFERENCES

(1) Ostwald, W., University Chronicle, Sept. 1903, p.1o1, Univ. of California 1903.
(2) Prigogine, I. and Wiame, J.M., Experimentia 2,451(1946).
(3) Glansdorff, P. and Prigogine, I., Structure, Stability and Fluctuations, Wiley, London 1971.
(4) Decker, P., Speidel, A. and Nicolai, W., in: Proc. 1st European Biophysics Congr. Baden/Vienna 1971, E.Broda et al. eds., Verl. Wiener Med. Akad., Vienna 1971, Vol. IV., p.515 (Bioids II).
(5) Decker, P., Ann.N.Y.Acad.Sci. 316,236(1979) (Bioids VII).
(6) Decker, P. and Speidel, A., Naturforsch. 27b, 257 (1972) (Bioids I).
(7) Decker, P. and Heidmann, W., in: Origin of Life, Proc. 2nd ISSOL Meetg., H. Noda ed., Jap.Sci.Soc.Press, Tokyo, pp. 617-623, 1978.
(8) Decker, P., Evolution in offenen Systemen: Prize essay submitted to the Bavarian Academy of Sciences; reprints (170pp) available from Docummentation Center of the Origin of Life,

Universitätsbibliothek, Ulm, FRG.,; cf. Decker, P., Nachr. Chem. Tech. 23, 167 (1975).
(9) Eigen, M., Naturwiss. 58, 465 (1971).
(1o) Buvet, R. in: The Origin of Life and Evolutionary Biochemestry, Dose, K. et al. eds., Plenum Press, New York 1974.
(11) Calvin, M., Chemical Evolution, Clarendon Press, Oxford 1969; Hoffmann-Ostenhoff, O., in: The Origin of Life on the Earth; Oparin, A.I. et al. ed., Pergamon Press, London 1959, p197.
(12) Oparin, A.I., The Origin of Life on the Earth, Oliver and Boyd, Edinburgh 1957.
(13) Kuhn, H., Naturwiss. 63, 68 (1976).
(14) Mitchell, P., Eur. J. Biochem. 95, 1 (1979).
(15) Saygin, Oe. and Decker, P., Naturforsch., (Bioids VIII) in press.
(16) Saygin, Oe., Z. Naturforsch., (Bioids IX) in press
(17) Saygin, Oe. and Decker, P., this Congress.
(18) Decker, P. in: Origins of Optical Activity in Nature, Walker, D.C. ed., Elsevier, Amsterdam 1979, pp.1o9-124.
(19) Krasna, A.I., Photochem.Photobiol. 31, 75 (1980).
(20) Decker, P., Pure and Appl. Geochemistry 112, 865 (1974).
(21) Decker, P., in: Orogin of Life, Proc. 2nd ISSOL Meetg., Noda, H. ed., Jap.Sci.Soc.Press Tokyo 1978, pp. 631-637.
(22) Decker, P., in: DFG-Kolloquium "Evolution in Planetenatmosphären", Schliersee, FRG, 2.-4.Okt. 1979, in press.
(23) Heinz, B., Ried, W. and Dose, K., Angew.Chem. 91, 510(1979).
(24) Kalyanasundaram, K. and Grätzel, M., Angew.Chem.91,759(1979).
(25) Thauer, R.K. and Fuchs, G., Naturwiss. 66, 89 (1979).
(26) Saygin, Oe., Schmidt, W. and Decker, P., in preparation.
(27) Haldane, J.B.S., The Inequality of Man, Chatto, London 1932.
(28) Lohmann, K., Naturwiss. 17, 624 (1929).
(29) Dauvillier, A., L'origine photochimique de la vie, Paris, Masson 1958; translation: The Photochemical Origin of Life Academic Press, New York 1965.
(30) Gaffron, H., in Horizons in Biochemistry, Kasha, M. and Pullman, B. ed., Academic Press, New York 1962, pp59-89.
(31) Granik, S., Ann.N.Y.Acad.Sci. 69, 292 (1957).

ULTRAVIOLET SELECTION PRESSURE IN THE PREPHANEROZOIC

Mitchell B. Rambler

Department of Biology, Boston University,Boston, Ma.

The late appearance of Metaphyta and Metazoa in the fossil record has often been attributed to low concentrations of ambient oxygen and restriction of organisms to shielded habitats due to intense ultraviolet selection pressure (Berkner-Marshall Hypothesis).

Microorganisms examined for survival as a function of UV dosage consistantly exhibit a capacity for radiation resistance and repair. Comparisons of lethal UV dosages of various microbes to calculations of present solar UV fluxes at 253.7 nm result in an average radiation tolerance of 30 minutes (4.65×10^3 ergs mm^{-2}). However, inspection of survival of protected cells(e.g. UV absorbants in the aqueous irradiation media or photoreactivation) increases radiation tolerance by three orders of magnitude or 24 hours of continual radiation exposure. The observation of protective mechanisms in extant microbes may be legacies from UV exposure in the Prephanerozoic. Radiation resistance taken together with the well preserved microbiota known from the Prephanerozoic suggests the Berkner-Marshall hypothesis to be indefensible.

The effects of unattenuated ultraviolet radiation during the Prephanerozoic aeon (4.5-0.6 billion years ago) as limiting terrestrial occupation and proliferation is a widely held concept. This idea largely originated by Berkner and Marshall (1,2,3) and since perpetuated by others(4,5) has only recently been questioned(6). A quotation representative of the acceptance is taken from a recently published monograph on atmospheric evolution(7):"Sagan (1973) has estimated that typical contemporary microorganisms would be killed in a matter of

Y. Wolman (ed.), Origin of Life, 537–542.

seconds if exposed to the full intensity of solar radiation in this wavelength"(2500 A).

There is presently a plethora of papers supporting either contention of atmospheric evolution: anoxic Archean conditions (4,8), microaerophilic(7), to aerobic conditions(9). In light of this controversy, we can assume a complete lack of atmospheric oxygen and no subsequent development of a protective ozone layer yielding a minimum of protective conditions against incoming solar UV radiation in the Archean.

The Archean fossil record, although certainly not continuous, atests to an abundant microflora(10,11). Land based microbial communities as set apart from aquatic bioherms may have existed during the Prephanerozoic(12,13). This evidence is suggestive of either atmospheric shielding of UV radiation or microorganisms evolved sufficient mechanisms of protection. If we assume the latter, the tolerance of microorganisms to UV light under varying non-protective and protective regimes can test this hypothesis.Although heterotrophs and chemoautotrophs could have remained sheltered from deleterious UV radiation in bottom sediments, photoautotrophs have an obligate light requirement and therefore would have needed efficient repair mechanisms due to the unavoidable exposure to UV light. By astronomical inference however, daylength during the Archean is thought to have been on the order of 5-6 hours(14) which would have permitted ample time for dark repair of UV damage. Furthermore, most estimates of solar luminosity suggest a lower solar flux in the Archean than at present(15).

Natural filters for protection against lethal UV radiation have been suggested. The shielding capacity by a vertical 10 meter column of water has been calculated to be protective against UV light(1). A microlayer of purines and pyrimidines has also been suggested to have protective capabilities(5). A microlayer of organic-rich metallo-complexes of dissolved and particulate chemical species has been reported to exist on present day oceanic surfaces(16). A comparable microlayer may have been present during the Archean, but the extent of UV absorption or protection of such a microlayer is not known. However, the coacervate theory (17), the abiotic synthesis of phospholipid monolayers(18), and proteinaceous spheroids(19), are all consistant with the idea that a surface layer of oceanic organic material was available during the Archean.

Microbial mats are precursors to lithified microbial communities, stromatolites. The matting habit of many cyanobacteria is a result of trapping and binding sedimentary particles(20). The formation of mats help protect underlying organisms from potentially lethal UV radiation(6,21). A UV protective protein was reported for the cyanobacterium,*Anacystis nidulans*(22). The fortuitous absorption of UV radiation by chemicals present in the irradiation media can prolong survival of various microbes for up to 24 hours of continual UV exposure(6,21,23). In addition, enzymatic repair processes are available to most organisms for correction of macromolecular damage induced by UV radiation(24).Photoreactivation, those enzymatic repair processes requireing light, has been observed in most cell types studied(25). This particular repair pathway has been shown to be 80% efficient in most bacterial strains(26). Furthermore, photoreactivation has been observed with a similar efficiency in obligate anaerobes(23).Sporulation, a response of microorganisms to suboptimal conditions, is another mechanism of radiation resistance(27).

Laboratory investigations usually study isolated aspects of radiation resistance or sensitivity. Several mechanisms interact in nature that produce increased tolerance to UV radiation not observed in the laboratory. For example,scattering and absorption of UV light due to organics and particulates in natural waters accordingly vary UV attenuation(29).How then can lethal radiation dosages for various microbes be determed? Is a lethal dose judged by a 10% or 99.9% reduction in the population under study? The following tables are presented to show the variation in tolerance to UV radiation(table 1) and the increased tolerance when a single protective parameter is added to the radiation study (table 2). Although the presence of nitrates in an anaerobic Archean environment is controversial, the protective capability of aqueous sodium nitrate serves to illustrate the ineffectiveness of UV light in this regime.

Solar flux calculations in an unattenuated atmosphere in the spectral interval of 253.3-256.4 nm for the present day solar luminosity values yield a UV flux of 2.26 ergs mm^{-2}(28). Based on this flux, for those microbial populations reduced by 10^{-3}(99.9%)in table 1, an average lethal value of 4.65×10^{3} ergs mm^{-2}(about 34 minutes of UV exposure) is produced. This evaluation although crude and incomplete, serves to illustrate the tolerance of unprotected bacterial strains. Survival values are increased further by imposing a single UV absorbant on the study(sodium nitrate 0.01-0.2 M). Those organisms presumed to be primitive with respect to metabolic capabilities, would be expected to be more resistant to UV radiation if they are decendants from an ancestor that lived in an environment of unattenuated solar radiation. That this is so, is demonstrated by the fact that an obligate anaerobe, Clostridium, requires an exposure of 1.2×10^{4} ergs mm^{-2}(60 minutes) for 10^{-3}reduction in an unprotected population(23). It is interesting to note, mammalian cell lines are reduced by 10^{-3} in less than two minutes exposure(6).

It is unlikely therefore, that UV radiation limited the

proliferation and diversification of life in the Prephanerozoic. Mechanisms of protection against UV light can be observed in extant microorganisms as possible legacies from Prephanerozoic UV exposure.

Table 1

Ultraviolet Dose Responses of Microorganisms @ 253.7 nm

Kingdom: Monera Phylum: Cyanobacteria	Dosage (ergs mm^{-2})	Surviving Fraction	Author
A. quadruplicatum	1.00×10^3	10^{-2}	30
A. flos-aquae	8.1×10^2	10^{-2}	31
G. alpicola	3.84×10^3	10^{-2}	32
Lyngbya sp.	1.46×10^3		6
Lyngbya sp. (+0.01 M NO_3)	3.1×10^4		6
Phylum: Aeroendospora			
Bacillus subtilis	5.7×10^3	10^{-3}	6
Bacillus subtilis (+0.01M NO_3)	6.7×10^4	10^{-3}	6
Bacillus cereus	5.5×10^2	10^{-2}	33
Phylum: Omnibacteria			
E. aerogenes	3.5×10^3	10^{-3}	6
E. aerogenes (+0.01M NO_3)	1.2×10^4	10^{-3}	6
E. coli B/r	1.2×10^3	10^{-4}	33
E. coli K12	1.3×10^3	10^{-3}	34
E. coli	4.0×10^3	10^{-3}	6
E. coli (+0.01 M NO_3)	2.0×10^4	10^{-3}	6
Klebsiella pneumoneii	4.0×10^3	10^{-3}	6
K. pneumoneii (+0.01 M NO_3)	2.2×10^4	10^{-3}	6
Phylum: Pseudomonads			
P. aerogenosa	2.08×10^3	10^{-3}	6
P. aerogenosa (+0.01M NO_3)	1.1×10^4	10^{-3}	6
Phylum: Fermenting Bacteria			
C. sporogenes	1.2×10^4	10^{-3}	6
C. sporogenes (+0.01 M NO_3)	3.1×10^4	10^{-3}	6

Classification (35)

Table 2

Ultraviolet Dose Responses of Microorganisms Irradiated in the Presence of 0.2 M Nitrate @ 253.7 nm (6)

Kingdom: Monera	Dosage (ergs mm^{-2}: hrs)	Surviving Fraction
Phylum: Omnibacteria		
E.aerogenes	1.75×10^5:15.5	10^{-3}
E.coli	1.70×10^5:15	10^{-3}
K.pneumoneii	1.85×10^5:17	10^{-3}
S.lutea	1.72×10^5:15	10^{-3}
Phylum: Cyanobacteria		
Lyngbya sp.	2.1×10^5 :24	?
Phylum: Aeroendospora		
B.subtilis	2.2×10^5: 24	10^{-3}
Phylum: Pseudomonads		
P.aerogenosa	1.7×10^5:15	10^{-3}
Phylum: Fermenting Bacteria		
C.sporogenes	2.3×10^5:24	10^{-3}

References

1.Berkner,L.V.,and Marshall,L.C. Origin and Evolution of the Oceans and Atmosphere,P.J.Brancazio and A.G.W.Cameron(eds),J.Wiley and Sons:N.Y.1964.102.
2.Berkner,L.V.and Marshall,L.C. J.Atm.Sci.22:225(1965).
3.Berkner,L.V.and Marshall,L.C.J.Atm.Sci. 23:133(1966).
4. Cloud,P. Econ.Geol. 68:1135(1973)
5.Sagan,C. J.Theor.Biol. 39:195 (1973).
6.Rambler,M.B.Ultraviolet Irradiation of Bacteria Under Anaerobic Conditions: Implications for Prephanerozoic Evolution,Ph.D.Thesis Boston University.
7.Walker,J.C.G. Evolution of the Atmosphere,Macmillan Pub.Co.:NY (1977)
8.Hart,M. Origins of Life 9:261 (1979).
9. Dimroth,E.and Kimberley,M. Can.J.Earth Sci. 13:1161(1976)
10.Barghoorn,E.S.andTyler,S. Science 147:(1965).
11.Schopf,J.W. Ann.Rev.Earth.andPlanetary Sci.3:(1975).
12.Campbell,S. Origins of Life 9:335(1979).
13.Margulis,L. Symbiosis in Cell Evolution,W.H.Freeman,San Francisco(1980).

14.Alfven,H.and Arrhenius,G. Evolution of the Solar System NASA Publication # SP-345(1976).
15.Newman,M.andRood,R. Science 198:1035(1977).
16.Chesselet,R. Abstract,Conference on Global Ecology,NASA, Washington,D.C.(1979).
17.Oparin,A.I. The Origin of Life,Macmillan:New York(1938).
18.Deborin,G.and Sorokina,A. The Origin of Life and Evolutionary Biochemistry,Dose,K.,Fox,S.,Deborin,G.,Pavlovskaya,T.(eds) Plenum Press:New York(1974).
19.Fox,S. The Origin of Life and Evolutionary Biochemistry, Dose,K.,Fox,S.,Deborin,G.,Pavlovskaya,T.(eds) Plenum Press: New York(1974).
20.Monty,C. Annales de la Soc.Geol.de Belgique 96:585(1973).
21.Rambler,M.B.,Margulis,L.and Barghoorn,E.S.Chemical Evolution of the Early Precambrian,Ponnamperuma,C.(ed.)Academic Press: New York(1977).
22.Bhattacharjee,S.K.Nature 269:82(1977).
23.Rambler,M.B.and Margulis,L. Science,in press,(1980).
24.Smith,K.C. Photochem.Photobiol. 28: 121(1978).
25.Sutherland,B.Nature 248: 109 (1974).
26.Bridges,B.and Munson,R. Biochem.Biophys.Res.Comm.30:620(1968).
27.Zamenhof,S.and Reddy,T.Rad.Res. 31:112 (1967).
28.Levine,J.S. Origins of Life,in press,(1980).
29.Smith,R.C.and Baker,K.S. Photochem.Photobiol. 29:311(1979).
30.Van Baalen,C.and O'Donnell,R.Photochem.Photobiol.15:269(1972).
31.Newton,J.,Tyler,D.and Slodki,M. Appl.Env.Micro. 38:347(1979).
32. Williams,E.,Lambert,J.,O'Brien,P.and Houghton,J. Photochem.Photobiol. 29:543 (1979).
33. Harm,W. and Haefner,K. Photochem.Photobiol. 8:179 (1968)
34. Tyrell,R. Photochem.Photobiol. 24:345(1976)

DNA-PROTEIN COMPLEX FROM AN EXTREME HALOPHILE : A HISTONE-LIKE PROTEIN IN ARCHAEBACTERIA

Masayuki OHBA and Tairo OSHIMA

Mitsubishi-Kasei Institute of Life Sciences,
Machida, Tokyo 194, Japan

ABSTRACT: A histone-like protein was found in an extract of halophilic archaebacteria, *Halobacterium halobium*. The protein is similar to histone in eukaryote nuclei in the following aspects: (i) The protein stabilized DNA. (ii) The protein-DNA complex formed nucleosome-like structure. (iii) The complex was stable in high salt concentrations which is similar to the physiological ionic strength. The presence of histone-like proteins in archaebacteria could be an important clue for studying the phylogenetic status of these unique microorganisms.

Based on nucleotide sequence homology in rRNAs from various organisms including prokaryotes and eukaryotes, Woese and his co-workers proposed a new kingdom named Archaebacteria (1-3). Extreme halophiles, acido-thermophiles and methanogens share with each other the common characters which differ from those of prokaryotes, and can be classified into Archaebacteria. The unique characters found in the archaebacteria include (i) the absence of peptideglycan cell wall, (ii) the presence of ether lipids with phytanyl chains, (iii) the presence of characteristic tRNAs and rRNAs.

Although Woese and Fox speculated that archaebacteria represent ancestral form of life (2,4), it has been reported that some of archaebacteria have several biochemical properties which is characteristics to eukaryotes. For example, extreme halobacteria lacks the Met-tRNA transformylase (5). A ribosomal protein (6) and 5S RNA (7) from the extreme halophile showed sequence homologies with the corresponding ones of eukaryotic cells. *Thermoplasma*, an acido-thermophilic archaebacteria has histone-like

Y. Wolman (ed.), Origin of Life, 543–549.

protein, actin- and myosin-like proteins, and respiratory enzymes similar to those of the eukaryotic microbodies (8). These findings are suggestive of phylogenetic relationship between archaebacteria and eukaryotes. However, further studies on biochemical characteristics of the archaebacteria would be necessary to draw the conclusion, since eukaryote-like properties are sometimes observed with some prokaryotes. In this paper, we will report the presence of histone-like protein in a halophilic archaebacteria, *Halobacterium halobium*.

The halobacteria, *H. halobium*, was grown under heterotrophic conditions (9), and the harvested cells were disrupted in a medium of low ionic strength consisting of 0.015M NaCl, 0.01M $MgCl_2$ and 0.01M Tris-Cl, pH 7.4. The lysate was homogenized with a blender for 10 min and then centrifuged at 100,000 g for 15 hours. The supernatant was applied on a Sepharose 2B column. The column was eluted with 0.01M Tris-Cl buffer, pH 7.4, containing 0.015M NaCl and 0.01M $MgCl_2$. A typical chromatogram is shown in Fig. 1. The DNA-protein complex was excluded from the gel and separated from most of the cellular proteins and metabolic intermediates. The fractions, of which absorbancy at 260 nm / 230 nm ratio was above 2.0, were collected and subjected to sucrose gradient centrifugation to remove free proteins and nucleic acids. The DNA-protein complex thus purified was used as the experimental material.

The DNA to protein ratio of the final preparation was 1 : 0.4

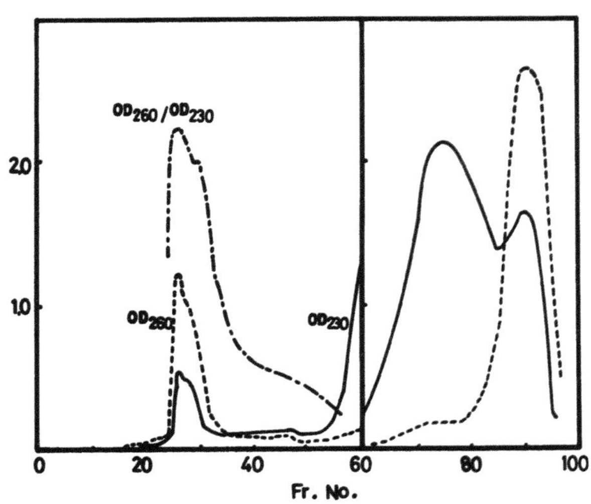

Fig.1 Sepharose 2B gel filtration chromatography of *H. halobium* supernatant. ---: absorbancy at 260 nm, ——: at 230 nm, —·—: ratio of A_{260}/A_{230}. The scale of vertical axis in the right half figure is reduced to 1/10 of that in the left half.

by weight. No RNA was detected by orcinol test, suggesting that the preparation was free from mRNA and ribosomes. On a SDS-disc gel electrophoresis, the protein dissociated from the DNA gave a single band. The molecular weight was estimated to be about 10,000.

The complex purified from the halobacteria is similar to the DNA-histone complex in the following three respects. (i) The DNA is stabilized by the protein under low ionic conditions. (ii) A nucleosome-like structure was seen under an electron microscope. (iii) The complex was stable in hight salt concentrations which is supposed to be similar to the intracellular conditions.

Figure 2 shows the thermal melting profiles of DNA in the low salt concentration. When the bound protein was removed by Pronase digestion, the thermal stability of DNA was lowered. After the digestion by 10 μg Pronase/ml at 30°C for 2 hours, the

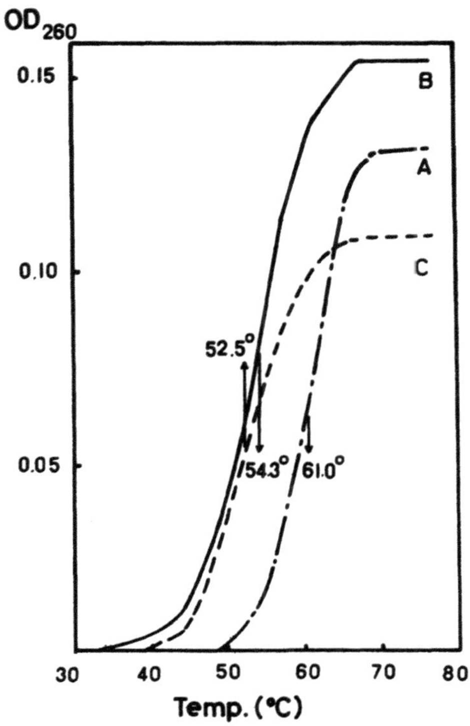

Fig.2 Thermal denaturation of the DNA-protein complex from *H. halobium*. A: The complex(11μg/ml) was heated in 0.25mM EDTA (pH 8.0). B: The complex was digested by Pronase (5μg/ml) at 30°C for 2 hours, followed by dialysis against 0.25mM EDTA (pH 8.0). C: Digested by 10μg/ml Pronase.

melting temperature observed (see curve (c) in Fig. 2) was identical to that of the protein free DNA extracted from the halobacteria.

The DNA-protein complex was contrasted with shadowing (10) and analyzed with an electron microscope. Nucleosome-like particles were observed on the electron micrograph as illustrated in Fig. 3. The particle has a diameter of 100Å, which is similar to that reported for nucleosomes (11). However, the distance between the particles was estimated to be about 1000Å, which is significantly longer than that of nucleosomes from eukaryotic cells. The longer distance between particles seems to be a characteristic feature of the halophile DNA-protein complex, and to account the low protein/DNA ratio of the complex compared with the eukaryotic DNA-histone complex.

Effect of the salt concentration on the complex was studied.

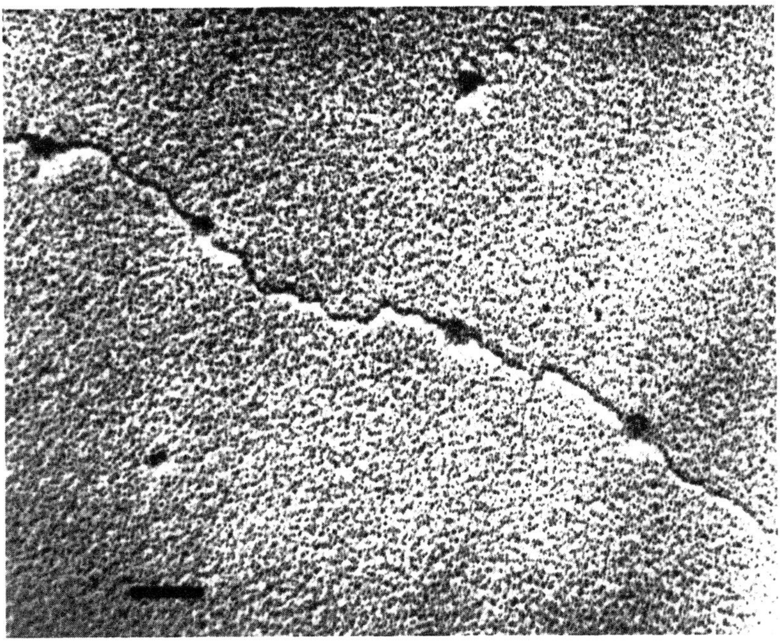

Fig.3 Electron micrograph of the DNA-protein complex. The sample was blotted on a carbon film supported on a mesh, and shadowed according to Ref. 10. The bar represents 500Å.

The complex was stable in the presence of 4M NaCl. A DNA-protein complex was also prepared from the halobacteria using a medium containing 4M NaCl, of which properties were identical with that prepared in the low ionic medium described above. Since it has been reported that the intracellular salt concentrations of the extreme halophiles are unusually high compared to those of normal cells (12,13), this observation suggests that the DNA-protein complex is stable in the cellular conditions and plays some roles in the regulation of gene expression *in vivo*.

Although the presence of histone-like proteins in some prokaryotic cells has been reported (14-16), Searcy and his co-workers (8) questioned about the physiological meaning of the complex, since the complex was stable only in lower ionic concentrations than the normal range of intracellular strengths. Recently it was reported (17) that no sequence homology was established between *E. coli* DNA-binding proteins and eukaryote

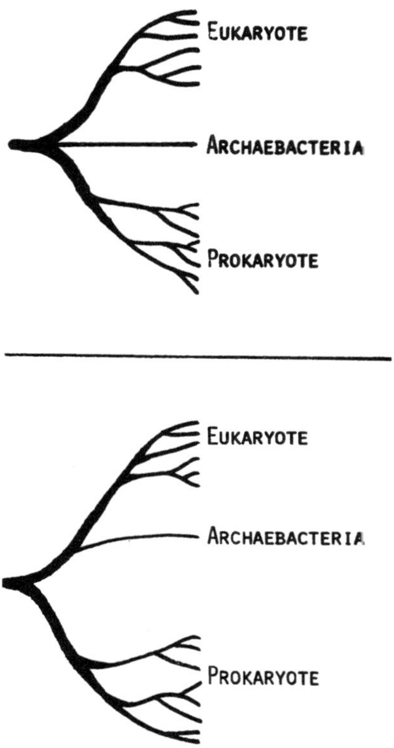

Fig.4 Models of phylogenetic relationships of three primary kingdoms.

histones. In contrast, the DNA-histone like complexes described here and reported for the acido-thermophilic archaebacteria are stable under ionic conditions similar to the physiological ones.

Two models have been proposed for the phylogenetic relationships among archaebacteria, prokaryotes (eubacteria) and eukaryotes as shown in Fig. 4. Woese and Fox (2) considered that three groups separated at an extremely primitive stage of life, and they are equidistant from one another. Hori and Osawa (7) suggested other model in which the primitive life separated initially into eubacteria and the common ancester of the extreme halophile and eukaryotes which was subsequently separated into two kingdoms. The model was also proposed by Yaguchi *et al.* based on the sequence homology in acidic ribosomal proteins from various organisms (8). To draw the final conclusion, further comparative studies will be necessary on the details of the machinery for macromolecule synthesis of archaebacteria with those of eubacteria and eukaryotes. The histone-like protein from the extreme halophile reported here, as well as that reported for *Thermoplasma* (8), could be an important clue for elucidating the phylogenetic relationship of archaebacteria to other groups. Work will continue to test the hybrid complex formation between the halophilic protein and the eukaryote DNA, and to determine the amino acid sequence of the protein.

REFERENCES

(1) Fox, G.E., Magrum, L.J., Balch, W.E., Woefe, R.S. and Woese, C.R.: Proc. Natl. Acad. Sci. USA, 74, 4537 (1977)
(2) Woese, C.R. and Fox, G.E.: Proc. Natl. Acad. Sci. USA, 74, 5088 (1977)
(3) Woese, C.R., Magrum, L.J. and Fox, G.E.: J. Mol. Evol., 11, 245 (1978)
(4) Woese, C.R.: J. Mol. Evol., 13, 95 (1979)
(5) White, B.N. and Bayley, S.T.: Biochim. Biophys. Acta, 272, 583 (1972)
(6) Yaguchi, M., Matheson, A.T., Visentin, L.P. and Zuker, M.: in "Genetics and Evolution of Transcriptional and Translational Apparatus" (S. Osawa *et al.*, eds.) to be published by Kodansha Scientific Press, Tokyo
(7) Hori, H. and Osawa, S.: Proc. Natl. Acad. Sci. USA, 76, 381 (1979)
(8) Searcy, D.G., Stein, D.B. and Green, G.R.: BioSystems, 10, 19 (1978)
(9) Gochnauer, M.B. and Kushner, D.J.: J. Microbiol., 15, 1157 (1969)
(10) Olins, A.L.: in "Methods in Cell Biology" Vol. 19 (G. Stein *et al.*, eds.) Academic Press, New York (1978) p. 61
(11) Thoma, F., Koller, T.H. and Klug, A.: J. Cell Biol., 83, 403 (1979)

(12) Lanyi, J.K. and Silverman, M.P.: Can. J. Microbiol., 18, 993 (1972)
(13) Ginzburg, M., Sachs, L. and Ginzburg, B.Z.: J. Gen. Physiol., 55, 187 (1970)
(14) Rouvière-Yaniv, J. and Gros, F.: Proc. Natl. Acad. Sci. USA, 72, 3428 (1975)
(15) Haselkorn, R. and Rouvière-Yaniv, J.: Proc. Natl. Acad. Sci. USA, 73, 1917 (1976)
(16) Rouvière-Yaniv, J.: Cold Spring Harbor Symp. Quant. Biol., 42, 439 (1978)
(17) Laine, B., Kmiecik, D., Sautiere, P., Biserte, G. and Cohen-Solal, M: Eur. J. Biochem., 103, 447 (1980)

THE EVOLUTION OF BLUE-GREENS AND THE ORIGINS OF CHLOROPLASTS

R.M. Schwartz and M.O. Dayhoff

National Biomedical Research Foundation and
Georgetown University Medical School,
Washington, D.C., 20007, USA

All of the available molecular data support the theory that the chloroplasts of eukaryote cells were originally free-living blue-greens. Of great interest is what the relationships are between contemporary types of blue-greens and eukaryote chloroplasts and whether the chloroplasts of the various eukaryotes are the result of one or more than one symbiosis. Combining information from phylogenetic trees based on cytochrome c_6 and 2Fe-2S ferredoxin sequences, we show that the chloroplasts of a number of eukaryote algae as well as the protist Euglena are polyphyletic; the chloroplasts of green algae and the higher plants may be the result of a single symbiosis.

The endosymbiotic theory for the origin of the eukaryote cell pictures the mitochondria and the chloroplasts as once having been free-living prokaryotes whose symbioses with a proto-eukaryote evolved into their current roles as cellular organelles (1). The chloroplasts are thought to have evolved from blue-greens (or cyanobacteria), whereas the mitochondria are thought to have evolved from aerobic bacteria. By combining information from different types of protein and nucleic acid sequences, we have shown that the eukaryote organelles share a recent common ancestry with bacteria and that this ancestry is indeed quite separate from the eukaryote host line (2). Recently elucidated sequences support this picture of early evolution and add exciting details. 5S ribosomal RNA (5S rRNA) sequences describe evolutionary histories for the eukaryote host and chloroplasts that are quite distinct and separate from one another. In phylogenetic trees based on ferredoxin and cytochrome c_6 sequences, branches leading to eukaryote chloroplasts are

Y. Wolman (ed.), Origin of Life, 551–558.

intermixed with the branches leading to blue-greens, suggesting an endosymbiotic, polyphyletic origin of these chloroplasts. These trees also treat the details of the phylogeny of contemporary blue-greens and their relationships to eukaryote chloroplasts.

The evolutionary trees presented here are based on complete nucleic acid or protein sequences and were derived by the least-squares matrix method (3). Branch lengths on all of the figures in this article are drawn proportional to the amount of inferred evolutionary change, and representative lengths are given in numbers of accepted point mutations per 100 residues.

5S rRNAs are relatively short molecules, about 120 nucleotides long, and are associated with the large ribosomal subunit. Fig. 1 pictures an evolutionary tree based on 5S rRNA sequences. These were elucidated by a number of workers, including T.A. Dyer, R.N. Nazar, and C.R. Woese (4-6). We placed the root of the

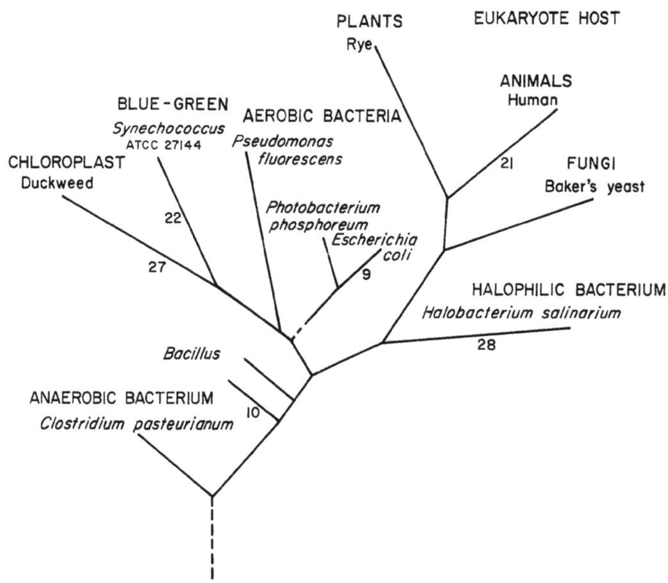

Figure 1. Phylogenetic tree based on 5S ribosomal RNA sequences. The branch leading to Escherichia coli and Photobacterium phosphoreum could have diverged as shown or from the Pseudomonas branch near its base. The two topologies have nearly the same overall length; the shorter one is shown. The first Bacillus divergence leads to B. stearothermophilus and the second to B. licheniformis (5S rRNA 1). Synechococcus ATCC 27144 is also called Anacystis nidulans. H. salinarium is called H. cutirubrum by the researchers who sequenced this molecule.

tree (or point of earliest time) on the clostridial branch in conformance with a tree based on bacterial ferredoxins and the strong evidence for a gene doubling in those sequences (2). Two branches of the tree are particularly noteworthy. The branch leading to animals, plants, and fungi is based on sequences from cytoplasmic ribosomes and, in our opinion, represents the evolution of the eukaryote cell that was the host for the symbioses that led to the establishment of the mitochondria and chloroplasts. The other branch of note is the one leading to the sequences from *Synechococcus* ATCC 27144 (a *Synechococcus* from subgroup 2 [7]) and from the chloroplast ribosome from duckweed; this branch represents the evolution of the blue-greens including the ancestor to the chloroplast of duckweed and the other higher plants. Clearly, the higher plants are composed of at least two kinds of bacteria: those that were ancestral to the eukaryote host and those that led to the chloroplasts. We expect that the plant mitochondrial 5S rRNA sequences will lie on yet a third branch, diverging from the *Pseudomonas* line. The evolution that we will discuss in the remainder of this article concerns the ancestry of eukaryote chloroplasts and can be pictured as having occurred in the neighborhood of the *Synechococcus*--duckweed chloroplast divergence on this tree.

Ferredoxins are a class of iron-sulfur proteins that act as electron carriers in a number of biochemical processes, including nitrogen fixation, nitrite reduction, and photosynthetic electron transport (8). The 2Fe-2S ferredoxins have been isolated from blue-greens, the chloroplasts of eukaryotes, and the bacterium *Halobacterium halobium*. Ferredoxin sequences were done by a number of workers, including H. Matsubara and K.T. Yasunobu (9-17). Chloroplast ferredoxins are translated on cytoplasmic ribosomes (18) and are therefore probably coded in the nucleus. However, like cytochrome c, they reflect the evolutionary history of the organelle in which they function. In the endosymbiotic theory, this is explained by postulating the transfer of genes from the organelle to the cell nucleus.

We have aligned the 2Fe-2S ferredoxins for which complete sequences have been determined (19). In sequence I from *Aphanothece*, there are deletions at positions 11 and 15 with respect to sequence II. (We believe that these are deletions in gene I rather than insertions in gene II because the sequence from *H. halobium* can be aligned better with sequence II without introducing gaps into the amino-terminal portion of either sequence.) Amino-terminal sequences of two ferredoxins from *Nostoc* MAC have been determined (20) and are consistent with this gene duplication having preceded the divergence of *Nostoc* and *Aphanothece*. Reports of two ferredoxins in both *Spirulina maxima* (21) and *Nostoc muscorum* (22) further suggest the antiquity of this duplication.

In the ferredoxin tree (Fig. 2), the two types of 2Fe-2S ferredoxin segregate as one would expect if a gene duplication preceded the blue-green divergences. Because the ancestral sequence lacked the two deletions, we can place the point of earliest time at the duplication as shown in Fig. 2 or on the Porphyra--Cyanidium branch. This is extremely important because it provides a temporal ordering to all the divergences. The topology of the tree in the immediate vicinity of the ancient gene duplication is not clearly resolved because the internodal distance between the two types of sequences is quite short in comparison with the branches surrounding it. Because the chloroplasts of Scenedesmus and the higher plants show no evidence of having the ancestral sequence without the two deletions, we presume that their ancestor lost the gene for this sequence.

If one scales branches on the right side of the tree so that the two Aphanothece branches are of equal length, the two halves of the tree can be superimposed to provide a description of the evolution of the organisms containing the sequences. Ignoring the very short internodal distance between the two types of sequences, these phylogenetic relationships are shown in Fig. 3. The earliest divergence on this tree was the one between prokaryotes that were ancestral to Synechococcus, to Aphanothece, to the chloroplasts of Porphyra and Cyanidium, and to those of the higher plants and green algae such as Scenedesmus. This last prokaryote may be related to Prochloron, a photosynthetic prokaryote containing chlorophylls a and b (23). Although we cannot be sure the chloroplasts of the green algae and the higher plants are monophyletic, there was at least one symbiosis on this line. At least one additional symbiosis led to the chloroplasts of Porphyra, a red alga, and Cyanidium, a eukaryote alga of uncertain affinity. These chloroplasts, like contemporary blue-greens, contain chlorophyll a and bilin pigments. Placing the root on the Porphyra--Cyanidium branch rather than at the gene duplication would simply mean that the ancestor of these chloroplasts diverged first, before the gene duplication--a possibility because only a single sequence without deletions has been reported from these species.

The radiation of blue-greens leading to Aphanothece and Mastigocladus suggests that an evolutionary schema that pictures the origin of filamentous blue-greens as a simple bifurcation of the coccoid and filamentous types is overly simplified. Aphanothece is coccoid, whereas Spirulina and Nostoc are filamentous.

If chloroplasts did not arise through symbiosis, then there are two alternative explanations of the nature of the ancestral forms on this tree. One is that the early organisms were prokaryotes and the eukaryote cell arose at least twice, once on

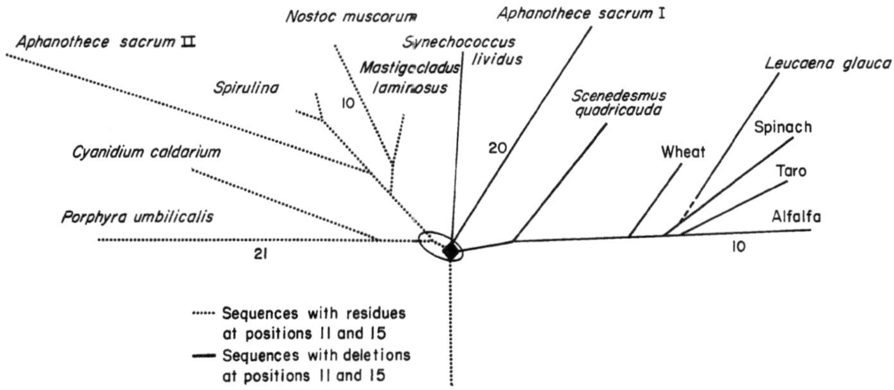

Figure 2. Evolutionary tree derived from 2Fe-2S ferredoxin sequences. Diamonds indicate gene duplications. The precise order of branching in the neighborhood of the gene duplication in Aphanothece sacrum is not clearly resolved computationally; the configuration among those with the shortest overall length that also segregates proteins on the basis of positions 11 and 15 in their alignment is shown.

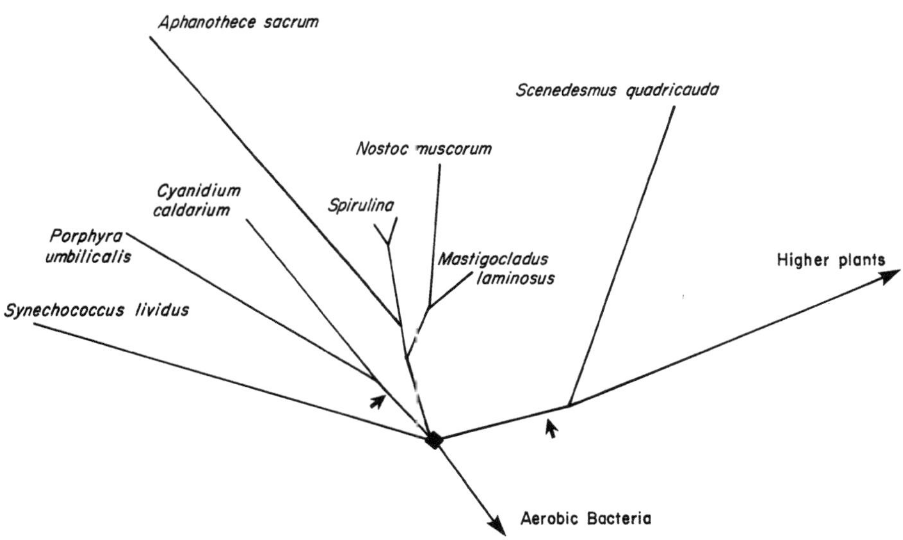

Figure 3. Phylogenetic tree based on 2Fe-2S ferredoxin sequences. Arrows indicate branches that independently gave rise to eukaryote chloroplasts.

the higher plant branch and once on the Porphyra--Cyanidium branch, a most unlikely coincidence. Alternatively, and equally unlikely, the early organisms were eukaryotes from which are descended a wide variety of contemporary blue-greens, including Aphanothece, Spirulina, Nostoc, and Mastigocladus.

Cytochrome c_6, also referred to as soluble cytochrome f or cytochrome $c_{552-554}$, is an electron transport protein that has been shown to substitute for plastocyanin in the electron transport chain between photosystems I and II depending upon the availability of copper to the organism (24). Sequences have been determined by a number of workers, including A. Aitken, R. Ambler, and M.V. Laycock (16, 25, 26). An evolutionary tree based on these sequences is shown in Fig. 4. The point of earliest time on the tree cannot be inferred from the sequences but was placed to conform with the gene duplication in the 2Fe-2S ferredoxins shown in Fig. 3. This tree has branches leading to Spirulina maxima, Synechococcus lividus, and the chloroplast of Porphyra in common with the ferredoxin tree. Like the Synechococcus on the 5S rRNA tree, Synechococcus ATCC 27167 is classified on the basis of DNA base ratio and genome size as belonging to subgroup 2 (7).

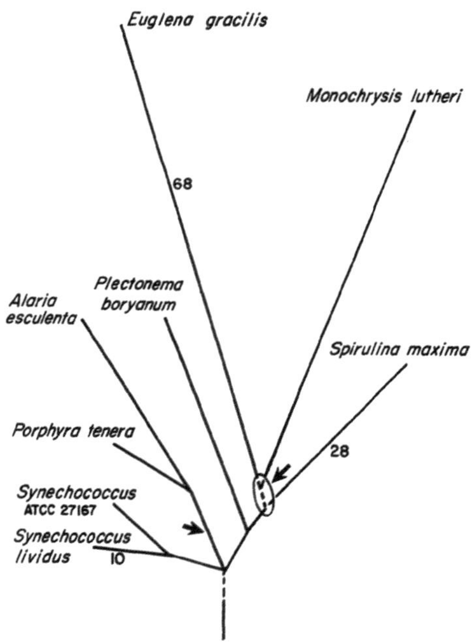

Figure 4. Phylogenetic tree derived from cytochrome c_6 sequences. Arrows indicate branches that independently gave rise to eukaryote chloroplasts. The order of divergence of the branches leading to Euglena, Monochrysis, and Spirulina could not be resolved.

In its overall configuration, this tree resembles the ferredoxin tree quite closely. The divergence of the branches leading to Spirulina, Plectonema, and the chloroplasts of Euglena and Monochrysis probably corresponds to the radiation of blue-green types on the ferredoxin tree that included Spirulina, Aphanothece, and Nostoc. Regardless of the exact position of the root of this tree, the intermixing of eukaryote and prokaryote branches here requires a minimum of two symbioses: one leading to the chloroplasts of Alaria and Porphyra and the other leading to those of Euglena and Monochrysis. The divergence of Monochrysis and Euglena being so close to what we believe was a radiation of blue-green types suggests that these chloroplasts may have arisen from different blue-greens; however, both were derived from a blue-green ancestor that was also common to Plectonema and Synechococcus.

Protein and nucleic acid sequence data offer distinct advantages in classifying prokaryotes and understanding their evolutionary relationships. What is true of free-living prokaryotes is even truer of their close relatives, the chloroplasts and mitochondria. In this area, where morphological characteristics, for example, have proved very inadequate, these gene products have provided unique insights. The elucidation of additional sequence data will undoubtedly provide exciting new details.

ACKNOWLEDGMENTS: We thank Drs. D.O. Hall and E. Margoliash for supplying us with sequences in advance of their publication, and K. Lawson for drafting the figures. This work was supported by NASA contract NASW-3317 and NIH grants GM-08710 and RR-05681.

REFERENCES

1. Margulis, L., Origin of Eukaryotic Cells, Yale Univ. Press, New Haven, 1970.
2. Schwartz, R.M., and Dayhoff, M.O., Science 199, 395 (1978).
3. Dayhoff, M.O., in Atlas of Protein Sequence and Structure, Dayhoff, M.O., Ed., Vol.5, Suppl.3, National Biomedical Research Foundation, Washington, D.C., 1979, pp.7-8.
4. Dyer, T.A., and Bowman, C.M., Biochem. J. 183, 595 (1979).
5. Nazar, R.N., Matheson, A.T., and Bellemare, G., J. Biol. Chem. 253, 5464 (1978).
6. Schwartz, R.M., and Dayhoff, M.O., in Atlas of Protein Sequence and Structure, Dayhoff, M.O., Ed., Vol.5, Suppl.3, National Biomedical Research Foundation, Washington, D.C., 1979, pp.327-336. 5S rRNA sequences and our alignment of them are collected here.
7. Rippka, R., Deruelles, J., Waterbury, J.B., Herdman, M., and Stanier, R.Y., J. Gen. Microbiol. 111, 1 (1979).
8. Rao, K.K., and Hall, D.O., in The Evolution of Metalloenzymes, Metalloproteins and Related Materials, Leigh, G.J., Ed., Symposium Press, London, 1977, pp.39-65.

9. Hase, T., Wada, K., and Matsubara, H., J. Biochem. 82, 267 (1977).
10. Hase, T., Wakabayashi, S., Wada, K., and Matsubara, H., J. Biochem. 83, 761 (1978).
11. Hase, T., Wakabayashi, S., Wada, K., Matsubara, H., Juttner, F., Rao, K.K., Fry, I., and Hall, D.O., FEBS Lett. 96, 41 (1978).
12. Hase, T., Wakabayashi, S., Matsubara, H., Rao, K.K., Hall, D.O., Widmer, H., Gysi, J., and Zuber, H., personal communication.
13. Takruri, I., Haslett, B.G., Boulter, D., Andrew, P.W., and Rogers, L.J., Biochem. J. 173, 459 (1978).
14. Wakabayashi, S., Hase, T., Wada, K., Matsubara, H., Suzuki, K., and Takaichi, S., J. Biochem. 83, 1305 (1978).
15. Takruri, I., and Boulter, D., Biochem. J. 179, 373 (1979).
16. Borden, D., and Margoliash, E., personal communication.
17. Schwartz, R.M., and Dayhoff, M.O., in Atlas of Protein Sequence and Structure, Dayhoff, M.O., Ed., Vol.5, Suppl.3, National Biomedical Research Foundation, Washington, D.C, 1979, pp.45-55. Ferredoxin sequences and our alignment of them are collected here.
18. Armstrong, J.J., Surzycki, S.J., Moll, B., and Levine, R.P., Biochemistry 10, 692 (1971).
19. Schwartz, R.M., and Dayhoff, M.O., Ann. N.Y. Acad. Sci., in press (1980).
20. Hutson, K.G., Rogers, L.J., Haslett, B.G, Boulter, D., and Cammack, R., Biochem. J. 172, 465 (1978).
21. Cammack, R., Rao, K.K., Bargeron, C.P., Hutson, K.G., Andrew, P.W., and Rogers, L.J., Biochem. J. 168, 205 (1977).
22. Hase, T., Wada, K., Ohmiya, M., and Matsubara, H., J. Biochem. 80, 993 (1976).
23. Lewin, R.A., Nature (London) 261, 697 (1976).
24. Wood, P.M., Eur. J. Biochem. 87, 9 (1978).
25. Aitken, A., Eur. J. Biochem. 101, 297 (1979).
26. Schwartz, R.M., and Dayhoff, M.O., in Atlas of Protein Sequence and Structure, Dayhoff, M.O., Ed. Vol.5, Suppl.3, National Biomedical Research Foundation, Washington, D.C., 1979, pp.28-43. Cytochrome c_6 sequences and our alignment of them are collected here.

EVOLUTION OF THE RHODOSPIRILLACEAE AND MITOCHONDRIA: A VIEW BASED ON SEQUENCE DATA

M.O. Dayhoff and R.M. Schwartz
Georgetown University and National Biomedical Research Foundation, Washington, D.C. 20007, USA

New sequence data from several protein families and from 5S ribosomal RNA confirm and elaborate our description of the phylogenetic connections between a variety of bacteria and the eukaryotes (1,2). Probably the first organisms were nonphotosynthetic anaerobic prokaryotes, which were followed soon by photosynthetic anaerobes. From this photosynthetic stock the aerobic line to Pseudomonadacae, Rhodospirillaceae, and blue-greens arose. The eukaryotes derived genetic material from the symbioses of at least three separate bacterial lines. Ancestors of Rhodopseudomonas globiformis gave rise to the eukaryote mitochondria, probably through at least three separate symbioses, one early on the flagellate line, one on the ciliate line, and one on the stem to the multicellular forms.

Many proteins and nucleic acids are "living fossils" in the sense that their structures have been dynamically conserved by evolution over billions of years: recognizably related forms are found in eukaryotes and prokaryotes. These are believed to be of common evolutionary origin, having evolved by a great number of small changes in the sequences (3,4). Recently, enough sequence information has become available from prokaryotes to permit us to build a biologically comprehensive phylogenetic tree (1,5,6). This composite tree, based on sequences of the c-type cytochromes, ferredoxins, and 5S ribosomal RNA molecules, is consistent with newer sequence data (see the paper by Barnabas et al. in this volume) (7). This tree was derived by objective mathematical methods from the chemical structures of the sequences, a line of reasoning independent of information from biological traits, metabolic capacities, or fossil evidence.

Y. Wolman (ed.), Origin of Life, 559–566.

On a phylogenetic tree, each point corresponds to a time, a species, and the macromolecular sequences within that species. There is one point representing the first divergence that corresponds to the earliest time and to the ancestral organism and sequences. Time advances on all branches of the tree emanating from this point. The topology of the branches gives the relative order of events.

We have found the least-squares matrix method most reliable for determining the best topology for distantly related sequences: it was used for constructing the trees in this paper (8,9). In this method, sequences are aligned to reflect the genetic changes in evolution and a matrix of percentage difference between them is calculated. These percentages are corrected for inferred superimposed mutations to yield the "observed" matrix of accepted point mutations per 100 residues (PAMs). The computer program determines branch lengths on a tree of given topology that provide a least-squares fit of the reconstructed matrix to the observed matrix. All reasonable topologies are examined; that one for which the sum of the absolute values of the branch lengths is smallest gives the best estimate of the true history. The least-squares matrix method makes no assumptions about the equality of evolutionary rates on the various branches and says nothing about the location of the point of earliest time.

In constructing the composite phylogenetic tree, our working hypothesis was that, for those sequences that perform basic metabolic functions, genetic transfer between major bacterial types followed by permanent acceptance in evolution is very infrequent. All such basic sequences within an organism share the same phylogenetic history, and any one of them can be used to infer its course.

The genetic doubling, most obvious in the clostridial-type ferredoxins but observed in all ferredoxin sequences, allows us to locate the base on the tree and therefore to infer the order of events. The hypothetical ancestral sequence that was formed just after the duplication was included with the present-day sequences in the calculations. The position of this ancestor identified the base of the composite tree. Because the species whose ferredoxin sequences have changed least since the doubling event were all anaerobic, heterotrophic bacteria, it is likely that the ability to live fermentatively is primitive. An early major innovation was the development of the ability to synthesize chlorophyll and perform bacterial photosynthesis.

The phylogenetic tree derived from selected c-type cytochrome sequences is shown in Fig. 1. In conformance with the ferredoxin doubling, the base occurs somewhere on the lines

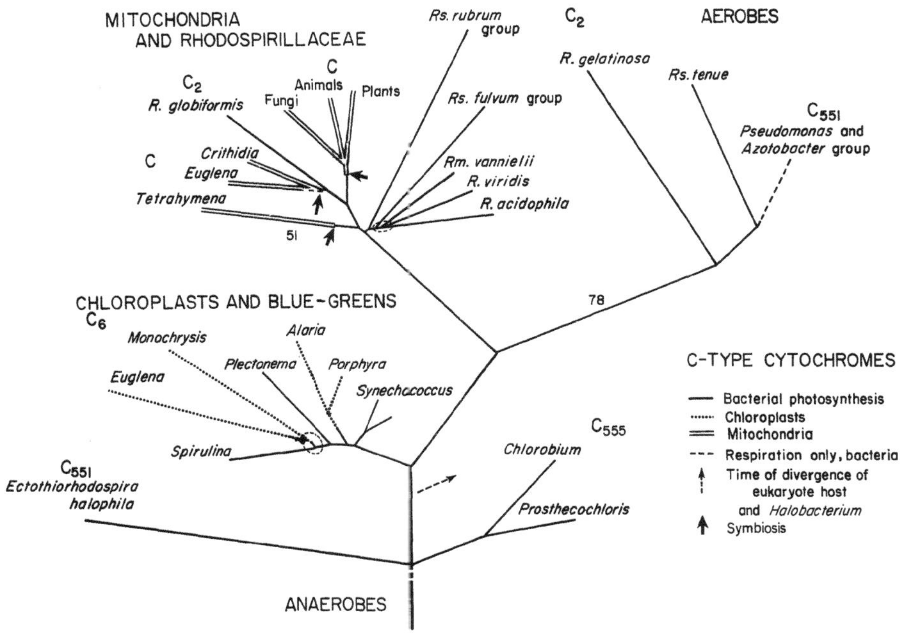

Figure 1. Phylogenetic tree of major groups from which c-type cytochrome sequences are known. The connections on the tree drawn in a single solid line show the divergences among the free-living prokaryotes. The parallel and dotted lines indicate the mitochondrion or the chloroplast, respectively. Solid arrows indicate branches on which mitochondrial symbioses have occurred.

leading back to Chlorobium and the other anaerobic photosynthesizers from the divergence of the blue-greens and pseudomonads, probably close to the position shown. The organism at the base of the c-type cytochrome tree was photosynthetic. The newly determined Ectothiorhodospira sequence (10) diverges near the bottom of the tree, as one would expect. No c-type cytochrome sequences are known that function in the cytoplasm of the eukaryote cell, so this molecule will not help us show the evolutionary history of the eukaryote host line. From the 5S ribosomal RNA tree shown in our other paper in this volume (11), it is clear that the eukaryote line, which also includes a branch to Halobacterium (12), came off just before the divergence of the blue-greens and Pseudomonas, as indicated by the dashed arrow. In Fig. 1, the blue-greens and chloroplasts clearly form a separate group. Their evolutionary history is described in detail in our other paper (11). The Rhodospirillaceae and the mitochondria also form a separate group.

There are two schools of thought concerning the origin of eukaryote organelles: one is that they arose by the compartmentalization of the DNA within the cytoplasm of an evolving protoeukaryote (13,14); the other is that they arose from free-living forms that invaded a host and established symbiotic relationships with it (15,16). In the nonsymbiotic theory, all genes arose within the organism and share the same phylogenetic history; homologues found in both the nucleus and the organelles arose by gene duplication. It is generally accepted that the eukaryotes arose only once, particularly in view of their highly complex mitotic apparatus. Organelles formed by rearrangement of parts of the nuclear material would probably not be capable of independent existence outside of the host cell, and their continued competitive extracellular existence in evolution would be most unlikely. Therefore, this theory would place the mitochondria, chloroplasts, and host from a species as one line on the phylogenetic tree; there would be no prokaryote descendents from the eukaryote lines. The symbiotic theory, on the other hand, proposes that the chloroplast was originally a free-living blue-green and the mitochondrion was originally a free-living aerobic bacterium. Their current status as organelles would have gradually evolved from these symbioses. In this theory, mitochondrial, chloroplast, and host genes are expected to show evidence of recent common ancestry with separate types of contemporary free-living prokaryote forms; the host and organelles would occur on different branches that could also contain free-living forms. The composite tree (7), as well as the c-type cytochrome tree in Fig. 1, clearly supports the symbiotic theory.

Of the several thousand genes that may have been present in the free-living form that became the mitochondrion, only a few remain in the genomes of contemporary organelles. These are transcribed and translated by the mitochondrial apparatus. An excellent review of these proteins is given by Tzagoloff et al. (17). Many other genes, including the one for cytochrome c, are thought to have originated in the free-living mitochondrial ancestor but been transferred to the nucleus. Their messengers are translated by the cytoplasmic apparatus and the mature proteins are transported to the inside of the mitochondrion or to its inner membrane where they function. Enough sequence information to be informative about mitochondrial evolution is presently known from the c-type cytochromes (10,18,19) and from the proteolipid subunit of ATP synthetase (20). Although phenylalanine and initiator methionine transfer RNA sequences are known from mitochondria, they have undergone so much change that they cannot be placed on the evolutionary trees for these molecules (21).

By far the most insight into the origin of mitochondria is obtained from the c-type cytochromes (Fig. 1). We have reviewed the literature on c-type cytochrome sequences (18) and presented detailed trees showing the evolutionary connections of all 12 species of purple, nonsulfur, photosynthetic bacteria, the Rhodospirillaceae, that are described in Bergey's Manual of Determinative Bacteriology (22) and of the major groups of eukaryotes from which mitochondrial sequences are known. In constructing the tree of Fig. 1, the matrix elements for closely related sequences have been averaged, and the group appears as a single line. Apparently, morphology is not an accurate guide to the evolution of these bacteria. Sequences from all three genera of Rhodospirillaceae are found intermixed on the tree. Further, the pseudomonads are shown diverging from one branch among these species and *Paracoccus denitrificans* from another.

Recently the cytochrome c_2 sequence from another species, *Rhodopseudomonas globiformis*, was determined (19). The computations indicate that this sequence clearly diverges from the *Euglena* and *Crithidia* branches close to the point of their divergence from one another. The best solution would have *R. globiformis* coming from the *Crithidia* branch. However, the cytochrome c sequences from *Euglena* and *Crithidia* share a rare substitution of an alanine for one of the heme-binding cysteines, an event that has not been observed in any other cytochrome c sequence examined. We presume that this change occurred once in an ancestor to *Crithidia* and *Euglena* just subsequent to the separation from *R. globiformis*. At the *R. globiformis* divergence, the ancestral cytochrome occurred in a free-living photosynthetic bacterium. Very soon thereafter, an endosymbiosis occurred to form the mitochondrion. The occurrence of a separate symbiosis among flagellates provides an explanation for the lack of mitochondria in the Trichomonadida, an anaerobic flagellate group. This group may have diverged before the endosymbiosis. If, as seems likely, the tree in Fig. 1 is correct, then there were at least two other symbioses established; these account for the *Tetrahymena* mitochondrion and that of the multicellular forms. Both could have been established by direct ancestors of *R. globiformis*, but more likely they were produced by related forms some time after their divergences from the *R. globiformis* line, as suggested in the figure. It is possible that these related forms have also given rise to other extant bacteria that are yet to be described. In each mitochondrial line photosynthetic ability was lost, either before or soon after the symbiosis. The connections on the tree drawn in a single line show the divergences among the free-living prokaryotes. The parallel lines indicate that the biological form is now the mitochondrion within the specified organisms. The solid lines need not reflect the evolutionary connections of the cytoplasmic constituents of the species shown.

Possibly, the mitochondria of animals, plants, and fungi originated from a single symbiosis preceding the divergences of these kingdoms as is shown in the figure, because the mitochondria of these lines show a great similarity in the genetic material that has been retained and in that transferred to the nucleus. There have been a considerable number of further independent changes in the genetic information in the mitochondria since the divergences of the kingdoms.

The evolutionary tree constructed from the sequences of the proteolipid subunit of ATP synthetase (20) (see Fig. 2) is consistent with that expected from our general scheme of Fig. 1. The point of earliest time and the connection of the eukaryote host should be prior to the chloroplast-mitochodrion divergence. The proteins that function in mitochondria appear together on a separate branch, showing the same recent evolutionary history as cytochrome c, even though the sequence from baker's yeast is coded in the mitochondrial genome and translated in the mitochondrion, whereas the sequence in Neurospora crassa is coded in the nucleus and translated in the cytoplasm. The close similarity of the two fungal sequences supports the view that the N. crassa gene was originally found in the mitochondrion. In this case, the place where the protein functions, rather than the location of its gene, is indicative of evolutionary history. Clearly, the amount of change in the mitochondrial line is greater than that in the chloroplast or the bacterial lines.

In conclusion, the sequence evidence overwhelmingly favors the origin of the eukaryote mitochondria by endosymbiosis. The mitochondria shown were all derived from ancestors of R. globiformis and their divergence times from these ancestors were more recent than their divergence times from any of the other 12 species of Rhodospirillaceae. At least three symbiotic events were required for the establishment of mitochondria in the eukaryote species examined here.

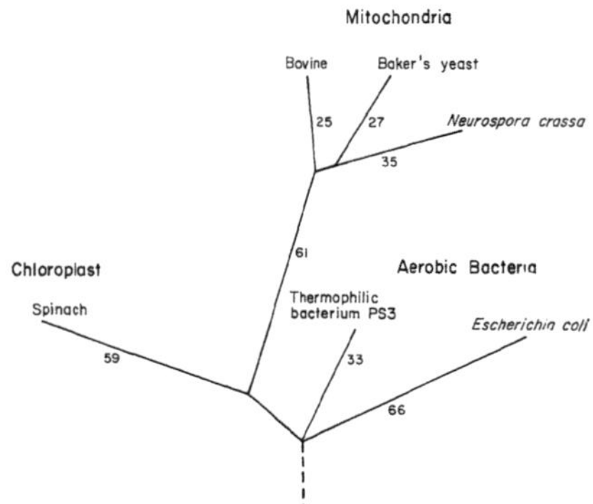

Figure 2. Evolutionary tree from the proteolipid subunit of ATP synthetase of membrane factor, F_0. The root of the tree has been placed in accord with that in Fig. 1. The mitochondrial, chloroplast, and bacterial sequences form three distinct groups, as expected.

ACKNOWLEDGMENTS: We would like to thank Dr. Richard Ambler for supplying us with the cytochrome c_{551} sequence from Ectothiorhodospira halophila in advance of its publication, and K. Lawson for drafting the figures. This work was supported by NASA contract NASW-3317 and NIH grant GM-08710.

REFERENCES

1. Schwartz, R.M., and Dayhoff, M.O., Science 199: 395 (1978).
2. Dayhoff, M.O., and Schwartz, R.M., in Proceedings in Endosymbiosis and Cell Research. In Press.
3. Bryson, V., and Vogel, H.J., Eds., Evolving Genes and Proteins, Academic Press, New York, 1965.
4. Dayhoff, M.O., Barker, W.C., and Hunt, L.T., in Atlas of Protein Sequence and Structure, Dayhoff, M.O., Ed., Vol. 5, Suppl.2, National Biomedical Research Foundation, Washington, D.C., 1976, pp.9-19.
5. Schwartz, R.M., and Dayhoff, M.O., in Comparative Planetology, Proceedings of the Third College Park Colloquium on Chemical Evolution, Sept. 1976, Ponnamperuma, C., Ed., Academic Press, New York, 1978, pp.225-242.
6. Dayhoff, M.O., and Schwartz, R.M., in Origin of Life, Proceedings of the Second ISSOL Meeting and the Fifth International Conference on the Origin of Life, April 1977, Center for Academic Publications Japan/Japan Scientific Societies Press, 1978, pp.547-560.
7. Barnabas, J., Schwartz, R.M., and Dayhoff, M.O. This volume.
8. Dayhoff, M.O., in Atlas of Protein Sequence and Structure, Dayhoff, M.O., Ed., Vol.5, Suppl.3, National Biomedical Research Foundation, Washington, D.C., 1979, pp.1-10.
9. Dayhoff, M.O., Fed. Proc. 35: 2132 (1976).
10. Ambler, R.P., sequence of cytochrome c_{551} from Ectothiorhodspira halophila, strain BN 9626. Submitted to the Atlas of Protein Sequence and Structure Reference Data Collection, August 1979.
11. Schwartz, R.M., and Dayhoff, M.O. This volume.
12. Nazar, R.N., Matheson, A.T., and Bellemare, G., J. Biol. Chem. 253: 5464, 1978.
13. Raff, R.A., and Mahler, H.R., Science 177: 575 (1972).
14. Uzzell, T., and Spolsky, C., Amer. Sci. 62: 334 (1974).
15. Margulis, L., Origin of Eukaryote Cells, Yale Univ. Press, New Haven, 1970.
16. Margulis, L., in Handbook of Genetics, King, R.C., Ed., Vol.1, Plenum Press, New York, 1974, pp.1-41.
17. Tzagoloff, A., Macino, A., and Sebald, W., Annu. Rev. Biochem. 48: 419 (1979).
18. Schwartz, R.M., and Dayhoff, M.O., in Atlas of Protein Sequence and Structure, Dayhoff, M.O., Ed., Vol.5, Suppl.3, National Biomedical Research Foundation, Washington, D.C., 1979, pp.28-43. The c-type cytochrome sequence literature is

reviewed here and our alignment and phylogenetic trees are shown.

19. Ambler, R.P., Abstr. Third Int. Symp. Photosynthetic Prokaryotes (Oxford), Nichols, J.M., Ed., Univ. Liverpool, 1979, Abstr. E17.
20. Sebald, W., Hoppe. J., and Wachter, E., in Function and Molecular Aspects of Biomembrane Transport, Palmieri, R. et al., Eds., Elsevier North-Holland, Amsterdam, pp.63-74, 1979. The sequences are reported in this paper.
21. Schwartz, R.M., and Dayhoff, M.O., in Atlas of Protein Sequence and Structure, Dayhoff, M.O., Ed., Vol.5, Suppl.3, National Biomedical Research Foundation, Washington, D.C., 1979, pp.313-325. The tRNA sequence literature is reviewed here and our alignments and trees are shown.
22. Buchanan, R.E., and Gibbons, N.E., Eds., Bergey's Manual of Determinative Bacteriology, 8th ed, Williams and Wilkins, Baltimore, 1974.

ON THE ORIGIN OF PHOTOSYNTHETIC EUKARYOTIC CELLS: CYANIDIUM CALDARIUM AS A "BRIDGE" ALGA BETWEEN PROKARYOTIC CYANOBACTERIA AND EUKARYOTIC RHODOPHYTES: EVIDENCE FROM ENVIRONMENTAL, POLYSACCHARIDE BIOCHEMISTRY AND ULTRASTRUCTURAL STUDIES.

Joseph Seckbach and Jerome F. Fredrick*

School for Overseas Students, Hebrew University of Jerusalem, Israel.
* The Research Lab. Dodge Chemical Comp.
Bronx, N.Y. 10469, U.S.A.

The similarities between Cyanidium caldarium and the rhodophytes are based on cellular ultrastructure and the site of storage glucan. In both the storage carbohydrate lies outside the chloroplast; their glucan is branched more than in the chlorophytes. The survival of Cyanidium in extreme-primeval atmosphere seems to be identical with the Cyanobacteria, as also the glucosyltransferases and the structure of storage glucan both algal types form. These isozymes of Cyanidium have structural and kinetic similarities with Cyanophyceae and rhodophytes. In line with all other observations, C.caldarium appears to be a transitional 'bridge' linking prokaryotic Cyanobacteria with the eukaryotic rhodophytes, this may obliterate the 'discontinuity' in the fossil record.

One of the most significant events in biological evolution was undoubtedly the transition from primitive prokaryotic cells to advanced eukaryotic cells. Unfortunately, data whereby this important step can be traced is not available and this constitutes one of the greatest 'discontinuities' in evolution between the anucleated and the nucleated cells. In algae, there are examples of both types of cells and, in addition, there are algae that are very suggestive of 'intermediate' stages. Therefore, it is among these 'primitive' plants that information should logically be found to overcome this 'missing link' in the fossil record between the Prokaryota and the Eukaryota. We believe that such an algal candidate for presenting the proposed bridge model should possess the following characteristics:
a) The alga should indicate phylogenetic antiquity such as being able to survive (or grow) in primitive atmosphere such as

Y. Wolman (ed.), Origin of Life, 567–574.

existed, for example, in the Precambrian or when the Eukaryota evolved. b) Because the first respirable substances of all cells were carbohydrates (1), and because cells possessing the capability for storing carbohydrates undoubtedly would have a selective advantage when the supply of sugars in the primordial 'soup' was depleted, it is logical to search for evolutionary trends among the storage polysaccarides of algae. At the same time, the storage glucans of thallophytes are the end result of enzymatic activities, and hence the mode of biosynthesis of these storage sugars, and the enzymes involved in their formation should also reflect such changes. c) At the ultramorphological level, this cell should possess eukaryotic characteristics preferably resembling the primitive Rhodophyceae (red algae) and at the same time retain cyanobacterial constituents.

We propose the unicellular eukaryotic alga *Cyanidium caldarium* as the most suitable organism for this transition linkage between the Cyanobacteria and nucleated primitive Rhodophyceae. In fact it may represent the only extant bridge in the development of the eukaryotic cell. For the establishment of this evolutionary link it is necessary to show that the features of *C. caldarium* and its biochemistry are shared by both these algal groups. This organism has not been placed in a fixed taxonomic position and its phylogenetic relationship to other algae is still debatable (1,2,3); Seckbach proposed recently (3) that it is a blue-green rhodophyte.

MILESTONES IN THE INVESTIGATION OF CYANIDIUM CALDARIUM

Recently there has been a great deal of attention devoted to *Cyanidium*, e.g., see bibliography collected by Fukuda (4), review by Brock (2) and ecological aspects elsewhere (5,6). However, despite the increase in studies, only a few research groups have dealt with the phylogenetic affinities of this anomalous alga with the thallophytes (2,7-14).

Tolerance to Extreme Environment

Cyanidium is a fascinating alga which inhabits very acidic and high temperature environments, i.e., thermal acidophilic springs and soils throughout the world (5). In fact it is the only photosynthetic organism existing above 40° in the acid thermal habitats (2). Allen (15) reported its culturing in the laboratory in 1N H_2SO_4 which is also closer to the acidic soil pH(5) and it grows on organic media in dark at the rate of growth of light (2). The maximum growth is at 55°C in the laboratory (15) which is the same as in nature (2,5,6). In this thermophilic demand *Cyanidium* resembles some species of the Cyanobacteria which thrive in higher temperature ranges (up to 74°C) in neutral or alkaline waters. Searcy et al. (16) pointed out that in Precambrian times the earth was much warmer than now and that the surface temperature on earth was about 50°-70°C during the period in which eukaryotic cells originated and the primordial atmosphere was more acidic. Thus conditions similar to those required by Cyanidium were believed to be common in Precambrian times. Furthermore this exotic alga also tolerates an extreme environment of elevated pressure (17,18), growth in lack of O_2 (19,20), and thrives under a pure atmosphere of CO_2 (19-21) which are conditions representing the stage of life evolution on earth. Stoecker (22) observed that some strains of cyanophytes (as Synechococcus, Synechosystis) survived under an atmosphere containing 50% or 60% of CO_2 similar to that which was assumed to be produced in the primeval atmosphere by anaerobic fermentation prior to CO_2 consumption by early photosynthetic organisms. The tolerance to the above environmental factors strongly suggests the phylogenetic antiquity of C. caldarium which is shared with the Cyanobacteria, both groups surviving atmospheres characteristic of primitive earth (17,21,22) which may indicate their close evolutionary connection.

The Glucans Enzymology and Cyanidium Transitional Position

Algae store sugars in the form of alpha-1,4 glucans (23). These glucans may exist as linear chains of glucose molecules with alpha-1,4-glucosidic bonds between residues such as the amylose component of 'starch'. This type of glucan is the common constituent of Chlorophyceae and Cryptophyceae. The other constituent of the 'starch' of both these groups is a moderately branched glucan called amylopectin. Amylopectin, in addition to the alpha-1,4-glucosyl bonds contains glucosyl residues branching from the main chains by alpha-1,6-glucosyl linkages.

The red algae form a storage glucan known as floridean starch. This is a single-component polysaccharide which resembles amylopectin but contains many more alpha-1,6 linkages, and is thus, much more highly branched than is amylopectin. The prokaryotic algae, i.e., the Cyanobacteria form a storage glucan which is very similar to the animal storage sugar, glycogen. The phytoglycogen of blue-green algae is very highly branched, more so than the floridean starch of red algae and the amylopectin of green algae.

Cyanidium caldarium forms a storage glucan which is identical with that found in cyanophytes. In fact, it is impossible to distinguish the storage glucan of Cyanidium from that of Oscillatoria and other Nostocales (10,24).

Two types of enzymes, glucosyltransferases, are able to synthesize alpha-1,4 glucosyl bonds between adjacent glucose molecules. These are the phosphorylases and the synthetases (E.C. 2.4.1.1. and E.C. 2.4.11). There is much controversy whether phosphorylases are physiologically active in forming alpha-1,4 bonds, or are degradative of these bonds (25,26). However, both do contribute to the alpha-1,4 bonding.
The other group of glucosyltransferases involved in storage glucan biosynthesis are the branching enzymes (E.C. 2.4.1.18). These are of two types (27,28), those that can branch linear

glucans such as amylose to form moderately-branched sugars like amylopectin (known as the 'Q' type of branching enzymes), and those that can introduce further alpha-1,6 linkages into already-branched sugars, and hence can convert amylopectin into phytoglycogens. This latter group of branching enzymes have been termed, the b.e. enzymes.

Incidentally, the active sites of both alpha-1,4 bond formers (the phosphorylases and the synthetases) show extensive similarities of the amino acid residues of the peptides forming these sites when sequenced (29,30). There is also evidence that the branching enzymes show immunological relatedness to phosphorylases (31). This has led to the hypothesis that probably in the primordial cell, the properties for forming both alpha-1,4 and alpha-1,6 bonds were incorporated into a single enzyme molecule (32).

TABLE 1

The Storage Polysaccharides of *Cyanidium* and Related Algae and their Glucosyltransferase Isozymes as Indicated by Two-dimensional Polyacrylamide Gel Electrophoresis

Group of Algae*	C.c.	Red Rhodym.	Blue-green Oscil.	Blue-green Anac.	Green Chl.	Green Spig.	Green Prt
Type of 'Starth'	Phyto-glycogen	Flori-dean	Phyto-glycogen		Amylose + Amylopectin		
Polysaccharides Stored in Cytoplasm	yes	yes	yes		no		
Enzymes Phosphorylases	1	1	2	2	2	2	2
Synthetases	2	2	2	2	2	2	2
Branching Isozyme Type	2 (b.e.)	2 (b.e.) 1(Q)	2 (b.e.)	3 (b.e.)	2 (Q)	3 (Q)	3 (Q)

* The algae are: C.c. = *Cyanidium caldarium*; Rhodym.= *Rhodymenia* Oscil. = *Oscillatoria*; Anac. = *Anacystis*; Chl. = *Chlorella*; Spig. = *Spirogyra*; Prt. = *Prototheca*

It should be mentioned that all three groups of enzymes responsible for storage glucan biosynthesis, have been shown to exist in algae as isozymes. First reported in the blue-green alga, Oscillatoria princepes (33), isozymes of these glucosyltransferases have also been found in other algae (34,35).

Two-dimensional polyacrylamide gel electrophoresis (P.A.G.E.) indicated the similarities of these glucosyltransferase isozymes in Cyanidium, rhodophytes and in cyanophytes (36). Table 1 indicates these properties as well as other polysaccharide features in these algae.

It should be noted that even the colorless Chlorophyte Prototheca zopfii, contains the Q type of branching enzymes so typical of Chlorophyceae.

Since the ultimate structure of the storage glucan is determined by the branching enzymes, it can be seen that those algae possessing only the Q branching enzymes form amylopectin as the branched glucan. Those algae containing the b.e. type, invariably form more highly branched glucans than amylopectin, such as the floridean starch of red algae and the phytoglycogens of blue-green algae. Note that Cyanidium caldarium contains b.e. branching enzymes exclusively, and thereby forms a storage glucan similar to that of Oscillatoria and Anacystis.

The red algae, particularly Rhodymenia pertusa contain three branching isozymes. Two of these are of the b.e. type while the third is a Q type. This alga forms floridean starch which is much more highly branched than the amylopectin of green algae, but not as highly-branched as the phytoglycogen of either Cyanidium caldarium or the blue-green algae, Oscillatoria princeps and Anacystis nidulans.

It has been found through P.A.G.E. studies that the Q enzyme of Rhodymenia is actually a charge isomer of its two b.e. isozymes, differing in only the substitution of a single amino acid residue (37). Boyer and Preiss have recently reported that the interaction of the three branching isozymes is responsible for the ultimate degree of branching in the phytoglycogen of maize (38). Surely, the lesser degree of branching in floridean starch as compared with the phytoglycogens of Cyanidium and Oscillatoria, is probably due to the interaction of the two b.e. isozymes with the lone Q isozyme of this alga.

The mode of action of the b.e. branching isozymes appears to be exactly the same for Cyanidium caldarium and the blue-green alga, Anacystis nidulans (39).

This similarity in the glucosyltransferase isozymes of Cyanidium caldarium, Oscillatoria princeps and Rhodymenia pertusa has been further borne out by double immunodiffusion studies in agar (12). When the purified phosphorylase of Cyanidium was used to immunize rabbits, the sera showed that there was indeed structural similarity between the phosphorylases of Cyanidium, Oscillatoria and Rhodymenia, by complete fusion of the precipitin lines. However, the phosphorylases from Chlorella and Spirogyra showed only partial fusion with extensive spur formation, indicative of structural differences between this glucosyltransferase of Cyanidium and green algae.

Recently, using the technique of tandem crossed immunoelectrophoresis (40), it has been shown that the Q isozyme of the red alga, Rhodymenia, is more closely related to the b.e. isozymes of Cyanidium than to the corresponding Q isozymes of Chlorophycean algae.

Cellular Ultrastructure of Cyanidium

Seckbach et al. recently compared the features of Cyanidium to other related algae (3). Because of its characteristics and a mixture of phylogenetic markers, this enigmatic alga has been called various names and has been placed in different taxonomic categories (1-3,5,7,10,12). Briefly, Cyanidium caldarium, the unicellular alga, possesses a true nucleus with a nucleolus, simply-developed organelles (mitochondria, chloroplasts), the simply constructed Golgi apparatus contains a few dictyosomes, and microbodies were also observed (3,7,20). Under certain conditions the chloroplasts are very big and may occupy most of the cytoplasmic volume (2,3,7,20), while in other instances these organelles are reduced in size (i.e., when grown in organic media) or even undeveloped (proplastids) as observed in heterotrophic and dark grown cells (3). The chloroplasts of C. caldarium are considered simple or primitive and are similar to the rhodoplasts; the thylakoid patterns are either single bands lying parallel across the plastid section or are nearly parallel, concentric, non-stacked rings of thylakoids (like onion rings). The non-concentric thylakoids are encircled by one or more girdle lamellae adjacent to the plastid envelope. No grana is formed with these chloroplasts. Phycobilisomes were detected in Cyanidium (3,7,8,41) on the external side of the photosynthetic bands, similar to the rhodoplast and the Cyanobacteria. In spite of the affinity of C. caldarium to the rhodophytes (which themselves have close phylogenetic relationships to the Cyanobacteria), there are some rhodophytic characters which have not been reported thus far for this fascinating cell; for example: sexual reproduction, mitotic apparatus, pyrenoid, phycoerythrin, α carotene. In most of these features (or the lack of them) Cyanidium is phylogenetically closer to the Cyanobacteria as it was considered in the past a member of the blue-green algae (5). The electron micrographs demonstrate the rhodophytic nature of Cyanidium cells, while these cells also have cyanobacterial traces like in the photosynthetic apparatus. Because of this, as well as other resemblances to the red algae, Cyanidium was placed among the Rhodophyta (2,3,5,7-9,12,14). In addition, Chapman (43) indicated that Cyanidium is the most primitive member of the Rhodophyceae.

In conclusion, because of the above-mentioned observations of this paradoxical anomalous rhodophytan, we are tempted to propose Cyanidium as the transitional link between the blue-green prokaryotes and the primitive rhodophytes. This idea is also supported by others, using other phylogenetic markers (1,3,9,13, 14,42). Cyanidium may be the only extant relic of the true bridge between the anucleated (Prokaryota) and the nucleated (Eukaryota) cell, and as such, may help to uncover the missing link in the fossil record.

REFERENCES

1) Klein, R.M. and Cronquist, A., The Quart. Rev. Biol. 42:105 (1967).
2) Brock, T.D., Thermophilic Microorganisms and Life at High Temperatures, Springer-Verlag, N.Y. 1978, pp 255-302.
3) Seckbach, J., Hammerman, I.S. and Hanania, J., Ann. N.Y. Acad. Sci., in press (1980).
4) Fukuda, I., A Possible Literature Survey of a Thermal Alga Cyanidium caldarium Geitler (I), Sci. Univer. Tokyo, 162 Japan, 1979.
5) Doemel, W.N., The Physiological Ecology of Cyanidium caldarium, Ph.D. dissertation, Indiana Univ. Bloomington, 1970.
6) Doemel, W.N. and Brock T.D., J. of General Microbiol. 67:17 (1971).
7) Seckbach, J., Microbios 5:133 (1972).
8) Seckbach, J. and Ikan R., Plant Physiol. 49:457 (1972).
9) Ikan, R. and Seckbach, J., Phytochem. 11:1077 (1972).
10) Fredrick, J.F., Phytochem. 7:1573 (1968).
11) Fredrick, J.F., Phytochem. 10:395 (1971).
12) Fredrick, J.F., Plant & Cell Physiol. 17:317 (1976).
13) Fredrick, J.F., Ann. N.Y. Acad. Sci., in press (1980).
14) Hase, T., Wakabayashi, S., Wada, K., Matsubara, H., Jüttner, F., Rao, K.K., Fry I. and Hall, D.O., FEBS Letters 96:41 (1978).
15) Allen, M.B., Arch. Mikrobiol. 32:270 (1959).
16) Searcy, D.G., Stein, D.B. and Green, G.R., BioSystems 10:19 (1978).
17) Seckbach, J. and Libby, W.F., Space Life Sci. 2:121 (1970).
18) Seckbach, J., Limnol. and Oceanog. 16:567 (1971).
19) Seckbach, J., Gross, H. and Nathan, M.B., Israel J. Bot. 20:84 (1971).

20) Seckbach, J., Israel J. Bot. 20:302 (1971).
21) Seckbach, J., Baker, F.A. and Shugarman, P.M., Nature 227:744 (1970).
22) Stoecker, R.R., Space Life Sci. 3:42 (1971).
23) Meeuse, B.J.D. and Kreger, D.R., Biochem. Biophys. Acta. 35:26 (1959).
24) Fredrick, J.F., Phytochem. 14:1911 (1975).
25) Leloir, L., DeFekete, M.A. and Cardini, C., J. Biol. Chem. 235:636 (1960).
26) Slabnik, E. and Frydman, R., Biochem. Biophys. Res. Com. 38:709 (1970).
27) Lavintman, N., Arch. Biochem. Biophys. 116:1 (1966).
28) Fredrick, J.F., Physiologia Plant. 24:55 (1971).
29) Larner, J. and Sanger, F., J. Mol. Biol. 11:491 (1965).
30) Nolan, C., Novoa, W., Krebs, E.G. and Fischer, E.H., Biochem. 3:542 (1964).
31) Fredrick, J.F., Ann. N.Y. Acad. Sci. 210:254 (1973).
32) Fredrick, J.F., Phyton 16:21 (1961).
33) Fredrick, J.F., Phytochem. 1:153 (1962).
34) Negashima, H., Nakamura, S. and Nisizawa, K., Bot. Mag. Tokyo. 81:411 (1968).
35) Mangat, B.S., Phytochem. 18:753 (1979).
36) Fredrick, J.F., Ann. N.Y. Acad. Sci. 151:413 (1968).
37) Fredrick, J.F., Phytochem. 7:931 (1968).
38) Boyer, C. and Preiss, J., Biochem. Biophys. Res. Com. 80:169 (1978).
39) Fredrick, J.F., J. Therm. Biol. 3:1 (1978).
40) Fredrick, J.F., Phytochem. 19: in press (1980).
41) Edwards, M.R. and Mainwaring, J.D., in "31st Ann. Proc. E.M. Soc. Amer." Arceneaux, C.J. (ed.) New Orleans, La. (1973).
42) Klein, R.M., Ann. N.Y. Acad. Sci. 175:623 (1970).
43) Chapman, D.J., Nova Hedwigia 25:673 (1974).

MICROBIAL LIFE IN COLD SALINE ENVIRONMENTS

S.B. Leschine*, K.J. Miller†, and R.L. Huguenin§

Departments of Microbiology*, Biochemistry†, Physics and Astronomy§, Univ. of Massachusetts, Amherst, USA

Microorganisms from Dry Valley (Antarctica) soils, from Great Salt Lake (Utah, USA) water, and from refrigerated salt codfish were enriched and isolated at 5°C in media containing high concentrations of salts. The microbial flora that developed varied considerably depending on the source of the inoculum. A variety of morphologically distinct microorganisms was isolated. Most isolated strains grew over a wide range of temperature (e.g., 0-25°C) and salt concentration (e.g., 0.1-2 M NaCl) suggesting that the evolutionary strategy for adaptation of these organisms to low temperature and high salt involved an increased tolerance of (rather than dependance on) these extreme conditions.

We have initiated a study of microbial growth at low temperature and high salinity. Our interest in the combined effect of these two environmental extremes was stimulated by recent reports of near-surface water at several locations on Mars, particularly the two large and temperate sites in Solis Lacus (-25, 85°) and Noachis-Hellespontus (-30, 315°) (1,2). A combination of Viking orbiter and earthbased remote sensing evidence indicated that H_2O vapor is anomalously outgassing from these regions, producing enhanced local concentrations of atmospheric water vapor, hazes, and frost throughout much of the martian year. Using Viking Infrared Thermal Mapper data and physical models of the soil column (based in part on Viking lander site measurements), it was deduced that the H_2O extends to within a few cm of the surface and undergoes a daily freeze/thaw cycle over at least half the martian year. Miller and Huguenin (3) have proposed that the water should range in composition from relatively pure liquid/ice to saturated brine/ice and be distributed heterogeneously in the

Y. Wolman (ed.), Origin of Life, 575–582.

"oases" with thicknesses ranging from relatively thin films to filled pore spaces and crevices. Independent corroboration of the occurrence of ground water within these two regions of Mars was recently revealed from earthbased radar data reported by Zisk and Mouginis-Mark (4). These data showed the oases regions to have anomalously high radar reflectivity (greater than 9%) and unusual smoothness (rms slopes less than 1%) that lead to the conclusion that ground water was indeed present and heterogeneously distributed within the top meter of soil. The radar data also showed a seasonal variation that was consistent with the seasonal variation and extent of thaw proposed by Miller and Huguenin (3).

The surface conditions that apparently exist within the oases are strikingly different from those encountered at the Viking lander sites. The Viking site environments were found to be highly desiccated, with detectable amounts of H_2O evolving from most samples only after heating to 350 and 500°C. There were other hostile elements as well, including the presence of oxidants, low temperatures and harsh UV radiation. With the occurrence of near-surface ground water, higher humidities, and the warmest temperatures on the planet (daytime high temperatures approach 20°C in summer), the oases environments would appear to be significantly less hostile. Indeed, even the oxidants found at the Viking sites would be unstable and should not form at the oases (3,5). Although conditions in the oases appear to be more suitable for life than the Viking lander environments, at least two factors, high salt and low temperature, would preclude the growth of most types of presently known microorganisms. However, the organisms inhabiting cold, hypersaline environments have not been well characterized possibly because these environments are relatively uncommon on Earth. It is not known whether these organisms could tolerate the various martian environmental extremes. Therefore, as a first step in assessing the plausibility of life in martian oases, we have isolated microorganisms from saline Antarctic soils, Great Salt Lake water, and refrigerated salt codfish at 5°C in media containing high concentrations of salts. In this paper we report initial studies aimed at characterizing these microorganisms.

Growth was enriched in liquid media. Medium TYE-HES contained HES salts (6) (0.85 M NaCl, 0.04 M $CaCl_2$, 0.2 M $MgSO_4$), 0.01% yeast extract (Difco) and 0.01% Trypticase (Baltimore Biological Laboratories); medium 5TYE-HES contained HES salts, 0.05% yeast extract and 0.05% Trypticase. In order to avoid precipitation of HES salts, concentrated solutions of each salt were prepared separately and then combined, adding the $MgSO_4$ solution last. The pH of the media was adjusted to 7.4 prior to sterilization by autoclaving. Tubes (16 x 150 mm) containing 10 ml of medium were inoculated with soil (≃0.1 g), water (≃0.5 ml), or fish (≃0.1 g) and incubated at 5°C in the light (approximately

TABLE 1
Sources of Inocula for Enrichment Cultures

Source	Description
A1	Soil collected[a] near algal mat west of hypersaline Don Juan Pond, Wright Valley, Antarctica[b]
A2	Soil collected[a] near fresh water entrance to Don Juan Pond[b]
A3	Soil from the south shore of Don Juan Pond[a,b]
A4	Soil and dried algal mat from Taylor Valley, Antarctica[a,c]
GS	Water from the southern arm of Great Salt Lake, Utah[d]
SC	Salt codfish[e]

[a]Antarctic soils collected by R.E. Benoit.
[b]Collected 1968-1969 austral summer; stored -40 to -80°C.
[c]Collected 1966-1967 austral summer; stored -40 to -80°C.
[d]Collected May 1979; stored 25°C.
[e]Purchased locally; stored 5°C.

35 cm from a 60-w bulb) without agitation. A description of the inocula is given in Table 1.

Growth was observed in enrichment cultures after 5 days to 2 weeks incubation except in medium TYE-HES inoculated with Taylor Valley soil which became turbid after 4 weeks. The microbial flora that developed in enrichment cultures varied considerably depending on the source of the inoculum. A great diversity of microorganisms was observed microscopically in cultures inoculated with Great Salt Lake water including bacteria (cocci, variously-sized motile and non-motile rods, helically-shaped cells), protozoa (ciliates, flagellates), and eukaryotic algae (diatoms, Dunaliella-like flagellated green algal cells). Similar microorganisms have been observed in water from the south arm of Great Salt Lake (7,8). In cultures inoculated with salt codfish, cocci predominated but rod-shaped cells and yeast were also observed. In cultures inoculated with Antarctic soil, we observed cocci and variously-sized motile and non-motile rods. The cocci were often in clumps of 4 or more cells. Microorganisms with similar morphologies have been isolated from Antarctic soils (9,10,11).

A variety of morphologically distinct microorganisms were isolated from enrichment cultures on 5TYE-HES agar medium (0.7% Difco Noble agar) at 5°C. Some of the isolated strains are described in Table 2 and the morphologies of some are illustrated in Fig. 1. Strains A3a (Fig. 1B) and A3c (Fig. 1D) formed spores that were resistant to heat and desiccation. Most strains were

TABLE 2
Description of Isolated Microorganisms[a]

Strain	Source[b]	Fig.[c]	Cell shape	Average cell size (μm)	Gram Reaction
A1a	A1	1A	pleomorphic	0.6 x 1.5	positive
A2b	A2	-	rod	0.8 x 2.0	positive
A3a	A3	1B	rod	0.6 x 4.0	variable
A3b	A3	1C	pleomorphic	0.6 x 3.0	positive
A3c	A3	1D	rod	0.8 x 2.5	positive
A4a	A4	1E	cocci	1 (diam.)	positive
GS1a	GS	-	rod	0.7 x 6.0	positive
GS1d	GS	-	rod	0.5 x 4.0	negative
SC1a	SC	-	yeast	5 (diam.)	-

[a]Microorganisms isolated from enrichment cultures when growth first appeared. Cells examined after 1-2 weeks growth at 5°C on 5TYE-HES agar medium.
[b]Sources of inocula for enrichment cultures described in Table 1.
[c]Numbers refer to figures in this article.

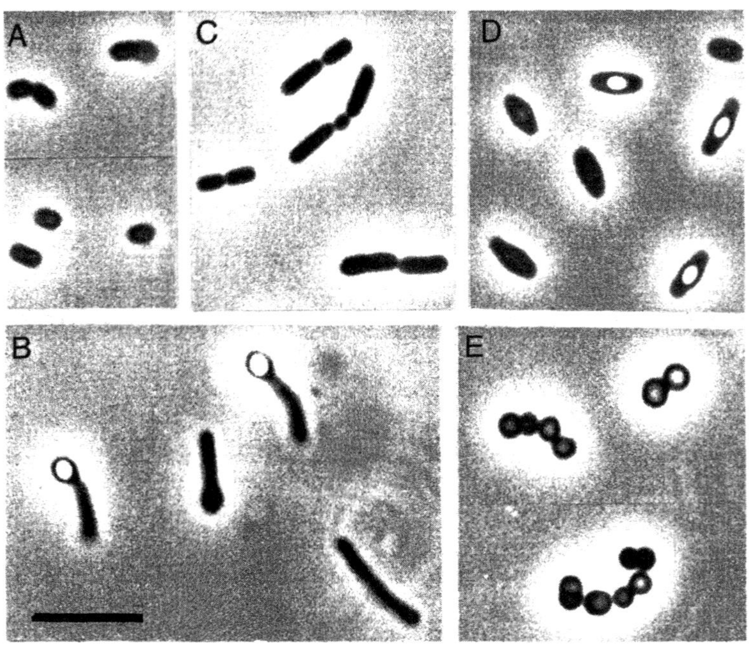

Fig. 1 (A-E). Phase contrast photomicrographs of some representative morphological types of isolated microorganisms. Sources and descriptions of isolates are given in Table 2. All micrographs are at the same magnification. Bar = 5 μm.

TABLE 3
NaCl Tolerance and Temperature Range for Growth of Microorganisms

Strain[a]	Temperature (^{o}C)[b]	NaCl concentration (M)[c]
A1a	0 - 25	0.1 - 2
A2b	0 - 25	0.1 - 2
A3a	0 - 25	0.1[d]- 2
A3b	0 - 25	0.1[d]- 2
A3c	5 - 25	0.1[d]- 2
A4a	0 - 35	0.1[d]- 2
GS1a	0 - 25	0.1[d]-0.85
GS1d	5 - 25	0.1 - 1
SC1a	0 - 25	0.1 - 1

[a]Strains described in Table 2.
[b]Range of incubation temperatures at which colony growth on 5TYE-HES agar medium was visible after 2 weeks incubation. Incubation temperatures used (^{o}C): 0, 5, 15, 20, 25, 35.
[c]Range of NaCl concentration supporting colony growth at 15°C after 2 weeks incubation. Cells grown on modified 5TYE-HES agar medium. NaCl concentrations used (M): 0.1, 0.2, 0.5, 0.85, 1, 2, 4.
[d]Growth also occured on TYE agar medium with no added salts.

motile and several formed pigmented colonies.

Salt tolerance and temperature range for growth of the isolated strains are shown in Table 3. All strains grew rapidly (visible growth $\lesssim$3 days) over a wide range of temperatures and tolerated relatively high concentrations of NaCl. Aerobic spore-forming rods (such as strains A3a & A3c) and Gram-positive, catalase-positive micrococci (such as strain A4a) are generally quite salt tolerant (12). In contrast, many other bacteria, including many marine bacteria (13), do not grow in the presence of 1 M NaCl. Larsen (12) reported that Gram-negative rods are generally completely inhibited by NaCl concentrations of between 0.8 and 1.7 M. Most Escherichia coli strains, for example, grow only in media containing less than 0.8 M NaCl.

Horowitz and coworkers (14) reported that spore-forming bacteria were uncommon in the Antarctic Dry Valleys and suggested that spore formation did not confer an advantage in cold, dry environments. Spore-formers found in these regions may not be indigenous members of the microflora (15). Nonetheless, the spore-forming strains we isolated grew rapidly at low temperature in the presence of high salt and, therefore, seem well adapted to cold, hypersaline environments.

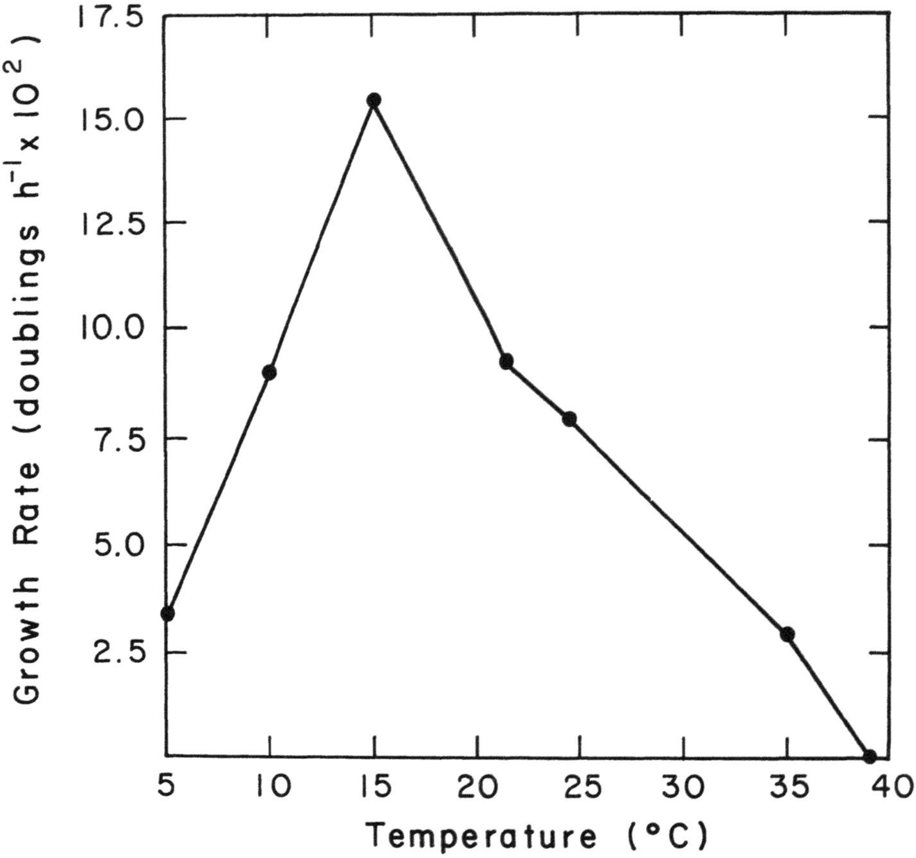

Fig. 2. Effect of temperature on the growth of strain A4a. Cells were cultured in 50 ml liquid 5TYE-HES medium in 250-ml sidearm flasks without agitation. Growth was monitored using a Klett-Sommerson colorimeter equipped with a 660 nm filter.

The method of Eimhjellen (16) was used to estimate the NaCl optima for growth of the isolates. Cells were streaked on modified 5TYE-HES agar media containing different concentrations of NaCl and constant concentrations of $CaCl_2$ (0.04 M) and of $MgSO_4$ (0.2 M). Generally, the most rapid appearance of colony growth was noted on media containing 0.1-0.2 M NaCl. The NaCl optima of strains A3a, A3b, and A3c were slightly higher. Temperature optima were estimated in a similar manner by determining the incubation temperature at which visible colony growth on 5TYE-HES agar medium occured most rapidly. The temperature optimum of all strains was between 15 and 25°C. A more accurate determination of the temperature optimum of strain A4a was made by measuring growth in liquid medium as shown in Figure 2. The temperature optimum of strain A4a was 15°C. Because this strain tolerates high salt concentrations, grows well at 0°C (visible colonies on solid media within one week), and has a temperature optimum below 20°C, it may be considered a "halotolerant psychrophile" (17, 18).

Previous studies of Antarctic organisms indicated that halotolerance declined as the temperature was lowered (9,14). Consistent with this idea, some strains (A1a, A3a, A4a, A4b) appeared to tolerate a lower concentration of NaCl at 5°C than at 15°C. Also, Kushner (17) noted that salt requirements may be reduced or abolished at low temperatures. Thus, organisms which are "halotolerant" at low temperature may actually require salt and therefore be considered "halophilic" at higher temperatures. Presently, we are investigating the effect of temperature on salt requirements and tolerance.

The evolution of adaptation to salt has been considered almost entirely in relation to the halobacteria (19,20). Halobacteria thrive in environments containing saturating concentrations of salts. They avoid loss of cellular water by accumulating KCl so that their intracellular water activity becomes equal to the water activity of their environment (reviewed in 17, 19). KCl is termed the "compatible solute" because cellular structures (the protein synthesizing apparatus, enzymes, etc.) must function in the presence of high concentrations of this solute. Indeed, several proteins from halobacteria require unusually high concentrations of salt for stability and/or activity (19). One apparent consequence of this evolutionary strategy is that the halobacteria tolerate only a limited range of salt concentrations. However, this is only one of several possible solutions to the problem of salt tolerance. Other organisms adjust their intracellular water activity by accumulating a "non-ionic compatible solute" (e.g., amino acids in bacteria (21) and glycerol in Dunaliella (22)). The proteins of these organisms did not have to evolve unusual salt tolerance for activity. Many of the microorganisms we have isolated not only tolerate high salt concentrations but also grow on either low concentrations of salts or in the absence of added salts at 5°C. This suggests that the evolutionary strategy for adaptation of at least some organisms to low temperature and high salt environments involved an increased tolerance to salt at low temperature (perhaps via the intracellular accumulation of a non-ionic compatible solute) rather than an over-all alteration in protein chemistry.

We are currently attempting to identify the special features that allow these microorganisms to tolerate such wide salt and temperature ranges and to thrive at 5°C in hypersaline environments. We are also continuing to characterize the oases environments through additional remote sensing observations and physical modeling. As our understanding of the microenvironment evolves, we will continue to assess the tolerance of these microorganisms to the various martian extremes. Ultimately we hope to be able to address a variety of questions related to whether indegenous or contaminant life could have adapted to the oases environments.

ACKNOWLEDGEMENTS. We thank Robert Benoit for the generous gift of the Antarctic soil samples and E. Canale-Parola, C. Harwood, and J. Barbieri for helpful discussions.

REFERENCES

1. Huguenin, R.L., Clifford, S.M., Sullivan, C.A., and Miller, K.J., NASA TM 80339, 208 (1979).
2. Huguenin, R.L. and Clifford, S.M., NASA TM 81776, 153 (1980).
3. Miller, K.J. and Huguenin, R.L., NASA TM 81776, 151 (1980).
4. Zisk, S.H. and Mouginis-Mark, Nature, in press (1980).
5. Huguenin, R.L., Miller, K.J., and Harwood, W.S., J. Mol. Evol. 14, 103 (1979).
6. Greenberg, E.P. and Canale-Parola, E., Arch. Microbiol. 110, 185 (1976).
7. Brock, T.D., Strategies of Microbial Life in Extreme Environments, ed., M. Shilo, Verlag Chemie, Weinheim, 1979, p. 29.
8. Post, F.J., Microbial Ecology 3, 143 (1977).
9. Benoit, R.E. and Hall, Jr., C.L., Antarctic Ecology, ed., M.W. Holdgate, Academic Press, New York, 1970, p. 697.
10. Cameron, R.E., Morelli, F.A., and Randall, L.P., Antarct. J. U.S. 7, 254 (1972).
11. Cameron, R.E., King, J., and David, C.N., Antarctic Ecology, ed., M.W. Holdgate, Academic Press, New York, 1970, p. 702.
12. Larsen, H., The Bacteria, vol. 4, ed., I.C. Gunsalus and R.Y. Stanier, Academic Press, New York, 1962, p. 297.
13. MacLeod, R.A., Bacteriol. Rev. 29, 9 (1965).
14. Horowitz, N.H., Cameron, R.E., and Hubbard, J.S., Science 176, 242 (1972).
15. Cameron, R.E., Honour, R.C., and Morelli, F.A., Extreme Environments: Mechanisms of Microbial Adaption, ed., M.R. Heinrich, Academic Press, New York, 1976, p. 57.
16. Eimhjellen, K., Zenter. Bakteriol. Parasitenk, Abt I, Suppl. 1, 126 (1965).
17. Kushner, D.J. Microbial Life in Extreme Environments, ed., D.J. Kushner, Academic Press, New York, 1978, p. 317.
18. Inniss, W.E., Ann. Rev. Microbiol. 29, 445 (1975).
19. Bayley, S.T. and Morton, R.A., Strategies of Microbial Life in Extreme Environments, ed., M. Shilo, Verlag Chemie, Weinheim, 1979, p. 109.
20. Lanyi, J.K., Strategies of Microbial Life in Extreme Environments, ed., M. Shilo, Verlag Chemie, Weinheim, 1979, p. 125.
21. Measures, J.C., Nature 257, 398 (1975).
22. Avron, M. and Ben-Amotz, A., Strategies of Microbial Life in Extreme Environments, ed., M. Shilo, Verlag Chemie, Weinheim, 1979, p. 83.

MUTAGENS AND CARCINOGENS: OCCURRENCE AND ROLE DURING CHEMICAL AND BIOLOGICAL EVOLUTION

A. Giner-Sorolla and J. Oró*

Sloan-Kettering Institute for Cancer Research, New York, N.Y. and *University of Houston, Houston, Texas, U.S.A.

Several products formed during biopoiesis were mutagenic. Since most mutagens are carcinogenic, their interaction with higher organisms may have led to the origin of cancer. Many elements of the periodic system are carcinogenic; extraterrestial mutagens or carcinogens include aromatic compounds from meteorites, nitrosamine precursors and formaldehyde (a biopoietic component) found in interstellar space. Volcanic activity, relevant to biopoiesis, is also a source of carcinogens. These findings indicate the ubiquitous, remote existence and cosmic origin of mutagens and carcinogens; the first needed for mutation processes in early evolution, the second leading to the appearance of malignancy.

Cancer cells might be envisioned as independent organisms which arise and proliferate in the susceptible host; they are tough and efficiently organized living systems with great adaptability and versatility which makes them quite resistant to attack or eradication. Although the gross development and clinical consequences of neoplasia are well known, the intimate nature and origin of malignant cells are still poorly understood. The Darwinian principle of "survival of the fittest" may well be applied to the cancer cells as they struggle to invade and subdue the organs of the body in a lethal competition for nourishment (1).

A model of malignancy in the form of a prebiotic neoplastic cell has been suggested as existing before the occurrence of normal cells and thus could represent the emergence of a "pre-life" (2). Current knowledge indicates, however, that cancer cells

Y. Wolman (ed.), Origin of Life, 583–588.

may have originated in a later stage of biological evolution from the interaction of chemical agents with normal cells because chemical carcinogens can induce malignant tumors in all vertebrates (3). It is likely that other factors, biological and physical, may have been contributory.

Several products which have been postulated as essential molecules for the origin and early evolution of life are mutagenic. Since most mutagens are carcinogenic, exposure of higher organisms in later stages of evolution to these agents or others produced during biological evolution, mainly by plants, might have led to the origin of cancer (4). In addition, physical agents such as ultraviolet radiation from the sun had the double role of catalyzing prebiotic synthetic processes and at the same time cause harmful effects in plants and animals.

Examination of the periodic system shows that in addition to radioactive elements, a proportion of metals and nonmetals exert carcinogenic or cocarcinogenic effects in animals and/or man by themselves or in simple combinations with variable degrees of intensity. Up to date, they account altogether for about half of the number of known elements (Fig. 1) and more than double the number of those which are components of living beings.

We and others have found in meteorites aromatic compounds such as benzene (5-7), a leukemogen, polycyclic hydrocarbons, among them, fluorene, phenanthrene, fluorantene and pyrene (5-10), formaldehyde (of relevance in the prebiotic syntheses of carbohydrates (11) and cyanuric acid (12), which are carcinogens (Table 1)). In addition, the existence of precursors of nitrosamines, which are potent carcinogens, such as methaneimine and nitroxyl radicals (13,14) as well as formaldehyde (15,16) have been detected in interstellar galactic space. Among the mutagens and carcinogens present during the early biopoietic stages are hydrazine (17), hydroxylamine (18), radioisotopes, ultraviolet radiation, and precursors of purine abiotic syntheses such as 5-aminoimidazole-4-carboxamide and its nitrosation product 2-azahypoxanthine (19). The prebiotic reaction leading to adenine, originally described by one of us (20), i.e., treating at 70° a concentrated aqueous ammonia solution of hydrogen cyanide, yields after 2 hr a mixture of products. These are highly mutagenic in the Ames Salmonella typhimurium test, strain TA98 indicating that the mutation is of the frameshift type. The test was negative in strain TA1535 (base substitution type). After a 48-hr reaction, the products obtained have a lower mutagenicity value (21). These results show that at the start of the reaction, ammonium cyanide is transformed into highly reactive and mutagenic oligomers such as aminomalononitrile, the corresponding amidines, aminoimidazolecarbonitrile, as well as polymers of low molecular weight. As the reaction proceeds, higher molecular

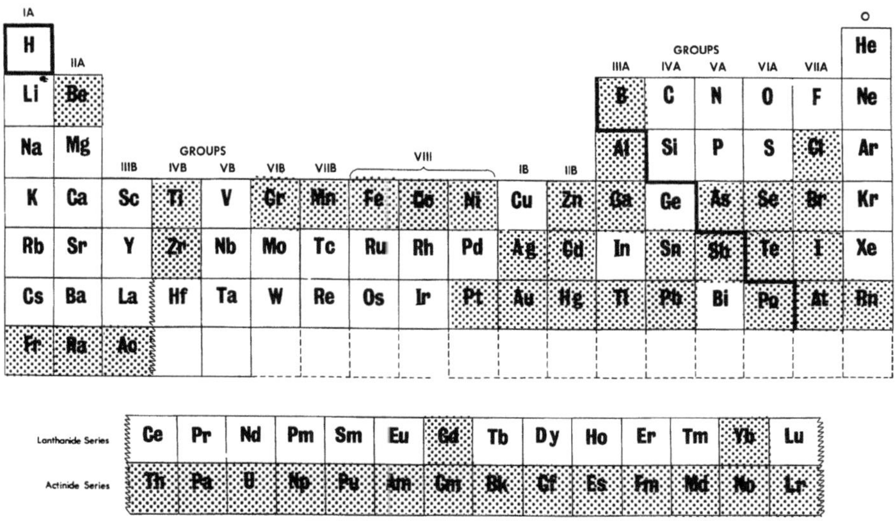

FIGURE 1. Periodic table of the elements: Carcinogenic or co-carcinogenic elements are shaded.

weight polymers are formed at the expense of the more reactive and mutagenic oligomers. Simple modifications (oxidation, nitrosation, hydroxylamine treatment) of nucleic acid components such as purines and pyrimidines obtainable by plausible prebiotic synthesis (20) lead to potent mutagenic and/or carcinogenic substances (22-25). The primitive Earth environment may well have contained compounds of this kind.

The appearance of an oxidizing atmosphere in the upper Precambrian was an essential factor to the further development of life on Earth, and also for the emergence of carcinogens. Most of the naturally-occurring, as well as synthetic, carcinogens either contain a reactive oxygen or require oxidation to exert their effects.

In early biological evolution, actinomycetales and other microorganisms may have produced the first biosynthetic carcinogenic substances in the form of certain antibiotics (paleocarcinogens) (4). In addition, there was an intense volcanic activity during early evolution and later periods, with subsequent discharge of carcinogenic aromatic polycyclic hydrocarbons,

TABLE 1

Synopsis of the Occurrence of Mutagens and Carcinogens During Cosmic and Planetary Evolution

Locale/Event	Years (B.P.)		Substrate
Interstellar matter	–	Formaldehyde, nitroxyl, methaneimine	-
Solar nebula	$> 4.5 \times 10^9$	Benzene, formaldehyde, polycyclic aromatic hydrocarbons, cyanuric acid	-
Prebiotic Earth	$< 4.5 \times 10^9$	Radioisotopes, ultraviolet radiation, hydrazine, hydroxylamine, purine precursors	-
Precambrian environment	$< 3.8 \times 10^9$	Secondary metabolites (antibiotics) polycyclic hydrocarbons (from volcanic activity)	-
Mesozoic flora	$\lesssim 4 \times 10^8$	Toxic products in ferns, cycads, fungi	Reptiles
Modern angiosperms	$\sim 1.5 \times 10^8$	Safrol, flavones	Mammalians
Industrial revolution	1760 (A.D.)	Metals, asbestos, aromatic amines, alkylating agents, nitrosamines, polycyclic hydrocarbons, radiation	Man

as have been found in ashes and volcanic debris (26). To this source and also to the continuous flow on the Earth's surface of meteorites may be attributed the ubiquitous presence of carcinogenic aromatic polycyclic hydrocarbons in soil and plants in the remote geological past, thus constituting a "background chemical carcinogenicity" analogous to that of background ionizing radiation (27).

Interaction between all these carcinogens and metazoans leading to the first neoplastic cell may have taken place in the early Paleozoic (28). A maximum of occurrence of carcinogens was reached in the Mesozoic when there was an abundance of plants (ferns, cycads) containing potent carcinogens which may have interacted with the predominant reptile population. In contrast, there is only an insignificant proportion of plants in present day flora containing weak carcinogens with negligible effect upon present fauna (4).

The fact that carcinogenic elements outnumber those which are biogenic suggests that the process of carcinogenesis is intrinsic in Nature. In addition, the presence of mutagenic and carcinogenic compounds in outer space and throughout evolution indicates the remote existence, cosmic origin and ubiquity of both mutagens and carcinogens. The first being necessary for the mutation processes essential for early biological evolution; the second might have led, at a later epoch in higher organisms, to the emergence of the first cancer cell.

ACKNOWLEDGEMENT - The authors wish to thank Professor H. S. Rosenkranz for mutagenesis assays, Dr. G. Stöhrer for valuable suggestions, and to Mrs. V. Rosso Taracido and Ms. M. Vicens-Ferragut for excellent technical assistance. (Supported in part by grants from the National Cancer Institute (CA 08748) and NASA (NCR 44 005-002)).

REFERENCES

1. Markert, C. L. Cell Differentiation and Neoplasia. Sanders (ed.), Raven Press, New York, 1978, p. 9.
2. Dawe, C. J. A Symposium on Neoplasms and Related Disorders of Invertebrates and Lower Vertebrate Animals. Dawe and Harshbarger (eds.), National Cancer Institute, Bethesda, MD, Monograph No. 31, 1969, p. 31.
3. Finstad, J. and Good, R. A. A Symposium on Neoplasms and Related Disorders of Invertebrates and Lower Vertebrate Animals. Dawe and Harshbarger (eds.), National Cancer Institute, Bethesda, MD, Monograph No. 31, 1969, p. 51.
4. Giner-Sorolla, A. and Bendich, A. Cosmochemical Evolution and the Origin of Life. Oró (ed.), Vol. 2, Reidel, Dordrecht, 1974, p. 315.
5. Studier, M. H., Hayatsu, R. and Anders, E. Geochim. Cosmochim. Acta 36, 189 (1972).
6. Simmonds, P. G., Bauman, A. J., Bollin, E. M., Gelpí, E. and Oró, J. Proc. Nat. Acad. Sci. 64, 1027 (1969).

7. Levi, R. L., Wolf, C. J., Grayson, M. A., Gilbert, J., Gelpí, E., Updegrove, W. S., Szlatkis, A. and Oró, J. Nature 227, 148 (1970).
8. Hayes, J. M. and Biemann, K. Geochim. Cosmochim. Acta 32, 239 (1968).
9. Anders, E., Hayatsu, R. and Studier, M. H. Science 182, 781 (1973).
10. Ponnamperuma, C. N.Y. Acad. Sci. 194, 56 (1972).
11. Breger, A., Zuboric, P., Chandler, J. C. and Clarke, R. S., Jr. Nature 236, 155 (1972).
12. Pliss, E. B. and Zabezinskii, M. A. Vopr. Onkol. 16, 82 (1970) (Chem. Abstr. 72, 99452v, 1970).
13. Pickles, J. B. and Williams, D. A. Nature 271, 335 (1978).
14. Mann, A. P. C. and Williams, D. A. Nature 283, 721 (1980).
15. Buhl, D. Nature 234, 332 (1971).
16. Wickramasinghe, N. C. Nature 252, 462 (1974).
17. Urey, H. C. The Planets, Yale University Press, New Haven, 1952, p. 152.
18. Oró, J., Kimball, A., Fritz, R. and Master, F. Arch. Biochem. Biophys. 85, 115 (1959).
19. Beal, D. D., Skiba, J. L. and Croft, W. A., Jr. Proc. Am. Assoc. Cancer Res. 15, 47 (1974).
20. Oró, J. and Kimball, A. P. Arch. Biochem. Biophys. 96, 293 (1962).
21. Rosenkranz, H. S., personal communication.
22. Brown, G. B. Prog. Nucleic Acid Res. Mol. Biol. 8, 209 (1968).
23. Teller, M. N., Giner-Sorolla, A., Stöhrer, G., Budinger, J. M. and Brown, G. B. Cancer Res. 38, 2229 (1978).
24. Giner-Sorolla, A., Greenbaum, J. H., Last-Barney, K., Anderson, L. M. and Budinger, J. H. Cancer Lett. 6, 79 (1979).
25. Anderson, L. M., Giner-Sorolla, A., Greenbaum, J. H., Last-Barney, K. and Budinger, J. M. Int. J. Cancer 24, 319 (1979).
26. Ilnitsky, A. P., Vinogradov, V. N., Riabchun, V. K., Mischenko, V. S., Gvildis, V. Y., Belitsky, G. A. and Shabad, L. M. Cancer Lett. 8, 51 (1979).
27. Grimmer, G. and Hildebrandt, A. Z. Krebsforsch. 67, 272 (1965).
28. Burnet, F. M. Prog. Exp. Tumor Res. 13, 1 (1970).

COULD THE BIOCHEMICAL METABOLISM BE DIFFERENT ?

BUVET René
Laboratoire d'Energétique Electrochimique et Biochimique.
Université PARIS VAL DE MARNE
F 94010 CRETEIL Cédex (France)

ABSTRACT

Starting from the experimental fact that biochemical reactions can generally be observed only in the presence of enzymic catalysts, it is commonly admitted that the chemical activity of living systems is determined as it were by the choice of available proteinic sequences, and that it could be different if chances of evolution had produced other sequences.

A thorough analysis of the set of known metabolic steps shows that all these reactions can be classified from only a few kinds of elementary processes. This classification noticeably differs from the enzyme classification, but presents the advantage of allowing easier comparisons of energetic and kinetic reactivities of the substrates involved.

In fact, it shows that in each group of elementary processes, the reactions occur only when the transformed groups of atoms in the reacting substrates have convenient substituents determining the energetic and kinetic reactivity of substrates. Examples are given mainly for redox reactions.

Reactions involved in the primeval chemical evolution should have been non-enzymic models of contemporary biochemical processes, involving more efficient substituents and higher energy balances, further eliminated by auto-catalytic evolution of the whole system.

It finally appears that the biochemical metabolism is built up from roughly the only set of reactions which can be set in place in aqueous solutions at ambient temperatures taking into account the physical-chemistry of carbonaceous compounds in aqueous media.

Y. Wolman (ed.), Origin of Life, 589–599.

Since a roundtable discussion will take place during this meeting on the subject of the origins of life as a curriculum in education, I would like to begin this speech by asking you to focus your minds for a while on the idea that perhaps the origin of life is still a problem only because our way to teach life sciences and particularly biochemistry makes it appear as such.

As a matter of fact, there are two ways of teaching biochemistry.

First the commonly used one, which consists in presenting the tools, i.e. proteins and enzymology, then the making of these tools, i.e. protein synthesis and the genetic code, and only finally the metabolic pathways. As if chronologically, as it appears experimentally in contemporary organisms, nucleic acids and the genetic code had determined proteins, which in their turn chose metabolic steps and pathways. Which implies that this set of reactions could have been nowadays different if the randomness of the formation of polynucleotidic sequences had been different at any time from the very beginning (1).

But, there exists another way for teaching biochemistry, that I do my best to emphasize in my courses. This way ultimately leads my students to the conclusion that biochemical metabolism is merely the regular and complete expression of the chemistry of carbonaceous substrates in aqueous solutions near room temperatures, and that it only progressively became sophisticated and oriented towards the selection of lower energy balances by autocatalytic evolution (2). However, even through the screen of this evolution, the general structure of the contemporary metabolism remains clearly, as we will see, determined by the reactivities of carbonaceous molecules and others such as derivatives of oxygen, sulfur, nitrogen and phosphorus and some metallic cations and complexes. Consequently, it is understandable from a sufficiently exhaustive and particularly physicochemical and energetically quantitative approach of this particular chemistry of aqueous media. In other words, it could not be different from what it is.

For reaching this conclusion, we must begin with reviewing classically the most common metabolic pathways, but instead of insisting upon the enzymes which catalyse the reactions, we will concentrate our attention on the kind of modifications which are implied at the level of substrates, searching step by step for a classification of these changes in categories, along a way which resembles the one which led to the classification of enzymes (3,4).

But apart from looking at substrates instead of tolls, we can go further in the reduction of the number of retained categories, by dividing some biochemical steps, which are in fact catalysed by one single enzyme, into two elementary processes assembled as it concerns mass and energy balances by stoechiometric addition in the considered step.

After having done all this work on a certain number of biochemical steps, indeed very limited regarding all the known ones, we are finally led to settle a very few kinds of elementary processes, which in fact permit furtherly to describe the entire metabolism, each category being defined clearly by the bond changes implied at the reacting site composed of no more than 3 to 4 atoms in the involved substrates.

TABLE 1

Main Kinds of Biochemical Elementary Processes

1) Redox reactions :

(n = 1 or 2)

$$Ox_1 + ne^- + p_1H^+ \rightleftharpoons Red_1 + q_1\ H_2O$$

$$Red_2 + q_2\ H_2O \rightleftharpoons Ox_2 + ne^- + p_2H^+$$

$$Ox_1 + Red_2 + (p_1 - p_2)H^+ \rightleftharpoons Red_1 + Ox_2 + (q_1 - q_2)H_2O$$

2) Condensations - hydrolyses

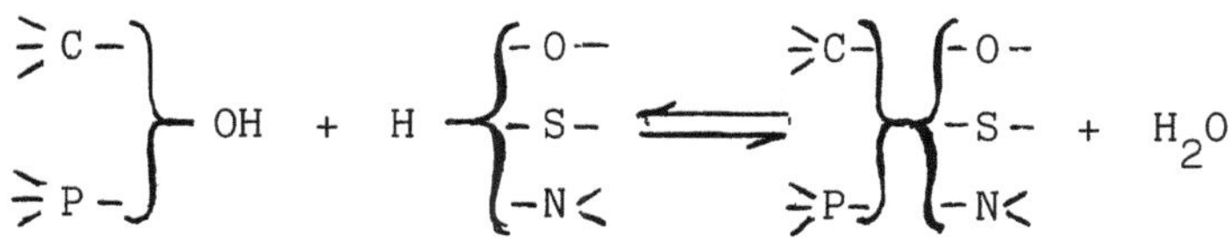

3) Additions - eliminations of C-H to C=O

4) Additions - eliminations of A-H (A=heteroatom O,S,N) to C=O

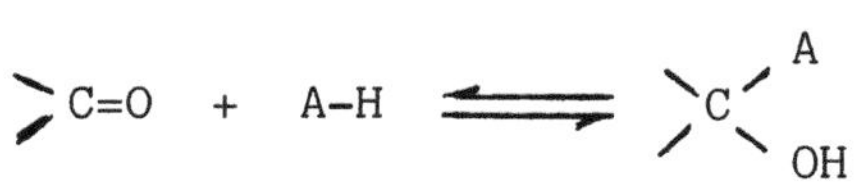

5) Additions - eliminations of A-H to C=C

6) Enol-oxo isomerisations

7) Methyl transfers (X,Y = C or A)

$$X\text{-}CH_3 + H\text{-}Y \rightleftharpoons X\text{-}H + H_3C\text{-}Y$$

8) Isoprenyl pyrophosphate polycondensations

For a first overview these categories are (table I) :
-Redox processes which contain all processes implying transfers of electrons from one reducing reagent to an oxidizing one and can be theoretically discussed in terms of electron transfers between two redox couples.
-Condensations, leading to the formation of condensed group, which can be in their turn hydrolyzed by the reverse reaction.
-Syntheses of C-C bonds by addition of C-H to C=O, which produce degradations of C-C bonds by the reverse reaction. To this group also belong degradative hydrolyses of C-C bonds.
-Additions of A-H to C=O bonds and the reciprocal elimination.
-Additions of A-H to C=C bonds and the reciprocal elimination.
-Enol-oxo isomerisations which are in fact intramolecular applications of the preceeding categories of reactions.
-Methyltransfers where the attaching atoms are either certain C or heteroatoms.
- to which I should add for completeness the isoprenyl pyrophosphate polycondensations.

Concerning the relationship between these elementary processes and biochemical steps two cases have to be considered.

A majority of biochemical steps can be simply identified with a given elementary process. Such are e.g. simple hydrolyses, oxydations of alcohols by NAD^+, $\alpha\beta$dehydrogenations, or the degradation of fructose diphosphate etc ... But some biochemical steps, named coupled reactions, cannot be so easily accounted for and their stoechiometric description implies the addition of two elementary processes. Such are e.g. transacylations and transphosphorylations, reactions implying the formation of condensed bonds from redox processes, thiolysis, etc ...(5)

I am afraid that as yet my talk seems to conclude that the enzyme classification should be only changed a little bit by reformulating it in terms of substrates transformations. But classifying is not explaining and we must go further.

If classifying is not explaining, it is a step and a tool towards this aim, because it permits us to put in the same box processes which are roughly similar and to compare them as for the remaining differences, and even much better here, to compare processes of a given category which do occur in living systems with processes obeying similar stoechiometries which are not known for participating to the life game.

Such analysis, which remains up to now purely empirical from biochemical facts, leads to very interesting conclusions concerning what enzymes can do and what lies apparently beyond their capabilities (4).

Let us e.g. consider redox couples involving carbonaceous substrates. Here two restrictive rules considerably limit the field of biochemistry (6).

First, although redox couples involving more than 2 electrons could be perfectly well written if we limit our judgment to mass and charge balances and should be even often energetically easier to set in action than couples exchanging only two electrons (figure 1), there is no biochemical one-step reaction which do so. In fact, considering now also non carbonaceous substrates, one only exception to this rule runs everywhere in treatises and publications of biochemistry. It concerns the case of the reduction of molecular oxygen. I am sorry to say, that we should consider it as a sin of temerity, and that the punishment which had to be paid for it by biology is the chemiosmotic theory of oxidative phosphorylations. In fact, many biochemical, chemical and electrochemical data clearly show that oxygen reduction does not escape to this common restriction and always occurs at least through the hydrogen peroxide intermediate, which must be considered as an essential step in the chemical coupling towards oxidative phosphorylations (5).

But let us leave aside this controversial point and focus our minds on the second restriction which apparently rules biochemical processes. Here the situation is a little bit more difficult to appreciate. Many, but not all redox reactions occur by mechanisms which imply intermolecular electron transfers between the reducing reagent and the oxidizing one without involving any known transient covalent association between both. The experimental evidence shows here that any time the mechanism of a redox process is of this kind, the electron transfer can eventually be accompanied by some changes of protonation states of the reagents, but never by any change of the covalent vicinity, or better topology, of relatively heavy atoms, such as C, O, N, S, etc .. in the involved molecules.

For clarifying it, let us give the simplest example where this rule applies (5). In the Krebs cycle succinate is oxidized to fumarate which in its turn is transformed by hydration into malate. Since this second process is weakly exergonic, it means that the direct oxidation of succinate to malate is in fact energetically easier than its oxidation to fumarate. We consider that the reason why the biochemical transformation does not take this direct route is not a matter of enzyme choice, but the simple consequence of the fact that relatively heavy atoms such as oxygen cannot be transported sufficiently fast to allow the oxidation of succinate to malate to occur in a single step, namely during the short time electrons take for being exchanged, even through the aid of enzymes which have themselves to cope with what is kinetically possible in this respect for the substrates.

But the same rule leads to a much more important consequence in several cases, of which thiolysis is an excellent example (5). Here we are primarily faced with the same situation. The βhydroxyacylcoA is biochemically oxidized to a βketo-derivative :

$$R\text{-}CHOH\text{-}CH_2\text{-}COScoA + NADH + H^+ \rightleftharpoons R\text{-}CO\text{-}CH_2\text{-}COScoA + NAD^+$$

but the keto group is in its turn the target of a nucleophilic attack by the thiol group of a coenzyme A, which leads to a transacylation producing a new acylcoA :

$$R\text{-}CO\text{-}CH_2\text{-}COScoA + HScoA \rightleftharpoons R\text{-}COScoA + CH_3\text{-}COScoA$$

But, since the degradation of the C_2-C_3 bond is so exergonic that it can even produce an energy-rich condensate, it means that the direct degradative oxidation of the βhydroxyacylcoA :

$$R\text{-}CHOH\text{-}CH_2\text{-}COScoA + H_2O \rightleftharpoons R\text{-}COO^- + CH_3\ COScoA + 2e^- + 3\ H^+$$

is here also energetically very much easier than the production of the βketoderivative. In fact, it does not occur directly in living systems and this must be related to the impossibility of changing the C-Cand C-O topology fast enough for performing it during the time electrons take to be exchanged. And the result is the production of a new energy-rich molecule which in fact stores a part of the redox energy involved in the sequence of reactions.

Let us now come to another group of restrictive rules which are revealed by the comparison of what occurs enzymatically and what is never observed in biochemistry, no more than in chemistry or electrochemistry (4).

To introduce it, we will e.g. consider the particular kind of redox couples which are named αβdehydrogenations. Only a few examples are known, mainly the reversible dehydrogenations of acylcoA or acylenzymes, of succinate and of dihydroorotate. On the contrary, this process never occurs (with flavinic or nicotinamidic coenzymes) in acyl chains of lipids, on free fatty acids, or generally from any CH-CH group which is not substituted at least either by one thioester or by two carboxyles or carboxylic derivatives.

Why not consider here, that the main reason for the reaction to occur in living systems is the presence of these substituents which do change something related to energetics and kinetics to the properties of the CH-CH group. This permits enzymes to finish the job instead of staying useless as if trying to cauterize a wooden leg !

This must be as much considered, that when looking at several other kinds of elementary processes, the same substituents present the same determining influence. This is e.g. the case of the additions of C-H to C=O which are well known to occur mainly from CH group placed in αposition from thioester functions, and from some other equivalent ones such as aldehyde functions, or of additions of H_2O to C=C bonds.

I cannot deal here with all such sensitization effects of reacting group related to the presence of substituents, but let me simply say that their enumeration evidently coincides with what is also empirically well- known about the interferences between neighbouring functional groups in organic molecules, and this must lead us now to look at physicochemical explanations.

Let us recall here that physicochemical explanation have to be developed in two steps.

The first question which needs to be asked is :
- what is energetically possible on the basis of the first and second principles of energetics and of available numerical data ?

Secondly, when some transformation is possible on this respect :
- does there exist a mechanism for performing it, without meeting any kinetic rule which forbids it and by using steps only involving intermediates of reasonable energy level ?

I cannot deal here with all the biochemical processes which can yet be explained in this way (4, 5). I only will focus on the use of one kind of numerical data for understanding why metabolism is built as it is, and in fact I will limit myself to a rapid reading of these data and, without demonstration, of their biochemically relevant consequences.

We will here consider as an exemple the values of the apparent standard potentials which define the transition from :
- a non oxidized carbon atom to an alcoholic or ethylenic group ;
- an alcoholic group to an aldehydic or ketonic one ;
- an aldehydic group to a carboxylic one.

On the diagrams (figure 1) which give the variations with pH of the apparent standard potentials of these redox couples, have been superimposed the apparent standard potentials involved in the reduction of water to hydrogen, and, to keep it in mind,the E°' values for the O_2/H_2O redox couple.

It clearly appears from these available numerical data that, whatever be the substituents placed in the vicinity of the considered carbon :,
- its transition from the hydrogenated redox level to the alcoholic redox level always occur between 0.05 and 0.20 V/SHE at pH 7, any change of carbonaceous substituent of the concerned carbon atom producing changes of potentials within this band ;
- the transitions from alcohols to aldehydes or ketones always occur between - 0.25 for isopropanol and - 0.10 V/SHE for glycolate at pH 7, i.e. the second electron pair is energetically easier to extract than the first one.
- and again, the transitions from aldehydes to carboxylic acids always occur at lower potential and here even below the potential of reduction of water to hydrogen, i.e. near -0.5 to - 0.6 V/SHE.

From these simple data, and using the restricting rules that we have introduced as axiomatically acceptable ones on the grounds of kinetics, it was possible through complete discussion of the concerned problems to conclude that :

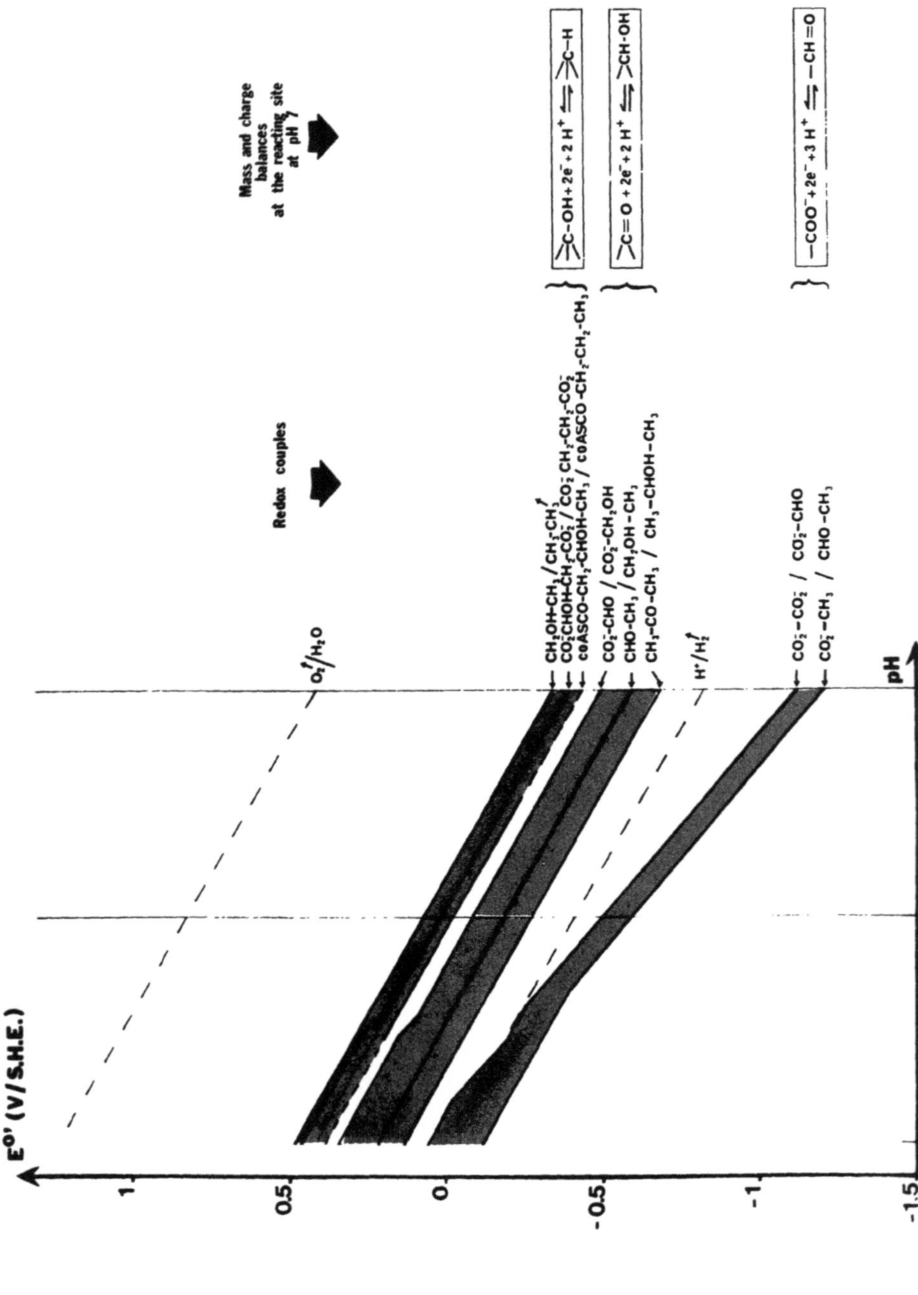
E°' (V/S.H.E.)
1
0.5
0
-0.5
-1
-1.5
0
7
14
pH
Redox couples
Mass and charge balances at the reacting site at pH 7
H⁺/H₂
O₂/H₂O

-The extraction of the first electron pair is only accessible to biochemically available electron exchangers, flavoproteins and quinones, if the substituents of the site are as revealed in biochemistry, and in these case it is only feasible in two steps, namely $\alpha\beta$dehydrogenation plus hydration of ethylenic bond, which is the case in biochemistry.

- In all other cases, it is only feasible by hydroxylations which occur through peroxydes produced by the bielectronic reduction of oxygen (8).

.-The oxydation of alcohols by nicotinamidic oxidants is only possible at weakly basic pH and, taking into account the substituent effects (which tend to increase the redox potential when the carbonaceous substituent is more oxidized) more and more alkaline if the carbon atom located in αposition of the one which has its level of oxidation changed is itself more oxidized, which is enzymatically the case in biochemistry.

-The alcohol function should always undergo dismutation as :

$$H{\gtrdot}C\text{-}OH + {\gtrdot}C\text{-}OH \rightleftharpoons {>}C{=}O + {\gtrdot}C\text{-}H + H_2O$$

but this process is only kinetically feasible without opposition of the second restrictive rule if both hydroxyles are located on two neighbouring C atoms :

$$>COH\text{-}CHOH\text{-} \rightleftharpoons\ >C{=}COH\text{-} + H_2O \rightleftharpoons\ >CH\text{-}CO\text{-} + H_2O$$

In this case it is strongly exoenergetic but occurs only if the substituants on the reacting site are conveniently selected and it is the origin of the hydrolysis energy of the phosphoenol pyruvate.

-Carboxylic acids cannot be reduced as such because this should imply the action of some reducing agent which should also reduce water into hydrogen and consequently disappear as soon as being produced. But their reduction can be completed from a sufficiently energy-rich condensed derivative of the carboxyl group, which is the case in biochemistry.

-Reciprocally, the oxidation of aldehydes being kinetically forbidden because of the second rule is only possible from hemiacetal forms and leads to condensed derivatives of acids, which is indeed the case.

- and so on...

Such examples could be multiplied and discussed more quantitatively and completely (5), but it is neither the place nor the time for doing it here, if you would like to consider that the elements for building a conclusion in connection with the topic of this conference are in our hands.

If one closes one's eyes to what is energetically and kinetically permitted in organic chemistry in aqueous solutions near room temperatures, i.e. if one takes only into account the possibility of writing developed formula and stoechiometries, then surely living systems seem to use only a very small part of what is admittedly possible.

But if one goes further in analyzing the criteria which determine the reactivities of substrates and the physicochemical conditions which permit these reactivities to effectively manifest themselves, then the set of reactions which make living systems live do seem to exploit all remaining possibilities (7).

This means that the biochemical metabolism is not determined by any random choice of enzymic sequences, or by any choice of enzymic sequences which could have evolved from any closed game between polypeptide and polynucleotide polymers. In fact, on the contrary we should consider that the metabolism is as it is because it could not be different owing to the physicochemical properties of carbonaceous compounds in aqueous media.

Evolution has systematically explored all possibilities of this particular chemistry, selecting all possible tools for fulfilling the job as well as possible, starting from the state of the art which have been just reached the day before, and making its choices for the maximum benefit of the whole system, which does not mean necessarily for the best benefit of each of the involved processes taken separately (2).

Exactly as men do, or at least should do, in their own works.

REFERENCES

(1) MONOD J.
Le hasard et la nécessité
Ed. Seuil publ. Paris (1970)

(2) BUVET R.
A simple analogical model for the selction of optical activity and of the most efficient catalysts in the course of molecular evolution.
Origins of Life 8 (1977) 267 - 270.

(3) BUVET R.
L'origine des êtres vivants et des processus biologiques.
Masson publ., Paris (1974)

(4) BUVET R.
Energetic structure of the metabolism.
Topics in Bioelectrochemistry and Bioenergetics
G.Milazzo ed., John Wiley publ. London (1976) p.107 - 177.

(5) BUVET R.
Energetics of coupled biochemical processes and of their chemical models.
Living systems as energy converters.
M.Allen, R.Buvet and J.P.Massué Eds., Elsevier/North Holland publ.Amsterdam (1977)

(6) BUVET.R
General energetical criteria for the implementation of electro-chemical, chemical and biochenical electron transfer processes.
to be published in Bioelectrochemistry and Bioenergetics.

(7) SCHOFFENIELS E.
L'anti hasard
Collection "Discours de la Méthode"
Gauthier Villars publ., Paris (1974)

(8) BUVET R. and LE PORT L.
Electrochemical energetics of biochemical hydroxylations and of their non-enzymic models.
Electrochimica Acta 25, 97 (1980)

AUTHOR INDEX

SUBJECT INDEX

GPSR Compliance
The European Union's (EU) General Product Safety Regulation (GPSR) is a set of rules that requires consumer products to be safe and our obligations to ensure this.

If you have any concerns about our products, you can contact us on

ProductSafety@springernature.com

In case Publisher is established outside the EU, the EU authorized representative is:

Springer Nature Customer Service Center GmbH
Europaplatz 3
69115 Heidelberg, Germany

www.ingramcontent.com/pod-product-compliance
Ingram Content Group UK Ltd.
Pitfield, Milton Keynes, MK11 3LW, UK
UKHW021029041026
15470UKWH00013B/363
* 9 7 8 9 0 2 7 7 1 2 2 9 5 *